THE ROUTLEDGE HANDBOOK OF SCIENTIFIC COMMUNICATION

Given current science-related crises facing the world such as climate change, the targeting and manipulation of DNA, GMO foods, and vaccine denial, the way in which we communicate science matters is vital for current and future generations of scientists and publics.

The Routledge Handbook of Scientific Communication scrutinizes what we value, prioritize, and grapple with in science as highlighted by the rhetorical choices of scientists, students, educators, science gatekeepers, and lay commentators. Drawing on contributions from leading thinkers in the field, this volume explores some of the most pressing questions in this growing field of study, including:

- How do issues such as ethics, gender, race, shifts in the publishing landscape, and English as the lingua franca of science influence scientific communication practices?
- How have scientific genres evolved and adapted to current research and societal needs?
- How have scientific visuals developed in response to technological advances and communication needs?
- How is scientific communication taught to a variety of audiences?

Offering a critical look at the complex relationships that characterize current scientific communication practices in academia, industry, government, and elsewhere, this *Handbook* will be essential reading for students, scholars, and professionals involved in the study, practice, and teaching of scientific, medical, and technical communication.

Cristina Hanganu-Bresch is Associate Professor of Writing and Rhetoric at the University of the Sciences, USA. She has authored a variety of books and articles in the area of the rhetoric of medicine, in particular psychiatry, and in scientific communication. Among her books is *Effective Scientific Communication: The Other Half of Science*, with Kelleen Flaherty (2020).

Michael J. Zerbe is Professor of Rhetoric and Writing Studies at York College of Pennsylvania, USA. He is the author of *Composition and the Rhetoric of Science: Engaging the Dominant Discourse*, and his work has also appeared in the *Journal of Technical Writing and Communication*, *Configurations*, and *POROI*.

Gabriel Cutrufello is Associate Professor of Composition and Rhetoric in the Department of Communication and Writing at York College of Pennsylvania, USA. His research interests include the rhetoric of science and the history of technical writing. His work has been published in *Rhetoric Review* and *Written Communication*.

Stefania M. Maci is Professor of English Language and Translation at the University of Bergamo, Italy. Her research is centered around ESP with a corpus linguistics approach, focusing, in particular, on medical and popularized discourse across genres (research articles, research letters, posters) from an intercultural perspective.

THE ROUTLEDGE HANDBOOK OF SCIENTIFIC COMMUNICATION

*Edited by Cristina Hanganu-Bresch, Michael J. Zerbe,
Gabriel Cutrufello, and Stefania M. Maci*

Routledge
Taylor & Francis Group

LONDON AND NEW YORK

Cover image © Getty Images

First published 2022
by Routledge
2 Park Square, Milton Park, Abingdon, Oxon OX14 4RN

and by Routledge
605 Third Avenue, New York, NY 10158

Routledge is an imprint of the Taylor & Francis Group, an informa business

British Library Cataloguing-in-Publication Data
A catalogue record for this book is available from the British Library

Library of Congress Cataloging-in-Publication Data
Names: Hanganu-Bresch, Christina, editor.
Title: The Routledge handbook of scientific communication / edited by Christina Hanganu-Bresch, Michael J. Zerbe, Gabriel Cutrufello, and Stefania M. Maci.
Description: 1 Edition. | New York, NY : Routledge, 2022. | Series: Routledge international handbooks | Includes bibliographical references and index.
Subjects: LCSH: Communication in science—Handbooks, manuals, etc. | Science—Language—Handbooks, manuals, etc. | Technical writing—Handbooks, manuals, etc.
Classification: LCC Q223 .R668 2022 (print) | LCC Q223 (ebook) | DDC 808.06/65—dc23/eng/20210927
LC record available at https://lccn.loc.gov/2021037531
LC ebook record available at https://lccn.loc.gov/2021037532

ISBN: 978-0-367-48979-3 (hbk)
ISBN: 978-1-032-19053-2 (pbk)
ISBN: 978-1-003-04378-2 (ebk)

DOI: 10.4324/9781003043782

Typeset in Bembo
by Apex CoVantage, LLC

CONTENTS

Contents

FIGURES

TABLES

CONTRIBUTORS

Moriah Ariely is a postdoctoral fellow and a teachers' instructor at the Department of Science Teaching, Weizmann Institute of Science, Israel, and a high school biology teacher. Her research focuses on developing and implementing new teaching strategies aimed to promote students' and teachers' scientific disciplinary literacy.

Ayelet Baram-Tsabari is a Professor at the Faculty of Education in Science and Technology at the Technion–Israel Institute of Technology and heads the Applied Science Communication research group. Her research focuses on the interactions between science education, science literacy, and public engagement with science in real-life scenarios.

Yael Barel-Ben David is Director of the Citizen Lab – a hub for citizen science in communities and in schools. Her research focuses on science communication training for scientists, their effectiveness, and how the public interacts with the products of these trainings as a way to examine their success.

Begoña Bellés-Fortuño is a senior lecturer in the Department of English Studies at Universitat Jaume I, Spain. She is Editor in Chief of *Language Value* journal and one of the executive directors of *IBERICA Journal* and Director of the Interuniversity Institute of Modern Applied Languages (IULMA) at Universitat Jaume I. Her research interests are focused on Discourse Analysis and, more concretely, academic discourse both written and spoken.

Marina Bondi is Professor of English Linguistics at the University of Modena and Reggio Emilia, where she coordinates the PhD program in Human Sciences. She has published extensively in the field of genre analysis, EAP, and corpus linguistics, with a focus on language variation across genres, disciplines, cultures, and media.

Raffaella Bottini is a doctoral researcher at the ESRC Centre for Corpus Approaches to Social Science, Lancaster University. Her research interests include corpus linguistics, statistics, second language acquisition, language teaching and testing. She focuses on the application of corpus methods to the analysis of vocabulary in spoken learner corpora.

Vaclav Brezina is Senior Lecturer at the Department of Linguistics and English Language and a member of the ESRC Centre for Corpus Approaches to Social Science, Lancaster

University. His research interests are in the areas of statistics, data visualization, corpus design and methodology. He is the author of *Statistics in Corpus Linguistics* (CUP, 2018).

Jonathan Buehl is an associate professor in the Department of English at the Ohio State University. He is the author of *Assembling Arguments: Multimodal Rhetoric and Scientific Discourse* and the coeditor (with Alan Gross) of *Science and the Internet: Communicating Knowledge in a Digital Age.*

Lauren E. Cagle is Assistant Professor of Writing, Rhetoric, and Digital Studies at the University of Kentucky. Cagle researches environmental, technical, and scientific rhetoric, often in collaboration with practitioner organizations, such as the Kentucky Division for Air Quality, the Kentucky Geological Survey, and The Arboretum, State Botanical Garden of Kentucky.

Lillian Campbell is Assistant Professor of English at Marquette University. Her research focuses on the rhetoric of health and medicine, technical and professional communication, and feminist rhetorics. Her publications can be found in *Technical Communication Quarterly*, *Written Communication*, *Rhetoric of Health and Medicine*, and *The Journal of Writing Research.*

James N. Corcoran is Assistant Professor of English as a Second Language and Applied Linguistics in the Department of Languages, Literatures, and Linguistics at York University. His research interests include language teacher education, (critical) English for specific/academic purposes, and relations of power in global academic knowledge production.

Danielle DeVasto is Assistant Professor in the Department of Writing at Grand Valley State University. Her research interests reside at the intersections of visual rhetoric, science communication, and uncertainty. Her work has been published in *Communication Design Quarterly*, *Community Literacy Journal*, *Present Tense*, and *Social Epistemology.*

Lisa DeTora is Associate Professor of Writing Studies and Rhetoric and Director of STEM Writing at Hofstra University. She is the editor of *Regulatory Writing: An Overview* (2017, 2020). Her scholarship bridges biomedical writing, rhetorics of health and medicine, technical communication, graphic narrative research, and the medical humanities.

Karen Englander is based at the International Foundation Program of the University of Toronto, bringing more than 15 years of working with and researching international scientists and scholars while Professor at the Universidad Autónoma de Baja California, Mexico. She analyzes these experiences through the lenses of policy, pedagogy, and critical theory.

Maurizio Gotti is Emeritus Professor of English Language and Translation at the University of Bergamo. His main research areas are the features and origins of specialized discourse, both in a synchronic and diachronic perspective. He is a member of the editorial boards of international journals and edits the Linguistic Insights series for Peter Lang.

Susanne Hall is Teaching Professor of Writing and Director of the Hixon Writing Center at Caltech. She is the editor of *Prompt: A Journal of Academic Writing Assignments.* Her work has appeared in *AILA Review*, *Journal of Technical Writing and Communication*, and *Journal of Microbiology and Biology Education.*

Glenn Hampson is the founder and executive director of the Science Communication Institute (SCI), a US-based nonprofit focused on improving the communication that happens inside science. He also directs the Open Scholarship Initiative (OSI), a UN-backed effort designing sustainable global policies for open science and other open solutions.

Ken Hyland is Professor of Applied Linguistics in education at the University of East Anglia. He is best known for his research into academic writing, having published over 260 articles and 28 books with 58,000 citations on Google Scholar. A collection of his work was published by Bloomsbury in 2018.

Natasha N. Jones is a technical communication scholar. Her research interests include social justice, narrative, and technical communication pedagogy. She holds herself especially accountable to Black women and marginalized genders and other systemically marginalized communities. She currently serves as the vice president for the Association of Teachers of Technical Writing and is Associate Professor at Michigan State University.

Tomás Koch Ewertz holds a PhD from Ghent University and is currently a faculty member at Universidad de Playa Ancha, Chile. His interests include scholarly communication, sociology of science, and higher education studies. He is also a member of the Observatorio de Participación Social y Territorio, Universidad de Playa Ancha.

Kalie Leonard is a lecturer in the English department at the University of Wyoming, where she primarily teaches technical and scientific writing courses. Leonard's primary research interests include the (changing) conventions of scientific posters, as well as the purpose of posters as a form of scientific communication.

Kate Maddalena teaches writing and media studies at the University of Toronto Mississauga. Dr. Maddalena's interests include media theory, science and technology studies (STS), and science communication. Her work has been published in journals such as *Theory, Culture, and Society*, the *International Journal of Communication, Canadian Journal of Communication*, and *Project on the Rhetoric of Inquiry (POROI)*.

Jennifer C. Mallette is an associate professor of English at Boise State University, where she collaborates with engineering faculty to support student writers. Her research builds on those collaborations, examining best practices for integrating writing into engineering curriculum; she also explores women's experiences in engineering settings through the context of writing.

Justin Mando is an associate professor of science and technical writing at Millersville University. His recent book is titled Fracking and the Rhetoric of Place: How We Argue from Where We Stand. He has published in Environmental Communication, Composition Studies, and Discourse & Communication. He received his Ph.D. in Rhetoric at Carnegie Mellon University.

Maureen A. Mathison is Associate Professor of Writing and Rhetoric Studies at the University of Utah. Her research interests include STEM studies and disciplinary rhetorics. Her publications include *Sojourning in Disciplinary Cultures: A Case Study of Teaching Writing in Engineering*, as well as multiple articles and book contributions.

Ashley Rose Mehlenbacher is Associate Professor in the Department of English Language and Literature at the University of Waterloo, author of *Science Communication Online: Engaging Experts and Publics on the Internet* (OSU Press, 2019), and coeditor, with Carolyn R. Miller, of *Emerging Genres in New Media Environments* (Palgrave, 2017).

Brad Mehlenbacher is Professor of Rhetoric and Communication in the Department of English Language and Literature at the University of Waterloo. His research focuses on communication and learning in technical, scientific, and engineering contexts. He is author of the award-winning *Instruction and Technology: Designs for Everyday Learning* (MIT Press, 2010).

Jenna Morton-Aiken is Assistant Professor of Humanities, Writing Program Administrator, and Director of Writing across the Curriculum at Massachusetts Maritime Academy. Her research has appeared in *Technical Communication Quarterly*, *WPA Writing Program Administration*, and other publications. She helped launch SciWrite@URI and is a recipient of the 2018–2019 CCCC Emergent Researcher Award.

Judy Noguchi is Professor Emerita of Kobe Gakuin University, Japan. She has been involved in research and teaching of English for specific purposes in science, engineering, and medical sciences. Her PhD work at the University of Birmingham was on the review article and its role in knowledge construction.

Mya Poe is Associate Professor of English at Northeastern University. Her books include *Learning to Communicate in Science and Engineering*, *Race and Writing Assessment*, and *Writing, Assessment, Social Justice, and Opportunity to Learn*. She is series coeditor of the Oxford *Brief Guides to Writing in the Disciplines*.

Amy D. Propen is Associate Professor of Writing at the University of California, Santa Barbara. She is author of *Visualizing Posthuman Conservation in the Age of the Anthropocene* (The Ohio State University Press, 2018) and *Locating Visual-Material Rhetorics: The Map, the Mill, and the GPS* (Parlor Press, 2012).

Gwendolynne Reid is Assistant Professor of English and directs the writing program at Oxford College of Emory University. Her research examines writing in disciplinary and professional contexts and can be found in *Science and Engineering Ethics*, *Written Communication*, and *Across the Disciplines*, as well as several edited collections.

Colleen A. Reilly is Professor of English at the University of North Carolina Wilmington. She teaches graduate and undergraduate courses in professional and scientific writing. She has published articles and book chapters related to scientific writing, privacy and surveillance in digital spaces, and genders, sexualities, and technologies.

Cheryl L. Sheridan, PhD English (Composition and TESOL), is Assistant Professor of National Chengchi University's Foreign Language Center in Taipei, Taiwan. Her research into academic writing, multilingual researchers' publishing experiences, and national scholarly journals has appeared in *ESP*, *Global Academic Publishing: Policies, Perspectives, and Pedagogies*, and *Written Communication*.

Michelle Sidler is Associate Professor of technical and professional communication at Auburn University. Her research interests include scientific publishing, grant writing, and technical editing. She currently serves on a university-sponsored grant with Rural Studio's Front Porch Initiative, an innovative architectural program dedicated to rural housing.

Sabrina Sobel is Professor of Chemistry and formerly chair of her department. She has served on three rounds of development of ACS standard undergraduate inorganic chemistry exams and has collaborated with Lisa DeTora on various pedagogical initiatives including Seminars in Chemistry and a summer precollegiate program for students interested in STEM fields.

Sarah Tinker Perrault is Associate Professor in the School of Writing Literature, and Film, and also Director of the Writing Intensive Curriculum program at Oregon State University.

Chad Wickman is Associate Professor in the Department of English at Auburn University. He teaches undergraduate and graduate courses in writing and rhetorical studies, and his current research explores intersections between technical discourse and knowledge-making in the life sciences.

Anat Yarden is Professor of Science Education, head of the department, and head of the Life Sciences Group at the Department of Science Teaching, Weizmann Institute of Science, Israel. Her group pioneered the adaption of primary scientific literature for the teaching and learning of biology in high schools.

ACKNOWLEDGEMENT

We are grateful to the York College of Pennsylvania Faculty Development Committee for a Research and Publication Grant to support the indexing of the book.

INTRODUCTION

Michael J. Zerbe, Gabriel Cutrufello, Cristina Hanganu-Bresch
and Stefania M. Maci

This handbook approaches the study of scientific communication from a primarily rhetorical perspective, though some chapters also contain some linguistic and narrative analysis as well. A rhetorical perspective is a form of textual analysis that focuses on the purpose(s) of a text, bearing in mind the text's effectiveness with respect to one or more target audiences. A rhetorical perspective, like other forms of textual analysis, is informed and shaped by organizational, national, and cultural contexts. Additionally, this handbook largely considers *scientific* communication—communication among scientists, including, in some cases, citizen scientists who participate in the scientific process—rather than *science* communication—communication between scientists and nonscientists, a topic that is covered in the *Routledge Handbook of Public Communication of Science and Technology*. Finally, this handbook considers science as the largely inductive, experimental process that is characterized in general by partition, measurement, and quantitative analysis and that has evolved since the Scientific Revolution, centered mostly in Europe, in the late 1600s.

It is fair to say that today no rhetoric defines our lives more than scientific rhetoric. As the form of rhetoric most commonly perceived as a source of knowledge, reality, and truth, scientific rhetoric occupies a dominant, privileged position among the types of rhetorics that shape human experience. Scientific rhetoric creates and consumes vast amounts of discursive energy for issues from the monumental—how will climate change impact our world, and does my child have autism, based on the authoritative *Diagnostic and Statistical Manual of Mental Disorders* (5th edition) criteria?—to the mundane—what kind of tomato should we purchase for today's meal? Given its enormous epistemological and ontological potential, then, scientific rhetoric deserves careful, continual analysis from scholars of rhetoric and communication. Additionally, scientists need to be aware of the powerful role that scientific rhetoric plays in our culture and attend to their work with this discourse assiduously and ethically.

Right from the start, scientific communication sought to distinguish itself from other forms of human communication. It aspired to rise above the fray of unending, intractable, and useless squabbles that, according to early proponents of science (or "natural philosophy" as it was called then), dominated intellectual rhetorics at the start of the Scientific Revolution in the late 1600s and for many centuries prior. Great Britain's Royal Society of London for Improving Natural Knowledge, one of the first scientific societies in the world, formed shortly after the Académie Montmor in France, immediately determined that one of its central tasks would be developing and describing a new kind of language that would create new knowledge and solve

DOI: 10.4324/9781003043782-1

problems rather than be bogged down in unceasing disputes that had not benefited humankind in any way for millennia. Indeed, the Royal Society's historian, Bishop Thomas Sprat, writing in 1667, laments "what wonders [science] would in all likelihood have produced before this, if [science] had been begun in the times of the Greeks, or Romans" (i.e. approximately 400 BCE to 100 CE) (p. 116), and he complains bitterly about rhetoric, describing it as "this vicious abundance of phrase, this trick of metaphors, this volubility of tongue" (p. 112). The language of the new science, Sprat contends, would be devoid of rhetoric because it would strive "to reject all the amplifications, digressions, and swellings of style: to return back to the primitive purity, and shortness, when [people] delivered so many things, almost in an equal number of words" (p. 113). Scientific communication would also, Sprat continues, be characterized by "a close, natural, naked way of speaking; positive expressions; clear senses, a native easiness: bringing all things as near the mathematical plainness as they can: and preferring the language of artisans, countrymen [sic], and merchants, before that, or wits, or scholars" (p. 113). The Royal Society, then, sought to initiate nothing short of a complete revolution of language in its efforts to establish science as an important and worthwhile enterprise for the betterment of humankind.

Sprat's goal of equating "things" and "words" was intended to eliminate the vagaries of interpretation in scientific communication and indeed all forms of communication. Sometimes referred to as the windowpane theory of language, the idea was that a word should mean exactly the same thing to everyone: in other words, metaphorically, the word is on one side of the transparent windowpane, the thing is on the other side, and one should always be able to think of the exact thing when the word is used because the thing is clearly visible through the window. Indeed, early proponents of science such as Francis Bacon were enamored of Chinese and other pictographic (or logographic as linguists would describe it) languages because they thought that a visual symbol was much more representative of the signified object than a phonetically based word. In *The Advancement of Learning* (published in 1605), for example, Bacon wrote, "[W]e understand further that it is the use of China and the kingdoms of the high Levant to write in Characters Real, which express neither letters nor words in gross, but Things or Notions" (p. 166). The goal was for phonetically based European languages, then, to achieve this same high—but utterly imagined—degree of correspondence between text and object.

The Royal Society's pronouncements about language, along with similar attempts to tame scientific rhetoric elsewhere in Europe, were not left uncontested. Italian humanist Giambattista Vico, in 1709, argued in *On the Study Methods of Our Time* that "Nature and life are full of incertitude" (p. 15) and that students, instead of speaking and writing only of absolutes, should be taught how to confront probability, opinion, and debate. In fact, Vico was alarmed by the prospect of students who were taught to communicate only complete truths or falsehoods associated with science, saying that such an approach would result in "odd or arrogant behavior" that would "render young people unfit" (p. 13) for human interaction because they would not know how to cope with difference or to negotiate with people. It would be far better, Vico continues, for students to learn about ethos and pathos appeals in addition to logos appeals and to think about audience. As Vico explains:

> Our young [people], because of their training [in science], are unable to engage in the life of the community, to conduct themselves with sufficient wisdom and prudence; nor can they infuse into their speech a familiarity with human psychology or permeate their utterances with passion. When it comes to the matter of prudential behavior in life, it is well for us to keep in mind that human events are dominated by Chance and Choice, which are extremely subject to change.
>
> *(p. 34)*

Ultimately, Vico's resistance proved futile as science embarked on an astonishingly prolific 300-year record of triumph in medicine, energy, transportation, communication, and agriculture that has utterly transformed the human race and the world around us. To be sure, along with this remarkable progress, science has also made war and the large-scale killing of human beings and the destruction of the environment not only possible but also prevalent. In any case, because of this chronicle of advancement, scientific discourse has solidified its reputation as truthful, neutral, objective, and universal, and, by the twentieth century, it largely replaced religious rhetoric as a dominant source of knowledge, reality, and truth.

Only in the past 75 years have questions about the infallibility of scientific rhetoric once again been effectively broached. Asking these questions has proved to be a challenge. Until the nineteenth century, science was a public spectacle and a public rhetoric. Scientific demonstrations took place in public venues and even in private homes as a form of entertainment for guests attending an afternoon tea or soirée. Scientific topics and debates were widely covered in the popular press and discussed by the (nonscientist) reading public (Shuttleworth and Cantor, 2004), and, like other prominent cultural institutions, science was a ripe target for satire (Paradis, 1997). By the beginning of the twentieth century, though, much science had moved into laboratories as its need for larger and more sophisticated instrumentation and controlled environments grew (Shapin, 1988). Now increasingly segregated from the public, science continued its record of success, but scientific communication disappeared from popular sources as the number of specialized scientific journals rapidly proliferated. Scientific rhetoric became steadily more characterized by highly technical terminology, Bishop Sprat's early pleas to use simple language notwithstanding, as scientists found that they needed a more precise lexicon to describe and interpret the phenomena they were observing. The use of passive voice and a restrained, obscured pathos were also progressively more common features of scientific rhetoric, leading to a perception of scientific discourse as neutral and objective but also highly complex and unable to be clearly understood by nonscientists; as a result, the general public no longer regularly participated in discussions about science. Although all of these changes in scientific communication potentially made its analysis more difficult, theorists such as Perelman and Olbrechts-Tyteca proceeded apace, in their *New Rhetoric* published in 1958, by questioning the scope of science and its ability to solve all human problems:

> The increased confidence thus brought about in the procedures and results of the mathematical and natural sciences went hand in hand with the casting aside of all the other means of proof, which were considered devoid of scientific value. Now this attitude was quite justifiable as long as there was the hope of finding a scientifically defensible solution to all actual human problems through an increasingly wide application of the calculus of probabilities. But if essential problems involving questions of a moral, social, political, philosophical, or religious order by their very nature elude the methods of the mathematical and natural sciences, it does not seem reasonable to scorn and reject all of the techniques of reasoning characteristic of deliberation and discussion—in a word, of argumentation.
>
> *(p. 512)*

So marked the beginning of attempts to vanquish Descartes's insistence, in *Discourse on Method* published in 1637, on mathematical certainty to answer every question posed by humankind, which had severed scientific communication from rhetoric for some three centuries. Similarly, at about the same time that Perelman and Olbrechts-Tyteca were publishing their work, Burke linked this preliminary critique of science more specifically to scientific rhetoric in *A Grammar*

3

of Motives by interrogating the existence of a neutral scientific vocabulary (pp. 284–285) and by questioning scientists' generalizing, via metaphor, of their results (pp. 510–511). These fledgling critiques of science and scientific rhetoric can be linked to mid-twentieth-century scientific advances such as the atomic bomb and the highly toxic pesticide DDT, which revealed that, while science was capable of curing long-fatal diseases and providing electricity and a means of worldwide transportation, it had reached a point that it could also end human life and make the planet uninhabitable. Scientists such as the physicist Robert Oppenheimer (in books such as *Science and the Common Understanding* and *The Open Mind*) and biologist Rachel Carson (in *Silent Spring*) publicly questioned science's role in society and its trajectory with their critical assessments of the atomic bomb (which Oppenheimer helped to develop) and DDT, respectively. Relatedly, in 1968, James Watson, who in 1953 along with Francis Crick had elucidated the structure of the DNA molecule, published a widely read autobiographical account of the race to ascertain DNA's configuration in *The Double Helix*, which significantly punctured scientists' carefully crafted and maintained ethos of seriousness and objectivity. The book revealed instead the ego-driven, competitive, and sexist nature of science with its description of the rivalry, which may have at times even approached deception, between Watson and Crick at the University of Cambridge and Rosalind Franklin and her colleagues at King's College to first establish the structure of the molecule and begin to learn how genetic information is transferred from one generation to the next. Indeed, *The Double Helix* was so controversial that its originally contracted publisher, Harvard University Press, refused to publish the book, a decision that drew withering scorn from the Harvard student newspaper; it was later published by Weidenfeld and Nicolson in the United Kingdom and by Atheneum Press in the United States instead. Finally, in his famous *Two Cultures* lecture and subsequent book, physical chemist and novelist C. P. Snow maintained that the humanities and sciences were so far apart that their practitioners could no longer talk to each other coherently and that this estrangement posed a serious obstacle to solving world problems. Thus did the intellectual winds begin to shift, and science and scientific communication became, for the first time, the subjects of critical analysis.

Scientists and their supporters worked diligently to maintain their dominance and perceived reputation for being able to avoid prejudice and emotion. Stories about science in the popular and influential American magazine *Life* (known for its photography), for example, portrayed science and scientists very positively during the magazine's heyday from about 1940 to 1960. Many of these stories included staged photographic portraits of scientists working with scientific equipment and instrumentation; these portraits were intended to enhance scientists' ethos (Gigante, 2015).

Nevertheless, by the 1970s, the idea that scientific discourse is as fully rhetorical as any other type of human communication reached critical mass. Theorists such as Lyotard (in *The Postmodern Condition*) and Žižec (in *Looking Awry: A Introduction to Jacques Lacan in Popular Culture*) described the narrative characteristics of scientific discourse. Foucault (in *The Order of Discourse*) recognized the epistemological dominance of scientific rhetoric, used as a method of controlling discourse by means of scientifically constituted distinctions between "reason" and "madness" and between truth and falsehood, as well as by means of disciplinarity (pp. 53–55). Additionally, scientific texts were, for the first time, analyzed rhetorically, as with Campbell's pioneering work (1970) on Darwin's *Origin of Species*. In 1976, an article entitled "The Rhetoric of Science," by Wander, was published in the journal *Western Speech Communication*, and rhetoric and science were finally recognized as inextricably intertwined, perhaps much to Descartes's and Bishop Sprat's everlasting dismay. Work on scientific discourse by rhetoricians and communication theorists thus began in earnest. More recently, connections between rhetoric and math/statistics have been established, such as describing statistics as the Toulminian warrant (i.e. the logical line of reasoning that enables evidence to support a claim) (Toulmin, Rieke, and Janik, 1979) of empirical, quantitative research that is the basis of most scientific research. Such warrants are integral to all branches of science except for descriptive sciences such as paleontology and theoretical work in math and physics. Building on this connection, a collection of essays entitled *Arguing with Numbers: The Intersections of Rhetoric and Mathematics*, edited by Wynn and Reyes, was released in 2021. Additionally, the arbitrary nature of the .05 *P* value threshold for statistical significance, a critically important gatekeeping constraint in scientific research, has been explored in a rhetorical history of this concept by Little (2001).

Through most of its 300-year history, science and scientific communication has largely been a Western European, male-dominated institution, and these biases are reflected in the institution's material manifestations and consequences. For example, during the late nineteenth and early twentieth centuries, French surgeon and anatomist Paul Broca and his colleagues removed brains from corpses that they dug up in Paris cemeteries and painstakingly measured the brains' weights and volumes. Finding that men's brains were, in general, larger and heavier than women's brains, Broca and his cohorts concluded that men were intellectually superior to women—a preposterous inference, admittedly, but one that matched the prevailing cultural attitudes toward women and that was therefore accepted as ironclad proof by nearly all white men, especially those in positions of power and authority (Gould, 1980). And Broca did not stop with gender. With no basis in data or logic, Broca and his colleagues extrapolated their findings to claim that Black men had brains that were similar in weight to women's brains and that thus Black men could not intellectually compete with white men either (Gould 1980). Though the prestigious scientific journal *Nature* conceded in a 1997 editorial that "fashionable ideas on the design of experiments to the negotiations that take place through the peer review process" do in truth influence science and scientific communication ("Science Wars," p. 373), these biases have not been eradicated. Indeed, the *Nature* editorial makes no mention of the fact that "fashionable ideas," a phrase that likely fits some instances of scientific research but trivializes the racist and misogynistic viewpoints in others, also shape the interpretation of data and the all-important articulation of conclusions, and science in the service of misogyny and racism continues to this day. In 1994, reprising Broca et al.'s work from a century earlier, social science researchers Herrnstein and Murray conducted a retrospective statistical analysis of standardized IQ test results in their book *The Bell Curve* to argue that Black people are less intelligent than white people, who are in turn less intelligent than Asian people, and that a reason for these different levels of intelligence is at least partially (between 40% and 80%) hereditary (i.e. racial) (pp. 22–23). Though a number of critics asserted that Herrnstein and Murray made questionable assumptions as well as dubious methodological and statistical decisions, the book's conclusions were publicly supported by other social scientists and still exert power as evidenced by continued structural and institutional racism. Currently, in 2021, scientific rhetoric on sickle cell trait, most commonly found in Black people, is used to excuse police brutality even though the condition is almost always benign because the individuals who have sickle cell trait only carry one of the two genes needed for full-fledged sickle cell anemia to occur (LaForgia and Valentino-DeVries, 2021). Scientific rhetoric is also used by an endowed professor of psychiatry, Paul McHugh at Johns Hopkins University, one of the most prominent medical schools in the United States, to argue against biological causation of sexual orientation and identity (Mayer and McHugh, 2016, pp. 7–9) and thus to legitimize ongoing discrimination toward the LGTBQ+ community. As a dominant cultural rhetoric, scientific rhetoric is, then, hijacked by those in power to consolidate and maintain their authority in the same way that religious rhetoric was used in ages past.

Scientific discourse started as a trickle in the 1700s but has exploded, exponentially, to a tsunami today. An estimated 30,000 peer-reviewed scientific journals publish some 2 million articles a year (Altbach and de Wit, 2018). Indeed, in *Laboratory Life: The Construction of Scientific Facts*, Latour and Woolgar describe scientists as "almost manic writers" (1979, p. 48) who spend a great deal of their time writing proposals, research reports, reviews, posters, and other kinds of texts. Indeed, writing is so important to a scientist's career that a great deal of effort is expended on scientific communication pedagogy and ongoing professional training. In terms of genre, scientific communication is dominated by the scientific research article, known colloquially as IMRAD for its Introduction, Methods, Results, and Discussion organizational scheme. Most professional scientists who have the responsibility of planning and managing scientific research must publish IMRAD articles regularly in peer-reviewed journals to advance professionally and maintain job security. At professional conferences, taking advantage of advances in communication technology, scientists now present their research in the form of scientific posters. Research proposals are another critically important genre and, chronologically, actually come before research articles and posters because most scientists must actively pursue funding from government sources and/or private charitable foundations to conduct their research.

Despite the now widespread acknowledgment that scientific discourse is rhetorical, it retains its distinction as a dominant rhetoric in terms of epistemological and ontological potential. Even so, scientific rhetoric is presently under assault in a number of areas in which conclusions formulated on the basis of scientific research and evidence are at odds with the religious, political, and/or otherwise ideological views of various population groups. The attacks on scientific rhetoric can be seen most clearly in the flashpoint issues of vaccine safety and effectiveness, climate change, evolution, and genetically modified foods (i.e. genetically modified organisms, or GMOs). Anti-vaxxers and those who are hesitant to use vaccines dispute their necessity and efficacy in large part because of the infamous (and since retracted) 1998 article by Wakefield et al. that attempted to link the administration of the measles, mumps, and rubella vaccine to autism (Wakefield et al., 1998). Many anti-vaxxers and vaccine hesitant individuals reject the finding that Wakefield's research was fraudulent, research that disproves Wakefield's theory, and any other scientific and medical work that demonstrates that vaccines are overwhelmingly safe and effective, and many also insist that vaccine mandates are a breach of freedom. Their distrust of any scientific discourse except that which supports their previously determined ideological views is a hallmark of the other flashpoints as well. For example, efforts to fight climate change are obstructed by wealthy, powerful energy corporations whose executives seek to protect profits, at least until they can determine how to make even more money in a post–fossil fuel world (Sikka, 2012), as well as by people who view recycling and attempts to reduce carbon use as an infringement of their individual rights. Religious fundamentalists continue to insist that evolution is a debatable theory that should compete on an equal footing with the belief that a deity created life. Anti-GMO activists contend that any food that is genetically modified potentially risks human health, the environment, and/or socioeconomic stability. In sum, what all of these flashpoint issues demonstrate is that scientists today face a potent rhetorical challenge in that many nonscientists rebuff logos and ethos appeals, the mainstay of persuasion in scientific research, out of hand. In fact, the tentative tone of the Discussion section in scientific research articles, with its de rigueur use of hedges such as "suggest," "may," and "possibly," is often used against scientists by those who seize on such phrasing to argue that the scientist's conclusions are not proven. Additionally, special interest groups who work to keep all of these flashpoint issues in the public spotlight, each of them a "manufactured scientific controversy" (Ceccarelli, 2011, p. 196), employ well -credentialed and occasionally prominent scientists who will, often (mis) using scientific rhetoric, say what the special interest groups want them to say.

Today, scientific communication is a behemoth. It is an enormous, powerful, profitable, worldwide institution that formidably influences nearly every person on Earth—the vast majority of them as stakeholders rather than as writers or readers/listeners—as well as the planet itself. It is now as much a visual rhetoric as a textual rhetoric, given the development of the Internet and social media platforms, the use of posters at conferences, and the advent of graphical abstracts. As with any large, complex institution, scientific communication faces a wide range of challenges that need to be addressed. First, the use of English as a lingua franca for scientific communication is perhaps advantageous in terms of efficiency and comprehension but hinders those for whom attaining English fluency is an obstacle. Additionally, given that different languages are different ways of thinking, English-only scientific communication limits the scope and understanding of science. Second, the evaluation of science and scientific communication has become far more complicated because of the sheer amount of research that is conducted and published and the potential for conflicts of interest that, to be fair, many editors of leading scientific journals and publishers work hard to prevent. Even so, retractions are now so common that a now popular website, www.retractionwatch.org, has been established to track them. Third, the pressure to publish is at times so intense that researchers will engage in dubious practices such as P-hacking (i.e. manipulating data until a statistically significant result is found) and author inflation (i.e. adding names of authors who contributed little or nothing to the research to the article so that they will get publication credit). Indeed, quantitatively scoring the impact of scientific journals and the number of publications a scientist has contributed to is now a major industry in and of itself, and P-hacking, also known as significance chasing and data dredging, is now more possible than ever with the arrival of big data (see the 2021 special issue of *POROI* for work on rhetorical approaches to big data). Fourth, access to scientific communication poses a lofty hurdle because of the skyrocketing cost of journal and database subscriptions. While open access endeavors such as PLoS (Public Library of Science) have gained a foothold, most scientific communication remains behind an expensive paywall. Fifth, both practitioners and scholars of scientific communication must consider the issue of science literacy and vigorously combat the notion that scientific information is too complex for nonscientists to understand. Ideas, even complex ones, can and should be explained clearly to any kind of audience by communicators willing to spend sufficient time and energy to determine the most rhetorically effective way to reach them. As a corollary, scientific communication should develop and implement rhetorical strategies to enable science to attract a more diverse population. Sixth, and perhaps most importantly, practitioners and scholars must ensure that the "ethic of expediency" (Katz, 1992, p. 255), the idea that human life is downplayed so drastically in scientific rhetoric that tragedies such as the Holocaust become possible, does not define scientific communication.

By addressing these challenges, scientific communication and science may become less isolated institutions that are more responsive to the needs of all humankind, not just those of the wealthy and powerful. A key to this more promising future is for practitioners and scholars of scientific communication to find a balance between supporting ethical, settled science and critiquing unethical and/or poorly executed science (Ceccarelli, 2011, p. 218). As evidenced by CRISPR, the recently developed method to target and manipulate DNA, science treats nature itself as a text, which it works to analyze and annotate, as with DNA's textual characterization in the form of letters to represent the four bases that constitute the molecule (Zerbe, 2019). As such, both practitioners and scholars of scientific communication have a lot to offer to this critically important field.

This book was completed during a very strange and challenging time in history, especially in medical and scientific history: the global coronavirus pandemic that spanned most of 2020 and 2021. During this time, scientific production and communication have been in overdrive, as researchers rushed to understand the origins, pathology, and course of treatment for the virus that paralyzed the world for an unprecedentedly long time in modern history. Many of the chapters in this collection have in fact started to address what that has meant for scientific communication and how this event will alter the pace of publication, the peer review processes, public understanding of science, and many other related issues (see, for example, Poe (Chapter 4), Mallette (Chapter 12), DeTora and Sobel (Chapter 11), Noguchi (Chapter 15), Maddalena (Chapter 27), and Reid (Chapter 29), this volume). While the definitive history of the COVID-19 pandemic and its implications for scientific communication remains to be written, we need to acknowledge that such events are going to happen more rather than less frequently and that science communication needs to remain extraordinarily supple and adaptable to these new challenges.

Section and chapter summaries

This handbook is divided into four sections: Scientific Research, Writing, and Publishing; Scientific Communication Genres; Scientific Visuals and Multimedia; and Scientific Communication Pedagogy. We have decided on this organization based on our perception of the current areas of interest in scientific communication. The topics in the first section aim to address some of the many challenges that confront scientific communication today—including the exponential growth of publishing and research, ethics, authorship, racism, sexism, predatory publishing, English as the de facto language of international science, measuring impact, assessing citation practices, scientific collaboration, and the emergence of citizen science. How we encode knowledge in genres, visualize it, and transmit it to others forms the focus of the remainder of the book. Methodologies used in the chapters that comprise the handbook include case studies, historical accounts, critical perspectives, and empirical approaches.

In **Part 1: Scientific Research, Writing, and Publishing**, the contributors explore the landscape of writing for publication in the sciences as a central activity to the work of scientific researchers. **Ken Hyland** describes scholarly publishing and, while discussing how academic publication has evolved over the last 50 years, explains the incredible rise of the number of journals, the Anglicization process of publication (cf. English as a T-Rex, Swales, 1997), the "publish or perish" pressure authors have had, the trend toward journal specialization, the relevance on citation and journal indexing, all of which, ultimately, have put an emphasis on the quantitative more than the qualitative aspects of research dissemination. **Sarah Tinker Perrault** details the ways that the concept of authorship has developed in order to assign credit, ownership, and responsibility for scientific work. **Michelle Sidler** draws a broad picture of open access's impact on scientific publishing, including a look at its history, rationale, evolution, and current international landscape. **Mya Poe** reviews how ethical standards are negotiated in regard to authorship, text recycling, data integrity, and peer review, which are to be understood as deeply influenced by the ways that scientists make knowledge in the global scientific community today. **Natasha N. Jones** examines the implications of racism and white supremacy in scientific communication scholarship and pedagogy, focusing in particular on testimonial quieting and self-censorship and their reverberations on how we practice, teach, and experience science and science communication. **Lillian Campbell** examines the relationship between feminism and scientific communication by recounting the history of women scientists and their rhetorical practices and discussing current challenges. **Brad Mehlenbacher** and **Ashley Rose Mehlenbacher** review the scholarship around scientific peer review, which they view as an imperfect "socially constructed knowledge-making activity" bound by historical and institutional constraints, among others. They focus in particular on the issue of expertise in peer review and identify and discuss current peer review trends such as training, transparency, and preprints, and data sharing. **Cheryl L. Sheridan**'s chapter on the editorial review process of Taiwan-based science journals investigates journal editors' approach and thoughts on their role in the development of scientists as communicators. **Tomás Koch Ewertz** looks at the history and role of citations in scientific publications, with special emphasis on the meaning and pitfalls of modern bibliometric indicators and how such tools influence citation practices. **Glenn Hampson** explores the perceptions and practices associated with the measurement of impact of scientific research articles. **Lisa DeTora** and **Sabrina Sobel** provide an overview of scientific collaborative practices in the context of various genres and settings in which scientists have to function, with special emphasis on regulatory practices. **Jennifer C. Mallette** examines the nature of citizen science and public science and what it means for the creation and communication of scientific knowledge. **Maurizio Gotti** explores the globalizing effects of English as the dominant scientific language in the context of a lengthy research project conducted at CERLIS, an institute devoted to the research of specialized discourse at the University of Bergamo. The research highlights the homogenizing and localizing trends in academic usage, which vary according to multiple factors including discipline, expertise, genre expectations, as well as local tradition and culture.

In **Part 2: Scientific Communication Genres**, the contributors investigate the various genres used in scientific communication to disseminate research to various audiences. **Marina Bondi** surveys the history and development of the all-important original scientific research report, now known almost universally by the acronym IMRAD (Introduction, Methods, Results, and Discussion), which is the genre's organizational structure. **Judy Noguchi** surveys the evolution and many forms and functions of the review article, arguing that the genre will undergo further evolution driven by the accelerated pace of publication and advances in data management software. **Begoña Bellés-Fortuño** provides a historical overview of proposals as research genres, summarizing multiple perspectives, such as the rhetorical moves of proposals, linguistic corpus analyses, and sociocognitive and dialogic dimensions of proposals. **Colleen A. Reilly** explores the evolution of grant proposals to the U.S. federal government funding agencies, National Institutes of Health and National Science Foundation, and comments on grant writing aspects of professional development programs for early career scientists. **Kalie Leonard** presents the results of an empirical study of scientists' attitudes toward the use of jargon in posters and their impressions of the future of poster design. **Chad Wickman** explores the genre of the medical case report, focusing on recent developments such as the CARE (CAse REport) guidelines and noting that case reports offer an opportunity for patient experience and agency. **Maureen Mathison** charts the history of scientific letters and commentaries and demonstrate that these genres of scientific communication are an understudied genre that help to illuminate the societal concerns of researchers. **Justin Mando** provides three case studies of nontraditional communication of science that demonstrate scientists acting in the role of citizens; the goal of such communication is to be both socially responsive and critically reflective.

In **Part 3: Scientific Visuals and Multimedia**, the contributors present research on the role of visual information (in print and digitally) in scientific communication. **Danielle DeVasto** explores the history of data visualization to demonstrate its long-standing centrality to scientific communication and provides 18 images as examples of the broad range of types of visuals and their uses in scientific communication. **Vaclav Brezina** and **Raffaella Bottini** provide a brief historical overview of data visualization with a focus on its core functions, interaction, and mutual replaceability for successful scientific communication. **Jonathan Buehl** examines the history, theory, and current usage of increasingly popular graphical abstracts via a series of case studies. **Lauren E. Cagle** helps to further refine the definitions of multimodality in light of interactive technologies intended to enhance scientific communication. **Amy D. Propen** investigates how citizen scientists use eBird to help develop visual information and other important data that aids in scientific communication.

In **Part 4: Scientific Communication Pedagogy**, the contributors detail a variety of pedagogical approaches and offer examples of classroom teaching, programmatic design, and institutional support for scientific communication learning. **Kate Maddalena** reviews three overlapping areas of scientific and science communication pedagogy and argues that teachers of scientific communication should focus on "embedded work," learning to communicate as part of the scientific research process. To help alleviate the biases and problems associated with English-dominant science, **James N. Corcoran** and **Karen Englander** propose an adaptable pedagogy for scientists in training that recognizes and draws on the strengths of their plurilingual abilities. **Gwendolynne Reid** proposes and explains a list of threshold concepts for scientific writing: (1) scientific writing is central to scientific inquiry; (2) scientific writing is rhetorical; (3) scientific genres serve distinct purposes in scientific genre ecosystems; and (4) scientific writing and language are contested and dynamic. These concepts can serve as stepping stones in all programs dealing with scientific literacy and communication and open a fruitful space of dialogue for any pedagogical practice in this field. **Moriah Ariely** and **Anat Yarden** introduce Adapted primary literature (APL) as an apprenticeship genre for promoting scientific literacy in high schools. **Yael Barel–Ben David** and **Ayelet Baram–Tsabari** look at several models of training programs in science communication for practicing scientists and assess these interventions in terms of efficacy. They conclude that such writing training is necessary and beneficial but that results may vary and long-term interventions may be considered to hone certain high-level communication skills. **Jenna Morton-Aiken** details the work undertaken to restart the WAC program at Massachusetts Maritime Academy. Using the Populations, Landscapes, Infrastructure, and Assessment framework, Dr. Morton-Aiken demonstrates how local context can be analyzed to help situate the work of WAC programs. **Susanne Hall** discusses the value of having STEM disciplinary experts working in the Writing Center. She argues that a tutor's expertise can and should be brought into the tutoring session to better help STEM writers.

This *Handbook* is directed at an audience of scientists, health care professionals, and other kinds of researchers who need to write and publish in their scientific, medical, or technological fields, as well as at professional science and technical writers and academics involved in the study, practice, and teaching of scientific, medical, and technical communication to undergraduate and graduate students. The exploration of scientific literature that the *Handbook* presents is a State-of-Scientific-Communication report for the early twenty-first century that offers scientists a critical look at the complex relationships characterizing current scientific practices in academia, industry, government, and elsewhere. The *Handbook* also identifies the drivers influencing domain-specific processes and discursive practices which seem more and more realized in rhetorical, pragmatic, and microlinguistic features indicative of (new) converging communicative practices. We hope that this volume is of use to scholars and teachers of scientific communication but also to practitioners of science, as it offers multiple, up-to-date perspectives (historical, theoretical, or empirical) on issues of heightened contemporary interest in this field.

References

Altbach, P. G., & de Wit, H. (2018) 'Too Much Academic Research Is Being Published', *University World News*.

American Psychiatric Association, DSM-5 Task Force. (2013) *Diagnostic and Statistical Manual of Mental Disorders: DSM-5* (5th ed.). Washington, DC: American Psychiatric Publishing.

Bacon, F. (2015) [1605] *The Advancement of Learning*. Oxford: Clarendon Press.

Burke, K. (1969) *A Grammar of Motives*. Berkeley: University of California Press.

Campbell, J. A. (1970) 'Darwin and *The Origin of Species*: The Rhetorical Ancestry of an Idea', *Speech Monographs*, 37, pp. 1–14.

Carson, R. (1962) *Silent Spring*. New York: Houghton Mifflin Company.

Ceccarelli, L. (2011) 'Manufactured Scientific Controversy: Science, Rhetoric, and Public Debate', *Rhetoric & Public Affairs*, 14(2), pp. 195–228.

Descartes, R. (2020) [1637] *Discourse on the Method of Rightly Conducting the Reason, and Seeking Truth in the Sciences*. Tr. John Veitch. Chicago: Open Court Publishing Company.

Foucault, M. (1970) 'The Order of Discourse', in *Untying the Text: A Post-Structuralist Reader*. Ed. Robert Young. Boston: Routledge & Kegan Paul, pp. 48–78.

Gigante, M. E. (2015) 'A Portrait of Exclusion: The Archetype of the Scientist at Work in *Life* Magazine', *Rhetoric Review*, 34(3), pp. 292–314, doi: 10.1080/07350198.2015.1040305

Gould, S. J. (1980) 'Women's Brains', in *The Panda's Thumb: More Reflections in Natural History*. New York: W. W. Norton & Company, pp. 152–159.

Herrnstein, R. J., and Murray, C. (1994) *The Bell Curve: Intelligence and Class Structure in American Life*. New York: Free Press.

Katz, S. B. (1992) 'The Ethic of Expediency: Classical Rhetoric, Technology, and the Holocaust', *College English*, 54(3), pp. 255–275.

LaForgia, M., and Valentino-DeVries, J. (2021) 'How a Genetic Trait in Black People Can Give the Police Cover', *New York Times* (May 15), https://www.nytimes.com/2021/05/15/us/african-americans-sickle-cell-police.html

Latour, B., and Woolgar, S. (1979) *Laboratory Life: The Construction of Scientific Facts*. Beverly Hills, CA: Sage Publications.

Little, J. (2001) 'Understanding Statistical Significance: A Conceptual History', *Journal of Technical Writing and Communication*, 31(4), pp. 363–372, doi: 10.2190/TUL8-X9N5-N000-8LKV

Lyotard, J.-F. (1984) *The Postmodern Condition*. Minneapolis: University of Minnesota Press.

Mayer, L. S., and McHugh, P. R. (2016) 'Sexuality and Gender: Findings from the Biological, Psychological, and Social Sciences', *The New Atlantis: A Journal of Technology & Society*, 50.

Oppenheimer, J. R. (1954) *Science and the Common Understanding*. Ann Arbor: University of Michigan Press.

Oppenheimer, J. R. (1955) *The Open Mind*. New York: Simon & Schuster.

Paradis, J. G. (1997) 'Science and Satire in Victorian Culture', in *Victorian Science in Context*. Ed. B. Lightman. Chicago: University of Chicago Press, pp. 143–178.

Perelman, C., and Olbrechts-Tyteca, L. (1969) *The New Rhetoric: A Treatise on Argumentation*. Tr. J. Wilkinson and P. Weaver. South Bend, IN: University of Notre Dame Press.

'Science Wars and the Need for Respect and Rigour' (1997) *Nature*, 385, p. 373.

Shapin, S. (1988) 'The House of Experiment in Seventeenth-Century England', *Isis*, 79, pp. 373–404.

Shuttleworth, S., and Cantor, G. (2004) 'Introduction', in *Science Serialized: Representations of Science in Nineteenth-Century Periodicals*. Cambridge: MIT Press, pp. 1–16.

Sikka, T. (2012) 'A Critical Discourse Analysis of Geoengineering Advocacy', *Critical Discourse Studies*, 9(2), pp. 163–175, doi: 10.1080/17405904.2012.656377

Snow, C. P. (1959) *The Two Cultures and the Scientific Revolution*. Cambridge: Cambridge University Press.

Sprat, T. (2014) [1667] *The History of the Royal Society of London, for the Improving of Natural Knowledge*. Columbus: Ohio State University.

Swales, J. (1997) 'English as Tyranosaurus Rex', *World Englishes*, 16(3), pp. 373–382.

Toulmin, S. E., Rieke, R. D., and Janik, A. (1979) *An Introduction to Reasoning*. New York: Macmillan.

Vico, G. (2018) [1709] *On the Study Methods of Our Time*. Tr. E. Gianturco. Ithaca, NY: Cornell University Press.

Wakefield, A. J., et al. (1998) 'Retracted: Ileal-lymphoid-nodular Hyperplasia, Non-specific Colitis, and Pervasive Developmental Disorder in Children', *The Lancet*, 351(9103), pp. 637–641.

Wander, P. C. (1976) 'The Rhetoric of Science', *Western Speech Communication*, 40(4), pp. 226–235, doi: 10.1080/10570317609373907

Watson, J. D. (1968) *The Double Helix: A Personal Account of the Discovery of the Structure of DNA*. New York: Atheneum Press.

Wynn, J., and Reyes, G. M. (eds.) (2021) *Arguing with Numbers: The Intersections of Rhetoric and Mathematics*. State College: Pennsylvania State University Press.

Zerbe, M. J. (2019) 'Toward a Rhetoric of DNA: The Advent of CRISPR', *POROI*, 14(2), Article 3, https://doi.org/10.13008/2151-2957.1276

Žižec, S. (1992) *Looking Awry: An Introduction to Jacques Lacan in Popular Culture*. Cambridge: MIT Press.

PART 1

Scientific research, writing, and publishing

1

THE SCHOLARLY PUBLISHING LANDSCAPE

Ken Hyland

Introduction

The last 50 years have seen, perhaps more than at any time since the invention of the printing press, massive changes in scholarly publishing. There has been an explosion of journals, papers, and researchers with the globalization of research and the insistent demands of publishing metrics on scholars across the planet. The period has also witnessed the increasing specialization of journals; the concentration of publishing into fewer corporate hands; the growth of multiple, even mega authorship; an increasing strain on the review system with a decline in the reviewer pool; the move to online publishing; the diversification of publishing genres; and the dominance of English as the international language of scholarship. This increasingly crowded and competitive academic marketplace has created an environment in which plagiarism, salami slicing of studies, and paper retractions have all increased. It has also generated a new breed of publisher, established on the basis of the 'writer pays' gold open access model, which threatens research standards and publishing ethics by guaranteeing publication following cursory 'peer review'. This chapter discusses this new landscape.

The massification of academic publishing

Once the preserve of gentleman scholars with private means and a desire for self-improvement, academic publication is now an enormous industry that dominates the professional lives of academics across the globe. It is thought that as many as nine million scholars are working in 17,000 universities seeking to publish in English-language journals each year (Björk et al., 2009) with some 10,000 publishers generating revenues of US$25 billion annually (Johnson et al., 2018). The number of papers produced is also staggering, with over 3 million new peer-reviewed articles each year (Johnson et al., 2018). When the US$3 billion scientific book market is added to this and the growing volumes of poor-quality research published in predatory (or fake) journals, we can see that academic publishing is a significant academic and financial force.

This is a time of unprecedented growth in academic publishing. There are now more journals, more scholarly papers, more publishers, more coauthorship and, crucially, more academics, many writing in a language which is not their native tongue (Hyland, 2015). Both the number of articles and the number of journals have grown steadily for the past 350 years,

 DOI: 10.4324/9781003043782-3

with the greatest increases in recent times. Bornmann and Mutz (2014), for instance, have identified three phases in the development of scientific communication, with growth rates tripling in each phase: from less than 1% per year up to the middle of the eighteenth century, to 2–3% up to the 1930s and 8–9% to 2012. The global scientific output is now doubling every nine years. In 2018, there were about 33,100 active scholarly peer-reviewed English-language journals plus 9,400 non-English-language journals (Johnson et al., 2018). One of the largest journal publishers, Elsevier, reported over 2 million articles submitted and 1 billion consumed in 2019 (Page, 2020).

One key factor in this expansion has been the migration to online publication and the retrospective digitization of earlier hard copy content, greatly increasing access to the scientific literature while reducing cost per unit. The main driving force, however, is the number of publishing academics worldwide. The latest UNESCO statistics report 7.8 million full-time equivalent researchers in 2013, an increase of 21% since 2007, or around 4–5% per year (UNESCO, 2017). The World Bank puts the total at 8.9 million, with most of the increase among emerging economies as countries recruit scholars to expand their higher education system to gain a foothold in the global 'knowledge economy'. But while publications in non-English-language journals count as part of a country's output, these have dramatically lower numbers of citations.

The fact that English is the language of the overwhelming majority of journals listed in prestigious bibliographic databases means that most publishing authors are now writing in a second language. This is a development viewed with alarm by some journal editors concerned with standards of written English. Visibility, of course, is all-important, and statistics show that academics all over the world are increasingly less likely to publish in their own languages and to find their English language publications cited more often. While many academics recognize that a lingua franca assists the exchange of ideas more effectively than a polyglot system, there is serious concern about the dominance of English on two main fronts: first, that it is accelerating the decline of alternative languages of scholarship, leading to domain loss for many languages, and second that it excludes many EAL (English as an additional language) writers from publishing, thus depriving the world of knowledge developed outside the Anglophone centers of research. This is, then, a politically charged and complex area.

The biggest driving force in the growth of academic publishing, however, has been the quantification of research outputs as a basis for funding and career advancement. The globalization and commercialization of the academy mean that researchers have found themselves in a culture which measures 'productivity' in terms of the number of papers they produce and the citations they receive on those papers. The promotion and career opportunities of scholars across the globe are increasingly tied to an ability to gain acceptance for work in high-profile journals. There are also financial rewards, with many universities in China and Iran offering substantial sums to encourage academics to publish in international journals indexed in the *Web of Knowledge* SCI databases, with inducements of up to $165,000 for a paper in *Science* or *Nature*, a sum 20 times that of a newly hired professor (Na & Hyland, 2016; Wei et al., 2017).

These pressures clearly impact academic life in important ways, reducing, for example, the time that can be devoted to editing journals, reviewing papers, mentoring students, and teaching. This intensive measurement regime also means that scholars must spend their time scrambling to publish whatever they can manage rather than in developing significant research agendas, sometimes leading them to sacrifice detailed, longitudinal, and novel studies for shallowness, repetition, and trending topics. This has also led to a process known as 'salami slicing', or fragmenting single coherent bodies of research into as many publications as possible. This sidesteps strict editorial policies against duplicate publication by breaking a single research paper into their 'least publishable units'.

Another strategy, growing in the sciences, is the use of 'publishing pacts' where researchers work as a team and add each other's names to papers which they may not even have read, let alone written. Interestingly, the number of authors per paper doubled from 1.8 to 3.7 between 1954 and 1998 (Mabe & Amin, 2002), and the Economist (2016) found the average number of authors per paper in Scopus grew from 3.2 in 1996 to 4.4 in 2015. There are numerous reasons for this, from increased specialization, technology, and social media links, but with all authors of a paper getting equal 'credit', pressures on academics to increase their outputs is certainly one of them. In addition to such 'guest authoring', the career rewards now associated with publishing have increased a variety of dubious practices such as plagiarism and a black market of 'paper selling' which offers authorship of papers written by ghost authors and accepted by SCI journals (Hvistendahl, 2013). Retractions of papers by journals are also increasing (retractionwatch.com), although they remain relatively rare, with only about four in every 10,000 papers affected.

Publishing practices: journals, books, and blogs

The appraisal culture and the massive increase in the number of journals have encouraged a shift away from book publishing, a relentless drive toward ever greater specialization, the strengthening of journal status hierarchies, and the imperative to reach new audiences.

The market for books in STEM subjects (science, technology, engineering, and mathematics) has shrunk dramatically in recent years, with income from growing but much smaller e-book sales yet to replace it. There has also been a decline in citations to books (RIN, 2009). While books are important in the humanities and social sciences, with up to 75% of citations to monographs in some fields (Zuccala & van Leeuwen, 2011), there has also been a decline in the popularity of books in the soft knowledge fields (RIN, 2009). So, in 1980 a scholarly publisher could expect to sell 2,000 copies of a history book, but this had declined to 200 by 2005 (Dalton, 2008). Today a book is doing well if it sells 200 copies in its first year.

The rapid pace of modern scholarship, as well as the desire to focus on very specific topics, means that shorter treatments become increasingly important. Classicists now have over 100 journals to choose from, for example, and historians well over 1,000 (SJR www.scimagojr.com/journalsearch.php). The culture of evaluation makes authoring monographs less attractive as not only are there considerable difficulties in measuring their 'quality', but institutional auditors typically value three or four articles more than a book. In addition, the fact that articles are online makes them both more available and more visible, especially as publishers now aggressively promote them through e-mail 'alerts' to readers. Many publishers, in fact, have moved to 'article-based publishing', putting papers online as they are ready without waiting for an issue to be compiled. Navigation to articles is increasingly driven by search rather than browsing, so, having found relevant content through a search engine, specialist database, or their library's online catalog, researchers spend very little time on publisher websites, dipping in to collect what they need for later reference.

Although journals are being undermined by article-based publishing, the popularity of their content remains undiminished. *The Review of Higher Education*, one of the field's most prestigious publications, for example, temporarily suspended submissions in 2018 due to a two-year backlog of articles awaiting publication. Journals not only remain the main vehicle for disseminating and archiving knowledge, but their position has been strengthened by the rise of evaluation systems which assign a value to an article by the quality of the journal in which it is published. There is, of course, a hierarchy of journals. Ninety percent of cited research is published in just 10% of journals, while 50% of all citations from the 11,877 journals in the 2019 *Journal Citations Report* came from only 400 journals (Clarivate, 2020). The disparity between top-ranked and other journals, moreover, may be growing as the ease of online access has led to a narrowing of scholarship in which researchers tend to cite the same small pool of more recent studies.

With increasing submissions, the journals market is becoming ever more specialized and fragmented, with almost every disciplinary niche congested with titles. Specialism generates new journals so that, for instance, there are now 61,300 dentistry journals listed in Pubmed and over 900 in language and linguistics on SJR. The journal brand is the badge of quality which researchers rely on when deciding what is worth reading from the mass of published material. Journal hierarchies have become the means of judging the integrity of individual research in a literature which has grown too large to judge in other ways. This hierarchy is influenced by a range of factors including the journal's timeliness of publication, peer review practices, and the reputation of its contributors, but most importantly is its impact factor.

There are several indices of journal impact, such as the Immediacy Index, Eigenfactor, or h-index; the most influential among evaluators, however, is the Journal Impact Factor (JIF). This measure reflects the yearly average number of citations to recent articles published in that journal but is controversial when extended to evaluate papers in a journal or to assess an individual researcher. This is because a few highly cited papers can dramatically increase the impact of average or less significant papers. Even disregarding editorial efforts to game the system, while highly prestigious journals publish many outstanding papers, they do not publish *only* outstanding papers, and neither do they have a monopoly on outstanding research. Academics and evaluators recognize this yet continue to use the measure as an indicator of research 'quality', strengthening the journal hierarchy in each field and increasing the competitive pressure on academics.

It is also important to mention new publishing venues beyond books and journals. The institutional imperative to reach new audiences and sponsors has encouraged scholars to engage new audiences through new genres. Funding bodies, governments, and cash-conscious universities are increasingly insistent on involvement with the private sector and in applied work. More scholarly publishing therefore takes the form of consultancy documents, patents, and reports. The mantra of 'knowledge exchange' also means taking research to wider publics, so that lay audiences can access the products of their tax dollars. The most significant means of facilitating this wider dissemination is the academic blog, an increasingly important genre to disseminate information, express academic views, and publicize research. By breaking down boundaries between public and private communication, they offer a space for scholars and interested publics to discuss and evaluate research through a more informal and accessible style of communication (Zou & Hyland, 2020).

Peer-review practices: challenges and innovations

The deluge in submissions to ever more journals poses serious challenges to a core principle of academic publication: its system of quality control. Peer review underpins how academia sees itself as a fair, consistent, and impartial enterprise. For readers overwhelmed by the volume of literature, it acts as a filter; for writers, it provides an indication of how other researchers understand their contribution. It is the guidance of useful feedback which is usually responsible for bringing 'revise and resubmit' papers to publication. Most centrally, however, peer review helps screen submissions for publication. Unreviewed research on personal websites is regarded with skepticism by the academic community, and a recent study shows that the more revisions a paper undergoes, the greater its citation impact (Rigby et al., 2018). Moreover, scholars seem committed to it. Eighty-four percent think that without peer review there would be no control in scientific communication, and virtually all believe that their own papers were enhanced by peer review (Publishing Research Consortium, 2016).

But while often seen as the cornerstone of academic credibility and fundamental to the development and integration of new research (Hyland, 2015), the massive scale of publishing is putting a huge strain on the creaking peer-review system. The journal *Nature* receives over 10,000 manuscripts a year, for instance, and Elsevier's online submission system processes 120,000 new manuscripts per month. This surge in submissions demands more reviewers. So just one publisher, Elsevier, made use of 700,000 peer reviewers in 2015 alone to conduct 1.8 million reviews (Reller, 2016). However, with the population of researchers growing, the *proportion* of reviewers is shrinking significantly. Reviewing is now a marginalized part of an academic's role, with increasing work demands allowing little time for it. The burden, moreover, falls particularly heavily on senior academics and U.S. researchers (Warne, 2016), and while Chinese authors are among the most prolific of submitters, they submit twice as much as they review (Warne, 2016).

While anonymity might help prevent personal bias, blind reviewing can make reviewers less accountable so although most reviews are collegial and supportive, some are caustic, rude, and unhelpful (Hyland & Jiang, 2020). Actual bias in peer review, however, is hard to prove and difficult to completely avoid as reviewers have pet theories and approaches. More serious, perhaps, is that reviewer agreement is little better than chance (Rothwell & Martyn, 2000), and it is just not feasible to get the six reviews for every paper needed for statistical reliability (Fletcher & Fletcher, 2003). Peer review is also poor at spotting fabricated data or plagiarism. With the deluge of papers, growing reviewer workloads, and increasing journal specialization, all journals encounter increasing difficulty in identifying willing readers with the time and expertise to review, and questionable work can slip through.

Various solutions have been suggested to address these problems. Some publishers, for example, reward reviewers with free online access to the journal or certificates for their work, while *Nature* and PLOS put reviewers' names on published papers to acknowledge their contribution. Additional training has also been initiated by several medical journals and recommended by the UK House of Commons Science and Technology Committee (2011). Certainly, junior scholars would like institutional recognition for reviewing (Warne, 2016), but studies suggest that training packages have little impact on the quality of reviews (e.g. Callaham & Tercier, 2007).

More dramatically, some journals have moved to an open peer-review system, where both authors and reviewers know the identity of each other. This might increase transparency and encourage reviewers to be constructive, but it comes with the potential to create animosity between authors and reviewers. *The British Medical Journal (BMJ)* adopted this system some 20 years ago, and all PLOS journals now offer authors the option to publish their accepted manuscript alongside the editor's decision letter, reviewer comments, and authors' responses. *Nature* and *PLOS Medicine*, however, have had problems with open peer review because of higher refusal rates, increased delays, and nonengagement of authors and reviewers (Lee et al., 2013). More radical is the idea of postpublication peer review where papers are published immediately after a light check by an editor, opening comments to the judgment of a broader audience. This deters authors from submitting low-quality manuscripts and allows reviewers to claim credit for their contribution (da Silva & Dobránszki, 2015), although engaging outsiders runs the risk of operating as social media 'likes' with only hot topics taken up positively and dissenting opinions voted down.

Business practices: writer pays, megajournals, and predatory publishing

Scholarly publishing is very big business. Profit margins of 35% at the top end of the market are common, so that Elsevier, for example, saw its operating profit rise to £982 million in 2019 (Page, 2020). These kinds of profits have seen publishers acquiring top-drawer journals from nonprofit societies over the past 30 years and consolidating their positions into fewer hands. Just three for-profit companies (Reed Elsevier, Springer, and John Wiley) now account for 42% of all the articles published. Savage cuts to library budgets have barely dented publishers' profits with the average price of an academic journal rising by 6% a year, faster than inflation and beyond library budgets. For libraries, a key complaint against the large publishers is not only the high prices but the practice of bundling subscriptions to lesser journals with valuable ones, forcing libraries to spend money on journals they don't want in order to get those they do want. These notorious 'Big Deals' lock libraries into long-term arrangements for more journals.

These huge profits are, of course, the result of the fact that the content of journals is provided free by unpaid authors, peer reviewers, and editors and are boosted by a reduction in publishing costs such as proofreading, typesetting, copy editing, printing, and worldwide distribution by online publishing. Journal publishing operates in a skewed market where, because every journal has a monopoly over the information of its field, journals compete more for authors than for subscribers. Without normal market mechanisms, publishers have few constraints on the prices they charge. This artificial inflation is shown most clearly in the trend of large publishers to increase prices more than small publishers, when in traditional markets, high volume and high sales enable cost savings and lower prices. This unsustainable economic model has created a backlash from both scholars and libraries. The 'Cost of Knowledge' boycott by academics against the business practices of Elsevier, for example, meant that over 5,000 academics worldwide refused to publish or perform peer review services for its journals. While the movement now seems to have run its course (Heyman et al., 2016), there is still considerable discontent about high subscription prices for individual journals and Big Deal bundles. As a result, many large university libraries in the United States and UK, as well as the British Library, have canceled their Big Deal agreements with the major publishers (Anderson, 2017).

The pricing crisis and desire for reform has helped fuel the Open Access (OA) movement, which makes published scholarly content (articles, monographs, conference proceedings, etc.) available online, free of charge to users, free of most copyright restrictions, and free of access barriers such as registration. OA has been the single biggest change in science publishing this century. Between 2000 and 2010, the number of OA journals grew at 18% a year and the number of articles by 30% a year (Laakso et al., 2011), as publishers 'backfilled' archived material on OA. More than 50% of the scientific papers published over the last decade can now be downloaded for free on the Internet (European Commission, 2015). The *Directory of Open Access Journals* lists nearly 5 million OA peer-reviewed papers and 14,430 journals (April 2020), while Scopus claims to have over 10 million peer-reviewed articles freely available. Sixty percent of conventional subscription journals are also now hybrid, offering an OA option to authors, and it is estimated that OA journals now make up about 26–29% of all articles published, with 5% more available via delayed access and another 12% via self-archived copies (Taylor, 2017).

OA has largely been driven by the idea that research funded by the public purse should be available to the public and that removal of access barriers allows the fruits of research to be exploited and knowledge to advance more quickly. In the UK, all authors funded by Research Councils are required to publish their work in open access, and the Research Excellence

Framework, the system for assessing the quality of research in UK higher education, will only accept submissions that are available by OA. The Chinese Academy of Sciences now requires its members to deposit articles in OA archives one month after their publication, and the US Office for Science and Technology Policy has directed federal agencies with R&D expenditures over $100 million to make research results freely available 12 months after publication. The European Commission has also announced an OA policy for the European Commission and recommended member states to adopt OA.

Essentially OA is actually three models of access: gold, delayed, and green. Gold open access is where the paper is made freely available immediately after acceptance by the publisher to whom it has been submitted, with the production costs recovered from the author's research budget or institution. Delayed, like gold is a subscription-based model but with an embargo on when the paper becomes available. In green open access, authors self-archive their work, either as 'gray' nonreviewed literature or as a version of a paper accepted by a regular peer-reviewed journal. Papers are hosted on the author's personal webpage or university repository, but there are an increasing number of specialized open-access websites. The Registry of Open Access Repositories (ROAR), for example, lists over 4,725 repositories in addition to academic databases such as *ResearchGate* and *Academia.com*.

Open access seems to reduce publication delays and has the potential to increase the reach and visibility of research, with some evidence that articles receive more downloads and citations than paywalled articles (e.g. Tang et al., 2017). But there are also problems. It is uncertain, for example, whether an author-pays model is sustainable for the humanities, where much longer articles and lower acceptance rates mean that it costs over three times as much to publish an article as it does in the sciences (Waltham, 2009). There is also concern that the model may freeze out authors in low-income countries and may jeopardize quality assurance by peer review (Taylor, 2017). Critically, OA does not provide a solution to scientific publishing's most serious funding problems. Several major universities ended their financial support of BioMed Central's Open Access Membership program in the late 2000s, for example, as article processing charges (APCs) soared. The aggregate cost to institutions rose by an average of 11% per annum between 2013 and 2017 (Pinfield & Johnson, 2018), and libraries have found themselves paying more than under the old subscription model.

Gold OA has also created two new breeds of publishers: the megajournal and predatory publishing. The megajournal sector, including platforms such as *PubMed Central* (PMC) for biomedical and life sciences and the *Public Library of Science* (PLoS) for the sciences more generally, has proved highly successful and represent a major innovation in scholarly journal publishing. The journals are characterized by a broad subject scope, encompassing either multiple disciplines or a single large discipline such as medicine or physics; a rapid "nonselective" peer review based on "soundness not significance"; and an Open Access business model based on quantity and competitive APCs (Wakeling et al., 2017).

There is evidence, however, that megajournals are in decline. *PLOS ONE* grew to become the world's largest journal, publishing more than 30,000 papers at its height in 2013, but its output fell by 44% by 2018. Another megajournal, *Scientific Reports*, saw its article count drop by 30% in 2018. With acceptance rates of over 50% and by publishing replication studies and negative results that might be rejected by traditional journals, megajournals could be expected to do better. It seems, however, that publishers have not yet persuaded the research community of their value, and megajournals collectively published only 3% of the total number of papers globally in 2018 (Brainard, 2019). They remain relevant as an option for European authors, however, due to low publishing fees and funders who require that their papers be free to read on publication.

The other major publishing innovation to emerge from the gold OA model is less benign: the massive increase in predatory, fraudulent, and mediocre journals. This unsavory parallel trade, which charges high APCs to authors and waives quality control, has been possible because of the large sums of money now available through research grants and publication allowances, particularly in the hard sciences, and the desperation of academics to get their research published. Unlike legitimate journals, they bombard academics with spam e-mails, misrepresent their country of origin, often fabricate their editorial boards, accept almost all submissions, and overstate the rigor of their peer-review processes, generally accepting anything submitted.

Shen and Björk (2015) estimate that there were 8,000 such journals in 2014, generating $75 million in revenues, while Cabells' Blacklist currently lists 12,000 journals with over 1,000 more under consideration for inclusion. Cabells employs 60 weighted criteria to identify predatory behavior covering integrity, peer review, vagueness over fees, and indexing and metrics (Toutloff, 2019). Red flags include the absence of a named editor or a claim of international scope that is contradicted by a lack of geographical diversity in the editorial board. The biggest red flag, however, is a promise to publish within a short time frame, such as a week, an impossibly short time for a comprehensive peer review. *Think. Check. Submit.* (http://thinkcheck-submit.org/), a cross-industry website devoted to helping researchers identify trusted journals, claims that about 1,000 new journals are launched each year, and most are 'predatory'.

These journals pose a serious threat to the integrity of scholarly communication by publishing work that is plagiarized, fabricated, or based on unsound methods. They also risk contaminating genuine open access journals by association. The presence of these journals on legitimate databases and platforms such as Scopus and PubMed does not help matters. The publishing industry has responded by tightening its codes of conduct and by collaborating to form *Think. Check. Submit.* as a resource for scholars to verify the credentials of publications. The Directory of Open Access Journals also responded by cleaning its database of over 900 suspect journals (Anderson 2014).

Conclusions and speculations

Scholarly publishing is currently in flux. The establishment of an institutional appraisal culture, by focusing on the number of papers published by individuals and universities, discourages the sharing of resources and ideas and replaces it with a highly competitive environment where each individual or research group is working for him- or herself. The structure has created a system in which research is booming, with the numbers of researchers, journals, and papers continuing to multiply; journals becoming increasingly specialized, hierarchical, and concentrated in fewer hands; and multiple authorship growing. On the other hand, traditional peer review is at risk due to the volume of submissions; publishing ethics are under pressure; libraries and scholars are despairing of a journal subscription model which seems exploitative; and publishers are in the midst of a transition to an uncertain alternative basis of funding.

It would, of course, be wonderful to end on a positive note, to be able to say that scholarly publishing is in the process of resolving these issues and moving toward a new age. In this imagined future, research is freely shared, technology is supportive of a reward system based on quality, peer review is transparent and rewarded, and evaluation criteria are independent of journal brands. Unfortunately, this is not the case, and what the future holds for publishing is anyone's guess.

Clearly any system needs to harmonize and coordinate four elements: quality control, certification and reputation, access and storage, and incentives for engagement. It is likely that the journal article will remain with us, in some form, for the foreseeable future. It has had incredible staying power and after 350 years is actually making serious inroads into disciplines where the monograph has traditionally occupied the same space, although new models are likely to gain acceptance and familiarity. Publication will always need to be wrapped in some form of quality assurance and validity measures, and we can expect networked technologies to increasingly offer innovative models in this regard. This could allow for multiple formats – videos, code, visualizations, text, data – to be published and for communication and peer review to become a combined, community-governed process. In this scenario, the quality of engagement, or how individuals make use of the material, rather than the impact factor, becomes the gold standard of success. Finally, gold open access has not been able to reform an unsustainable funding model. Scholarly collaboration networks may offer a viable alternative in empowering librarians and facilitating shared content, as do initiatives such as the Global Sustainability Coalition for Open Science Services (SCOSS).

A recent report to the European Commission (2019) sees strengths in the current system and proposes a vision for the future of scholarly communication with collaboration among all stakeholders. The report, moreover, places changes to the research evaluation system at the heart of reform, seeing 'scholarly/learned societies and other researcher communities' as 'best positioned to affect change across all aspects of scholarly communication' (ibid., p. 6). Clearly, concerted efforts by publishers, institutions, academics, funders, and editors will be needed to effect the reforms needed, but it is certain that scholarly publishing is going to see even more changes in the next decade.

References

Anderson, R. (2014) 'Housecleaning at the directory of open access journals', http://scholarlykitchen.sspnet.org/2014/08/14/housecleaning-at-the-directory-of-openaccess-journals/

Anderson, R. (2017) 'When the Wolf finally arrives: Big deal cancellations in North American libraries', *The Scholarly Kitchen*, https://scholarlykitchen.sspnet.org/2017/05/01/wolf-finally-arrives-big-deal-cancelations-north-american-libraries/

Björk, B.-C., Roos, A., & Lauri, M. (2009) 'Scientific journal publishing: Yearly volume and open access availability'. *Information Research*, vol. 14, no. 1, http://InformationR.net/ir/14-1/paper391.html

Bornmann, L., & Mutz, R. (2014) 'Growth Rates of modern science: A bibliometric analysis based on the number of publications and cited references', Eprint arxiv:1402.4578.

Brainard, J. (2019) 'Open-access megajournals lose momentum as the publishing model matures', *Science*, www.sciencemag.org/news/2019/09/open-access-megajournals-lose-momentum-publishing-model-matures

Callaham, M.L., & Tercier, J. (2007) 'The relationship of previous training and experience of journal peer reviewers to subsequent review quality', *PLoS Med*, vol. 4, no. 1, p. e40.

Clarivate. (2020) 'Journal citations report', https://jcr.clarivate.com/

Dalton, M.S. (2008) 'The publishing experiences of historians', *Journal of Scholarly Publishing*, vol. 39, no. 3, pp. 197–240.

da Silva, J.A., & Dobránszki, J. (2015) 'Problems with traditional science publishing and finding a Wider Niche for post-publication peer review', *Accountability in Research*, vol. 22, no. 1, pp. 22–40, doi:10.1080/08989621.2014.899909, PMID: 25275622

The Economist. (2016) 'Why research papers have so many authors', *The Economist*, www.economist.com/science-and-technology/2016/11/24/whyresearch-papers-have-so-many-authors

European Commission. (2015) 'Proportion of open access papers published in peer-reviewed journals at the European and world levels – 1996–2013', http://science-metrix.com/sites/default/files/science-metrix/publications/d_1.8_sm_ec_dg-rtd_proportion_oa_1996-2013_v11p.pdf

European Commission. (2019) *Future of scholarly publishing and scholarly communication: Report of the expert group to the European Commission*, https://op.europa.eu/en/publication-detail/-/publication/464477b3-2559-11e9-8d04-01aa75ed71a1

Fletcher, R.H., & Fletcher, S.W. (2003) 'The effectiveness of editorial peer review', in F. Godlee & T. Jefferson (eds.), *Peer review in health sciences*, 2nd edition, London: BMJ Books, pp. 62–75.

Heyman, T., Moors, P., & Storms, G. (2016) 'On the cost of knowledge: Evaluating the Boycott against Elsevier', *Frontiers in Research Metrics and Analytics*, vol. 1, doi:10.3389/frma.2016.00007

House of Commons Science and Technology Committee. (2011) *Peer review in scientific communications*, Eighth Report of Session 2010–12, London: The Stationary Office Limited.

Hvistendahl, M. (2013) 'China's publication Bazaar', *Science*, vol. 342, no. 6162, pp. 1035–1039, doi: 10.1126/science.342.6162.1035

Hyland, K. (2015) *Academic publishing: Issues and challenges in the construction of knowledge*, Oxford: Oxford University Press.

Hyland, K., & Jiang, K. (2020) '"This work is antithetical to the spirit of research": An anatomy of harsh peer reviews', *Journal of English for Academic Purposes*, vol. 46, doi:10.1016/j.jeap.2020.100867

Johnson, R., Watkinson, A., & Mabe, M. (2018) *The STM report 5th ed: International association of scientific*, Holland: Technical and Medical Publishers.

Laakso, M., Welling, P., Bukvova, H., Nyman, L., Björk, B.-C., et al. (2011) 'The development of open access journal publishing from 1993 to 2009', *PLoS One*, vol. 6, no. 6, p. e20961, doi:10.1371/journal.pone.0020961

Lee, C.J., Sugimoto, C.R., Zhang, G., & Cronin, B. (2013) 'Bias in peer review', *Journal of the American Society for Information Science and Technology*, vol. 64, pp. 2–17, https://doi.org/10.1002/asi.22784

Mabe, M.A., & Amin, M. (2002) '"Dr Jekyll and Dr Hyde: Author-reader asymmetries in scholarly publishing', *Aslib Proceedings*, vol. 54, no. 3, pp. 149–157, https://doi.org/10.1108/00012530210441692

Na, L., & Hyland, K. (2016) 'Chinese academics writing for publication: English teachers as text mediators', *Journal of Second Language Writing*, vol. 33, pp. 43–55, https://doi.org/10.1016/j.jslw.2016.06.005

Page, B. (2020) 'Elsevier sees 2019 profits and revenues lift', *The Bookseller*, www.thebookseller.com/news/elsevier-sees-profit-and-revenue-lift-1192787

Pinfield, S., & Johnson, R. (2018) 'Adoption of open access is ring – but so too are its costs', *LSE Impact Blog*, https://blogs.lse.ac.uk/impactofsocialsciences/2018/01/22/adoption-of-open-access-is-rising-but-so-too-are-its-costs/

Publishing Research Consortium. (2016) *Peer review survey 2015*, Mark Ware Consulting, www.publishingresearch.org.uk.

Reller, T. (2016) 'Elsevier publishing – a look at the numbers, and more', *Elsevier Connect*, www.elsevier.com/connect/elsevier-publishing-a-look-at-the-numbers-and-more.

Rigby, J., Cox, D., & Julian, K. (2018) 'Journal peer review: A bar or bridge? An analysis of a paper's revision history and turnaround time, and the effect on citation', *Scientometrics*, vol. 114, no. 3, pp. 1087–1105, https://doi.org/10.1007/s11192-017-2630-5

RIN. (2009) 'E-journals: Their use, value and impact', *Research Information Network*, http://rinarchive.jisc-collections.ac.uk/our-work/communicating-anddisseminating-research/e-journals-their-use-value-and-impact

Rothwell, P.M., & Martyn, C.N. (2000) 'Reproducibility of peer review in clinical neuroscience. Is agreement between reviewers any greater than would be expected by chance alone?' *Brain*, vol. 123, pp. 1964–1969, doi:10.1093/brain/123.9.1964

Shen, C., & Björk, B.C. (2015) '"Predatory" open access: A longitudinal study of article volumes and market characteristics', *BMC Medicine*, vol. 13, no. 1, https://doi.org/10.1186/s12916-015-0469-2

Tang, M., Bever, J.D., & Yu, F.-H. (2017) 'Open access increases citations of papers in ecology', *Ecosphere*, vol. 8, no. 7, p. e01887, https://doi.org/10.1002/ecs2.1887

Taylor, S. (2017) 'The rise of open access', *Royal Society*, https://blogs.royalsociety.org/publishing/the-rise-of-open-access/.

Toutloff, L. (2019) 'Cabells blacklist criteria v 1.1. the source', https://blog.cabells.com/2019/03/20/blacklist-criteria-v1-1/

UNESCO (2017) 'Science report: Towards 2030', https://en.unesco.org/unesco_science_report/

Wakeling, S., Spezi, V., Fry, J., Creaser, C., Pinfield, S., & Willett, P. (2017) 'Open access megajournals: The publisher perspective (Part 1: Motivations)', *ALSP*, doi:10.1002/leap.1117

Waltham, M. (2009) 'The future of scholarly journal publishing among social science and humanities associations', *Journal of Scholarly Publishing*, vol. 41, no. 3, pp. 257–324, doi:10.3138/jsp.41.3.257

Warne, V. (2016) 'Rewarding reviewers – sense or sensibility? A Wiley study explained', *Learned Publishing*, vol. 29, pp. 41–50.

Wei, Q., Chec, B., & Shu, F. (2017) 'Publish or impoverish: An investigation of the monetary reward system of science in China (1999–2016)', *Aslib Journal of Information Management*, vol. 69, no. 5, pp. 1–18, doi:10.1108/AJIM-01-2017-0014

Zou, H., & Hyland, K. (2020) ' "Think about how fascinating this is": Engagement in academic blogs across disciplines', *Journal of English for Academic Purposes*, vol. 43, https://doi.org/10.1016/j.jeap.2019.100809

Zuccala, A., & van Leeuwen, T. (2011) 'Book reviews in humanities research evaluations', *Journal of the American Society for Information Science and Technology*, vol. 62, no. 10, pp. 1979–1991, https://doi.org/10.1002/asi.21588

2

THE EVOLUTION OF AUTHOR FUNCTIONS IN SCIENTIFIC COMMUNICATION

Sarah Tinker Perrault

Introduction

The authorship concept links a text to an author – a person or a group – and gives them certain rights and responsibilities. These links work only for certain kinds of texts: writing notes on authorship did not make me an author, but the published chapter will count. Authorship involves locating a text in a discursive context and having it be recognized as part of a discourse (Foucault, 1998).

Because authorship is a function of discourses, authorship criteria vary across time and contexts (national, disciplinary, and institutional), as do the rights and responsibilities of authorship (Biagioli, 2012; Sutherland-Smith, 2016), resulting in there being "little standardization . . . as to what should constitute authorship on a scientific paper" (Osborne and Holland, 2009, p. 2). However, criteria, rights, and responsibilities do share an underlying set of values that arose in early modern England and Europe, in the same time and contexts as natural philosophy (the precursor to today's science). Constructs of authorship rooted in that intellectual heritage dominate the globalizing world of English-language science today but are increasingly challenged by changes in and beyond scientific domains.

To explain today's authorship concepts, this chapter explores the underlying values that inform current concepts of authorship, these concepts' origins, and what kinds of things people can think and do because of the currently dominant concept of authorship.

Authorship criteria

Looking at themes that arose in early modern England and Europe reveals "a number of related histories, all of which point toward the same general constellation of meanings" (Ede, 1985, p. 6) used today.

To understand modernist authorship, it helps to start with the medieval and Renaissance ideas that modernist ideas grew and deviated from. Before the early modern period, writers occupied two roles, and authorship was an "unstable marriage" between "two distinct concepts" (Woodmansee, 1984, p. 426): the craftsperson and the amanuensis. The craftsperson adapted existing ideas and techniques to meet specific needs, or "culturally determined ends" (Woodmansee, 1984, p. 428), using "rules extrapolated from classical literature" (p. 430). In contrast,

DOI: 10.4324/9781003043782-4

the amanuensis served as a vehicle for expressing divine inspiration (Woodmansee, 1984, p. 427) and was less a creator than a tool for a divine being's act. The premodern individual was not considered a source of knowledge, not "distinctly and personally responsible for his creation" (Woodmansee, 1984, p. 427); that role belonged to the ancients, or the god(s).

Three factors in the early modern period created a new conception of writers as authors. First was recognition of the individual; the author is a "modern figure" that developed when modernist thinkers "discovered the prestige of the individual, or, as it is more nobly put, the 'human person'" (Barthes, 1977, pp. 142–143). Second, creativity shifted from outside to inside the person as the period saw "the privileging of a single authorial mind, rather than community wisdom, as the source of invention and the concomitant privileging of texts as reflections of the workings of this sovereign authorial mind" (Crowley, 1990, p. 12). These changes, combined with the third factor – shifting ideas about social status and knowledge production – underpin the three values underlying the modernist construct of authorship: individuality, originality/creativity, and social status.

During this same time, what we now call science was developing in the form of natural philosophy. In that domain, the values underlying authorship developed in conjunction with empiricism. Empiricism created new ideas of epistemic validity, replacing trust in ancient wisdom with trust in experiences described by qualified reporters. In the late 1600s, as the Scientific Revolution "reached a stage of consolidation" via institutions such as scientific societies, practitioners developed new forms of authority (Dear, 2015, p. 53), leaving behind practices in which "attaching the name of an authority to a statement of experiential fact rendered it probable and hence suitable for use in argument" (Dear, 2015, p. 58).

The new form of knowledge-making involved two shifts. First, God's creation (nature), rather than traditional authoritative texts, became the proper source of knowledge (Dear, 2015, p. 61). Second, instead of individuals working with ancient texts, it involved a "cooperative inquiry" (Dear, 2015, p. 55) – not yet today's collaborative work (researchers working together) but cooperation between "an association of individuals with common interests who were engaged in work that was commonly perceived to be of value" (Dear, 2015, p. 57).

This cooperative inquiry required two things: a "tacit agreement on the nature of knowledge" and, in keeping with that agreement, "a common standard of authority" (Dear, 2015, p. 57). However, while leading thinkers at the time "agreed that the authority of the ancients should be rejected, they did not agree on what should replace it" (Dear, 2015, p. 61). What developed was *direct experience* as a source of authority and "the rooting of claims about the natural world in discrete events" (Dear, 2015, p. 66)

As this brief review of empiricism shows, both individuality and originality were vital to the development of natural philosophy. Along with social status, they are the values underlying scientific authorship today.

Modern authorship requires an identifiable entity – a person or group – who is linked to a text. Although in the medieval and renaissance periods the individual was subordinated to collective concepts of knowledge, there gradually emerged "a belief in possessive individualism which is the belief that individuals are entitled to protect themselves and the products of their labors" (Sutherland-Smith, 2016, pp. 576–577). This idea, initially about physical labor, was "soon extended to the labors of an individual's mind" and supported ideas of "the author as an individual creator of text" (p. 577) and of a text as "a unique individual creation – unlike any preceding it" (p. 577).

In sciences, then as now, the goal was to add to existing stores of knowledge, holding "originality as a supreme value" (Merton, 1957, p. 640). Early scientific societies had a "pronounced dedication to the discovery of new experimental data – as opposed to the exposition of texts, which had characterised scholastic science" and a "new ethic" in which science "was conceived of as a *venatio*, or a hunt, an aggressive pursuit after the deepest secrets of nature" (Eamon, 1985, p. 334). But if the hunt is collective, authorship is not; historian Pamela O. Long (1991) distinguishes novelty (marking something as new) from originality (marking something as having been developed by the specified individual) (p. 847). Originality in science is thus about both new ideas and also about the person who comes up with them, "the belief in the importance of individual ingenuity and genius in creation and invention" (Long, 1991, p. 881).

The third value is social status. To be an author, one must have the recognized right to speak in the given context. In early modern England and Europe, people had to be upper-class males to participate in education at all, let alone to become an author.

In early modern science, social status especially mattered because empiricism required new relations of trust which depended on a person's status as a gentleman and ability to join in discussions modeled on gentlemanly social norms. Natural philosophers knew one another and trusted one another's gentlemanly probity, and authorial presence mattered because personal authority still mattered. This shows in writing that Atkinson (1999) describes as "involved" (p. 152), meaning it resembled the in-person interactions it was based on and reflected.

These restrictions show in the Royal Society of London. In theory, knowledge was welcomed from many kinds of correspondents. Thus, Thomas Spratt's 1667 history of the society "emphasized the diversity of those contributing to the Royal Society's work" (Dear, 2015, p. 63). Spratt welcomed knowledge not only from "Learned and profess'd Philosophers; but from the Shops of *Mechanicks*; from the Voyages of *Merchants*; from the Ploughs of *Husbandmen*; from the Sports, the Fishponds, the Parks, the Gardens of *Gentlemen*" (Spratt, cited in Dear, 2015, p. 62). Accordingly, knowledge from any "plain, diligent and laborious observers" (Spratt cited in Dear, 2015, p. 62) was allegedly welcome. In reality, the trust given to information varied depending on the source. Dear (2015) explains that the "identity and status of those involved" was important, that despite Spratt's "boasting of the wide diversity of people who contributed . . . it should not be imagined that all were therefore equal" (p. 68).

The value of status is highlighted by the fact that the person presenting work and viewed as its creator often had not done it; members presented so-called firsthand results, even though technicians had done the hands-on work. Shapin (1989) repeatedly distinguishes between "mere skill (or hand knowledge) and truly reflective and rational philosophical knowledgeability" (p. 561), parallel to the organization of early modern society around "the distinction between those who worked and those who thought or fought" (p. 561). Shapin's distinction explains Dear's (2015) point that "[t]he humble could play their part, but not in quite the same way as the mighty" (p. 68), and it held even when technicians did more than manual labor – up to and including designing and carrying out experiments and writing up the experimental reports, which was still recognized as done by the "master-scientist" (Shapin, 1989, p. 558).

Who can have a voice has shifted over time. Amateurs ("the humble") were squeezed out as sciences professionalized (Atkinson, 1999; Bensaude-Vincent, 2001), while the educational franchise slowly opened to those traditionally excluded on the bases of gender, race, ethnicity, and socioeconomic status. Overall, however, the distinction between the humble and the mighty, the doers and the thinkers, persists in the structures of scientific work, for example in counting intellectual but not manual work as worthy of authorship.

Shapin (1989) finds commonality in "the sort of work that both seventeenth-century and present-day scientists tend to regard as of no importance to the production of knowledge" even when that work is essential (p. 557). Some workers are seen as unskilled and replaceable, such that "certain fundamental elements in the scientist–technician relationship" are echoed in and probably connected to arrangements today (p. 562). The "evaluative distinction between skill and knowledgeability" endures, as shown in Osborne and Holland's (2009) list of common authorship criteria:

- conception or design,
- data collection and processing,
- analysis and interpretation of the data, and
- writing substantial sections of the paper (p. 4).

Overall they recommend assigning authorship based on "conceptualization and design" and on "data collection, data analysis, and creation of the manuscript" (p. 7) but not on "acquisition of funding, general supervision, and clerical or mechanical contributions" (p. 4).

Rights of authorship

Linking people/groups to texts as authors gives those people/groups certain rights, most significantly the right to credit for the work, to ownership of or control over the work, and to determine who else can have the first two rights. This section describes the development of each right over time.

Perhaps the most important right of academic authorship is the right to receive credit for one's work (Cronin, 2001, p. 559). Despite appeals to the Mertonian norm of disinterestedness – which says that scientists work only for knowledge, not for personal credit, fame, or reward – credit has been part of scientific endeavors since the early modern era. Textual gestures that established an account's credibility by linking it to a person – the author – also served to credit that person with the finding or idea being shared. Then, as now, authorship was partly about making priority claims. Alan Gross (2006) explains that even in the Royal Society's early years, priority claims were an issue and spurred the development of "dated journal publication" as a way to ensure that whoever first made a knowledge claim received credit for having done so (pp. 175–176).

The emphasis on "claim-staking and priority" that started in the seventeenth century (Cronin, 2001, p. 559) now undergirds the academic economy. The adage "publish or perish" holds true; since hiring, promotions, and grants all depend on the publication record, "credit for authorship is essential for academic survival" (Sutherland-Smith, 2016, p. 585). (See Tomás Koch Ewertz [Chapter 9] and Hampson [Chapter 10] this volume.)

While credit is the basis of academic rewards systems, authorship also confers rights to ownership and control. Ownership rights exist legally via copyrights and patents and socially via institutional norms (e.g. expectations that scholars give credit via citations, refrain from stealing others' ideas, and so on). Academic contexts use systems of determining priority to give credit and ownership to whoever first shared an idea. These legal and social systems show that ownership, like authorship, depends on concepts of individuality and originality. For example, the first English copyright case, in 1741, emphasized individuality, as "the Lord Chancellor . . . determined writing a text was 'a solitary and self-sufficient act of creation'" (Sutherland-Smith, 2016, p. 577). Likewise, the case emphasized originality, as "the court decided that the concept of originality was a key element" (Sutherland-Smith, 2016, p. 577).

Ownership of ideas was part of science from its start. Eamon (1985) links the idea of "public knowledge" with the sixteenth- and seventeenth-century development of "institutional mechanisms to protect the interests of discoverers" (pp. 321–322), including patents and copyrights. Sciences were key to the institutionalization of intellectual property, as the idea of intellectual property, and its codification in legal rights, "emerged in response to a growing awareness that scientific knowledge could be put to practical use, and that as long as new discoveries were kept secret, the advance of knowledge, and hence profit, would be retarded" (Eamon, 1985, p. 328). While the idea of intellectual property was not new with modernism, the shift to individualism enabled "the emergence of the priority dispute" as "a direct result of this enhanced view of individual genius as intrinsic to discovery and invention" (Long, 1991, p. 881).

Alongside legal protections, natural philosophers developed systems for giving credit and recognizing knowledge claims. For example, the Royal Society's secretary kept an extensive correspondence, and "correspondents understood that the contents of their letters were likely to be read or summarized at meetings of the Royal Society, and would be entered in abstract in its register" to "provide a means for establishing priority in scientific discovery" and to ensure they would receive "recognition and status" (Eamon, 1985, p. 344). After the correspondence became *The Philosophical Transactions of the Royal Society*, publication became a means to create ownership claims. Merton (1957) explains, "Then, as now, complex ideas were quickly published in abstracts" to ensure credit, and gives the example of Halley telling Newton to publish an abstract "in order to secure 'his invention to himself till such time as he would be at leisure to publish it'" (p. 654).

Perhaps nothing shows the importance of priority more than the passionate fights it has fueled. Merton's (1957) account reads like a who's who of early modern science: it includes Newton versus Leibniz (the famous question of who invented calculus); Newton versus Hooke; Hooke versus Huygens; Cavendish, Watt, and Lavoisier in "a three-way tug of war" (p. 636); Jenner versus Pearson and Rabaut over smallpox; Lister versus Lemaire; "several of the Bernoullis" (p. 636) (including against each other); and so on. Fights over priority, "far from being a rare exception in science, have long been frequent, harsh, and ugly" (Merton, 1957, p. 636), so common that Darwin's and Wallace's mutual civility was considered remarkable, and Euler is still recognized for his generosity (Merton, 1957, p. 637). Linking these fights and new constructs of authorship, Long (1991) explains that "disagreements among contemporaries concerning who first invented or discovered something seem to have appeared initially in the 16th century" (p. 883). She names them as part of the birth of science: "these conflicts . . . were at the center of the development of early-modern scientific and technological thought and practice" (p. 884).

The priority aspect of ownership continues today through credit systems, as just described, and through copyright and patents. Scientists use copyright in ways both like and unlike other authors. Like all authors, scientists own the copyright to work they produce. However, it is not a primary consideration since scientists, like other academics, "typically sign copyright away to publishers at no cost in exchange for the reputational and career benefits that will accrue from the broad circulation of their work" (Birnholtz, 2006, pp. 1759–1760). (For more about copyright, scholarship, and control over access once something is published, see Sidler [Chapter 3] this volume.)

Another increasingly important form of ownership in some sciences is patenting. Although the "kind of credit generated by scientific authorship is categorically distinct from the rights granted by either copyright or patent law" (Biagioli, 2012, p. 458), there are now overlaps, resulting in the rise of patenting and in patent fights. Authorship now holds the potential for actual as well as symbolic capital for scientists in fields with potential for commercially appealing intellectual property.

Social status also affects not only who can claim authorship but also who determines who else has that right. While especially clear in big science, where research teams with dozens or even hundreds of members are growing more common, this has been part of science since its outset. In early science, gentlemanly standing defined who could speak and write with authority and who could authorize others do so. Boyle had the right to publish findings from his lab and to decide who else could write about his lab's findings, as well as who could be credited by name. Like other natural philosophers, he "enjoyed authority over those whose labor he engaged" (Shapin, 1989, p. 560), which was entirely consistent with the relative roles of gentlemen-employers and technicians – the "'servants' or, at times specifically, 'chemical servants'" of the time (Shapin, 1989, p. 554).

Over the centuries, especially with the professionalization of sciences in the late nineteenth and early twentieth centuries, the right to determine insider status has shifted from gentlemen to scholars. Today gatekeeping of both people and knowledge is firmly in the hands of those with academic credentials, especially those in more prestigious PhD-granting institutions. Students and sometimes postdocs now fill most roles formerly occupied by servants but have, like them, "often been viewed as cheap labor rather than as part of an intellectual mentorship model" (Osborne and Holland, 2009, p. 7). While this varies across disciplines and even individual research groups (Shapin, 1989, p. 562), the underlying facts of stratification and gatekeeping are universal.

Responsibilities of authorship

Concomitant with the right of authorship come responsibilities. When authorship developed as a way to link a text and a person/group, it made authors accountable in a way they were not before (Foucault, 1998). Birnholtz (2006) points out that "ownership," for example via copyright, comes with "taking responsibility for one's work, including error or controversy that might lie within it" (p. 1760). Specific reasons for and forms of accountability vary over time and across contexts. In sciences, they tend to fall into two areas: accuracy and ethics.

While responsibility for the overall quality of knowledge in sciences rests with peers (Biagioli, 2012, p. 465) – via quality control measures such as peer review and replication – authors are individually responsible for making sure their accounts are accurate to the best of their knowledge. Parallels to this personal responsibility for accuracy show in the natural philosophers' activities, situated within the seventeenth and eighteenth centuries' precursors to peer review. Peer review – subjecting work to the judgment of trusted community members – began with witnessing. Shapin (1988) describes how an idea or claim about nature became a fact only by undergoing a process of "disciplined witnessing" (p. 366–367) in which the Royal Society's members discussed experiments or observations. Thus, a fact came into being not when it was displayed or reported but "Only when there was clear agreement" among society members (Shapin, 1988, p. 399). In texts, witnessing appeared via the practice of mentioning who was present, "that is, the naming of (typically important) persons who were present at the scientific event being reported" (Atkinson, 1999, p. 77) and via rhetorical strategies that so vividly depicted activities they allowed for "virtual witnessing" (Shapin and Schaeffer, 1985, pp. 60–65).

However, as important as this process was, presenters also had to ensure that witnesses received only information or witnessed only experiments that had already been tested as thoroughly as possible. Shapin distinguishes between the trials natural philosophers did on their own, many of which were expected to fail, and their demonstrations before the society. The latter were expected not to fail but instead were to provide evidence for members to use in actually vetting the scientific knowledge via their "experimental discourses," that is, the "expatiatory and interpretative verbal behaviors" they engaged in collectively (Shapin, 1988, p. 400). For example, Robert Hooke's job was to conduct experiments and show the successful ones to the society, and when a demonstration failed, "Hooke's wrist was smartly slapped," and he was told to try again when he could make the experiment work (Shapin, 1988, p. 402).

The assumption that a researcher will adequately test an idea or experiment before submitting work holds today, though obviously it is not enforced in the same way. Generally, problems not caught via peer review (see Mehlenbacher and Mehlenbacher [Chapter 7] of this volume) are supposed to be addressed by authors or editors through corrections and/or retractions.

Another issue is responsibility for ethics. Like accountability, ethics began in the gentlemanly tradition and its "rhetorical *technology of trust and proximity*" (Atkinson, 1999, p. xxvii; emphasis in original). Historians emphasize how trust was a direct function of social standing. In explaining that knowledge-making was done in gentlemen's homes, Shapin (1988) says, "What underwrote assent to knowledge claims was the word of a gentleman, the conventions regulating access to a gentleman's house, and the social relations within it" (p. 404). Likewise, Dear (2015) says, "Gentlemen were trustworthy just because they *were* gentlemen" (p. 69).

Authorship ethics today fall into two categories: those related to data and those related to textual production. Although ethical issues in scientific publication (such as plagiarism) are described elsewhere (see Poe [Chapter 4] of this volume), two specifically related to authorship are worth mentioning here: guest authorship and its opposite, ghost authorship.

With guest authorship, or honorary authorship, a person receives authorship credit without contributing to a paper. This can lead to problems such as those identified by O'Brien et al. (2009) in medical research, including "dilution of relative contribution for their coauthors" and "a negative effect on patient care" (p. 231) when a respected author's name means a study is believed and affects clinical decisions more than it would otherwise (p. 233). With ghost authorship, a contributor's name is left off the publication, which allows for "[u]nacknowledged industry influence over publications" (Lo et al., 2009, pp. 5–26).

Both are fairly common; cross-disciplinary studies show "evidence of honorific authorship in 18 to 19 percent of sampled articles and of ghost authorship in 8 to 11 percent of articles" in the Public Library of Science (PLOS) (Larivière et al., 2016, p. 419; see also Rennie et al., 1997; Cronin, 2001). Sometimes the two combine, as when drug companies write papers and "offer honorary authorships to 'opinion leaders' in order to influence clinicians" (Vaux, 2016, p. 901).

Another ethics issue is that conflicts of interest can arise when scientists are publishing as allegedly disinterested scholars in areas where they also have financial interests (Biagioli, 2012; Cohen, 2017). (See Poe [Chapter 4] this volume.)

Problems with dominant concepts of authorship

The current concept of authorship has limitations, some already addressed. Additional problems fall into three areas: interdisciplinarity, hyperauthorship, and noncanonical knowledge systems.

Authorship credit can challenge research teams encompassing fields with different authorship criteria. Author order illustrates the problems; in some fields, the first author is the team leader, while in others the team lead goes last. Some areas, in attempts to resolve authorship disputes, list authors alphabetically, but this is not consistent across journals and disadvantages people with names late in the alphabet (Osborne and Holland, 2009, p. 2). And name order is only one facet of authorship differences; each field also has its own norms for author presence (e.g. self-mentions), for how others' work is integrated via citations, etc.

Another challenge is the increasing number of authors in individual papers, or "hyperauthorship" (Cronin, 2001). As early as 1979, Beaver and Rosen (1979) found a "marked increase in the growth of collaboration" (p. 243), and numbers have since grown. In high-energy physics, each experiment has "contributions by nearly 2,000 individual physicists, all of whom will be listed . . . on each article published by any member of the experiment" (Birnholtz, 2006, p. 1759). Although this example is extreme, other areas face challenges as "the number of collaborators on each paper skyrockets" (Larivière et al., 2016, p. 429). Hyperauthorship also creates challenges to academic hiring and promotion decisions as evaluators struggle "to discern the nature and extent of individual contributions to the work" (Birnholtz, 2006, p. 1758) and ethical challenges as it is hard to know "whom to hold liable in the event of errors or other controversy" (Birnholtz, 2006, p. 1758). Since responsibility is unclear, "the ensuing tendency is to give credit, but to withhold blame" (Beaver and Rosen, 1978, p. 69).

These problems have prompted two reactions. One is attempts to pin down criteria that will work across contexts; these are mostly unsuccessful. Another is a contributorship model analogous to movie credits, in which each person involved is listed with their role clearly defined (Rennie et al., 1997), to help with assigning credit and responsibility (Vaux, 2016, p. 902). However, it is not a panacea. For example, Vaux (2016) mentions a paper that has 30 authors listed but only describes contributions by the three co–first authors (p. 902). Furthermore, because "these statements remain self-declared," readers must watch for variations in how contributorship is decided and described in different contexts (Larivière et al., 2016, p. 429). Overall, multiauthorship makes assigning credit and responsibility harder than it was when texts were mainly written by individuals or by a small number of easily identifiable people.

The third challenge is knowledge arising from noncanonical epistemologies, that is, knowledge systems that exist separately from and do not conform to the knowledge-making systems of science. This is particularly true of systems whose values do not align with those of the "moral economy" (Daston, 1995) that characterizes science and that includes the authorship criteria described here. Jaszi and Woodmansee (2003) list groups disadvantaged by the current system, which "has functioned to marginalize or deny the work of many creative people: women, non-Europeans, artists working in traditional forms and genres, and individuals engaged in group or collaborative projects, to name but a few" (p. 195).

The case of indigenous and traditional knowledge (ITK) demonstrates how some kinds of knowers do not have the right to control if and how their knowledge is used. Looking at copyright shows how a legal system "promotes particular cultural interpretations of knowledge, ownership, authorship and property" (Anderson, 2010, p. i) and how this disadvantages communities with differing interpretations of these concepts. With copyright, "the solitary and sovereign 'author' holds clear sway: copyright cannot exist in a work produced as a true collective enterprise; copyright does not hold in works that are not 'original'" (Ede and Lunsford, 2001, p. 359). For holders of ITK, legal expectations for individuality and originality can bar their right to control over collective knowledge.

Legal scholar Jane Anderson (2010) reviews legal problems and proposed solutions and distinguishes solutions that attempt to fit ITK within dominant traditions from those that recognize the validity of "the manifold indigenous laws and governing structures that historically and contemporarily exist for regulating knowledge use" (p. 33). The National Congress of American Indians (2019) has worked for years to increase the participation of Indigenous governments in United Nations decisions about issues "of particular concern to Indigenous peoples, including climate change and the protection of Indigenous traditional knowledge and genetic resources" (p. 47). When culturally specific and historically based authorship systems are taken as natural and neutral, people and knowledge that don't fit those systems are disadvantaged.

Concluding remarks

Despite challenges, the authorship paradigm that has underpinned scientific knowledge-making for nearly 400 years remains largely intact. As it did in early modern England and Europe, it continues to prize individuality, originality, and social status as determinants of who can be an author. Nevertheless, some scholarship recognizes the drawbacks to this system. Looking ahead, work on authorship must accept the current system's failings and seek to remedy them. One way to do this is to look to other knowledge-making systems for other ways to create links between texts and knowledge, and the people involved in their development.

References

Anderson, J. (2010) *Indigenous/traditional knowledge & intellectual property*, Issues Paper, Durham, NC: Center for the Study of the Public Domain, Duke University.

Atkinson, D. (1999) *Scientific discourse in sociohistorical context: The philosophical transactions of the royal society of London, 1675–1975*, Mahwah, NJ: L. Erlbaum Associates (Rhetoric, knowledge, and society).

Barthes, R. (1977) 'The death of the author', in S. Heath (trans.), *Image, music, text*, London: Fontana, pp. 142–148.

Beaver, D. deB., & Rosen, R. (1978) 'Studies in scientific collaboration: Part I. The professional origins of scientific co-authorship', *Scientometrics*, vol. 1, no. 1, pp. 65–84.

Beaver, D. deB., & Rosen, R. (1979) 'Studies in scientific collaboration: Part III. Professionalization and the natural history of modern scientific co-authorship', *Scientometrics*, vol. 1, no. 3, pp. 231–245, https://doi.org/10.1007/BF02016308

Bensaude-Vincent, B. (2001) 'A genealogy of the increasing gap between science and the public', *Public Understanding of Science*, vol. 10, pp. 99–113, https://doi.org/10.3109/a036858

Biagioli, M. (2012) 'Recycling texts or stealing time?: Plagiarism, authorship, and credit in science', *International Journal of Cultural Property*, vol. 19, no. 3, pp. 453–476, https://doi.org/10.1017/S094 0739112000276

Birnholtz, J.P. (2006) 'What does it mean to be an author? The intersection of credit, contribution, and collaboration in science', *Journal of the American Society for Information Science and Technology*, vol. 57, no. 13, pp. 1758–1770, https://doi.org/10.1002/asi.20380

Cohen, J. (2017) 'How the battle lines over CRISPR were drawn', *Science*, www.sciencemag.org/news/2017/02/how-battle-lines-over-crispr-were-drawn (Accessed 31 January 2021).

Cronin, B. (2001) 'Hyperauthorship: A postmodern perversion or evidence of a structural shift in scholarly communication practices?', *Journal of the American Society for Information Science and Technology*, vol. 52, no. 7, pp. 558–569, https://doi.org/10.1002/asi.1097

Crowley, S. (1990) *Methodical memory: Invention in current-traditional rhetorics*, Carbondale, IL: Southern Illinois University Press.

Daston, L. (1995) 'The moral economy of science,' *Osiris*, vol. 10, pp. 2–24.

Dear, P. (2015) 'Totius in verba: Rhetoric and authority in the early royal society', in T. Skouen & R. Stark (eds.), *Rhetoric and the early royal society: A sourcebook*, Leiden: Koninklijke Brill, pp. 53–76.

Eamon, W. (1985) 'From the secrets of nature to public knowledge: The origins of the concept of openness in science', *Minerva*, vol. 23, no. 3, pp. 321–347, https://doi.org/10.1007/BF01096442

Ede, L. (1985) 'The concept of authorship: An historical perspective', Annual Meeting of the National Council of Teachers of English, Philadelphia, 22 November.

Ede, L., & Lunsford, A. (2001) 'Collaboration and concepts of authorship', *PMLA*, vol. 116, no. 2, pp. 354–369.

Foucault, M. (1998) 'What is an author?', in J.D. Faubion (ed.), *Michel Foucault: Aesthetics, method, and epistemology*, New York: New Press.

Gross, A.G. (2006) *Starring the text: The place of rhetoric in science studies*, Carbondale, IL: Southern Illinois University Press.

Jaszi, P., & Woodmansee, M. (2003) 'Beyond authorship: Refiguring rights in traditional culture and bioknowledge', in M. Biagioli & P. Galison (eds.), *Scientific authorship: Credit and intellectual property in science*, London: Routledge, pp. 195–223.

Larivière, V., Desrochers, N., Macaluso, B., Mongeon, P., Paul-Hus, A., & Sugimoto, C. (2016) 'Contributorship and division of labor in knowledge production', *Social Studies of Science*, vol. 46, no. 3, pp. 417–435, https://doi.org/10.1177/0306312716650046

Lo, B., Field, M., Committee on Conflict of Interest in Medical Research, Education, and Practice, & Board on Health Sciences Policy (2009) *Conflict of interest in medical research, education, and practice*, Washington, DC: National Academy of Sciences.

Long, P.O. (1991) 'Invention, authorship, "intellectual property," and the origin of patents: Notes toward a conceptual history', *Technology and Culture*, vol. 32, no. 4, pp. 846–884, https://doi.org/10.2307/3106154

Merton, R.K. (1957) 'Priorities in scientific discovery: A chapter in the sociology of science', *American Sociological Review*, vol. 22, no. 6, pp. 635–659, https://doi.org/10.2307/2089193

National Congress of American Indians. (2019) *Policy update*, Albuquerque, www.ncai.org/attachments/PolicyPaper_zuufQisBiFmwncntutmICsEJXHWPReKwjpbpYsxALTXBECBsNDS_2019%20NCAI%20Annual%20Convention%20Policy%20Update.pdf (Accessed 31 January 2021).

O'Brien, J., Baerlocher, M.O., Newton, M., Gautam, T., & Noble, J. (2009) 'Honorary coauthorship: Does it matter?', *Canadian Association of Radiologists Journal*, vol. 60, no. 5, pp. 231–236, doi:10.1016/j.carj.2009.09.001

Osborne, J.W., & Holland, A. (2009) 'What is authorship, and what should it be? A survey of prominent guidelines for determining authorship in scientific publications', *Practical Assessment, Research & Evaluation*, vol. 14, no. 15, pp. 1–19, https://doi.org/10.7275/25pe-ba85

Rennie, D., Yank, V., & Emanuel, L. (1997) 'When authorship fails: A proposal to make contributors accountable', *JAMA*, vol. 278, no. 7, pp. 579–585.

Shapin, S. (1988) 'The house of experiment in seventeenth-century England', *Isis*, vol. 79, no. 3, pp. 373–404, https://doi.org/10.7275/25pe-ba85

Shapin, S. (1989) 'The invisible technician', *American Scientist*, vol. 77, no. 6, pp. 554–563.

Shapin, S., & Schaeffer, S. (1985) *Leviathan and the air-pump: Hobbes, Boyle, and the experimental life*, Princeton, NJ: Princeton University Press.

Sutherland-Smith, W. (2016) 'Authorship, ownership, and plagiarism in the digital age', in T. Bretag (ed.), *Handbook of academic integrity*, Dordrecht, The Netherlands: Springer, pp. 575–589.

Vaux, D.L. (2016) 'Scientific misconduct: Falsification, fabrication, and misappropriation of credit', in T. Bretag (ed.), *Handbook of academic integrity*, Dordrecht, The Netherlands: Springer, pp. 895–911.

Woodmansee, M. (1984) 'The genius and the copyright: Economic and legal conditions of the emergence of the "author"', *Eighteenth-Century Studies*, vol. 17, no. 4, pp. 425–448, https://doi.org/10.2307/2738129

3

THE NEW NORM

Open access publishing and scientific research

Michelle Sidler

In his "Afterword" for *Science and the Internet: Communicating Knowledge in the Digital Age*, Charles Bazerman (2016) lauds the myriad rhetorical analyses of networked science presented in the edited collection, from participatory science to postpublication evaluation of science to enhanced scientific collaboration. But he ends his piece with an observation about what is missing:

> [W]hile some chapters in the volume take examples from open access academic publishing, this volume does not examine how digital production and distribution through the Internet are changing the economics of academic publishing, particularly in a time when university libraries are challenged financially by corporate acquisition and marketization of traditionally boutique academic publishing that was culturally congruent with academic culture. All this is challenging the information chain that supported the emergence of the encapsulated scientific article.
>
> *(p. 281)*

This chapter begins such a discussion, though it is just that: a beginning. Open access (OA) publishing is a moving target and has been for almost two decades. Publishing options for scientific researchers vary, and the politics and rhetoric behind those options are even more complex. The chapter provides a brief overview of OA as a way to navigate current trends, drawing a broad picture of its impact on scientific publishing, including a look at its history, rationale, evolution, and current international landscape.

Bazerman ends the "Afterword" by musing about digital communication's impact on the "encapsulated article," which has been the generic currency of scientific communication since the Royal Society first published *Philosophical Transactions* in 1665. The chapter shows how OA has not so much changed the genre of the scientific article, but it has exposed its commercial monetization. Ownership of and access to scientific publishing have become a complex question of global equity.

In the beginning: the Budapest, Bethesda, and Berlin open access meetings

In late 2001, the Open Society Foundations (known at the time as the Open Society Institute) gathered a group of OA advocates in Budapest, Hungary, for a meeting to discuss an international,

DOI: 10.4324/9781003043782-5

coordinated approach to OA publishing. They composed a petition called the Open Access Initiative that was both a mission statement and a strategic plan to make research articles in all fields openly available online. The petition, whose signatories represented many areas of scholarship including the hard sciences, medicine, social science, and the humanities, laid out a two-part strategy that still provides a framework for achieving OA. The first component is self-archiving, wherein researchers deposit their own articles into OA databases with coding standards that make them easy to locate and access. The second component is OA journals, which are journals that are freely accessible and do not charge subscription fees that result in paywalls for those who cannot afford to pay. Instead, OA journals are funded by alternative sources such as foundations, government agencies, money derived from "add-on" premium content, and payments from authors.

In 2003, the Howard Hughes Medical Institute sponsored another meeting of OA advocates in Bethesda, Maryland, to discuss the impact of OA on the biomedical community and to further refine the terms of OA. Their definition required publishers to grant the "perpetual right of access to, and a license to copy, use, distribute, transmit and display the work publicly and to make and distribute derivative works" (Suber, 2003). In October of that same year, the Max Planck Society convened a group of more than 200 researchers, funders, publishers, and other learned societies in Berlin, Germany, for the first of several conferences about OA. The Berlin Conferences, as they came to be known, codified two forms of OA publication, self-archiving through OA repositories and publication through OA journals. These two methods came to be known as "green OA" and "gold OA," respectively.

Science publishing and the public good

The Open Access Initiative's first sentence reads: "An old tradition and a new technology have converged to make possible an unprecedented public good" (Open Society Foundations, 2002). The old tradition is the doctrine followed by scientific communities – dating back to the Royal Society, publisher of the first scientific journal still in existence and developer of the peer review process – that obligates scientists to publish their research without payment for the advancement of scientific inquiry. The Internet is the new technology, digitally affording scientists the opportunity to make their research widely available, instantaneously. Advocates of OA research argue that such access is crucial to the advancement of knowledge and desirable for the social good as well. With faster, easier access to the latest research, advancements are achieved more quickly, with a wider array of researchers who contribute. Moreover, public access to research fosters a better informed citizenry by widely disseminating information like medical advances and reinforcing the value of science to the larger society.

The Open Access Initiative's emphasis on the public good stands in stark contrast to the late twentieth-century privatization and commodification of scientific research. Academic publishing in the sciences has become big business. Increasing corporate control resulted in the five largest publishing companies owning the rights to 50% of the world's research by 2015 (*Science-Daily*, 2015). Most of this research resides behind paywalls, available only to subscribers and to those willing to pay substantial fees for access to one article. With such control, publishers can charge increasingly high prices for subscriptions, even in the information age, when most of the cost of print production and distribution has given way to digitized formats. Major publishers like Elsevier make profits of 30–40% every year; those returns rival the most lucrative industries like oil. (Buranyi, 2017; *NewScientist*, 2018)

Publishers exploit the Royal Society tradition that obligates researchers to offer their written work for free, greatly reducing the costs of publishing compared to industries that pay authors for their work like trade publishing and journalism. Those same researchers also volunteer to serve as editors of journals, further reducing the cost of journal production. Moreover, academic publishers have a ready-made customer base – academic libraries – that is obligated to pay large subscription rates in order to support academic research. For-profit publishers reap incommensurate monetary benefits from the academic discourse loop that was intended to foster the creation of knowledge and human innovation through the free flow of ideas.

The NIH OA policy and other early U.S. government intervention

By the turn of the twenty-first century, the conflation of high publishing costs and ease of online publication led scientists and foundations like those in attendance at the Budapest, Bethesda, and Berlin meetings to call for change. Government funding agencies soon realized that they have power to make change as well. Although public funding amounts have fallen over the last ten years, the majority of noncorporate scientific research is still funded by governments or private foundations. In 2015, for example, 44% of all research in the United States was funded by government agencies (Mervis, 2017). That level of support allowed the National Institutes of Health (NIH), the largest single public funder of biomedical research in the world ($26 billion in 2016), to embark on an OA initiative of its own. In 2004, it began requesting that all publications based on NIH-funded research should be made publicly available (National Institutes of Health, 2004). Grantees were asked to deposit their published journal articles in PubMed Central (PMC) within 12 months of publication.

Although this request was groundbreaking at the time, it was also incomplete and tentative. The NIH did not mandate that authors provide the articles, so the policy was not enforceable. Many researchers chose to ignore the request, knowing that commercial publishers would not allow their published works to be distributed freely. One study found that of 3,578 NIH-funded articles published in 2006, only 9.4% were deposited in PMC (DeGroote et al., 2015).

In response, the NIH implemented a more forceful policy in 2006, which mandated that all NIH-funded research articles published after April 7, 2008, be deposited into PMC (National Institutes of Health, 2016) by the end of the 12-month moratorium. Commercial publishers began taking the mandate more seriously, lobbying Congress to support legislation that would forbid such policies (Fair Copyright in Research Works Act of 2009, H.R. 801; Research Works Act of 2011). In response, OA advocates and members of Congress sponsored bills that would expand access to scientific publishing (Fair Access to Science and Technology Research Act of 2013, S.350; Fair Access to Science and Technology Research Act of 2017, S.1701). However, no bills were successful.

Becoming the new norm: the Obama administration's OA memorandum

Impatient with the impasse in Congress, OA advocates began petitioning the White House in 2012 "to implement open access policies for all federal agencies that fund scientific research" (J.W., 2012). In response, the Obama administration's Office of Science and Technology Policy (OSTP) released an executive order on February 22, 2013, directing all federal agencies with an annual budget of more than $100 million to develop a plan that would increase public access (Office of Science and Technology Policy, 2013). The policy memorandum offered two main areas of guidance: each agency should make the published results of federally funded research freely available to the public within one year of publication, and each should develop a public repository (similar to PMC) for archiving the publications. The Executive Order was positively received by publishers and other copyright advocates because it espoused a 12-month mandate rather than the six-month mandate proposed in earlier Congressional bills. As a result, the policy has endured through both the Obama and the Trump administrations. Agencies have spent the ensuing years devising and implementing OA plans and adhering to the 12-month moratorium. Even without the permanent, more easily enforced mandate of a legislative act, the executive order quietly set into motion a culture of open science.

In 2013, the same year that the Obama administration mandated compliance from the highest levels of government, the statewide University of California (UC) system committed to a bottom-up OA approach for its universities. It started with an open access policy for scholarship wherein all UC-employed researchers were obliged to deposit their publications in the university open access repository (University of California Office of Scholarly Communication, n.d.). UC accounts for approximately 10% of all research published in the United States (Smith, 2019), making this policy particularly notable and influential.

In 2019, UC began a new phase of its commitment to OA when it attempted to renegotiate library subscription fees with the three largest scientific publishing companies: Springer, Wiley, and Elsevier. UC mandated transformative agreements, which are contracts that allow libraries to shift some amount of their publisher fees away from the paywalled subscription model and toward more open access publishing. UC wanted a share of its subscription fees to be diverted to cover publishing fees for UC researchers' journal articles, enabling scholarship to be published as green OA in the publishers' journals.

Springer and Wiley reacted favorably to transformative agreement negotiations, but Elsevier was less receptive. The negotiations eventually broke down, and in February 2019, UC announced that it was terminating all subscriptions with Elsevier. Although smaller institutions had previously made similar decisions, UC was by far the largest and most influential to do so. The schism between a major university system and a publishing giant garnered national attention and caused Elsevier's parent company, RELX, to lose 6% of its stock price (Nilsson, 2019). UC's libraries had been paying Elsevier $10 million a year in subscription fees, and its researchers published roughly 10,000 articles in Elsevier-owned journals, making this a consequential decision for both the system's libraries and its researchers (Fox and Brainard, 2019; McKenzie, 2019). Negotiations between UC and Elsevier stalled for over a year, but currently, the two organizations are said to be in discussions again. In the meantime, Elsevier has faced similar pressure from other academic institutions and has, at least tentatively, dropped its resistance to transformative agreements (Elsevier, n.d.).

The academic community followed this story closely, wondering if the lack of access to so much research would harm UC's research productivity and frustrate its research faculty. After a year, though, UC reported minimal negative reaction (at least publicly) (Hwang, 2020). Before the 2019 publisher negotiations, UC had spent years boosting faculty buy-in for open access, both through its open repository program and by educating faculty about alternate OA sources for accessing scholarly publications. They also spearheaded the Pay it Forward Project, a multiuniversity study of open access scholarship costs and sustainability in North America (University of California Libraries, 2016). The Pay it Forward Project concluded that through careful planning, a distribution of costs, and a multiplayer approach to publishing (involving faculty, universities, libraries, and publishers), green OA was affordable for large research universities.

The UK's Wellcome Trust OA policy and other early global interventions

Although the NIH was the first major agency to propose an OA policy in 2004, the agency did not make the policy mandatory until 2008. Globally, the first science funder to do so (in 2006) was the Wellcome Trust, the largest philanthropic funder of biomedical research and the sixth largest overall ($909 million in 2016) (Viergever and Hendriks, 2016). Like the NIH, the Wellcome Trust initially had difficulty achieving compliance from authors, so in 2012, it imposed sanctions on authors who did not comply. With the threat of losing future funding, compliance increased to 65% by 2014 (Wellcome Trust, 2014). Unlike the NIH, the Wellcome Trust only allowed a six-month moratorium on research published behind paywalls before it had to be deposited in either PMC or the UK PubMed Central.

Progress toward OA in other European countries was less swift than that of the UK's Wellcome Trust. Most of it has occurred since 2013, when the Global Research Council (GRC), a voluntary board composed of research councils and government funding agencies from around the world, published its *Action Plan Towards Open Access to Publication*. This document is a statement of support for, and commitment to, OA for all member nations. The Action Plan was groundbreaking, serving as global recognition that the Internet changed the way science is practiced and its results are published.

The action plan encouraged nations to promote what had become the three pillars of OA action: publishing in OA journals (gold OA), self-archiving in repositories (green OA), and the development and interconnection of institutional repositories. The GRC recognized that individual countries have different resources and priorities, so the action plan does not mandate a specific OA timeline or methodology but instead acts as a global commitment that each country will work toward that goal (Global Research Council, 2013).

cOAlition S and the globally groundbreaking Plan S

The GRC's action plan was ratified three months after the Obama administration announced its OA policy, making 2013 a global turning point in attitudes about the necessity of OA research. In that same year, the European Research Council (ERC) and the European Commission's (EC) consortium of research agencies (that together make up the world's second largest medical research funding organization [Viergever and Hendriks, 2016]), as well as the UK Medical Research Council (RCUK), the world's third largest public medical research funder, announced OA policies similar to those of Wellcome Trust, mandating public access to research articles within six months of publication (University of Cambridge, n.d.; European Research Council, 2013).

Although other individual European nations – including France and Germany – have pursued various OA policies since the Budapest, Bethesda, and Berlin meetings, the most significant OA advancement for the European Union (EU) came in 2018. At that time, Science Europe, a nonprofit association of 36 European research funders and national research councils (including ERC and RCUK) sponsored a program called cOAlition S to create a more unified standard for OA within the EU and around the world. cOAlition S aims to make science as open as possible, so the organization developed a broad set of principles, coupled with specific and groundbreaking policies, called Plan S. Plan S is said to stand for "science, speed, and shock." "Science," obviously, refers to the belief that OA research benefits the pursuit of science; "speed" refers to a short timeline for the policy's implementation; and "shock" acknowledges the radically transformative impact of the broad coalition's policy.

Plan S articulates 10 broad principles for implementation, but its basic mission is both straightforward and potentially groundbreaking:

> With effect from 2021, all scholarly publications on the results from research funded
> by public or private grants provided by national, regional and international research

councils and funding bodies, must be published in Open Access Journals, on Open Access Platforms, or made immediately available through Open Access Repositories without embargo.

(cOAlition S, 2020b)

This mandate has several implications that make it transformative. First, the large number of funders that back the policy (a number which, to date, has grown to 15 member organizations, including both national funding agencies and private foundations, as well as about two dozen other organizations who have expressed support) provide further impetus to shift both authors and publishers toward OA, while at the same time encouraging other funders outside the EU to participate. Second, the timeline was ambitious because participating researchers, institutions, and publishers only had until 2021 to comply (cOAlition S, 2019).

The most contentious elements of Plan S are made clear in its underlying principles. Most notably, the mandate calls for *immediate* OA for all publications, eliminating any time embargo, a requirement that could potentially disrupt the academic publishing industry. Major publishers have adapted to embargoed OA policies by charging author fees, by raising subscription fees, and by creating "hybrid journals" – journals that feature some OA research and some that are closed and proprietary. Plan S would eliminate publishers' advantages in the first two of those three strategies. OA embargos – whether they were for six months or a year – provide an incentive for researchers, institutions, and libraries to maintain their subscriptions to journals so that researchers can have access to the most up-to-date scientific information. With no embargo, consumers of scientific research can access once-proprietary information immediately and without paying a subscription or fee.

This immediate, free, and open access is precisely the goal of cOAlition S. In an essay for *Frontiers in Neuroscience* that reads more like a manifesto than a scientific article, Mark Schiltz, president of Science Europe, makes this case unequivocally. Schiltz (2018) draws on the tradition of the scientific enterprise – one that values collaboration and communal contributions to scientific progress – to argue that the scientific method itself is at odds with pay-for-access approaches to scientific knowledge. He justifies Plan S (a proposal that would soon be called "radical" by authors in both the journal *Science* and the journal *Nature*) by arguing that closed access is antithetical to the very foundations of science (Else, 2018; Rabesandratana, 2019).

Plan S is radical. For the first time, a large and ambitious consortium of major funding agencies and research councils have overtly decried monetized academic publishing and its model of propriety and paywalls, with the intent to eliminate closed access altogether. Plan S strives for universal inclusion, allowing all researchers (and citizens as well) equal access to the foundations of knowledge, regardless of their ability to pay publisher subscription prices.

Global inequality challenges plan S

The radical goals of Plan S are laudable and have the potential to move science forward – faster and with more equity. Unfortunately, the plan's feasibility and its equitable goals are not guaranteed. One only needs to read the comments section at the end of Schiltz's article to see multiple obstacles, both systemic and pragmatic, laid bare. Resistance from both publishers and researchers has already slowed its progress (the timeline for Plan S was pushed back by one year in response to public concern), and portions of the plan may actually contribute to new inequities. The range of challenges is too vast to cover in their entirety, but one issue, in particular, has incited the most resistance and offers a glimpse into the unintended consequences of progress toward complete and immediate OA: the rise of hybrid journals.

Hybrid journals are subscription journals that offer authors the option to transfer their articles to OA for a price, called an article processing charge (APC). This model has been popular with publishers because it accommodates traditional publishing practices, wherein institutions, libraries, and/or individuals pay for subscriptions to journals (or access individual articles for an access fee) while at the same time, allowing authors to make their work OA for an additional fee (the standard APC charge is approximately $3,000) (Björk, 2017). Because hybrid journals do not require a major overhaul of publishers' subscription and fee-by-view models, publishers have been quick to embrace this approach as a way to accommodate governmental and institutional OA policies while showing at least some commitment to OA. As a result, one study from the University of Utrecht Libraries estimated that over 75% of major, high-ranking journals (including those in the sciences, social sciences, and arts and humanities – the last of which has been slow to join the OA movement) now offer some form of OA publishing. Notably, hybrid journals constitute the vast majority of those journals (Universiteit Utrech, 2019). On its surface, this progress toward OA would appear to be a strong indication of OA's rising influence on academic publishing, and certainly it shows a recognition on the part of publishers that the twentieth-century exclusive, subscription-only model is no longer tenable.

However, the reality of hybrid journals is more complex than these promising numbers show. Many OA advocates, including cOAlition S, argue that hybrid journals do more harm than good. cOAlition S deems most articles published in hybrid journals to be noncompliant with Plan S, even if the authors pay APCs to make the articles OA (European Research Council, 2020). The most obvious reason for the resistance is a term often referred to as "double-dipping," wherein the publisher of a hybrid journal gathers revenue from two sources: the subscription fees from libraries, institutions, and researchers who want access to the full journal (not just the OA articles) and APCs charged to authors who want OA status granted to their articles (Research Libraries UK, 2014). In addition, APC rates tend to be higher in hybrid journals than those fully OA journals that do not charge subscriptions (Rettberg, 2018). So, not only do publishers keep much of the status quo – the commercial subscription model – intact, but they also gather more revenue from those authors (and their funders) who advocate OA.

When given the choice, most authors choose the status quo as well. This same University of Utrecht Libraries study found that of those articles published in hybrid journals, the vast majority are not OA. Without a mandate for full, immediate OA, most authors choose to skip the APCs and publish their works behind paywalls.

For some authors, the reluctance to pay APCs stems from a lack of resources. Funding agencies in North America and northern Europe can provide funds to cover APCs, but many researchers are in less wealthy countries without the resources to pay $3,000 fees. Ironically, although the goal of Plan S is to make research more accessible and equitable, its guidelines tend to exacerbate the financial disparities among different countries, particularly between wealthy countries of the Global North and poorly resourced countries of the Global South (Debat and Babini, 2019). The limitations of paying APCs are reflected in many of the comments at the end of Schiltz's article introducing Plan S. One comment by Manfred Fehr, a Brazilian scientist, is fairly representative:

> It is so easy for Science Europe to ask me that I pay OA fees for all my publications. I am not a millionaire. They simply tackled the problem from the wrong end. I join the quest for a more realistic procedure. In the meantime, I keep publishing in journals that do not charge me.
>
> *(Schiltz, 2018)*

cOAlition S has acknowledged the disparities and is committing set-aside money for researchers from the most financially disadvantaged countries (cOAlition S, 2020a) – but those funds cannot possibly cover the cost of all articles currently being published under closed licenses in hybrid (and fully closed) journals.

Researchers from the UC-led Pay it Forward Project study (2019) also recognize this disparity. In their final report, the authors suggest that the "transformative agreement" model might be a feasible strategy because institutions in developing countries could redirect funds from journal subscriptions to APC fees (p. 130). This strategy assumes that poorly resourced institutions are currently subscribing to all the journals in which their researchers are publishing, which may not necessarily be the case. In addition, the transformative agreement model sustains hybrid journal publishing, and some advocates like cOAlition S disparage that approach because it is not fully OA.

The extent of this equity quagmire is evident in the latest major change to cOAlition S: on July 7, 2020, the ERC, a founding partner of cOAlition S, pulled its support from the consortium and from Plan S. In a press release, they echoed Fehr's concerns about the high cost of APCs and added that Plan S will serve as a barrier to entry for early career researchers, who often do not have the grant writing expertise and supportive resources of more experienced faculty. Implicit in this quote is the reality that funding amounts are tight, both in the EU and worldwide. Public funding for scientific research is already scarce – NIH only funded 18% of applications in 2014, for example (Rockey, 2015) – and as more developing nations join the scientific community, the competition for resources and for publication in the highest-ranked journals will only increase.

With its strict hybrid journal policy, it is no surprise that countries in the Global South have not signed on to Plan S. One prominent example is Latin America, which already has a preexisting, centrally controlled OA ecosystem for scholarly publishing. Plan S's reliance and emphasis on individual authors paying APCs could threaten this ecosystem (López and García, 2019). Other global science leaders have been reluctant to fully commit to Plan S as well. At the Open Access 2020 conference, officials from China's Ministry of Science and Technology stated that they "will now urge Chinese funders, research organizations and academic libraries to make the outcomes of publicly funded research free to read and share as soon as possible" (Schiermeier, 2018), but they did not commit (and still have not committed) to full partnership with cOAlition S. India also offered initial support, but in December 2018, K. VijayRaghavan, the principal governmental scientific adviser, announced that India was pulling back its support (Mukuntth, 2019). And so, with the exception of a few African nations and Jordan, cOAlition S remains primarily a consortium of EU nations, further fueling concerns about inequality and the privilege of the Global North as well as the questionable potential for Plan S to become a worldwide OA initiative.

Global open access and the race for the new norm

Conspicuously absent from most media and discussions about cOAlition S and Plan S is the status of academic publishing in the United States. U.S. policy has remained the same since the Obama administration's executive order in 2013, a tenuous policy that could easily be overturned by subsequent administrations and one that now seems conservative, overly generous to publishers, and potentially a hindrance to scientific progress, especially in comparison to European efforts. On the other hand, the United States has, at least in the short term, chosen a path that seems less disruptive to what Bazerman calls the "information chain," the circulation of academic journal articles that has for centuries created a global culture of knowledge sharing.

In the long term, the information chain is threatened by the polarization of scientific publishing: on one end is the consolidation of copyrights and ownership by a small group of powerful publishers, and on the other is the well-intentioned but potentially explosive guidelines espoused by OA provocateurs like cOAlition S. If the goal of scholarly publishing is participation by all of those who wish to contribute to the advancement of knowledge, any OA plan must adhere to that principle. Open access is only open if it makes both the reading and writing of published research accessible. The future of the scientific journal article will be determined less by changes to its generic, "self-contained" form and more by the ways in which it is accessed. Academic publishing has become commodified, and as with all commodities, the scientific journal article is now traded internationally, wielded for economic power, and deeply embedded in complex global inequalities.

References

Bazerman, C. (2016) 'Afterword', in A.G. Gross & J. Buehl (eds.), *Science and the internet: Communicating knowledge in a digital age*, Amityville, NY: Baywood Publishing, pp. 267–282.

Björk, B. (2017) 'Growth of hybrid open access, 2009–2016', *PeerJ*, vol. 5, p. e3878, https://doi.org/10.7717/peerj.3878

Buranyi, S. (2017) 'Is the staggeringly profitable business of scientific publishing bad for science?', *The Guardian*, 27 June, www.theguardian.com/science/2017/jun/27/profitable-business-scientific-publishing-bad-for-science

cOAlition S (2020a), *Part I: The Plan S principles*, www.coalition-s.org/addendum-to-the-coalition-s-guidance-on-the-implementation-of-plan-s/principles-and-implementation/

cOAlition S (2020b) 'What is cOAlition S?', www.coalition-s.org/about/

cOAlition S (2019) 'Rationale for the revisions made to the Plan S principles and implementation guidance', www.coalition-s.org/rationale-for-the-revisions/

Debat, H., & Babini, D. (2019) 'Plan S in Latin America: A precautionary note', *PeerJ Preprints* vol. 7, p. e27834v2, https://doi.org/10.7287/peerj.preprints.27834v2

DeGroote, S.L., Shultz, M., & Smalheiser, N.R. (2015) 'Examining the impact of the national institutes of health public access policy on the citation rates of journal articles', *PLoS One*, vol. 10, no. 10, 8 October, https://doi.org/10.1371/journal.pone.0139951

Else, H. (2018) 'Radical open-access plan could spell end to journal subscriptions', *Nature*, vol. 561, pp. 17–18, 4 September, www.nature.com/articles/d41586-018-06178-7

Elsevier. (n.d.) 'Elsevier and open access', www.elsevier.com/about/elsevier-and-open-access

European Research Council (2013) 'Open access guidelines for researchers funded by the ERC', https://erc.europa.eu/sites/default/files/document/file/ERC_Open_Access_Guidelines-revised_2013.pdf

European Research Council (2020) 'ERC scientific council calls for open access plans to respect researchers' needs', https://erc.europa.eu/news/erc-scientific-council-calls-open-access-plans-respect-researchers-needs

Fair Access to Science and Technology Research Act of 2013, S.350, 113th Cong. 2013 (2013–2014) www.congress.gov/bill/113th-congress/senate-bill/350

Fair Access to Science and Technology Research Act of 2017, S.1701, 115th Cong. 2017 (2017–2018) www.congress.gov/bill/115th-congress/senate-bill/1701

Fair Copyright in Research Works Act of 2009, H.R. 801, 111th Cong. 2009 (2009–2011) www.congress.gov/bill/111th-congress/house-bill/801

Fox, A., & Brainard, J. (2019) 'University of California boycotts publishing giant Elsevier over journal costs and open access', *Science*, 28 February, www.sciencemag.org/news/2019/02/university-california-boycotts-publishing-giant-elsevier-over-journal-costs-and-open

Global Research Council (2013) 'Action plans towards open access to publication', www.globalresearchcouncil.org/fileadmin/documents/GRC_Publications/grc_action_plan_open_access_FINAL.pdf

Hwang, I. (2020) 'UC's termination of Elsevier contract has had limited negative impact', *Daily Bruin*, 27 February, https://dailybruin.com/2020/02/27/ucs-termination-of-elsevier-contract-has-had-limited-negative-impact

J.W. (2012) 'Require free access over the Internet to scientific journal articles arising from taxpayer-funded research', *The White House President Barack Obama*, https://petitions.obamawhitehouse.archives.gov/response/increasing-public-access-results-scientific-research/

López, E.A., & García, A.B. (2019) 'Latin America's longstanding open access ecosystem could be undermined by proposals from the Global North', *London School of Economics and Political Science*, https://blogs.lse.ac.uk/latamcaribbean/2019/11/06/latin-americas-longstanding-open-access-ecosystem-could-be-undermined-by-proposals-from-the-global-north/.

McKenzie, L. (2019) 'UC drops Elsevier', *Inside Higher Ed*, 1 March, www.insidehighered.com/news/2019/03/01/university-california-cancels-deal-elsevier-after-months-negotiations

Mervis, J. (2017) 'Data check: U.S. government share of basic research funding falls below 50%', *Science*, 9 March, www.sciencemag.org/news/2017/03/data-check-us-government-share-basic-research-funding-falls-below-50

Mukuntth, V. (2019) 'India will skip plan S, focus on national efforts in science publishing', *The Wire*, 26 October, https://thewire.in/the-sciences/plan-s-open-access-scientific-publishing-article-processing-charge-insa-k-vijayraghavan

National Institutes of Health. (2004) 'Notice: Enhanced public access to NIH research information', https://grants.nih.gov/grants/guide/notice-files/not-od-04-064.html

National Institutes of Health. (2016) 'NIH public access policy details', *Public Access Policy*, https://publicaccess.nih.gov/policy.htm

NewScientist. (2018) 'Time to break academic publishing's stranglehold on research', 21 November, www.newscientist.com/article/mg24032052-900-time-to-break-academic-publishings-stranglehold-on-research/

Nilsson, P. (2019) 'Relx his as University of California cancels $11m contract', *Financial Times*, 1 March, www.ft.com/content/cc135e32-3c31-11e9-b72b-2c7f526ca5d0

Office of Science and Technology Policy. (2013) 'Increasing access to the results of federally funded scientific research', https://obamawhitehouse.archives.gov/sites/default/files/microsites/ostp/ostp_public_access_memo_2013.pdf

Open Society Foundations. (2002) 'Budapest open access initiative', www.budapestopenaccessinitiative.org/read

Rabesandratana, T. (2019) 'Will the world embrace Plan S, the radical proposal to mandate open access to science papers?' *Science*, 3 January, www.sciencemag.org/news/2019/01/will-world-embrace-plan-s-radical-proposal-mandate-open-access-science-papers

Research Libraries UK. (2014) 'Fair prices for article processing charges (APCs) in hybrid journals', www.rluk.ac.uk/wp-content/uploads/2014/02/RLUK-stance-on-double-dipping-Final-November-2013.pdf

Research Works Act of 2011, H.R. 3699, 112th Cong. (2011) www.congress.gov/bill/112th-congress/house-bill/3699

Rettberg, N. (2018) 'The worst of both worlds: Hybrid open access', *OpenAire*, 26 June, www.openaire.eu/blogs/the-worst-of-both-worlds-hybrid-open-access

Rockey, S. (2015) 'What are the chances of getting funded?', *National Institutes of Health Office of Extramural Research*, https://nexus.od.nih.gov/all/2015/06/29/what-are-the-chances-of-getting-funded/

Schiermeier, Q. (2018) 'China backs bold plan to tear down journal paywalls', *Nature*, vol. 564, pp. 171–172, www.nature.com/articles/d41586-018-07659-5

Schiltz, M. (2018) 'Science without publication paywalls: cOAlition S for the realisation of full and immediate open access', *Frontiers in Neuroscience*, vol. 12, p. 656, https://doi.org/10.1371%2Fjournal.pbio.3000031

ScienceDaily. (2015) 'Five companies control more than half of academic publishing', 10 June, www.sciencedaily.com/releases/2015/06/150610143624.htm

Smith, M. (2019) 'University of California's break with the biggest academic publisher could shake up scholarly publishing for good', *The Conversation*, 7 March, https://theconversation.com/university-of-californias-break-with-the-biggest-academic-publisher-could-shake-up-scholarly-publishing-for-good-112941

Suber, P. (2003) 'Bethseda statement on open access publishing', http://legacy.earlham.edu/~peters/fos/bethesda.htm

Universiteit Utrech. (2019) 'Open access potential and uptake in the context of plan S – a partial gap analysis', *Innovations in Scholarly Communication*, https://101innovations.wordpress.com/2019/11/21/open-access-potential-and-uptake-in-the-context-of-plan-s-a-partial-gap-analysis/

University of California Libraries. (2016) 'Pay it forward: Investigating a sustainable model of open access article processing charges for large North American research institutions', www.library.ucdavis.edu/wp-content/uploads/2018/11/ICIS-UC-Pay-It-Forward-Project-Final-Report.pdf

University of California Office of Scholarly Communication. (n.d.) 'Participate in the UC open access policies', https://osc.universityofcalifornia.edu/for-authors/open-access-policy/

University of Cambridge. (n.d.) 'A brief history of OA', https://osc.cam.ac.uk/open-access/brief-history-oa

Viergever, R., & Hendriks, T.C.C. (2016) 'The 10 largest public and philanthropic funders of health research in the world: What they fund and how they distribute their funds', *Health Research Policy and Systems*, vol. 14, no. 12, https://doi.org/10.1186/s12961-015-0074-z

Wellcome Trust. (2014) 'Independent review of the implementation of RCUK policy on open access', https://wellcome.ac.uk/sites/default/files/wtp057467.pdf

4

ETHICAL ISSUES IN SCIENTIFIC PUBLICATION

Rapidly evolving practices in the global scientific community

Mya Poe

Introduction

The global impact of SARS–CoV-2 has made visible the many pressures that scientists face in the development of drug technologies and the dissemination of scientific knowledge. From getting reliable data on the disease's origin to securing funding for vaccine development, the pandemic has highlighted how scientific research and publishing can swiftly change. The rate of "pandemic publishing" alone has been remarkable; according to a December 2020 review by *Nature*, more than 100,000 articles on PubMEd had been published about the coronavirus pandemic, and more than 200,000 coronavirus papers could be identified through the Dimensions database on the coronavirus pandemic (Else, 2020). Of those papers traced by PubMed and Dimensions, "between 17% and 30% of total COVID-19 research papers" were preprints, meaning that they were made public in advance of peer review (Else, 202, para. 7). The average time for peer review was a speedy six days (Palayew et al., 2020, p. 668). In short, the current "pandemic publishing" has demonstrated the frantic pace by which scientific discovery can be circulated through a network of preprint servers and digital journals.

Of course, pandemic publishing has also come with ethical challenges. As Retraction Watch, which is hosted by the Center for Scientific Integrity, reported this year, "[I]f Covid were an author, it would be fifth on our leaderboard [of retracted papers], with 72 so far. We're certain that's not the high-water mark for Covid retractions given the haste with which scientists have churned out papers about the disease and the virus behind it" (2020, para. 2). Such concerns over ethical issues have led the European Association of Science Editors to release a statement on quality standards that encourages editors "to ensure that reports of research on COVID-19 meet required standards and comply with agreed guidelines, and that any limitations are clearly stated" (2020, para. 1).

I draw on the SARS–CoV-2 pandemic because it has laid bare the many ethical issues in scientific publishing. At the time that I composed this chapter, more than 3,600,000 lives have been lost to the disease. When the loss of life is so great, it is tempting to ignore an overstated finding, rush informed consent, or accept materials that have not been peer reviewed. As I explain in this chapter, which focuses on four ethical challenges that are central in scientific publishing, guidelines for research misconduct are insufficient to address these concerns. Ethics is not just

DOI: 10.4324/9781003043782-6

about ferreting out misconduct or thinking in mechanistic ways about publishing processes. Ethical standards must be updated to reflect changing practices and conditions under which science is conducted. Moreover, research on ethics in scientific publishing must attend to vulnerable groups who are most likely to be affected by misconduct while also understanding how interventions designed to address misconduct can misfire and harm the very researchers they were designed to protect.

Several caveats are important to acknowledge here. First, my use of the term "science" in this article is purposely broad and meant to include a range of scientific disciplines. I use the term "science" broadly because research on ethical issues spans a range of scientific disciplines, and researchers interested in publishing ethics can learn from the variety of perspectives represented across those disciplines.

Second, it is beyond the scope of this chapter to discuss topics like consortia licensing, changes in copyright laws, and the emergence of massive, global research teams. A discussion of some of these may be found at "STM Report: An Overview of Scientific and Scholarly Publishing" (Johnson et al., 2018). Likewise, I do not have the space to discuss in-depth the way that impact factors and publishing metrics are driving some of the ethical issues discussed in this chapter. Entire books have been recently published on this topic: see *Gaming the Metrics: Misconduct and Manipulation in Academic Research* (Biagioli & Lippman, 2020).

Third, the norms for ethics in science are described within Western philosophical frameworks with concepts such as virtue ethics, deontology, and consequentialism. Values like objectivity, honesty, carefulness, efficiency, openness, universalism, freedom, credit, and accuracy are meant to follow from that orientation (Resnik, 1998). For example, as explained by Bevilacqua et al. (2001), when Robert Merton argued that race, nationality, religion, class, or other personal or social attributes of the researcher should not matter to the validity of conclusions, he was espousing a value of "universalism" (p. 186). The scientific method, then, follows in the insistence on research that is repeatable, predictable, quantitative, theory driven, based on observable natural phenomena.

Finally, science itself may be viewed as a colonial enterprise, meaning that no amount of ethical action can undo unjust legacies of genocide and what Sousa Santos (2018) calls "epistemicide" (p. 8). A decolonial perspective on scientific publishing, thus, would demand a reexamination of taken-for-granted processes such as peer review, publishing platforms, and ownership of manuscripts and potentially yield different interventions than the ones I describe in this chapter (Macaulay et al., 1998). Further discussions of gender inequality as well as race/racism may be found in Chapters 5 and 6 in this book.

Framing ethics in scientific publishing

Standards for research misconduct arose after World War II. The atrocities of World War II, as well as changes in funding and shifting academic standards, drew attention to the need to articulate standards for ethical practice.

In the United States, the federal government passed a series of laws, including the 1974 National Research Act (PL 93–348), that addressed the use of humans and animals in research as well as research misconduct and provided regulation of scientific research through the federal government's spending powers in awarding grants for scientific research. Together, these principles, policies, and federal regulations comprise what is known as Responsible Conduct in Research. As outlined by the Office of Research Integrity (ORI), the principles of Responsible Conduct in Research are driven by the belief that public support for scientific research demands that researchers act in ways that ensure "researchers have clear obligations to conduct their research in a responsible manner" (Steneck, 2007, p. xi). Research misconduct is defined as the "fabrication, falsification or plagiarism in proposing, performing or reviewing research, or in reporting research results. . . . Research misconduct does not include honest error or differences of opinion" (Office of Research Integrity, 2000, 42 C.F.R. Part 93). To be considered research misconduct, actions must "represent a "significant departure from accepted practices"; have been "committed intentionally, or knowingly, or recklessly"; and be "proven by a preponderance of evidence" (Steneck, 2007, p. 20). Internationally, standards for responsible conduct in research have also been articulated by, among others, the European Code of Conduct for Research Integrity, the Danish Committee on Scientific Dishonesty, and the Australian Research Integrity Committee.

Beyond standards for research misconduct, there are also specific standards for scientific publishing. The Council of Science Editors (CSE), a U.S.-based nonprofit, for example, has published the "White Paper on Promoting Integrity in Scientific Journal Publications since 2006." The White Paper includes an overview of research ethics organizations in other countries as well as a process for addressing potential cases of misconduct and the potential parties affected. The White Paper, now updated "on a rolling basis as new sections are added and/or existing sections are updated to reflect new information or best practices," provides publication guidance for editors, authors, and reviewers (Council of Science Editors, 2018).

In addition to the CSE, the International Committee of Medical Journal Editors (ICMJE) gained recognition in 1978 when it produced the Uniform Requirements for Manuscripts Submitted to Biomedical Journals, making it one of the first organizations to advanced publication guidelines for scientific research as "a way of standardizing manuscript format and preparation across journals" (International Committee of Medical Journal Editors, "History", 2021). Since then, the ICMJE has continued to provide leadership on scientific publication through its "Recommendations for the Conduct, Reporting, Editing, and Publication of Scholarly Work in Medical Journals," which provides guidance on roles and responsibilities for authors, editor, reviewers, and journal owners, as well as manuscript preparation and submission guidance. Unlike the CSE, the ICMJE is not an open membership group, but the guidelines may be adopted voluntarily by other journals.

Finally, the Committee on Publication Ethics (COPE), founded in 1997, is a British organization that provides detailed guidance for journal editors on issues related to scientific publishing. In 2017, the organization outlined ten core "robust and well described publicly documented practices" that should "be considered alongside specific national and international codes of conduct for research and are not intended to replace these" (Committee on Publication Ethics, 2020b, para. 1). The ten practices include advice on how to develop policies related to:

1   allegations of misconduct;
2   authorship and contributorship;
3   complaints against the journal, staff, or editorial board;
4   potential conflicts of interest;
5   data availability and reproducibility;
6   informed consent, including vulnerable populations, as well as policies related to ethical conduct of research using animals and use of confidential data;
7   intellectual property, including copyright and publishing licenses;
8   journal management;
9   peer review;
10 postpublication discussions and corrections. (Committee on Publication Ethics, 2020a, "Core Practices")

The organization has also produced additional guidance, for example, the Principles of Transparency and Best Practice in Scholarly Publishing; Text Recycling Guidelines for Editors; and Sharing of Information Among Editors-in-Chief Regarding Possible Misconduct. By 2020, COPE membership exceeded 12,500 in 103 countries, and its publications had been translated into Spanish and Chinese editions (Committee on Publication Ethics, 2020b, "History").

Efforts by professional organizations such as the CSE, ICMJE, and COPE have been integral to establishing widely held understandings of what constitutes appropriate roles for individuals involved in scientific publishing. In doing so, they have drawn attention to the complicated relationships in the production of scientific knowledge. However, as I will explain in the following sections, the practical orientation of such guidelines still elides many deeper questions about power, prestige, and the making of scientific knowledge.

The negotiation between ethics guidelines and everyday practice

Guidelines for publishing ethics cover a small portion of the many ethical and legal issues that scientific authors, editors, and reviewers face. In what follows, I examine four areas – authorship, text recycling, data integrity, and peer review – where professional guidelines have offered advice. I argue that guidelines are useful starting points for ethical discussions, but they are retrospective documents. As the production of scientific knowledge itself changes, discussions about ethics must evolve too, and there are many areas of negotiation that remain unresolved as scientific practices change.

Ethics of authorship

Mario Biagioli, Distinguished Professor of Law and Communication at UCLA, has written extensively about the complexity of scientific authorship:

> Authorship is a particularly thorny issue in science because of the specific logic of its reward system. . . . A scientific claim is not rewarded as the material inscription of the scientist's personal expression, but a nonsubjective statement about nature. Consequently, it cannot be the scientist's property. This means that he or she does not have inherent rights in a scientific claim in the way a "normal" author has rights in the product of his or her personal expression simply by virtue of being the creative producer of that inscription. From this, it follows that *unless it is published and evaluated by peers, a scientific claim does not count as such and does not bring rewards to the scientist who produced it*. In sum, scientific authorship is not a right but a reward.
>
> *(2003, pp. 253–254, 256; emphasis added)*

As Biagioli suggests, authorship is power and prestige in scientific publishing because of the type of knowledge claims made by scientists. Recognition of authorship not only means that a researcher made a meaningful contribution to a project but also provides a means to claim ownership of the idea itself. Yet there is no agreement within the scientific community about a threshold for recognition of authorship, and there are various definitions regarding the distinction, for example, between an author and a contributor who is acknowledged but not given authorial recognition.

Issues of authorship are interwoven with the collaborative nature of scientific research. Few researchers work independently, and collaborative authorship is critical to career success in that "junior researchers who coauthor work with top scientists enjoy a persistent competitive advantage throughout the rest of their careers, compared to peers with similar early career profiles but without top coauthors" (Li et al., 2019, p. 1). Tensions over authorship are further exacerbated by "gift authorship" or "honorary authorship" in which an individual is added to the author's list, even though they may have contributed little or nothing to the research being reported. The collaborative, power-differential nature of scientific publishing, thus, gives rise to tensions about who should be given credit and how much credit should be given. Such tensions surface, among other ways, in relation to unequal professional status, gender discrimination, linguistic discrimination, and global "parasitic" publishing inequalities (Rees et al., 2017).

In an attempt to offer some guidance on the matter of authorship, the ICMJE recommends that authorship be based on four criteria:

1. Substantial contributions to the conception or design of the work; or the acquisition, analysis, or interpretation of data for the work; AND
2. Drafting the work or revising it critically for important intellectual content; AND
3. Final approval of the version to be published; AND
4. Agreement to be accountable for all aspects of the work in ensuring that questions related to the accuracy or integrity of any part of the work are appropriately investigated and resolved. (2019, p. 2, emphasis in original)

The ICMJE definition links effort, process, responsibility, and accountability. What is not addressed in this definition, however, are questions about order of authorship and corresponding authorship.

Order of authorship is an important marker of recognition across the sciences (Parish et al., 2018), even though individual disciplines differ in their approach to assigning authorship order (Dance, 2012). Securing the position of first author typically confirms prestige. The position of last author on a publication also implies significance because the position is often "assumed to be an individual (often in a more advanced career stage) who is providing overarching intellectual questions, funding, and writing, but has not necessarily been directly involved in data collection" (Faulkes, 2018, "Background," para. 3). Order of authorship thus conveys power and prestige, and it matters for the purposes of job security.

Lack of recognition in authorship extends beyond order of authorship to recognition as the corresponding author. Women, for example, are often not equally recognized in authorship (Macaluso et al., 2016), and even when women are recognized as authors, they are less likely to be the corresponding author. A "corresponding author" is not merely the contact person for questions related to the publication but also "assume[s] responsibility for all aspects of a publication" (Steneck, 2007, p. 136). Not being credited as the corresponding author further undermines professional recognition for their work (Fox et al., 2018; Hagan et al., 2020). That authorship and recognition gap translates into a citational gap and ultimately into a gap in symbolic recognition of the production of scientific knowledge (and potential revenue claims that might come later with patents, capital ventures, and further innovation). Journals are loath to mediate such disputes, and research misconduct rules often do not apply to such cases. As a result, as Faulkes (2018) points out, "having no way of resolving disputes in progress has corrosive effects on communities, particularly its most vulnerable members" (p. 3).

In the end, ethical issues related to authorship are multifaceted and must be recognized as including recognition for authorship, order of authorship, and corresponding authorship. Research on authorship must take this multidimensional approach in order to understand the nature and extent of authorship concerns as well as how local solutions can address such inequities (Wininger et al., 2017). Contribution statements and badging may help address some of the tensions surrounding authorship, although there is little sense of how authorship disputes will play out in the move to increasing preprint publication or how changing publication metrics may exacerbate such tensions. What is clear is that not attending to who is being affected by these changes will likely simply further marginalize certain members of the community.

Ethics of text recycling

In contrast to plagiarism, which is defined as "both the theft or misappropriation of intellectual property and the substantial unattributed textual copying of another's work" (Office of Research Integrity, 1994, para. 2), text recycling or the reuse of "the author's own original material" across subsequent texts (sometimes called self-plagiarism) is less clear (Anson and Moskovitz, 2020, p. 3). While it is widely accepted that boilerplate language may be reused across publications and that authors may reuse text from unpublished work, such as grants, in their published findings, "there is also no consensus as to how much text an author may recycle from his/her previous writings" (Roig, 2015, p. 21). Text recycling thus has become an area of scholarly research as well as debate. For example, the Text Recycling Research Project, a group based out of Duke University, seeks to "understand text recycling, to help build consensus among stakeholders, and to promote ethical and appropriate practice" (n.d., "About the TRPP"). To these ends, the group is studying professional researchers' beliefs on text recycling, text recycling in practice through corpus analysis, and legal issues surrounding text recycling, such as author rights and copyright law. Recent research by the group has included a survey of 200 STEM (science, technology, engineering, and math) researchers, including 130 "novices" and 70 "experts" with the purpose of understanding "how STEM researchers view factors such as the source of recycled material, the rhetorical purpose and amount of that material, and configurations of authorship in deciding whether any instance of text recycling is appropriate" (Moskovitz and Hall, 2020, p. 2). They found "broad agreement among both experienced STEM researchers and those at the start of their careers about the line between text recycling and plagiarism, [but they] saw considerably less agreement regarding the conditions under which recycling is acceptable" (Moskovitz and Hall, 2020, p. 16).

In another study by the Text Recycling Research Group, Anson and Moskovitz (2020) studied "verbatim and altered (or 'patch-written') forms of recycling" in four disciplines from articles linked to NSF grants (p. 6). They found that "reusing some amount of material from one's prior published articles appears to be fairly common" but that "verbatim reuse of entire sentences is not ubiquitous" (p. 17). Finally, they found that "all four disciplinary areas exceeded three recycled sentences per paper, corresponding to around 1.2% of the average 250-sentence paper" (p. 11), meaning that text recycling was consistent across the disciplines and confirming other research that showed "opinions on the acceptability of text recycling were largely independent of discipline" (p. 17).

Text analysis software has added an algorithmic dimension to the detection of plagiarism and text recycling, such that some journals now require authors to submit their papers to software analysis before being published (or even before an article is sent out for peer review). While such software provides the appearance of algorithmic ethics, it cannot address how such information should inform decision making. Furthermore, in a forthcoming critique, Moskovitz and Colton (2021) argue that iThenticate's widely cited white paper, "The Ethics of Self-Plagiarism" is "a disguised marketing ploy that misrepresents the realities and ethics of academic research."

For multilingual researchers, many of the questions surrounding text recycling "are not just situated in the linguistic abilities of the second language English researchers but also deeply situated within the structural components of publication," including lack of access to research writing resources, lack of access to professional networks, and lack of access to publication support (Hanauer et al., 2019, p. 137). Compounding these challenges, multilingual writers also report greater anxiety in publishing across languages. According to a study by Hanauer and Englander (2011), "multilingual scientists perceived English science writing as 24% more difficult, generating 21% more anxiety and 11% less satisfaction than science writing using their L1" (p. 138). Given these concerns, questions surrounding text recycling need to be addressed more clearly by the scientific community (Corcoran, 2019).

In sum, while definitions of plagiarism suggest a rather straightforward ethics issue, the ambiguity surrounding text recycling and the changing nature of who is conducting science – that is, international, multilingual researchers – suggest more nuance. Research by groups such as the Text Recycling Research Project and by scholars studying the publishing practices of international researchers provides such nuance to conversations about plagiarism and text recycling. Such work helps us understand not just scientific community practices but also who is most affected by changes to these practices.

Ethics of data integrity

Beyond authorship, plagiarism, and text recycling, the manipulation of visual evidence is a topic that has garnered much attention over the last decade (Rossner and Yamada, 2004). For example, Bik et al. (2016)'s research on the manipulation of scientific images has revealed the extent of data manipulation. In a 2016 study, "A total of 20,621 research papers containing the search term 'western blot' from 40 different journals and 14 publishers were examined for inappropriate duplications of photographic images, with or without repositioning or evidence of alteration" (p. 3). They found that "3.8% of published papers contained problematic figures, with at least half exhibiting features suggestive of deliberate manipulation" (p. 2). They concluded that the problem was likely much more widespread than they discovered, the problem of manipulation of visual images has increased extensively in the last decade, and the problem seems to increase when authors publish multiple papers.

Like plagiarism, data manipulation can include a number of intended and unintended practices, ranging from deliberate fabrication and rerendering of visual images to produce effects that were not observed to more subtle manipulations from cropping, low image resolution, and color enhancing images (journals dissuade "beautifying" images). In response, journals like *Nature*, *Science*, and the *Journal of Cell Biology* now provide expectations on data integrity as well as specific guidelines for types of visual evidence (microscopy, electrophoretic gels and blots, etc.), use of software, and statements about "spot-checking" images for manipulation (*Nature Research*, 2021). (See also Council of Science Editors, 2012; Office of Research Integrity, 2017.) Other journals now require authors to upload their raw data into a data repository (Matheson, 2018), or they have turned to software for prepublication image screening to identify figures that have been reused within and across publications (Acuna et al., 2018).

In the end, the main ethical concern today surrounding data integrity is the manipulation of data. More powerful software does not just allow researchers to capture better images, it also comes with more capabilities to manipulate those images. How algorithmic solutions will fare in addressing data manipulation remains to be seen. What is clear is that the same pressures to publish quickly and frequently that give rise to ethical concerns like plagiarism also drive the pressure to produce clean-looking data.

Ethics of peer review

A final issue of concern with the rapidly changing nature of scientific publishing is peer review. Peer review has long been the metric to assess the quality of research and can be delivered in a number of different ways, ranging from double-blind to open review. Furthermore, peer review may occur prior to publication or after publication, as well as being facilitated by the journal, a third party, or individual reviewers (Committee on Publication Ethics, 2017a, 2017b).

The vast changes in the publishing world, however, including the introduction of preprint servers such as arXiv, bioRxiv, BioRN, ChemRxiv, or SSRN, have reshaped the traditional role of peer review. For example, preprint servers do not rely on peer review. As explained on the bioRxiv website, which is operated by the Cold Spring Harbor Laboratory:

> Because [the peer review] process can be lengthy, authors use the bioRxiv service to make their manuscripts available as "preprints" *before* completing peer review and consequent certification by a journal. This allows other scientists to see, discuss, and comment on the findings immediately. Readers should therefore be aware that articles on bioRxiv have not been finalized by authors, might contain errors, and report information that has not yet been accepted or endorsed in any way by the scientific or medical community.

Even before the COVID-19 pandemic, the relatively slow pace of scientific peer review was driving the popularity of preprint servers. The pandemic has accelerated this trend.

In tandem with the growth of preprints, the scientific community has embraced the postpublication peer review (PPPR) process. Postpublication review has been a long-standing feature of many journals in which readers submit letters or other commentary to the journal. Much of this work now occurs in digital formats on websites such as PubPeer and F1000. PPPR provides open and even fully attributable reviews. Moreover, unlike traditional peer review that relies on double-blind conditions in which neither the reviewer nor the author is aware of the other's identity, open peer review, such as PPPR, is an attempt to recognize the value of quality peer discussion, spur higher-quality reviews, and make scientific discussions about current research more transparent.

With traditional peer review no longer a gatekeeper for academic publishing and the print medium no longer being the single venue for the dissemination of scientific research, it has been argued that the sciences should move from the curate-publish approach to the publish-curate approach in which "authors decide when and what to publish; peer review reports are published, either anonymously or with attribution; and curation occurs after publication, incorporating community feedback and expert judgment to select articles for target audiences and to evaluate whether scientific work has stood the test of time" (Stern and O'Shea, 2019, p. 1).

The popularity of preprint servers and postpublication peer review is leading to seismic changes in scientific peer review. It remains to be seen how these changes will be felt in different scientific disciplines and whether this change will result in other consequences, such as a certain number of postpublication peer reviews being factored into tenure decisions or whether postpublication reviews themselves may count as publications.

Conclusion

The SARS-CoV-2 pandemic has fundamentally changed scientific publishing. From accelerating the move to sharing data on open-access preprint servers to changing the kinds of publications that are being embraced by the scientific community. These changes intersect with long-standing ethical concerns about authorship, plagiarism, and text recycling while also being shaped by technological advances that make data manipulation easier. The pandemic has demonstrated that the tradition of scientific publishing that occurred primarily through peer-reviewed print publications will not be the same and that the ethical guidelines for publishing that were based on such models are insufficient. Instead, ethical guidelines for scientific publishing must evolve as scientific research itself evolves, and the scientific community must recognize that the very pressures to publish – pressures the community puts on itself – has led to the very conditions that result in ethical misconduct.

It remains to be seen whether these changes will ripple outside the sciences into publishing in the social sciences and, potentially, even the humanities. It also remains to be seen whether concerns about gender inequality and linguistic bias in conversations about changing publishing ethics will be extended with conversations about racial discrimination and accessibility for readers with disabilities. In conclusion, it is exciting to think that the long-standing gatekeeping mechanisms of scientific publishing that has restricted many struggling scientists might finally be open to a broader range of scientists in the postpandemic world.

References

Acuna, D.E., Brookes, P.S., & Kording, K.P. (2018) 'Bioscience-scale automated detection of figure element reuse (preprint)', *Scientific Communication and Education*, https://doi.org/10.1101/269415 (Accessed 15 February 2021)

Anson, I.G., & Moskovitz, C. (2020) 'Text recycling in STEM: A text-analytic study of recently published research articles', *Accountability in Research*, https://doi.org/10.1080/08989621.2020.1850284 (Accessed 15 February 2021)

Bevilacqua, F., Giannetto, E., & Matthews, M. (eds.) (2001) *Science education and culture: The contribution of history and philosophy of science*, New York: Springer.

Biagioli, M. (2003) 'Rights or rewards: Changing frameworks of scientific authorship', in M. Biagioli & P. Galison (eds.), *Scientific authorship: Credit and intellectual property in science*, New York: Routledge, pp. 253–279.

Biagioli, M., & Lippman, A. (2020) *Gaming the metrics: Misconduct and manipulation in academic research*, Cambridge: MIT Press.

Bik, E.M., Casadevall, C., & Fang, F.C. (2016) 'The prevalence of inappropriate image duplication in biomedical research publications', *bioRxiv*, http://dx.doi.org/10.1101/049452 (Accessed 15 February 2021)

Committee on Publication Ethics (2017a) 'Guidelines: Ethical guidelines for peer reviewers', https://publicationethics.org/files/cope-ethical-guidelines-peer-reviewers-v2_0.pdf (Accessed 15 February 2021)

Committee on Publication Ethics (2017b) 'Who "owns" peer reviews?' https://publicationethics.org/files/Who_owns_peer_reviews_discussion_document.pdf (Accessed 15 February 2021)

Committee on Publication Ethics (2020a) 'Core practices', https://publicationethics.org/core-practices (Accessed 15 February 2021)

Committee on Publication Ethics (2020b) 'History of COPE', https://publicationethics.org/about/history (Accessed 15 February 2021)

Corcoran, J. (2019). 'Addressing the "bias gap": A research-driven argument for critical support of plurilingual scientists' research writing', *Written Communication*, vol. 36, pp. 538–577. [online]. https://doi.org/10.1177/0741088319861648 (Accessed 15 February 2021)

Council of Science Editors. (2012) 'Digital images and misconduct', www.councilscienceeditors.org/resource-library/editorial-policies/white-paper-on-publication-ethics/3-4-digital-images-and-misconduct/ (Accessed 15 February 2021)

Council of Science Editors. (2018) 'CSE's white paper on promoting integrity in scientific journal publications', CSE-White-Paper_2018-update-050618.pdf (Accessed 15 February 2021)

Dance, A. (2012) 'Authorship: Who's on first?' *Nature*, vol. 489, pp. 591–593, https://doi.org/10.1038/nj7417-591a (Accessed 15 February 2021)

Else, H. (2020) 'How a torrent of COVID science changed research publishing – in seven charts', *Nature*, www.nature.com/articles/d41586-020-03564-y (Accessed 15 February 2021)

European Association of Science Editors. (2020) 'EASE statement on quality standards', https://ease.org.uk/publications/ease-statements-resources/ease-statement-on-quality-standards/?fbclid=IwAR088siNa8HSjypWxO78CfZ-j0sr-cOUpacZ9IdtNmR_YSzTckYIzKQkmzE (Accessed 15 February 2021)

Faulkes, Z. (2018) 'Resolving authorship disputes by mediation and arbitration', *Research Integrity and Peer Review*, vol. 3, no. 12, https://doi.org/10.1186/s41073-018-0057-z (Accessed 15 February 2021)

Fox, C., Ritchey, J., & Paine, C.E. (2018) 'Patterns of authorship in ecology and evolution: First, last, and corresponding authorship vary with gender and geography', *Ecology and Evolution*, vol. 8, pp. 11492–11507, doi:10.1002/ece3.4584

Hagan, A., Topçuoğlu, B., Gregory, M., Barton, H., & Schloss, P. (2020) 'Women are underrepresented and receive differential outcomes at ASM journals: A six-year retrospective analysis', *mBio*, vol. 11, no. 6, https://10.1128/mBio.01680-20, https://doi.org/10.1128/mBio.01680-20 (Accessed 15 February 2021)

Hanauer, D.I., & Englander, K. (2011) 'Quantifying the burden of writing research articles in a second language: Data from Mexican scientists', *Written Communication*, vol. 28, no. 4, pp. 403–416, doi:10.1177/0741088311420056

Hanauer, D.I., Sheridan, C.L., & Englander, K. (2019) 'Linguistic injustice in the writing of research articles in English as a second language: Data from Taiwanese and Mexican researchers', *Written Communication*, vol. 36, pp. 136–154, https://doi.org/10.1177/0741088318804821

International Committee of Medical Journal Editors. (2019) 'Recommendations for the conduct, reporting, editing, and publication of scholarly work in medical journals', www.icmje.org/recommendations/ (Accessed 7 May 2021)

International Committee of Medical Journal Editors. (2021) 'History', www.icmje.org/recommendations/browse/about-the-recommendations/history-of-the-recommendations.html (Accessed 15 February 2021)

Johnson, R., Watkinson, A., & Mabe, M. (2018) *STM report: An overview of scientific and scholarly publishing*, The International Association of Scientific, Technical and Medical Publishers, www.stm-assoc.org/2018_10_04_STM_Report_2018.pdf (Accessed 15 February 2021)

Li, W., Aste, T., Caccioli, F., & Livan, G. (2019) 'Early coauthorship with top scientists predicts success in academic careers', *Nature Communications*, vol. 10, p. 5170.

Macaluso, B., Larivière, V., Sugimoto, T., & Sugimoto, C. (2016) 'Is science built on the shoulders of women? A study of gender differences in contributorship', *Academic Medicine*, vol. 91, no. 8, pp. 1136–1142, doi:10.1097/ACM.0000000000001261

Macaulay, A.C. et al. (1998) 'Participatory research with Native community of Kahnawake creates innovative code of research ethics', *Canadian Journal of Public Health*, vol. 89, pp. 105–108, https://doi.org/10.1007/BF03404399

Matheson, S. (2018) 'Why you should deposit your raw data', http://crosstalk.cell.com/blog/why-you-should-deposit-your-raw-data (Accessed 15 February 2021)

Moskovitz, C., & Colton, A. (2021) 'iThenticate, self-plagiarism and fraud: A complicated relationship', *Inside Higher Education*, www.insidehighered.com/views/2021/03/05/resources-avoiding-self-plagiarism-are-scarce-and-problematic-opinion

Moskovitz, C., & Hall, S. (2020) 'Text recycling in STEM research: An exploratory investigation of expert and novice beliefs and attitudes', *Journal of Technical Writing and Communication*, vol. 51 no. 3, pp. 252–272, https://doi.org/10.1177/0047281620915434

Nature Research. (2021) 'Image integrity and standards', www.nature.com/nature-research/editorial-policies/image-integrity (Accessed 15 February 2021)

Office of Research Integrity (1994) 'ORI policy on plagiarism', https://ori.hhs.gov/ori-policy-plagiarism (Accessed 7 May 2021)

Office of Research Integrity (2000) 'Federal research misconduct policy', http://ori.hhs.gov/policies/fed_research_misconduct.shtml (Accessed 15 February 2021)

Office of Research Integrity (2017) 'Tips for presenting scientific images with integrity', https://ori.hhs.gov/sites/default/files/2017-12/6_Image_Manipulation_scalable.pdf (Accessed 15 February 2021)

Palayew, A., Norgaard, O., Safreed-Harmon, K., Andersen, T.H., Rasmussen, L.N., & Lazarus, J.V. (2020) 'Pandemic publishing poses a new COVID-19 challenge', *Nature Human Behavior*, vol. 4, pp. 666–669, https://doi.org/10.1038/s41562-020-0911-0

Parish, A.J., Boyack, K.W., & Ioannidis, J.P.A. (2018) 'Dynamics of co-authorship and productivity across different fields of scientific research', *PLoS One*, vol. 13, p. e0189742, https://doi.org/10.1371/journal.pone.0189742

Rees, C.A., Lukolyo, H., Keating, E.M., Dearden, K.A., Luboga, S.A., Schutze, G.E., & Kazembe, P.N. (2017) 'Authorship in paediatric research conducted in low- and middle-income countries: parity or parasitism?', *Tropical Medicine & International Health*, vol. 22, no. 11, pp. 1362–1370, doi:10.1111/tmi.12966

Resnik, D. (1998) *The ethics of science: An introduction*, London and New York: Routledge.

Retraction Watch. (2020) 'A look back at retraction news in 2020 – and ahead to 2021', 30 December, https://retractionwatch.com/2020/12/31/a-look-back-at-retraction-news-in-2020-and-ahead-to-2021/ (Accessed 15 February 2021)

Roig, M. (2015) 'Avoiding plagiarism, self-plagiarism, and other questionable writing practices: A guide to ethical writing', *Office of Research Integrity*, https://ori.hhs.gov/sites/default/files/plagiarism.pdf (Accessed 15 February 2021)

Rossner, M., & Yamada, K.M. (2004) 'What's in a picture? The temptation of image manipulation', *Journal of Cell Biology*, vol. 166, pp. 11–15, https://doi.org/10.1083/jcb.200406019

Sousa Santos, B. (2018) *The end of the cognitive empire: The coming of age of epistemologies of the south*, Durham, NC: Duke University Press Books.

Steneck, N. (2007) 'Introduction to the responsible conduct of research', *Office of Research Integrity*, viewed 15 February 2021, https://ori.hhs.gov/sites/default/files/rcrintro.pdf

Stern, B.M., & O'Shea, E.K. (2019) 'A proposal for the future of scientific publishing in the life sciences', *PLoS Biol*, vol. 17, p. e3000116, https://doi.org/10.1371/journal.pbio.3000116

Text Recycling Project. (n.d.) 'About the TRPP', https://textrecycling.org/ (Accessed 15 February 2021)

Wininger, A.E. et al. (2017) 'Bibliometric analysis of female authorship trends and collaboration dynamics over *JBMR's* 30-year history', *Journal of Bone and Mineral Research*, vol. 32, no. 12, pp. 2405–2414, doi: 10.1002/jbmr.3232. Epub 6 September 2017. PMID: 28777473; PMCID: PMC5732055.

5

MY TESTIMONY

Black Feminist Thought in scientific communication

Natasha N. Jones

Introduction: how to read this chapter

I blend personal narrative and academic genre conventions to critique "recognize[d] scientific communication itself as a site of ongoing racial basis" (Hanganu-Bresch et al., personal communication, 2020). Drawing on scholarship and personal dialogues, I interrogate silencing in scientific communication scholarship. I center my critique on the discipline itself. Science communication, in this chapter, refers specifically to the academic discipline and to scholarship dedicated to the study of communication of and about science. I do this in order to note that communication by scientists to other scientists or to lay audiences (how scientific communication is often defined) sets up a false dichotomy between who can *do* and *communicate about* science. This definition obscures that academia, as a gatekeeping mechanism, decides who is knowledgeable, respectable, and "educated" enough to do science and communicate about it and often ignores how marginalized populations communicate *about* (and do) science in their own communities. Thus I focus on academic contexts as a space for examining how racism and white supremacy can shape our understandings of scientific communication. Finally, my critique acknowledges that sites of academic work (research, scholarship, pedagogy, and practice) are interrelated and interdependent.

My testimony

I often hesitated when writing this chapter. I stopped and started. I added text only to highlight it all and send my words back into the ether. I talked through ideas with myself over coffee. Some of those thoughts drifted away. The ideas that I was able to grasp and hold onto, proved to be a feeble attempt to capture what I wanted to say. Finally drafted, I sent the chapter to the editors. Weeks later, the editors replied, requesting revisions that asked me to disappear more of my ideas, my knowledge. I'd worked hard to find words to express the ways that silencing stifled knowledge-making. Choking out each phrase, I'd corralled and pinned down and penned down each word with the force of my fingers and pulled in and pushed out my argument onto the screen with the raced, gendered weight of my embodied ways of knowing. I felt unseen, unvoiced. I thought of Serena Williams.

DOI: 10.4324/9781003043782-7

In 2017, tennis superstar Serena Williams birthed her first child. Williams was candid about her experiences of childbirth as a Black woman and revealed in an interview (Haskell, 2018) that she nearly died after giving birth because medical professionals ignored her concerns and dismissed her requests for diagnostic tests. As it turns out, Williams was suffering from a hematoma in her abdomen. This life-threatening condition would have been discovered earlier if Williams' health care team had listened and taken her concerns seriously. Williams joined the ranks of Black women who are routinely ignored, undertreated, and labeled as noncompliant as medical patients, their words disappeared and their voices silenced.

A Centers for Disease Control (CDC) press release (2019) noted that Black women and Alaskan Native women are "two to three times more likely to die from pregnancy-related causes than white women" (Centers for Disease Control, 2019). For Black women, maternal and childbirth-related deaths can be traced back to U.S. chattel slavery. In "Black Maternal and Infant Health: Historical Legacies of Slavery," Owens and Fett (2019) draw connections between slavery and Black women's birthing experiences. "Legal and medical attention to enslaved women's bodies played an especially important role in the entrenchment of American racism and its manifestation as a public health crisis today" (2019, p. 1342). Blatant disregard for Black women's lives persists in subtle and overt ways in the medical practices of today. Chadwick points out that "intersecting social inequalities mean that poor black women and girls are more likely to be the recipients of coercive and persistent forms of silencing during labour/birth" (2019, p. 33). This silencing is pervasive and is evidence of the failure to address systemic racism that is embedded in the ways our academic programs – medical programs, scientific and technical communication programs, rhetoric of science programs – are hesitant to address racism and white supremacy in pedagogy, subsequently impacting how professionals and practitioners apply their knowledge in their various workplace contexts.

Maternal health is only one area where we see disparity that illuminates the vestiges of anti-Black racism playing out in scientific practices. We have historical and contemporary evidence that racism is a factor across the spectrum of science- and health-related disciplines. For example, Black and Indigenous populations are at a greater risk of health issues and have higher mortality rates due to illness and disease (Serchen et al., 2020), are more likely to live in and be impacted by environmental pollution and air quality (Robinson et al., 2018), are less likely to receive vaccines and pain medications needed to treat illness (Hollingshead et al., 2016; Quinn et al., 2017), are less likely to have access to a variety of nutritious foods (Holt-Giménez, 2018), are less likely to receive care-oriented assistance for addiction (Hart and Hart, 2019), are less likely to have inclusive access to museums (Dawson, 2018), are less likely to be represented in science, technology, engineering, and mathematics (STEM) fields at both the K–12 and college levels, and are less likely to be represented in STEM field and computer technology professions (AAUW, 2015). These are all evidence of systemic racism that is intertwined in science-related disparities of education, public policy, and communication.

Though many of these disparities are tracked, traced, and surveilled,[1] academic disciplines present a resistance to addressing issues of white supremacy directly. To be clear, it is disingenuous of academic disciplines to focus on the critique of practice and practitioners without reflectively and reflexively examining the ways that our pedagogy and scholarship influences practices and practitioners. I argue that how we engage with, teach, and publish scientific communication scholarship engages in racist practices that silences the voices of marginalized groups, ignoring intersectional and overlapping oppressions. This silencing in academic spaces enables the silencing and dismissal of the voices and experiences of marginalized people in practice and in communities. Scholarship always informs pedagogy, which informs practice, which informs scholarship, and on and on. Because of the interdependent and dialogical nature of scholarship, pedagogy, and practice, the field of scientific communication as an academic enterprise cannot assume that scientific communication itself will not replicate unaddressed racial bias. I advocate for explicitly and directly interrogating the structure of the field *alongside* practitioners, addressing white supremacist ideas and racial bias. Instead, it seems that much science communication scholarship relies on the vague language of diversity and inclusion when pointing to disparities and inequities that are based in racial oppression and bias. In fact, it is not until recently, as we are seeing the social justice turn in related fields like technical communication, that some scholars are becoming more willing to specifically use the term "white supremacy." Understanding that white supremacy is, according to Ansley, a "political, economic, and cultural system in which whites overwhelmingly control power and material resources, conscious and unconscious ideas of white superiority and entitlement are widespread, and relations of white dominance and non-white subordination are daily reenacted across a broad array of institutions and social settings" is a start (1989, p. 1024). "White supremacy undergirds the way we organize our society, and the ways in which we distribute resources and power" – that is, the preceding "definition of white supremacy focuses primarily on the institutional arrangements" with "individual race-based animus" being a secondary concern (Wilson, 2018, p. 3). As Wilson further explains, there is a clear resistance to talk about white supremacy that is not about "individual race-based animus" (2018, p. 3), which points to an indication of one of the reasons why vague language may be used instead. Fundamentally, however, word choice matters and has rhetorical implications in disciplinary silencing practices.

Silencing in scientific communication: a site of racial bias

Types of silencing occur for a variety of reasons. Silencing is defined broadly as "not being heard or understood, not being included or represented, being ignored or delegitimized, not being valued, or . . . marginalized" (Jones, 2016, p. 478). Feminist conceptualizations of silence interrogate how power is maintained and manifested, understanding that "power can silence or support the voices of others" (Jones, 2016, p. 478). When considering silence from a Black Feminist perspective, "silencing is often systemic" and "can occur without malicious intent, ill will, or even active engagement," with complicity finding a path through "heteronormative, patriarchal, eurocentric" ideologies that go unacknowledged and unchallenged in any explicit way (Jones, 2016, p. 478). Using Black Feminist Thought as a frame, I examine two of the ways that silencing in science communication occurs. Silence can be enacted in oppressive ways that devalue Black and marginalized knowledges and knowledge-makers, while, in other contexts, silence is taken up in agentive ways when authoritative voices or experts in science communication choose complicity in the form of self-censorship in order to avoid surrounding sociopolitical understandings of racism and white supremacy. The differences between what may appear to be self-silencing similar to self-censorship by Black and other marginalized groups can be understood outside of the bounds of complicity because of systemic and institutional power dynamics. That is, "power relations and other contextual factors" impact if, how, and when a marginalized individual may feel protected in speaking up or speaking out (Dotson, 2011, p. 239). What may be seen as complicity via self-censorship in marginalized groups can be "testimonial quieting" (discussed later in this chapter) or "testimonial smothering," in that the individual or group is "coerced" into silence because of the *risk* of "social, political, and/or material harm" (Dotson, 2011, p. 244).

The words we value

Early in my career, I struggled to adjust to a predominately white program in a predominantly white institution located in a predominately white city. Years later, I realized that I did not yet have the language to speak out about what it was like for me to be a Black woman in spaces that did not or could not acknowledge that my experiences were largely different from those around me. Even as a scholar who studied the impact of language in constructing realities, I just did not have the words. I was effectively silenced. Fricker (2009) calls the phenomenon that I experienced "hermeneutic injustice," which occurs when "an agent lacks the ability to understand a key aspect of her lived experience, due to the fact that conceptual resources for her community have been undercut by systemic oppression" (Droira, 2019). The systemic failures of our institutions to address the impact of white supremacy on the work we do and the lives we live, both within and outside the academy, have an impact. Often, there are no conversations about representation, inclusion, racial equity, anti-Black racism, or oppression. In not talking about these things, in the quiet surrounding of these things – the ways that white supremacist ideals are infused in every aspect of academia – the folks most impacted by the oppressive systems are not only silenced but separated from the very knowledge creation and legitimization functionings of our scholarly disciplines.

It is also important for me to acknowledge the difficulty of "valuing" the utterance of the words "white supremacy." Most likely those words conjure an immediate and visceral reaction from folks reading this text. Whether this reaction is recoiling in defensiveness, righteous anger, fear, or solemn acknowledgment, I assume that it is partially due to these emotionally tethered, reactive responses that scholars in science communication (and related fields) often refrain from saying these words out loud in our academic spaces. Yet the more I research issues of social justice and oppression in scientific and technical communication, the more I am convinced that, without a direct engagement with white supremacist ideals and their impact, we are poorly equipped to do the type of work that we claim to want to do as scholars – which, as Ceccarelli details, includes work that seeks to understand, gain insight, illuminate, show, reveal, consider, broaden our thinking in order to become more attuned and heighten our awareness (Ceccarelli, 2013, p. 2). We are even more stifled in the scholarly goal of disrupting norms and conventions if we refuse to address the effect of systematic white supremacy on the sciences and science communication (Ceccarelli, 2013, p. 2; Majdik & Platt, 2012, p. 138).

From the outset of writing this chapter, I had some idea that scholarship explicitly naming white supremacy and anti-Black racism in science communication would be scarce. Dawson (2018) lamented that, even generally, "questions of social justice, equity and exclusion are rarely discussed" (2018, p. 774). Still, out of curiosity, as I conducted a keyword search of a few science communication journals, I was hard-pressed to find the term "racism," let alone "white supremacy," in many of the published articles. My quick keyword search was limited to three journals (*Science Communication*, the *Journal of Science Communication*, and *Public Understandings of Science*). I restricted my search to the past five years (2015–2020). Results revealed that the term "racism" had been used within the text of published articles a total 14 times across all the three journals (that is, 14 articles used the term racism at least one time in the main text). Results for the term "white supremacy" were even more revealing, with the term being used in only one article across all three journals. This is not to say that no work is being done in this area. To the contrary, I believe work *is* being done related to racism, oppression, and marginalization in science communication. Again, I was looking for the explicit use of the terms (a calling out, if you will). In a few notable exceptions, scholars specifically use the word "racism" or "white supremacy," for example, Dawson's (2018) work on excluded publics in science communication (though Dawson calls out cultural imperialism, the author does *not* specifically use the words "white supremacy," but it can be argued that white supremacy is an instantiation of cultural imperialism). But this is the nature of locating silence. Locating silence requires paying attention to what is *not* there, to what is obfuscated, opaque, hidden. Silence looks at what is marginalized and makes an assessment about what *could* be. This understanding led me to ask myself: Is this calling out, this naming in the form of explicitly using the terms "racism" and "white supremacy," truly not happening in science communication? Why or why not? Moreover, the results of this search made me, as a scholar, ask, Is explicitly using the words "racism" and "white supremacy" key (or necessary) for *doing* science communication work? What work does silence preclude us from doing in science communication?

To aid my reflections, I turned to a mentor, another Black woman scholar in science and technical communication. She posed a provocative question. She asked me to be specific about *who* was not using the terms "racism" and "white supremacy" to talk about and engage in science communication scholarship. Our dialogue shifted the focus of the questions that I was grappling with. My questions became more about whose voices, perspectives, and knowledges we value as a field. Who gets published and why? Whom do we not hear from, and why are those that are not published in our academic venues and speaking at our academic conferences ignored or not acknowledged as science communicators? So when I ask whether racism and white supremacy are key to *doing* science communication work, whose ideals of *doing* am I considering? This required me to sit with the tension of locating silence and the uncertainty of locating work. In the following sections, I interrogate two types of silencing that are tied to (1) knowledge legitimization and (2) the privilege of remaining apolitical.

Testimonial quieting and expertise

Dotson describes testimonial quieting as "an audience fail[ing] to identify a speaker as a knower" (2011, p. 242). This idea is intricately connected to knowledge creation and knowledge legitimization. According to Dotson, the "disappearing of knowledge" occurs when "local or provincial knowledge is dismissed due to privileging alternative, often Western, epistemic practices" (Dotson, 2011, p. 236). Drawing on the theoretical frame of Black Feminist Thought (Patricia Hill Collins, 1990), Dotson notes how the presumption of incompetence constrains Black women's ways of learning and knowing and restricts the spaces in which Black women can engage in epistemic and knowledge-making practices. Collins demonstrates how Black women's knowledge is undervalued and devalued.

Our academic institutions are not exempt from this indictment (see Muhs et al., 2012). Science communication, in drawing together disciplinary knowledge work, also falls prey to the devaluing and undervaluing of the contributions of marginalized populations. In examining Fricker's definition of epistemic injustice ("wronging an agent in her capacity as a knower"), Droira reiterates how epistemic injustice aligns with power (Fricker, 2009). Droira asserts that these injustices are "pervasive wherever social power interacts with economies of knowledge, and science is no exception" (2019, para. 1). Academic disciplines themselves (yes, even academic disciplines like science communication that often engage in public-facing knowledge work) often gatekeep by defining the boundaries of what is "validated" knowledge(s) and by deciding what counts as disciplinary work in complicity with ways that academia (writ large) marginalizes certain groups, certain knowledges, and certain ways of knowing. This reproduction of marginalization includes testimonial quieting of individuals or groups of individuals who are not valued as knowers or experts. We see an example of testimonial quieting in the Flint

Michigan Water crisis when the overwhelmingly Black and impoverished community of Flint was disregarded and ignored as they tried to raise awareness of the polluted water source that they were forced to rely on by city and state governments. Ultimately, this testimonial quieting has had long-standing impacts, resulting in the death of at least 12 Flint citizens due to a Legionnaire's outbreak (Childress, 2019).

As Yu and Northcut indicate, in science communication, the distinction between expert (who is engaged in "real" science communication) and nonexpert is fraught, and some of this tension stems from fragmented definitions of what science communication is. In fact, Yu and Northcut advise that "we will do well to recognize that the supposed demarcation between "experts" and "non-experts" is situation-dependent, politically charged, and potentially problematic" (2018, p. 12). Further, "the boundary between insiders and outsiders is blurred as science becomes a social enterprise, a cultural phenomenon," and localized, culturally grounded knowledge is valid and valuable in relation to science communication (Yu and Northcut, 2018, p. 12). It stands to reason that systemically marginalized groups are likely to experience silencing in a disproportionate manner, simply by virtue of their positionalities and identities. Recognizing this, this discipline has moved to include citizen science and other participatory models in their research and scholarship.

Even as scholars in science communication turn to citizen science as a way to address bias about who does science and who communicates about science, there are still questions about how marginalized communities are valued and how knowledges are reframed as expertise in science. Are marginalized communities actually "empowered" by citizen science? Some contend that "traces of this elitist attitude[s] continue to make it hard for marginalized communities with problems and no answers to find a science-enabled voice" (https://phys.org/news/2016-01-citizen-science-empower-disenfranchised.html). This elitism is inherently tied to racism and white supremacy when considering that the communities in question are often raced communities that experience overlapping and intersectional oppressions. Even as more participatory approaches push back against the deficit model as a frame for understanding how and why a community might engage with science, there remain persistent ideals about who are considered experts and capable science communicators. This persistence can be illustrated by how science communication scholarship can cast marginalized individuals solely as advocates and activists or as extraordinary examples but not as science communicators in their own right.

Race talk

The relationship between testimonial quieting and knowledge creation and legitimization reveals the pervasiveness of marginalization and its limiting effects on the capability of institutions (in general) and academic disciplines (specifically) to be able to get out of their own way and do the knowledge work that they seek to do. However, the silencing of marginalization communities is only one consideration of the way that racism and white supremacy affect disciplinary science communication. Next, I consider the use of vague language and self-censorship by those positioned as experts.

In recent years, academic institutions and academic disciplines have forged ahead with espoused missions of diversity, inclusion, and equity. Aside from the prolific creation of offices of diversity and inclusion within our institutions (Parker, 2015), between the years of 2005 and 2015, approximately 60 chief diversity officer positions were created at colleges and universities. Overwhelmingly, these terms ("inclusion," "diversity," and "equity") have been picked up and used in vague, undefined ways. Parker details, for example regarding the term "diversity," that "there is little uniformity in the prior scholarship regarding the term, *diversity*, in organizations and institutions, as well as the underlying definition of diversity" (2015, p. 27). Further, terms like "diversity" and "inclusion" have been used as stand-ins for anything ranging from "inclusive ideals" to a "diversity of methodological approaches," in part allowing for the illusion of an apolitical approach to racism in academia. Vague words like "diversity" offer cover for scholars and scholarship that seeks to avoid public conflict for calling out white supremacist ideals (including anti-Black racism) as such.

In writing this chapter, I went back and forth with myself about "calling a thing a thing" and the importance of using specific words like "white supremacy." But ultimately, and in relation to science communication and the public understanding of science, it is no secret that anti-Black racism and white supremacy are inescapable factors that can be historically traced to and through universities and academic research and scholarship (for example, the cruelty of J. Marion Sims, the "father of gynecology," and the Tuskegee "study" on syphilis). What we choose to acknowledge about the impact of racism on the ways that we talk about, explain, engage with, and do science necessarily shifts the focus of what interventions, responses, resolutions, and advancements we choose to consider and employ. Language is complex and messy, and avoiding conflict (or the appearance of conflict) and examining how, when, and why controversy plays a role in science communication and the public understanding of science is simultaneously a point of privilege and a point of contention. Thus race talk, rather than racism and white supremacy itself, is often seen as controversial, making discourse about racism and white supremacy all too easily pushed aside by science communicators as they "self-censor" and instead choose not to talk directly about white supremacy at all.

Harvin explains how "race talk" has been rhetorically framed as impolite in the U.S. public sphere. Moreover, as Harvin acknowledges, people tend to talk about race among people with whom they are most alike. This allows people to be less likely to be "challenged" (1996, p. 15). More recently, scholars have theorized about the unequal labor (emotional and intellectual) that talking about race and white supremacy requires of Black and Indigenous people, while white people who "evidently don't feel pressed to think about how considerations of race play out in their own lives, at home, at work" (1996, p. 15) can exert less effort engaging in complicated dialogues about race. Further, discussion and dialogue about racism and white supremacy have long been points of political divisiveness in the United States.

Barbara Smith reminds us that, despite long-standing historical evidence, "millions deny the existence of systemic racism, including a cohort that enthusiastically supports white supremacy" (Smith, 2020). In 2019, Black Americans recognized 400 years of white supremacy and systemic racism. An entire issue of *The New York Times* was dedicated to Nikole Hannah-Jones's "The 1619 Project" that marks late August 1619 as a "defining moment" in U.S. history and the very "point of origin" of the nation. The compilation of literary works and essays traced the ways that Americans have been asked to sit with this historical framing that points out that slavery and anti-Black racism was fundamental in seeding "everything that has made America exceptional" (Hannah-Jones, 2019). Effects of anti-Black racism are also evidenced in science and technology developments and advancements (for example, innovation around cotton and agricultural products driven by the need to make slave owners' plantations more profitable). And yet, despite historians' work that traces how chattel slavery and colonization in the service of white supremacy anchor every aspect of this country (see also Kendi, 2016), this assertion is considered a "controversial" idea in the public and political spheres. In fact, this "controversy" has been the focus of an effort to suppress Hannah-Jones's work as not "patriotic."

This more recent framing of race talk as unpatriotic and anti-American is particularly kairotic as attitudes about racism have illustrated disparity between how white populations and minority groups *acknowledge* the existence of racism and racial progress (or the lack thereof) in the United

States. In a poll by Pew Research Center (Horowitz et al., 2019), 78% of Black respondents said that the United States "hasn't gone far enough toward equal rights" for Black people while only 37% of white respondents shared that same opinion. Dei, Karumanchery, and Karumanchery-Luik argue, "It should be realized that racism and oppression are such an integral part of our history in Western context that they have attained an almost accepted/acceptable station in our cultural/political/social milieu" (2007, p. 9). This passive acceptance empowers those who may be motivated to maintain white supremacist views, and this subtle positioning of what is accepted/acceptable "is solidified through the commonsense beliefs, rhetoric, and practices that seek to mute and downplay the impact and scope of racism and oppression" (2007, p. 9).

What does this mean for science communicators who already struggle with issues of uncertainty and even manufactured scientific controversy? "Scientific controversy is 'manufactured' in the public sphere when an arguer announces that there is an ongoing scientific debate in the technical sphere about a matter for which there is actually an overwhelming scientific consensus" (Ceccarelli, 2011, p. 196). And though tensions around acknowledging the impact of white supremacy and anti-Black racism on science are not a manufactured scientific controversy, the manufactured *public* controversy has an impact on how science communicators do their work and whether or not science communicators explicitly call out racism and white supremacy. In fact, we see some of the same argumentative appeals to "open-mindedness, freedom of inquiry, and fairness" nurture discursive silence in disciplinary science communication (Ceccarelli, 2011, p. 198). Further, these argumentative appeals, deployed in a manufactured scientific controversy as a way to goad debate on scientific issues and position the scientific mainstream as "opposed to democratic values" should they refuse to engage, have the opposite effect in the manufactured public controversy about white supremacy. The manipulated and deceptive rhetorical frame then becomes an argument that to *engage* in debate and to *talk* about racism and white supremacy in explicit ways in relation to science (and in other contexts) *is* antidemocratic. And, because "science communication practices are shaped by structural inequalities" that are raced and gendered, favoring dominant ideals about who engages with and participates in science-related activities (Dawson, 2018, p. 773), the "safe play" then becomes for science communicators to self-censor in order to appear unproblematic in the public sphere.

While this self-censorship allows individuals to enact their rhetorical agency in order to avoid "controversy" in the public and political sphere, these methods of disengagement contribute to "color blind racism" that feeds off the refusal to acknowledge that persistent racial inequity is founded in white supremacist ideals. As Bonilla-Silva writes, contemporary (or "new racism") is "subtle, institutional, and apparently nonracial," and this "aids in the maintenance of white privilege without fanfare, without *naming* [emphasis added] those who [sic] it subjects and those who [sic] it rewards" (2006, p. 3). The discursive and rhetorical strategies of naming and defining (or, in this case, the refusal to do such) recreate and reinstantiate oppression and marginalization. In this way, self-silencing and self-censorship has a larger consequence than individuals simply avoiding a controversy. Instead, silence on behalf of those deemed as "experts" surreptitiously works to further silence the already marginalized and to further invalidate (by way of refusal to name) lived experiences and different knowledges.

Pedagogical suggestions

I focused on the often invisible, macro-level racism that is frequently unexamined in our disciplines, many times with critiques jumping straight into a focus on practice and practitioners rather than looking reflectively at what we publish and subsequently at how our research influences pedagogy that we then go on to implement into curriculum that we use to teach the future practitioners who filter into and out of our academic spaces and places. To me, this is disingenuous. How do we critique science writers without critiquing the scholarship that teaches and trains these writers? Admittedly, it's hard for me to wrap my head around clear divisions between practice and scholarship because I see these as always interconnected. To that end, I offer a few brief pedagogical suggestions in an acknowledgment that critique should be a beginning and not an end in recursively and iteratively seeking change.

- Address oppressions (racism, sexism, homophobia, transphobia, etc.) in our classrooms and pedagogies, highlighting connections between how we educate and train professionals to the application of that education in workplace contexts
- Engage with multiply marginalized scholars in our classrooms, exposing our students to a multiplicity of perspectives and lived experiences
- Make our students aware of the precarity involved in addressing and redressing oppressive practices in workplace contexts (Our students should be aware that their resistance may come at a cost.)
- Use language purposefully in our classroom and in our scholarship

Conclusion: testifying with Black Feminist Thought

I return to my question: is explicitly using the words "racism" and "white supremacy" necessary for doing science communication work? I think so. I have argued that systemically marginalized groups (like Black women) experience testimonial quieting while some experts in science communication self-censor. What holds true across both of these concepts (testimonial quieting and self-censorship) is the separation of people from the symbolic expression that enables knowledge sharing. Just as I returned to my framing question for this chapter, I also return to my acknowledgment of the affective and emotional reactions to hearing, speaking, and using the words "white supremacy" in disciplinary science communication scholarship and in other contexts. I recognize that some of the resistance and silencing may be attributed to uncertainty and even to fear – fear that we won't be heard, fear of retribution, fear of getting it wrong, fear of offending. In this realization, I turn to the words of Black Feminist scholar, writer, and poet Audre Lorde: "For we have been socialized to respect fear more than our own needs for language and definition, and while we wait in silence for that final luxury of fearlessness, the weight of that silence will choke us" (Lorde, 1977).

Note

1 The CDC's Pregnancy Mortality Surveillance System (PMSS gathers data annually from all 50 states, New York City, and DC. www.cdc.gov/reproductivehealth/maternal-mortality/pregnancy-mortality-surveillance-system.htm?CDC_AA_refVal=https%3A%2F%2Fwww.cdc.gov%2Freproductivehealth%2Fmaternalinfanthealth%2Fpregnancy-mortality-surveillance-system.htm

References

American Association of University Women (AAUW). (2015) 'Solving the equation: The variable for women's success in engineering and computing', www.aauw.org/research/solving-the-equation/

Ansley, L. (1989) 'Stirring the ashes: Race, class, and the future of civil rights scholarship,' *Cornell Law Review*, vol. 74, no. 6, pp. 993–1077.

Bonilla-Silva, E. (2006) *Racism without racists: Color-blind racism and the persistence of racial inequality in the United States*, Lanham, MD: Rowman & Littlefield Publishers.

Ceccarelli, L. (2011) 'Manufactured scientific controversy: Science, rhetoric, and public debate', *Rhetoric & Public Affairs*, vol. 14, no. 2, pp. 195–228, doi:10.1353/rap.2010.0222

Ceccarelli, L. (2013). To Whom Do We Speak? The Audiences for Scholarship on the Rhetoric of Science and Technology, *Poroi*, vol. 9, no. 1, pp. 1–7, https://doi.org/10.13008/2151-2957.1151

Centers for Disease Control. (2019) 'Racial disparities continue in pregnancy-related death [press release]', www.cdc.gov/media/releases/2019/p0905-racial-ethnic-disparities-pregnancy-deaths.html

Chadwick, R. (2019) 'Practices of silencing: Birth, marginality and epistemic violence,' in *Childbirth, vulnerability and law*, London: Routledge, pp. 30–48.

Childress, S. (2019) 'We found dozens of uncounted deaths during the Flint water crisis: Here's how', *PBS, Frontline*, www.pbs.org/wgbh/pages/frontline/interactive/how-we-found-dozens-of-uncounted-deaths-during-flint-water-crisis/

Collins, P.H. (1990) *Black feminist thought: Knowledge, consciousness, and the politics of empowerment*, London: Routledge.

Dawson, E. (2018) 'Reimagining publics and (non) participation: exploring exclusion from science communication through the experiences of low-income, minority ethnic groups', *Public Understanding of Science*, vol. 27, no. 7, pp. 772–786, https://doi.org/10.1177/0963662517750072

Dei, G.J.S., Karumanchery, L.L., & Karumanchery-Luik, N. (2007) *Playing the race card: exposing white power and privilege*, New York: Peter Lang Publishers.

Dotson, K. (2011) 'Tracking epistemic violence, tracking practices of silencing,' *Hypatia*, vol. 26, no. 2, pp. 236–257, https://doi.org/10.1111/j.1527-2001.2011.01177.x

Droira, J. (2019) 'Caring to ask: A new picture of Inquisitiveness,' in C. Deane-Drummond, T. A. Stapleford & D. Narvaez (eds.), *Virtue and the practice of science: Multidisciplinary perspectives*, Notre Dame: Center for Theology, Science, and Human Flourishing, University of Notre Dame.

Fricker, M. (2009) *Epistemic injustice: power and the ethics of knowing*, New York: Oxford University Press.

Hannah-Jones, N., & Elliott, M.N. (eds.) (2019) 'The 1619 project', *The New York Times*, 18 August.

Hart, C.L., & Hart, M.Z. (2019) 'Opioid crisis: Another mechanism used to perpetuate American racism,' *Cultural Diversity and Ethnic Minority Psychology*, vol. 25 no. 1, p. 6, https://doi.org/10.1037/cdp0000260

Harvin, C.B. (1996) 'Conversations I can't have', *On the Issues,* vol. 5, no. 2, p. 15. Retrieved from https://search.proquest.com/docview/221140015?accountid=6724

Haskell, R. (2018) 'Serena Williams on motherhood, marriage, and making her comeback,' *Vogue Magazine*, www.vogue.com/article/serena-williams-vogue-cover-interview-february-2018

Hollingshead, N.A., Meints, S.M., Miller, M.M., Robinson, M.E., & Hirsh, A.T. (2016) 'A comparison of race-related pain stereotypes held by White and Black individuals,' *Journal of Applied Social Psychology*, vol. 46, no. 12, pp. 718–723, https://doi.org/10.1111/jasp.12415

Holt-Giménez, E. (2018) 'Overcoming the barrier of racism in our capitalist food system', *Food First Backgrounder*, vol. 24, no. 1, pp. 1–4.

Horowitz, J.M., Brown, A., & Cox, K. (2019) *Race in America 2019*, Pew Research Center, www.pewsocialtrends.org/2019/04/09/race-in-america-2019/

Jones, N.N. (2016) 'Narrative inquiry in human-centered design: Examining silence and voice to promote social justice in design scenarios', *Journal of Technical Writing and Communication*, vol. 46, no. 4, pp. 471–492, https://doi.org/10.1177/0047281616653489

Kendi, I.X. (2016) *Stamped from the beginning: The definitive history of racist ideas in America*, New York: Random House.

Lorde, A. (1977) 'The transformation of silence into language and action,' in B. Ryan (ed.), *Identity politics in the women's movement*, New York: New York University Press, pp. 81–86.

Majdik, Z.P., & Platt, C.A. (2012) Selling certainty: Genetic complexity and moral urgency in Myriad Genetics' BRAC Analysis campaign. *Rhetoric Society Quarterly*, 42, 120–143.

Muhs, G.G., Niemann, Y.F., González, C.G., & Harris, A.P. (eds.) (2012) *Presumed incompetent: The intersections of race and class for women in academia*, Louisville: University Press of Colorado.

Owens, D.C., & Fett, S.M. (2019) 'Black maternal and infant health: Historical legacies of slavery', *American Journal of Public Health*, vol. 109, no. 10, pp. 1342–1345, https://doi.org/10.2105/AJPH.2019.305243

Parker, E.T. (2015) 'Exploring the establishment of the office of the chief diversity officer in higher education: A multisite case study', PhD (Doctor of Philosophy) thesis, University of Iowa, https://doi.org/10.17077/etd.irgq8poj

Quinn, S.C., Jamison, A., Freimuth, V.S., An, J., Hancock, G.R., & Musa, D. (2017) 'Exploring racial influences on flu vaccine attitudes and behavior: Results of a national survey of White and African American adults', *Vaccine*, vol. 35, no. 8, pp. 1167–1174, doi: 10.1016/j.vaccine.2016.12.046

Robinson, T.M., Shum, G., & Singh, S. (2018) 'Politically unhealthy: Flint's fight against poverty, environmental racism, and dirty water', *Journal of International Crisis and Risk Communication Research*, vol. 1, no. 2, p. 6, DOI: 10.30658/jicrcr.1.2.6

Serchen, J., Doherty, R., Atiq, O., & Hilden, D. (2020) 'Racism and health in the United States: A policy statement from the American college of physicians', *Annals of Internal Medicine*, vol. 7, no. 173, pp. 556–557, doi: 10.7326/M20-4195

Smith, B. (2020) 'How to dismantle white supremacy', *The Nation*, www.thenation.com/article/politics/how-to-dismantle-white-supremacy/

Wilson, E. (2018) 'The legal foundations of white supremacy', *DePaul Journal for Social Justice*, vol. 11, no. 2, Art. 6.

Yu, H., & Northcut, K. (eds.) (2018) *Scientific communication: Practices, theories, and pedagogies*, New York: Routledge.

6

GENDER AND SCIENTIFIC COMMUNICATION

Lillian Campbell

While public interest in women scientists is on the rise, scholars continue to discover a leaky pipeline for women pursuing careers in the sciences (Wang and Degol, 2017). Women begin studying science at the same rates as men but are rarely found in the upper echelons of scientific research. According to feminist philosophers of science, by limiting representation in the sciences, scientific research perpetuates gendered biases (Haraway, 1997; Barad, 2007). Meanwhile, some point to the masculine nature of scientific communication as contributing to this problem (Gonsalves, 2014).

This chapter briefly overviews the question of whether women's language exists. Next it highlights key moments in the history of women's scientific communication practices, including public communication of science in the 1700–1800s, early women doctors and patient communication in the late nineteenth century, and technical writing by women scientists during World War II. It discusses appropriations of scientific discourse during second-wave feminism, such as the feminist pamphlet *Our Bodies, Ourselves*, and concludes by identifying ongoing challenges for women scientists. These include self-perceptions, treatment by peers, and stereotyping during the publication process.

Women's language

Virginia Woolf was an early advocate of the distinctly female sentence in the early 1900s, echoed by French feminists Hélène Cixous and Luce Irigaray in the 1970s and 1980s who argued that systems of language were man-made and thus inscribed with oppressive views of women (Ritchie and Ronald, 2001). In order to overcome this oppression, Cixous modelled poetic approaches to language that relied on unexpected pairings and nontraditional uses. Feminist linguist Robin Lakoff carried these ideas into linguistics in the 1970s with an assimilationist focus, arguing that women's speech operates within a male institution that disqualifies them from positions of power. She identified a range of linguistic patterns like tag questions and hedging that rendered women tentative and trivial in conversation. Lakoff's work was a precursor to feminist linguistics scholarship, including Deborah Tannen's research on the differences between men and women's conversational styles (1996, 2006) and Deborah Cameron's examinations of language and sexual politics (2006, 2007). However, Lakoff received methodological backlash for relying on casual

 DOI: 10.4324/9781003043782-8

observation and others went on to argue that strategies Lakoff identified with oppression could also be used as power plays (Lakoff and Lakoff, 2004).

Discussing scientific communication, feminist scholars have argued that a discourse that values "dichotomy and hierarchy" is hostile to women's language and worldview, which is "governed by a logic of connection and relationship" (Wells, 2001, p. 141). Indeed, feminine writing is often associated with genres (like personal narrative) and traits (like emphasis on interaction and relational connection) that are diametrically opposed to scientific discourse (Black et al., 1994). For feminist science philosopher Donna Haraway (1997, p. 24), the figure of the "modest witness" so encapsulates the problems of technical scientific language that it obscures people and actions with its focus on objects. This object-focus is enhanced by grammatical metaphor, a tenet of technical discourse that transforms actions into nouns that can then become actors in a sentence, a process called nominalization (Martin et al., 1993). Grammatical metaphor is one linguistic strategy that enables the author to remove themselves from scientific narratives, in direct contrast to feminist research approaches that call for transparency about one's positionality and relationship to the research subjects and environment (Haraway, 1997; Barad, 2007).

Linguists have responded to public critiques of nominalization, arguing that these strategies serve important functional and structural roles that are fundamental to science's taxonomizing work. While Martin et al. (1993) agree that technical jargon can be mobilized to obscure powerful actors, they reject the possibility of eliminating grammatical metaphor and critique the misguided assumption that it is too difficult for children (especially girls) to learn. Other linguistic studies of scientific articles have provided a more complex picture of the genre, pointing to the use of politeness strategies (Myers, 1989) and hedging (Hyland, 1996) in order to establish deference to scientific colleagues. Meanwhile, other feminist science scholars extend far beyond grammatical metaphor in their critiques of scientific communication and practice, calling for more intersectional approaches to data gathering and analysis (Bright et al., 2016), mixed methodologies that can center rather than obscure smaller populations and individuals (Longino, 2013), and centering women and women's health in medical research (Morris, 2020).

Meanwhile, the prospect of rejecting scientific discourse as not aligned with women's worldviews carries far more risks than rewards. Haraway (1988, p. 578) warns against a feminist impetus to dismiss scientific discourse: "We ended up with one more excuse for not learning any post-Newtonian physics and one more reason to drop the old feminist self-help practices. . . . They're just texts anyway, so let the boys have them back." In contrast, recent feminist philosophy of science like Barad's *Meeting the Universe Halfway* engages with scientific principles while still unpacking the sexist material conditions that undergird scientific practice. Meanwhile, Butler's theories of gender performativity (1990) highlight the potential for flexibility and play within gendered norms of communication, emphasizing that every discursive act is a performance. Moving into a discussion of women scientists' communication practices throughout history, the focus will be on the rhetorical strategies they used to balance their role as women and their participation in scientific practice as well as how their work influenced scientific perspectives on gender.

History of women's scientific communication

This section overviews some of the key historical moments in women's rhetoric of science. It does not undertake a complete history of women's role in the sciences; such comprehensive accounts can be found elsewhere (Herzenberg, 1986; Ogilvie, 1993). Attention is on scholarship on women scientist communicators, so that readers can understand the rhetorical strategies that have been affiliated with this group over time.

Public communication of science in the eighteenth and nineteenth centuries

While women were largely excluded from scientific research in the 1700s and 1800s, they played a critical role as mediators of knowledge who studied and then reinscribed scientific research for children and adult lay audiences. Indeed, Gates and Shteir (1997) argue that more historical women scientists would be found by expanding our archives to include scientific popularizations like children's books, public talks, and illustrations. These popularizations often featured a wise mother or governess figure and took the form of a letter, dialogue, or conversation in which she instructed children about the wonders of the natural world. Other women science popularizers took to the stage, lecturing to audiences of primarily women about topics ranging from the mysteries of Africa (Mary Kingsley, 1862–1900) to female anatomy and physiology (Mary Gove Nichols, 1810–1884).

While clear and accurate representations of scientific knowledge were of paramount importance to these popularizers, they frequently used their public roles to revise scientific research and connect it to calls for social change. For example, natural historian Mary Roberts (1788–1864) produced traditionally feminine accounts that highlighted connections between the natural world and Christianity. However, she occasionally opposed male scientists, which Gould argues in his essay "The Invisible Woman," has been rendered invisible in previous accounts of her work (Gates and Shteir, 1997). Similarly, Jane Marcet's (1769–1858) *Conversations on Chemistry* featured a woman teacher and two girl students conducting scientific demonstrations in their home, using the environment to counter the radical prospect of a woman doing scientific research. However, throughout the narrative the teacher, Mrs. B., is pointedly limited by her access to materials. In this way, she both uses the domestic setting as a way to normalize her foray into scientific research while also taking a critical stance (Gates and Shteir, 1997, p. 57).

In public scientific performances, women similarly relied on familiar and comfortable feminine roles to contrast their more radical personas as women with scientific knowledge. For example, Julie Early describes how Mary Kingsley shared her anthropological fieldwork in Africa during public speeches where she dressed in somber silks, taking on the persona of the "maiden aunt" (Gates and Shteir 1997). Meanwhile, Skinner (2009, p. 242) argues that Mary Gove Nichols used the content of her lectures to the ladies to construct her audience as already aware of her facts and sharing her values for reform. As one of the first women to lecture to audiences of women about anatomy and physiology, this made it difficult "to condemn the character of the speaker, because in the course of the speech the audience begins to see themselves as sharing her motivations for engaging in behavior they might otherwise denounce as radical."

Toward the end of the nineteenth century, evolutionary theory became a primary focus for many women science writers. Some, like Arabella Buckley (1840–1929), focused on unpacking these theories and applying them to children's daily experiences. Her illustrated book popularizations of Darwin's theories focused on parenting as a dominant metaphor and made mutuality a key evolutionary principle (Gates and Shteir, 1997). Others used popularizations even more explicitly to transform scientific claims to reflect female power and feminine values. Eliza Gamble's popular book *The Evolution of Woman*, published in 1894, identified strategic discontinuities in Darwin's evolutionary theory and argued that altruism rather than control was a force for evolution. She used this interpretation to advocate for female power over sexuality as a missing link in Darwin's work. In this way, Gamble's engagement with evolutionary theory became a grounds for countering essentialized views of women's maternal, nurturing role and calling for social action even while she sought to inform and educate (Gates and Shteir, 1997).

At the turn of the twentieth century, cookbooks also became a venue in which women were engaging in public scientific communication. The rise of domestic science that followed the passage of the Land Grant Act in 1862 and the spread of the Progressive movement in the United States caused shifts in the rhetorical strategies and aims of women's cookbooks. Form and content changed as cookbooks moved from a narrative of the recipe to a list of ingredients and standardized measurements (Walden, 2018, p. 115). Moving out of a Victorian model that emphasized ideal mothers as spiritual stewards for a growing nation, Progressive-era cookbooks grounded claims about taste and morality in scientific standards and sought to professionalize women's domestic work: "No longer a conversational or participatory narrative, the modern recipe indicated a scientific authority, rather than an experienced housewife, at its helm" (Walden, 2018, p. 115). However, as Walden (2018, p. 118) points out, these scientific arguments still addressed questions of citizenship and morality that were inseparably tied to fears about poor and minority Americans and immigrants. One leader of the domestic science movement, Ellen Richards, even coined the term "euthenics," which combined the era's interest in eugenics with a focus on health and morality through domestic science.

Despite the expansion of the domestic sciences, women's claim over the domain of scientific popularization began to wane early in the twentieth century due to the increasing complexity of the science taught in schools as well as the rising competition from male popularizers writing for adult audiences (Farkas, 1998). Women were also beginning to gain footholds as professional scientists – a trend that would be accelerated by the World Wars in the early twentieth century, as I discuss later in the chapter. Today, studies find that male scientists are more likely to engage in public outreach (Crettaz von Roten, 2011; Amarasekara and Grant, 2019). However, just as popularizers are often forgotten in historical accounts of scientific work, there is limited current research on gender distribution among individuals pursuing full-time careers in public communication of science.

Women doctors, medical journals, and patient communication in the nineteenth century

In the United States, women began formally practicing medicine in the mid-1800s, facing significant barriers because their gender was perceived to be in conflict with the profession. Male physicians would reference both biological models of female weakness and social views that the women's domain was in the home to oppose women's admission to medical schools and societies (Skinner, 2012). This presented significant rhetorical challenges for women physicians, who had to counter mismatches between "proper" feminine communication and effective medical discourse, which was "expected to be assertive, analytical, and direct" (Skinner, 2012, p. 309). Still, the latter half of the 1800s saw a sharp increase in women's presence in the field – from Elizabeth Blackwell becoming the first women's medical school graduate in 1849 to the nearly 7,000 women physicians in the United States at the turn of the century (Morantz-Sanchez, 2000, p. 92).

Scholars have identified a range of strategies leveraged by women physicians who were publishing their research in nineteenth-century male-dominated medical journals. These included "cross-dressing," where they rhetorically presented themselves like male physicians, or using literary language and a humorous tone to present medical content (Wells, 2001; Skinner, 2012). Skinner (2012) finds that Julia Carpenter's attempts at cross-dressing were not well received, with participants criticizing the omission of details from her report as well as her description of a patient's tumor. Meanwhile, her artistic account of hay fever, which was lighthearted and included character sketches, was praised, at least in part because it refrained from analysis and left discussion to the men.

Both Wells (2001) and Skinner (2012) also argue that women physicians could take an approach to medical writing that leveraged more traditional research styles but foregrounded gendered content and perspectives. One way to do this was to integrate the voices of patients through survey or interview data. For example, Wells (2001) identifies Mary Putnam Jacobi as the first individual to publish survey research in a medical journal, using women's accounts on hysteria to challenge Weir Mitchell's advocacy of the rest cure. Similarly, Theriot (1993) discusses women doctor's surveys of female asylum patients used to counter dominant views in gynecology that hysteria was caused by a disease of the reproductive organs.

Meanwhile, in woman-dominated contexts, women physicians had more rhetorical flexibility. Examining the theses of students at the Woman's Medical College of Pennsylvania in the 1860s Wells (2001) identified opportunities for subversion. These included the use of satire and irony to negotiate the role of both female and expert in the text, a topical focus on gendered diseases, and the use of lay opinion to build authority. In 1893, female physicians created their own journal, the *Woman's Medical Journal*, which was both edited by women and focused on representing women's interests in medicine. Fancher et al. (2020) demonstrate how the journal was leveraged to provide social networking for women in the profession and challenge sexist institutions but rarely included black female physicians as authors or readers.

Finally, women intervened not only in formal medical communication but also in their day-to-day conversations with patients. Typical studies find that in the course of a diagnostic interview, doctors frequently interrupt and ask questions to redirect a patient's narrative and determine both the blame and illness of the patient (Segal, 2002). Wells (2001) found that historically women physicians intervened in this medical discourse through the recording of "heart histories," i.e. patient's narratives of their illness experiences. While the physicians in Wells's study took dutiful accounts of the details of their patient's experiences, Wells (2001, p. 54) points out that these narratives also allowed for even greater intervention into patients' lives. Thus a more empathetic exchange did not necessarily translate to more ethical action.

Much like women's declining presence in the public communication of science, their participation as medical practitioners also waned with the turn of the century. The Flexner Report, published in 1910, provided a survey of medical training practices and standards arguing for greater standardization in curriculum and scientific training requirements. Many women's medical colleges and African American medical schools were forced to close their doors or drop enrollments as a result, so that between 1904 and 1915 women graduates declined from 198 to 92 (More, 1990, p. 166). It would take nearly another century to recoup these losses and see more gender parity in the profession again.

Technical writing during the early twentieth century

While histories of feminism frequently ignore the decades between the passage of women's suffrage in 1920 and the second-wave feminist movement in the 1960s, a range of social, political, and economic forces conspired to increase women's participation in scientific activity during this time. The interwar period saw both a renewed investment in the hard sciences as well as a rising interest in the social sciences like psychology, anthropology, and economics (George et al., 2013, p. 3). During World War II, women were encouraged to participate in scientific study both to replace men who had gone to war and also to boost scientific efforts in America (Jack, 2009). However, this influx of women into the field was unbalanced and short-lived, with most women taking on semiskilled positions and leaving scientific work at the war's end.

In their introduction to women's interwar rhetorics, George et al. (2013, p. 7) argue that, having gained a foothold in professional spheres, "the woman of the 1920's spent less time demanding a right to speak and more time proving her right to act (and be judged) in accordance with the norms of her chosen field." Thus this period saw a rise in women's participation in technical genres in the disciplines including the publication of scholarly articles. An examination of articles coming out of a 1941 lobby for greater professional opportunities from the National Council of Woman Psychologists, Jack (2009, p. 15) finds, however, that data-driven arguments often reinforced gender-neutral perspectives. She demonstrates how authors used statistical analysis to identify underrepresentation of women in the field, for example, but ultimately attributed differences to innate ability, dedication, or personal choice.

In cases where women scientists did work to challenge the status quo with data, Jack (2009, p. 94) finds they were often limited by the cultural valuing of technical rationality. Scientist Katherine Way, who worked on the Manhattan Project, tried to justify concerns about safety hazards using quantitative arguments. However, her appeals were constrained both by the use of data as well as by the politeness strategies required by the technical memo. Similarly, responding to calls for national nutritional guidelines in 1940, nutritionist Lydia Roberts found numerous gaps in expert research and turned to informal reports to supplement the data, much like Theriot's (1993) findings on women doctors and asylum patients. Still, Roberts had to defend her standards to a panel of experts and thus relied on a discourse of expertise (Jack, 2009). Resonating with Frost's (2016) theory of apparent feminism, the technical writing of these women scientists demonstrates how they could often be drawn into masculinist cultural logics like efficiency and rationality. These values seem tacit and natural but can undermine a scientific woman's potential to create change.

However, women scientists' "backstage" and popular writing in the early twentieth century created more space for reform. Applegarth (2015, p. 195) argues that the rise in the social sciences in the interwar period and in anthropology specifically opened up opportunities for women's participation but also created a lot of "boundary work" within the field. Thus women anthropologists used letters and private correspondence to bolster one another's careers, including naming gender discrimination, promoting one another's work and accomplishments, and providing feedback on drafts. Similarly, she demonstrates how the vocational autobiography – a popular genre of the time – worked to promote women's participation in the sciences. Applegarth's (2015, p. 205) examination of Ann Morris's archaeological autobiographies shows how they were recruitment tools, promoting a life of adventure while also sharing access to technical knowledge. In Applegarth's view, these autobiographies challenge exceptionalist rhetoric that was common in career guides for women in the sciences during the interwar period.

As women scientists became more visible in society, their rhetorical strategies were not only textual but also embodied. Applegarth (2015, p. 533) shows how authors of vocational autobiographies frequently depicted themselves in "professional spaces: behind news desks, in surgical wards, laboratories, station houses, and so on." Hillin makes a similar argument about women pilots in the early twentieth century, arguing that they created an "aerocyborg" identity in their public writing that allowed them to display their mechanical talents without the limits imposed by gender or race (George et al., 2013, p. 178). Overall, the 1920s and 1930s were a contradictory time for women in the sciences; as they overcame visible barriers to entry, the discrimination against them became more subtle. They responded rhetorically by speaking to the technical and rational values of the times in their research, while also finding ways to challenge the status quo, whether through the use of data-driven arguments that critiqued discrimination, backstage, and popular writing that provided support or simply through a physical presence that countered gendered assumptions.

The women's health movement of the 1970s

The women's health movement began alongside other second-wave feminist movements as a radical, grassroots initiative aimed at providing lay women with techniques for accessing both medical and embodied knowledge. Journalist and activist Barbara Seaman's 1969 book, *The Doctors' Case Against the Pill*, is often referenced as a starting point for the movement. Like women doctors publishing in the late 1800s, Seaman incorporated testimony from women about their dangerous health experiences with legal birth control while also drawing extensively on interviews with physicians and medical researchers (Dubriwny, 2012, p. 17). These moves – elevating women's bodily experiences and putting them in conversation with medical discourse – were characteristic of the women's health movement's rhetorical strategies as a whole.

Sobnosky (2013, p. 219) describes three modes of storytelling that characterized the rhetoric of the second-wave women's health movement: (1) linking current medical care to biased and unscientific practices of the past, (2) using "doctor stories" to demonstrate how these practices negatively impacted care, and (3) positioning laywomen as the real experts. Feminist consciousness-raising was at the heart of this narrative structure. As Campbell (1973, p. 79) explains, consciousness-raising succeeds by bridging the personal and the political and demonstrating that "what were thought to be personal deficiencies and individual problems are common and shared, a result of their position as women." Another practice that was central to the women's health movement was embodied self-help. Describing the iconic vaginal self-exam, Murphy (2004, p. 119) explains how individual women learned to "produc[e] the evidence of experience" through self-help practices. These practices enabled them to value and claim their embodied knowledge and to leverage it as evidence for change in the scientific community.

The feminist health book *Our Bodies, Ourselves* (*OBOS*) provides a specific example of the rhetorical strategies of the women's health movement. *OBOS* was born out of a workshop on women and their bodies at a women's conference in Boston in May 1969. Interested in questions about access to health care and doctor relationships, a group of women began working to compile a list of good doctors. This group ultimately became the Boston Women's Health Collective and shifted their focus to learning for themselves. They designed a laywoman's course which they ultimately revised to create "Women and Their Bodies," a 138-page booklet published for $0.30 by the New England Free Press (Wells, 2010). Enormous demand led to a publication deal with mainstream publisher, Simon & Schuster in 1973, which Musser (2007) argues also marked a shift away from collective health goals toward a more essentialist, individual vision. Since its move to Simon & Schuster, *OBOS* has been revised for eight editions and translated into 29 languages.

Wells' (2010, p. 698) book on *OBOS* draws on numerous interviews with the book's authors and readings of their early drafts, notes, and revisions to argue that it was a "labor of language." She sees the text as strategically written and revised to simultaneously empower women and to provide an epistemological framework for their health experiences. Wells (2010, p. 177) attributes much of *OBOS*'s success to its unique rhetoric, which integrated personal anecdotes from members of the collective with excerpts from medical texts and conversational discussions of health topics: "[*OBOS*] transformed medical discourse into something colloquial, metaleptic, and mobile. It entered the medical register, took it over, and also maintained critical distance from it." This method had its pitfalls: it essentialized both women and the medical institution.

African American women, Latina women, lesbians, and women with disabilities gradually received individualized attention in later editions as the authors struggled with maintaining inclusive rhetoric while also acknowledging difference.

Conservative backlash in the 1980s led to a shift toward "unobtrusive mobilization" by insiders, i.e. women working to create change as employees in the medical system (Katzenstein, 1990). Unfortunately, what looks like a still flourishing women's health movement today – in the form of nationwide campaigns to support breast cancer awareness, research on women's heart disease, or attention to postpartum depression – has lost many of its connections to feminist aims. Instead, these movements promote a vision of what Dubriwny (2012, p. 9) calls "the vulnerable empowered woman" who supports neoliberal structures like traditional gender roles "through her various practices of risk management and consumption." In addition, antiscience rhetoric today has appropriated some of the iconic rhetorical moves of the women's health movement, especially their elevation of embodied knowledge and experience (Campbell, 2019; Whidden, 2012). In considering this problematic uptake, it is important to highlight that the early women's health movement sought only to reclaim and challenge scientific knowledge but never to outright reject scientific research.

Ongoing challenges

As each of these sections suggest, women across time have found ways to participate in, challenge, and even change scientific discourse. However, the work of equalizing gender participation in scientific communication is by no means complete. Today, while women enter many scientific disciplines at the same rates as men, they continue to be underrepresented in scientific leadership (De Welde et al., 2007). Scientific communication is at the heart of these struggles. And while this chapter has focused on women's experiences historically with scientific communication, more recent scholarship also notes the ongoing challenges for other gender minorities including trans- and gender-fluid individuals (Pérez-Bustos, 2014).

As practicing scientists, women today face internal challenges with self-identification as scientists and scientific writers. Falconer's (2019) case study of an African American woman undergraduate demonstrates how the student's perception that scientific discourse was not for people like her initially caused her to be alienated from scientific practices. Close mentors showed her that scientific discourse was a foreign language for any newcomer to the discipline that helped her gain the confidence to enter into research writing. Similarly, Smith et al. (2015) show the role that stereotype threat plays in undergraduate women's perceptions of their scientific fields. Comparing male-dominated physics with female-dominated biology, the authors find that students are less likely to take on scientific identities in these male-dominated contexts because of concerns about confirming negative gender group stereotypes.

Once in the workplace, strength as a scientific writer is not always perceived as an asset for women. Mallette's (2017, p. 419) study of a woman engineer finds that her skill and attention to writing at the firm serve to isolate her from the perceived "real work" of engineering. Meanwhile, barriers to women publishing in the sciences continue to be steep. One study found that readers would rate "female-typed" scientific abstracts lower and have less interest in collaborating with the author during a blind assessment (Knobloch-Westerwick et al., 2013). Thus systems of promotion in the sciences often necessitate enculturation, leaving little room for critique or innovation until a scientist has achieved significant notoriety and status (Wang and Degol, 2017).

Overall, negative self-perception and outward resistance combine with the continued expectation that women take on a greater role in child rearing and family life to limit women scientists (Ledin et al., 2007). However, the breadth and complexity of women's contributions to scientific communication over the last two centuries demonstrates that their participation is vital. By helping children and laypeople access scientific knowledge, incorporating the voices of suffering patients, leveraging the body for communicative power, and much more, women have pushed the boundaries of scientific communication again and again. Looking toward the future, expanding their participation and supporting greater access for gender-non-conforming voices will offer valuable new perspectives and contributions to the scientific community.

References

Amarasekara, I., & Grant, W.J. (2019) 'Exploring the YouTube science communication gender gap: A sentiment analysis', *Public Understanding of Science*, vol. 28, no. 1, pp. 68–84, https://doi.org/10.1177/0963662518786654

Applegarth, R. (2015) 'Personal writing in professional spaces: Contesting exceptionalism in interwar women's vocational autobiographies', *College English*, vol. 77, no. 6, pp. 530–552.

Barad, K. (2007) *Meeting the universe halfway: Quantum physics and the entanglement of matter and meaning*, Durham, NC: Duke University Press.

Black, L., Daiker, D., Sommers, J., & Stygall, G. (1994) *Writing like a woman and being rewarded for it: Gender, assessment, and reflective letters from Miami University's student portfolios*, Heinemann: New Directions in Portfolio Assessment, pp. 235–247.

Bright, L.K., Malinsky, D., and Thompson, M. (2016) 'Causally interpreting intersectionality theory', *Philosophy of Science*, vol. 83, no. 1, pp. 60–81, www.jstor.org/stable/10.1086/684173

Butler, J. (1990) *Gender trouble: Feminism and the subversion of identity*, New York: Routledge.

Cameron, D. (2006) *On language and sexual politics*, New York: Routledge.

Cameron, D. (2007) *The myth of Mars and Venus: Do men and women speak different languages?* Cambridge: Oxford University Press.

Campbell, K.K. (1973) 'The rhetoric of women's liberation: An oxymoron', *Quarterly Journal of Speech*, vol. 59, no. 1, pp. 74–86, doi: 10.1080/00335637309383155

Campbell, L. (2019) 'Rhetorically framing the "inside woman": Female healthcare workers across editions of *Our Bodies, Ourselves*', *Peitho*, vol. 21, no. 3, pp. 610–625.

Crettaz von Roten, F. (2011) 'Gender differences in scientists' public outreach and engagement activities', *Science Communication*, vol. 33, no. 1, pp. 52–75, https://doi.org/10.1177/1075547010378658

De Welde, K., Laursen, S., & Thiry, H. (2007) 'Women in science, technology, engineering and math (STEM)', *Advance Library Collection*, Paper 567, https://digitalcommons.usu.edu/advance/567

Dubriwny, T.N. (2012) *The vulnerable empowered woman: Feminism, postfeminism, and women's health*, New Brunswick, NJ: Rutgers University Press.

Falconer, H.M. (2019) '"I think when I speak, I don't sound like that": The influence of social positioning on rhetorical skill development in science', *Written Communication*, vol. 36, no. 1, pp. 9–37, https://doi.org/10.1177/0741088318804819

Fancher, P., Kirsch, G., & Williams, A. (2020) 'Feminist practices in digital humanities research: Visualizing women physician's networks of solidarity, struggle and exclusion', *Peitho*, vol. 22, no. 2, https://cfshrc.org/article/feminist-practices-in-digital-humanities-research-visualizing-women-physicians-networks-of-solidarity-struggle-and-exclusion/

Farkas, C.A. (1998) '*Natural Eloquence: Women reinscribe science* edited by Barbara T. Gates and Ann B. Shteir', *Victorian Review*, vol. 24, no. 1, pp. 82–85.

Frost, E.A. (2016) 'Apparent feminism as a methodology for technical communication and rhetoric', *Journal of Business and Technical Communication*, vol. 30, no. 1, pp. 3–28, https://doi.org/10.1177/1050651915602295

Gates, B.T., & Shteir, A.B. (eds.) (1997) *Natural eloquence: Women reinscribe science*, Madison, WI: University of Wisconsin Press.

George, A., Weiser, M.E., & Zepernick, J. (eds.) (2013) *Women and rhetoric between the wars*, Carbondale, IL: SIU Press.

Gonsalves, A. (2014) 'Physics and the girly girl – there is a contradiction somewhere: Doctoral students' positioning around discourses of gender and competence in physics', *Cultural Studies of Science Education*, vol. 9, no. 2, pp. 503–521, DOI:10.1007/S11422-012-9447-6

Haraway, D. (1988) 'Situated knowledges: The science question in feminism and the privilege of partial perspective', *Feminist Studies*, vol. 14, no. 3, pp. 575–599, https://doi.org/10.2307/3178066

Haraway, D. (1997) *Modest−witness@second−millennium.femaleman−meets− oncomouse: Feminism and technoscience*, East Sussex: Psychology Press.

Herzenberg, C.L. (1986) *Women scientists from antiquity to the present*, West Cornwall, CT: Locust Hill.

Hyland, K. (1996) 'Writing without conviction? Hedging in science research articles', *Applied Linguistics*, vol. 17, no. 4, pp. 433–454, https://doi.org/10.1093/applin/17.4.433

Jack, J. (2009) *Science on the home front: American women scientists in world war II*, Champaign, IL: University of Illinois Press.

Katzenstein, M.F. (1990) 'Feminism within American institutions: Unobtrusive mobilization in the 1980s', *Signs: Journal of Women in Culture and Society*, vol. 16, no. 1, pp. 27–54, https://doi.org/10.1086/494644

Knobloch-Westerwick, S., Carroll, J.G., & Huge, M. (2013) 'The Matilda effect in science communication: An experiment on gender bias in publication quality perceptions and collaboration interest', *Science Communication*, vol. 35, no. 5, pp. 603–625, https://doi.org/10.1177/1075547012472684

Lakoff, R., & Lakoff, R.T. (2004) *Language and woman's place: Text and commentaries*, Cambridge: Oxford University Press, vol. 3.

Ledin, A., Bornmann, L., Gannon, F., & Wallon, G. (2007) 'A persistent problem: Traditional gender roles hold back female scientists', *EMBO Reports*, vol. 8, no. 11, pp. 982–987, doi: 10.1038/sj.embor.7401109

Longino, H. (2013) *Studying human behavior: How scientists investigate aggression and sexuality*, Chicago, IL: Chicago University Press.

Mallette, J.C. (2017) 'Writing and women's retention in engineering', *Journal of Business and Technical Communication*, vol. 31, no. 4, pp. 417–442, https://doi.org/10.1177/1050651917713253

Martin, J.R., Martin, J., & Halliday, M.A.K. (1993) *Writing science: Literacy and discursive power*, New York: Routledge.

Morantz-Sanchez, R. (2000) *Sympathy and science: Women physicians in American medicine*, Chapel Hill, NC: University of North Carolina Press.

More, E.S. (1990) 'The American medical women's association and the role of the woman physician, 1915–1990', *Journal of the American Medical Women's Association (1972)*, vol. 45, no. 5, p. 165.

Morris, K.K. (2020) 'Women and bladder cancer: Listening rhetorically to healthcare disparities', in E. Frost & M. Eble (eds.), *Interrogating gendered pathologies*, Logan, UT: Utah State University Press, pp. 157–170.

Murphy, M. (2004) 'Immodest witnessing: The epistemology of vaginal self-examination in the US feminist self-help movement', *Feminist Studies*, vol. 30, no. 1, pp. 115–147.

Musser, A.J. (2007) 'From our body to yourselves: The Boston women's health book collective and changing notions of subjectivity, 1969–1973', *Women's Studies Quarterly*, vol. 35, no. 1–2, pp. 93–109.

Myers, G. (1989) 'The pragmatics of politeness in scientific articles', *Applied Linguistics*, vol. 10, no. 1, pp. 1–35, DOI:10.1093/APPLIN/10.1.1

Ogilvie, M.B. (1993) *Women in science: Antiquity through the nineteenth century*, Cambridge, MA: MIT Press.

Pérez-Bustos, T. (2014) 'Of caring practices in the public communication of science: Seeing through trans women scientists' experiences', *Signs: Journal of Women in Culture and Society*, vol. 39, no. 4, pp. 857–866, https://doi.org/10.1086/675540

Ritchie, J., & Ronald, K. (eds.) (2001) *Available means: An anthology of women's rhetoric(s)*, Pittsburgh, PA: University of Pittsburgh Press.

Segal, J.Z. (2002) 'Problems of generalization/genrelization: The case of the doctor-patient interview', in R.M. Coe, L. Lingard & T. Teslenko (eds.), *The rhetoric and ideology of genre: Strategies for stability and change*, Cresskill, NJ: Hampton Press, pp. 171–184.

Skinner, C. (2009) '"She will have science": Ethos and audience in Mary Gove's lectures to ladies', *Rhetoric Society Quarterly*, vol. 39, no. 3, pp. 240–259, DOI: 10.1080/02773940902766730

Skinner, C. (2012) 'Incompatible rhetorical expectations: Julia W. Carpenter's medical society papers', *Technical Communication Quarterly*, vol. 21, no. 4, pp. 307–324, DOI:10.1080/10572252.2012.686847

Smith, J., Brown, E., Thoman, D., & Deemer, E. (2015) 'Losing its expected communal value: How stereotype threat undermines women's identity as research scientists', *Social Psychology of Education*, vol. 18, no. 3, pp. 443–466, https://doi.org/10.1007/s11218-015-9296-8

Sobnosky, M.J. (2013) 'Experience, testimony, and the women's health movement', *Women's Studies in Communication*, vol. 36, no. 3, pp. 217–242, https://doi.org/10.1080/07491409.2013.835667

Tannen, D. (1996) *Gender and Discourse*, Oxford: Oxford University Press.

Tannen, D. (2006) *You're wearing that?: Mothers and daughters in conversation*, New York: Ballantine Press.

Theriot, N.M. (1993) 'Women's voices in nineteenth-century medical discourse: A step toward deconstructing science', *Signs: Journal of Women in Culture and Society*, vol. 19, no. 1, pp. 1–31, https://doi.org/10.1086/494860

Walden, S.W. (2018) *Tasteful domesticity: Women's rhetoric and the American cookbook, 1790–1940*, Pittsburgh, PA: Pittsburgh University Press.

Wang, M.T., & Degol, J.L. (2017) 'Gender gap in science, technology, engineering, and mathematics (STEM): Current knowledge, implications for practice, policy, and future directions', *Educational Psychology Review*, vol. 29, no. 1, pp. 119–140, doi: 10.1007/s10648-015-9355-x

Wells, S. (2001) *Out of the dead douse: Nineteenth-century women physicians and the writing of medicine*, Madison, WI: University of Wisconsin Press.

Wells, S. (2010) *Our bodies, ourselves and the work of writing*, Palo Alto, CA: Stanford University Press.

Whidden, R.A. (2012) 'Maternal expertise, vaccination recommendations, and the complexity of argument spheres', *Argumentation and Advocacy*, vol. 48, no. 4, pp. 243–257, https://doi.org/10.1080/00028533.2012.11821775

7

PEER REVIEW IN SCIENTIFIC PUBLISHING

Brad Mehlenbacher and Ashley Rose Mehlenbacher

Introduction

Peer review is central to the academic publishing enterprise and has shaped scientific knowledge for hundreds of years. Yet what constitutes peer review and how it is managed have notable distinctions by field and even by venue. Variation across disciplines is an important feature as distinct disciplines have distinguishable epistemologies, and honoring them by conducting reviews that understand those commitments is essential to producing meaningful contributions in a field or area of research. In addition to competency in one's own discipline, field, or specialization and the attendant difficulty of defining exactly what we mean by a discipline or field (Scholz, 2016), peer review requires expertise in the assessment of a study as well as the compositional activities to recommend study improvements. Gaining expertise in peer-review activities, however, can be challenging as the genre is occluded. John Swales explains that "occluded genres" (1996, p. 46) of scientific communication are those that are "out of sight" (2004, p. 18). In this chapter, we argue that peer review is an important occluded genre-ing activity that requires continued attention so that we can better understand and improve the peer-review process.

The general model of review used in scholarly publishing has a long history. Henry Oldenburg, the first secretary of the Royal Society of London (founded in 1660), integrated expert consultation or peer review into the publishing process of research articles being submitted to *Philosophical Transactions*, and by 1752, the Royal Society had established a formal approach to the important process. Scientific journals in general did not routinely employ peer review until sometime during the nineteenth century (Baldwin, 2015, p. 338), and it is not entirely clear how, in the twentieth century, peer review "came to assume a much more central part of modern science than it had . . . before World War II" (Baldwin, 2015, p. 348), although Burnham (1990) notes that the growth in peer review can be linked generally to the increasing specialization of science and technology (p. 1325).

Not all researchers agree that *Philosophical Transactions* was the first scientific journal to integrate peer-review processes. Kronick (1990), in his review of the use of peer review by eighteenth-century scientific societies, argues that the Royal Society of London had peer review as part of its publication management process for its *Medical Essays and Observations* more than 20 years earlier. The journal even provided a disclaimer, noting that "peer review did not guarantee truthfulness or accuracy" (Benos et al., 2007, p. 145). Baldwin (2015) has documented

DOI: 10.4324/9781003043782-9

the history of the weekly scientific journal, *Nature*, and its anomalous and sometimes outright controversial peer-review history, with early editors famously having an incredible influence on what was published and not published by the prestigious journal. Over time, perhaps following the general historical increase in journals that relied more and more on peer review, the editors of *Nature* integrated review into their publication management process. Today, peer review is a common feature of most scientific journals' publication management processes, a feature that many journals view as critical to the scientific community's goals for rigor, accountability, and progressive knowledge-building. It is, as Richard Smith, editor of the *British Medical Journal* for 13 years, has argued, "at the heart of . . . science" and "the method by which grants are allocated, papers published, academics promoted, and Nobel prizes won" (2006, p. 178).

The contemporary publishing cycle of the scientific research article commonly entails scientists submitting manuscripts to journals or to the journal editor who then, first, decides whether a manuscript is suitable for review, of interest to the journal audience, and "of adequate scientific quality to merit peer review" (Yates, 2017, p. 869). If a manuscript is suitable for the peer-review process, editors select (on average between two and five) reviewers who they think will be knowledgeable about the research literature and the manuscript's topic, methods, or argument. Some journals have automated systems that allow submitting authors to suggest possible reviewers for their article or to disqualify reviewers for personal reasons or methodological ones (Yates, 2017), and general editors may select some peer reviewers based on these suggestions or on algorithms that match article keywords to research keywords that peer reviewers have provided to the journal. Journals can differ in their application of blinding to the peer-review process; that is, journals can practice single-blind reviews, where the reviewers know the names of the authors of the manuscript, or double-blind reviews, where neither the reviewers' nor the authors' names are known as part of the review process. Journals have been exploring more transparent modes of peer review over the last 30 years to varying success (Tennant and Ross-Hellauer, 2020, p. 5).

Whichever system of review is used, peer reviewers then assess the manuscript generally with the goal of assessing whether the research paper is publishable as-is, it is publishable with limited or extensive revisions, or it should be rejected. Many journal editors and journals have criteria for reviewing submitted manuscripts; for example, is the experiment well designed; are the results understandable; are the conclusions reasonable, grounded in the data, interesting, technically accurate; and so on? At this point, the peer reviewers send their feedback (sometimes called "reports") to the journal editor, who, either alone or with the help of associate editors or through a formal editorial meeting (Weller, 1990), uses the reviews to inform their final decision to accept or reject the manuscript. Peer review serves multiple scientific constituents: peer review serves the manuscript writers by double-checking their claims, methods, evidence, and so on; the journal editors by turning the decision-making process into a collaborative enterprise; the greater scientific community which relies on shared responsibility for quality research, careful research design, and researcher accountability (Bakanic et al., 1987); and the general publics who look to scientific societies and peer-reviewed journals as trustworthy, legitimate, and authoritative bodies of knowledge-makers (Bedeian, 2004, p. 198).

Scientific peers are not always, however, equal in ability or expertise (Resnik et al., 2008), and exemplary scientific researchers may fall prey to the Dunning-Kruger effect, wherein "peer" reviewers may not be capable of assessing the quality of the research given that their knowledge is less than the research being reviewed (Kruger and Dunning, 1999). Bedeian (2004) describes, too, how peer reviewers can sometimes police low-level stylistic or conceptual issues in submitted manuscripts or, worse, how highly critical reviews can sometimes lead to HARKing (hypothesizing after the results are known) in subsequent revisions. Benos et al.

(2007) and Tamblyn et al. (2018) further describe challenges to the peer-review process, including bias, for example, related to issues like institutional affiliation, gender, ethnicity, race, age, and other demographic characteristics; bias related to studies reporting significant or positive results (Dwan et al., 2013); fraud involving dishonesty, problematic ethics, and misrepresented data (Rennie, 2016); delays caused by slow response and the publishing process (Das, 2016); and the enormous amount of work put into reviewing manuscripts for uncertain reward (Yankauer, 1990). Still, Benos et al. (2007) argue that, despite its haphazard and inconsistent development in scientific research publishing, peer review has become the "imprimatur" for research articles (p. 145). That is, peer review is currently critical to scientific work, often determining who is published, funded, and promoted; indeed, peer review often determines which research will be carried out in different disciplines and how well supported institutions and laboratories are to do that work (Souder, 2011).

Peer review is very much a socially constructed knowledge-making activity that is far from perfect. The goal of this imperfect system is to balance scientific authors' desires to make claims to research originality while addressing reviewers' steps to ensure that the authors' claims are framed carefully in a developing literature with vetted methods and shared notions of what is acceptable and what is unacceptable in terms of research design, data collection, and analysis (Myers, 1990; Bedeian, 2004). Up until the 1990s, most journals adhered to what Walker and Rocha da Silva (2015) have referred to as the "classical peer review" (p. 1), but since then increased research on the peer-review process and numerous calls for improvement have encouraged explorations of alternative ways to increase transparency in peer review. Efforts to make reviews more transparent include setting up triage reviews where both referees and editors work together to review manuscripts, paying referees for their time and energy reviewing manuscripts, allowing preposting and open exchange repositories for research manuscripts in process, and so on (Sub and Martin, 2009; Couzin-Frankel, 2013).

Although calls to improve peer review are many, there are challenges to accomplishing change, including understanding fully the nature of the problems in current peer-review models. One of the greatest criticisms of peer review is that it has been historically difficult to study, in large part because scientific journals have been somewhat secretive or at least nontransparent about their peer review and internal editorial processes (Brownlee, 2006; Squazzoni et al., 2020). This criticism was levied at journals even before editors integrated peer review into their review processes (Baldwin, 2015), but it is even more problematic in terms of peer review because the scientific community relies on it to ensure that scientific research is vetted for rigorousness, thoughtful design, and ethical implementation. To this end, cases of publication bias abound in the peer-review literature (Chalmers et al., 1990): for example, biases based on cronyism (despite objections by some researchers, e.g. Travis and Collins, 1991); biases based on institutional and national affiliation (e.g. the Vine and Matthews team from England versus Morley from Canada, publishing on plate tectonics in the 1970s) (Baldwin, 2015, pp. 342–345); biases based on editor selection of peer reviewers (Kassirer and Campion, 1994, p. 96); and biases based on editor interpretation of controversial or contested issues and trying to represent different "sides" of the issue in their reviewer selection process (Bower, 1991). Bedeian (2004), indeed, notes, "Empirical studies do, in fact, suggest that biases associated with various social, intellectual, and political considerations enter into referee recommendations" (p. 199). Resnik et al. (2008) reported that over 50% of the 283 people at the National Institute of Environmental Health Sciences believed that reviewers had exhibited bias during the peer-review process (p. 307). Additionally, gender bias in the review process has received considerable research attention (Souder, 2011), and affiliation and geographic bias has also been scrutinized (Wellington and Nixon, 2005).

Further, Bazeley (1998) found that funding outcomes for Australian Research Council Large Grants were influenced by whether the authors held research-only positions and were affiliated with more prestigious institutions. Other, subtle issues can evoke reviewer bias. For example, in examining the peer reviews of manuscripts submitted to the *American Sociological Review* between 1977 and 1981, Bakanic et al. (1987) found that reviews tended to favor quantitative data analyses, that assigning editors factored highly into the decision-making outcome, and that mean recommendations of the reviewers and revision numbers influenced the editors' final decisions significantly.

Kassirer and Campion (1994) note that although considerable research has been done on peer review as part of the publication management process, few researchers have looked at "its most important aspect – the cognitive task of the reviewer assessing a manuscript" (p. 96). We expect that peer review is undertaken by our *expert* colleagues in a field of research, our peers. But we tend not to carefully examine the kinds of expertise that we expect our peers to embody when they go about reviewing our scientific research manuscripts or grant proposals (Park et al., 2014). We rely on expert judgment, yet we understand that "[f]or objective judgment to be judgment, it must have as its basis subjective, interested, informed expert opinion" (Fitzpatrick, 2011, p. 198). For this reason, we would argue that expertise in a specific domain and expertise in the process of peer review require distinct ways of practicing expertise.

Expertise in peer review

Broadly, in the matter of becoming expert, it is important that one engage not only with the research findings in a field but with the assortment of social and rhetorical activities associated with becoming "enculturated" as a member of the discipline and its discourse (Bazerman, 1992) through, for example, attending conferences and engaging in other face-to-face and online interactions. Indeed, this understanding of expertise is supported by Collins and Evans' (2007) model of expertise that requires such engagement with a discipline to become expert given that *only* reading in a field may leave out important socially recognized features of a discipline. This interaction between the processes of science and reading and writing activities was reported by Florence and Yore (2004), as well, when they concluded that "novice scientists came to appreciate that the writing, editing, and revising process influenced the quality of the science as well as the writing" (p. 637).

Schriver (2012) has reviewed research on expertise in professional communication that is applicable here, given that we situate peer review as a form of professional communication and, further, focus on the question of expertise in the refereeing of scientific materials. Making Schriver's framework of expertise in professional communication particularly useful is its rhetorical comportment. She explains that professional communication relies not merely on other experts to evaluate performance but also on the stakeholders. Consider the example of a scientist conducting a peer review. In this example, we take the domain knowledge to be some area of scientific research. The particulars are not important for our argument, and our scientist could be an entomologist studying infectious disease vectors, a physicist studying photonics, or a geophysicist studying plate tectonics. In any case, scientists have expertise in their domain of science, but in the rhetorical activities of peer review, they must also be expert professional communicators. Here we might imagine the peer reviewers are the expert professional communicators and the stakeholders are the authors and perhaps editors. However, we can also understand that editors must undertake expert professional communications to help the authors interpret referee feedback. In this respect, the expert community requires what Luntley (2009) calls "activity-dependence," given that "the opportunity to communicate, let alone account for, one's expert judgement will be limited by the extent to which one's audience shares the appropriate capacities for activity on which the concepts depend" (p. 369).

Thus, while on the one hand we are arguing that scientists are engaged not only in scientific processes but also in discursive ones, we are also acknowledging the tight relationship between the two activities or processes. As Ericsson et al. (1993) point out, "during the process of writing scientists develop and externalize their arguments. The written products can be successfully criticized and improved by the scientists themselves, even after long delays, and also easily shared with other scientists for evaluation and comments" (p. 391). In this way, peer review can be viewed as a knowledge-generating activity where reading and writing as a scientist are as central to the enterprise of science as, strictly speaking, the science itself.

As a rhetorical activity or as a form of professional communication, the peer-review process involves numerous different roles and communicative moments. First, let us consider the role of journal editors as they commonly undertake a form of broad peer review and decide which articles are suitable for full peer review and which will receive a desk rejection. In addition to automated systems helping the editor think about potential reviewers for the manuscript, the editor must also consider how germane the research is to the journal's contents and audience's interests, and how contested the content or arguments in the manuscript are. In short, the editor has numerous issues to juggle in selecting peer reviewers for the work. If an article is sent for review, journal editors must have an expertise in professional communication that allows them not only to understand and interpret the comments made by the reviewers but also to synthesize comments from multiple reviewers and effectively share those with the authors. This is an especially important skill that requires more than the ability to operate within a domain of a specialist expert's scientific thinking. This activity also requires that editors be able to anticipate the needs of the authors. When conveyed to the author of the manuscript, feedback from the reviewers should not merely be summarized by the journal editor but instead synthesized into an account that helps authors advance their research, even when their manuscript is as yet not suitable for publication. Inexpert editorial responses may be partially to blame for the feeling some authors relate when they are given contradictory feedback by multiple reviewers. To push beyond simplistic summary, we believe that what Schriver (2012) describes as complex communication challenges must be responded to creatively through "high-wire rhetorical acts of interpretation, anticipation, production, and reflection" that we might call "expert performance" (p. 280).

Peer reviewers, then, need what Schriver (2012) calls "persuasion expertise" (p. 283; cf. Scardamalia and Bereiter's [1991] "Literate expertise") to assess a particular study and effectively communicate the needed changes to editors and, in turn, authors. For the purposes of our discussion here, we will take it to be true that, often, reviewers will have relevant expertise in scientific and technical subject matter. Then we might focus on the ways that a failure to consider persuasive elements can undermine the review process and science itself, especially given that scientists themselves sometimes downplay such elements in scientific communication (Yore et al., 2004). Yet learning to make persuasive arguments within a disciplinary framework is a key part of socializing an individual into a discipline. The persuasive arguments that must be made for peer review are also shaped by the discipline as well as the norms of the genre. First, they must be capable of writing knowledge-transforming text that would normally be appropriate to a scholarly journal. Also, once they have submitted their work and it has been passed on to their peers for review, it is important that they are able to interpret and respond to reviews when they come in. Interpreting involves understanding the editor and reviewer feedback, making a revision plan, executing that plan in your manuscript (which may sometimes involve reframing your research in terms of your literature review or even reexamining your data or results), and then drafting a rhetorically appropriate letter in response. Inexpert authors tend to have several immediate difficulties with peer reviews: they may assume they must modify their manuscript to address every issue the reviewers raise; they may assume that significant suggestions for revision are akin to having the reviewers outright reject their manuscript; and they may assume that inconsistent feedback from reviewers means that there is a problem with the peer-review process. After Shanteau (2015), we see that the "experts-should-agree perspective of expertise" is problematic, and instead we emphasize the complex nature of the problem space, in this case the peer-review process, where reviewers guide authors rather than making single-answer decisions and work with "multiple, constantly changing, and dynamic factors" as part of the review process (p. 172). The goal of peer review is, as an ideal, a negotiated, knowledge-making and -sharing activity among journal editors, authors, and peer reviewers. A successful peer-review encounter normally improves the author's manuscript as part of this collaboration.

Trends and challenges in improving peer review

Numerous other topics than those we have been able to raise here might be considered when investigating how to improve peer review. Scholarly venues and associations continue to debate these subjects heatedly. A range of interrelated challenges and changes for peer review exist on the immediate horizon, and we briefly outline and explore some of them here.

Training peer reviewers, we argue, is a central concern that prefigures other challenges in peer review (Smith, 2006). Different fields have different processes for teaching junior scholars how to produce quality peer reviews, including having graduate students perform reviews in apprenticeship scenarios, sharing other peer reviews with researchers new to peer review, providing formal training and informal heuristics or evaluation checklists for performing effective reviews, and so on (Bower, 1991). Bruce et al. (2016) list various types of training such as structured workshops, feedback, mentoring, self-taught or informal face-to-face instruction (p. 4). Blockeel et al. (2017) describe how numerous biomedical societies, such as the American Society for Reproductive Medicine (ASRM), are designing and providing courses on peer review for referees and editors, although much of the instruction on scientific research publishing is geared more toward authors than reviewers. We believe that, in such training, attending to the complex rhetorical activities required of peer reviewers might help address challenges that arise in the activity of reviewing as well as communicating recommendations and engaging in dialogue with the editors and authors of scientific journals. As well, we hope that the training would promote antiracist practices and awareness of biases in review, such as affiliation, geographic, and gender bias. It is important to remember, as Bazerman (1998) observes, "Within the bounded discursive world of science, an intertext of cited works or a literature defines a gradually transforming discursive space within which new claims vie for acceptance, judged by an epistemic court of insider specialists" (p. 383).

Transparency in peer review has long been an issue that complicates our understanding of the process and its shortcomings (Wicherts, 2016; Horbach and Halffman, 2018; Bravo et al., 2019). Almost 75% of 1,340 internationally distributed biomedical researchers, for example, suggested that the peer-review process in their disciplines was not transparent, and approximately 50% indicated they believed that the process was neither fair nor scientific (Ho et al., 2013). Some experiments in transparency have been applied over the last 30 years and can be connected to movements toward open peer review or open science, including open review participation, open reports, and so on (Tennant, 2018). Examples include articles in *Frontiers* sharing the names of editors and reviewers with authors and the *British Medical Journal* in 1999 disclosing the names of peer reviewers with authors, an activity that, when compared to unsigned reviews, made no significant difference in the quality of the reviews (Henderson, 2010). An additional mode of review involves no blinding, where the reviewers and the authors know each other's names, a method that some researchers have suggested increases transparency and accountability among editors, authors, and reviewers, while others have noted that it may further complicate the review process by reducing candid and sometimes negative reviews (Wendler and Miller, 2014). Isenberg et al. (2009) found, for example, that fewer manuscripts were published, and manuscript recommendation scores were significantly lower when the reviewer did not know the names of the authors. Not surprisingly, calls for transparency and for more research on the peer-review process recently culminated in *The Journal of the American Medical Association* (JAMA) releasing a Call for Research on Peer Review and Scientific Publication in 2019 for its Ninth International Congress being held in 2022 in Chicago, Illinois; suggested topics include bias, peer-review decision making, research and publication ethics, and models for peer review and scientific information dissemination (Ioannidis et al., 2019). Numerous editors of science journals have lamented the "woeful state of affairs" (Friedberg, 2010, p. 477) in the current research on the peer-review process (Smith, 2006), and Larson and Chung (2012) note that their "systematic review" of research on peer review "consisted primarily of editorials and commentaries" (p. 43).

Registered reports are an emerging model in scientific publishing characterized by a bipartite model of the scientific research article where, normally, the first version of the manuscript involves study design, including the rationale and methods proposed, and the procedures for analysis prior to any data collection (beyond, say, a pilot study). The second version of the manuscript completed once data are collected and analyzed can be thought of as a rather standard representation of the scientific research article genre (Mehlenbacher, 2019). Registered reports account for the intentions of an author or authors in their study in a manner that a traditional scientific research article submission and subsequent review cannot. In the first version of the manuscript, the intentions of the author to conduct a study in a particular manner are explicitly stated, and authors will be held accountable for following this plan and explaining any diversions, which are to be quite minor, if at all. For the reviewer, the focus is shifted from the significance of the findings to the methodological approach and design.

Preprints and data sharing servers are another issue challenging the current model of peer-review system. Servers like *ArXiv.org*, launched in 1991, make e-prints and data available to scientific audiences without having gone through the peer-review process or with very limited review (Mayernik et al., 2015; Smith, 2016). These manuscripts are sometimes called "rogue reports," that is, published manuscripts that have not been "properly" peer reviewed but that are being shared prior to publication in "classic" journals (Sipido et al., 2017). The risk of accelerated scientific publication without peer-review raises new challenges, as Abbas and Lamb (2020) discuss, especially when public and policy need for scientific answers are high (as we are experiencing now, with the race to understand and address the COVID-19 crisis). So, while preprints successfully speed the circulation of scientific data and knowledge, researchers have only begun to examine their influence on peer-review processes; *Lancet, British Medical Journal*, and other BioMed journals, for example, have been allowing authors to post preprints of their manuscripts, while they are being reviewed (Sub and Martin, 2009). Further, who has sufficient scientific expertise to assess preprints both within disciplines and in interested fields (e.g. journalism) is an important question requiring continued attention (Yan, 2020).

References

Abbas, N., & Lamb, S. (2020) 'A little science is a dangerous thing', *Healthy Debate*, 3 July, https://healthydebate.ca/2020/07/topic/little-science-is-dangerous/

Bakanic, V., McPhail, C., & Simon, R.J. (1987) 'The manuscript review and decision-making process', *American Sociological Review*, vol. 52, no. 5, pp. 631–642, https://doi.org/10.2307/2095599

Baldwin, M. (2015) 'Credibility, peer review, and "Nature", 1945–1990', *Notes and Records of the Royal Society of London*, vol. 69, no. 3, pp. 337–352, https://doi.org/10.1098/rsnr.2015.0029

Bazeley, P. (1998) 'Peer review and panel decisions in the assessment of Australian research council project grant applicants: What counts in a highly competitive context?', *Higher Education*, vol. 35, no. 4, pp. 435–452, https://doi.org/10.1023/A:1003118502318

Bazerman, C. (1992) 'From cultural criticism to disciplinary participation: Living with powerful words', in A. Herrington & C. Moran (eds.), *Writing, teaching and learning in the disciplines*, New York: Modern Language Association of America.

Bazerman, C. (1998) 'The production of technology and the production of human meaning', *Journal of Business and Technical Communication*, vol. 12, pp. 381–387, https://doi.org/10.1177/1050651998012003006

Bedeian, A.G. (2004) 'Peer review and the social construction of knowledge in the management discipline', *Academy of Management Learning & Education*, vol. 3, no. 2, pp. 198–216.

Benos, D.J. et al. (2007) 'The ups and downs of peer review', *Advances in Physiology Education*, vol. 31, no. 2, pp. 145–152, https://doi.org/10.1152/advan.00104.2006

Blockeel, C. et al. (2017) 'Review the "peer review"', *Reproductive Biomedicine (RBM) Online*, vol. 35, no. 6, pp. 747–749, https://doi.org/10.1016/j.rbmo.2017.08.017

Bower, B. (1991) 'Peer review under fire', *Science News*, vol. 139, no. 25, pp. 394–395, https://doi.org/10.2307/3975252

Bravo, G. et al. (2019) 'The effect of publishing peer review reports on referee behavior in five scholarly journals', *Nature Communications*, vol. 10, no. 1, pp. 1–8, https://doi.org/10.1038/s41467-018-08250-2

Brownlee, C. (2006) 'Peer review under the microscope', *Science News*, vol. 170, no. 25, pp. 392–393

Bruce, R. et al. (2016) 'Impact of interventions to improve the quality of peer review of biomedical journals: A systematic review and meta-analysis', *BMC Medicine*, vol. 14, no. 85, pp. 1–16, https://doi.org/10.1186/s12916-016-0631-5

Burnham, J.C. (1990) 'The evolution of editorial peer review', *JAMA: The Journal of the American Medical Association*, vol. 263, no. 10, pp. 1323–1329, https://doi:10.1001/jama.1990.03440100023003

Chalmers, T.C. et al. (1990) 'Minimizing the three stages of publication bias', *JAMA: The Journal of the American Medical Association*, vol. 263, no. 10, pp. 1392–1395, https://doi:10.1001/jama.1990.03440100104016

Collins, H., & Evans, R. (2007) *Rethinking expertise*, Chicago: University of Chicago Press.

Couzin-Frankel, J. (2013) 'Secretive and subjective, peer review proves resistant to study', *Science (American Association for the Advancement of Science)*, vol. 341, no. 6152, p. 1331, https://doi:10.1126/science.341.6152.1331

Das, A.K. (2016) '"Peer review" for scientific manuscripts: Emerging issues, potential threats, and possible remedies', *Medical Journal Armed Forces India*, vol. 72, pp. 172–174, https://doi.org/10.1016/j.mjafi.2016.02.014

Dwan, K. et al (2013) 'Systematic review of the empirical evidence of study publication bias and outcome reporting bias – An updated review', *PLOS one*, vol. 8, no. 7, pp. 1–37, https://doi.org/10.1371/journal.pone.0066844

Ericsson, K.A., Krampe, R.T., & Tesch-Römer, C. (1993) 'The role of deliberate practice in the acquisition of expert performance', *Psychological Review*, vol. 100, no. 3, pp. 363–406, https://doi.org/10.1037/0033-295X.100.3.363

Fitzpatrick, K. (2011) 'Peer review, judgment, and reading', *Profession: The Modern Language Association*, vol. 2011, no. 1, pp. 196–201

Florence, M.K., & Yore, L.D. (2004) 'Learning to write like a scientist: Coauthoring as an enculturation task', *Journal of Research in Science Teaching*, vol. 41, no. 6, pp. 637–668, https://doi.org/10.1002/tea.20015

Friedberg, E.C. (2010) 'Peer review of scientific papers – A never-ending conundrum', *DNA Repair*, vol. 9, pp. 476–477, https://doi.org/10.1016/j.dnarep.2010.03.003

Henderson, M. (2010) 'End of peer review show?', *BMJ: British Medical Journal*, vol. 340, no. 7749, pp. 738–740

Ho, R. C-M. et al. (2013) 'Views on the peer review system of biomedical journals: An online survey of academics from high-ranking universities', *BMC Medical Research Methodology*, vol. 13, no. 74, pp. 1–15, https://doi.org/10.1186/1471-2288-13-74

Horbach, S.P.J.M., & Halffman, W. (2018) 'The changing forms and expectations of peer review', *Research Integrity and Peer Review*, vol. 3, no. 8, pp. 1–15, https://doi.org/10.1186/s41073-018-0051-5

Ioannidis, J.P.A. et al. (2019) 'The Ninth International Congress on peer review and scientific publication: A Call for Research', *The Journal of the American Medical Association*, vol. 322, no. 17, pp. 1658–1660, https://DOI:10.1001/jama.2019.15516

Isenberg, S.J. et al. (2009) 'The effect of masking manuscripts for the peer-review process of an ophthalmic journal', *British Journal of Ophthalmology*, vol. 93, no. 7, pp. 881–884, https://doi:10.1136/bjo.2008.151886

Kassirer, J.P., & Campion, M.D. (1994) 'Peer review: Crude and understudied, but indispensable', *JAMA: The Journal of the American Medical Association*, vol. 272, no. 2, pp. 1321–1322, https://doi:10.1001/jama.272.2.96

Kronick, D.A. (1990) 'Peer review in 18th-Century scientific journalism', *JAMA: The Journal of the American Medical Association*, vol. 263, no. 10, pp. 96–97, DOI:10.1001/JAMA.1990.03440100021002

Kruger, J., & Dunning, D. (1999) 'Unskilled and unaware of it: How difficulties in recognizing one's own incompetence lead to inflated self-assessments', *Journal of Personality and Social Psychology*, vol. 77, no. 6, pp. 1121–1134, https://doi.org/10.1037/0022-3514.77.6.1121

Larson, B.P., & Chung, K.C. (2012) 'A systematic review of peer review for scientific manuscripts', *HAND*, vol. 7, pp. 37–44, https://doi:10.1007/s11552-012-9392-6

Luntley, M. (2009) 'Understanding expertise', *Journal of Applied Philosophy*, vol. 26, no. 4, pp. 357–370, https://doi:10.1111/j.l468-5930.2009.00468.x

Mayernik, M.S. et al. (2015) 'Peer review of datasets: When, why, and how', *American Meteorological Society (BAMS)*, pp. 191–201, https://doi:10.1175/BAMS-D-13-00083.I

Mehlenbacher, A.R. (2019) 'Registered reports: Genre evolution and the research article', *Written Communication*, vol. 36, no. 1, pp. 38–67, https://doi.org/10.1177/0741088318804534

Myers, G. (1990) *Writing biology: Texts in the social construction of scientific knowledge*, Madison: University of Wisconsin Press.

Park, I-U. et al. (2014) 'Modelling the effects of subjective and objective decision making in scientific peer review', *Nature*, vol. 506, pp. 93–98, https://doi:10.1038/nature12786

Rennie, D. (2016) 'Let's make peer review scientific', *Nature*, vol. 535, pp. 31–33, https://doi.org/10.1038/535031a

Resnik, D.B., Gutierrez-Ford, C., & Peddada, S. (2008) 'Perceptions of ethical problems with scientific journal peer review: An exploratory study', *Science and Engineering Ethics*, vol. 14, no. 3, pp. 305–310, https://doi:10.1007/s11948-008-9059-4

Scardamalia, M., & Bereiter, C. (1991) 'Literate expertise', in K.A. Ericsson & J. Smith (eds.), *Toward a theory of expertise: Prospects and limits*, Cambridge: Cambridge University Press, pp. 172–194.

Scholz, O.R. (2016) 'Symptoms of expertise: Knowledge, understanding and other cognitive goods', *Topoi*, vol. 37, no. 1, pp. 29–37, https://DOI:10.1007/s11245-016-9429-5

Schriver, K. (2012) 'What we know about expertise in professional communication', in V.W. Berninger (ed.), *Past, present, and future contributions of cognitive writing research to cognitive psychology*, New York: Psychology Press, pp. 275–312.

Shanteau, J. (2015) 'Why task domains (still) matter for understanding expertise' *Journal of Applied Research in Memory and Cognition*, vol. 4, pp. 169–175, https://doi.org/10.1016/j.jarmac.2015.07.003

Sipido, K.R. et al. (2017) 'Peer review: (R)evolution needed', *Cardiovascular Research*, vol. 113, pp. e54 – e56, https://doi:10.1093/cvr/cvx191

Smith, D.R. (2006) 'Peer review: A flawed process at the heart of science and journals', *Journal of the Royal Society of Medicine*, vol. 99, pp. 178–182, https://doi:10.1258/jrsm.99.4.178

Smith, D.R. (2016) 'Will Publons popularize the scientific peer-review process?', *BioScience*, vol. 66, no. 4, pp. 265–266, https://doi.org/10.1093/biosci/biw010

Souder, L. (2011) 'The ethics of scholarly peer review: A review of the literature', *Learned Publishing*, vol. 24, no. 1, pp. 55–74, https://doi:10.1087/20110109

Squazzoni, F. et al. (2020) 'Unlock ways to share data on peer review', *Nature*, vol. 578, pp. 512–514, https://doi:10.1038/d41586-020-00500-y

Sub, J., & Martin, R. (2009) 'The air we breathe: A critical look at practices and alternatives in the peer-review process', *Perspectives on Psychological Science*, vol. 4, no. 1, pp. 40–50, https://doi:10.1111/j.1745-6924.2009.01105.x

Swales, J.M. (1996) 'Occluded genres in the academy: The case of the submission letter', in E. Ventola & A. Mauranen (eds.), *Academic writing: Intercultural and textual issues*, Amsterdam: John Benjamins, pp. 45–58.

Swales, J.M. (2004) *Research genres: Explorations and applications*, Cambridge: Cambridge University Press.

Tamblyn, R. et al (2018) 'Assessment of potential bias in research grant peer review in Canada', *CMAJ*, vol. 190, no. 16, pp. E489 – E499, https://doi.org/10.1503/cmaj.170901

Tennant, J.P. (2018) 'The state of the art in peer review', *FEMS Microbiology Letters*, vol. 365, pp. 1–10, https://doi.org/10.1093/femsle/fny204

Tennant, J.P., & Ross-Hellauer, T. (2020) 'The limitations to our understanding of peer review', *Research Integrity and Peer Review*, vol. 5, no. 6, pp. 1–14, https://doi.org/10.1186/s41073-020-00092-1

Travis, G.D.L., & Collins, H.M. (1991) 'New light on old boys: Cognitive and institutional particularism in the peer review system', *Science, Technology, & Human Values*, vol. 16, no. 3, pp. 322–341, https://doi.org/10.1177/016224399101600303

Walker, R., & Rocha da Silva, P. (2015) 'Emerging trends in peer review – A survey', *Frontiers in Neuroscience*, vol. 9, no. 169, pp. 1–18, https://doi.org/10.3389/fnins.2015.00169

Weller, A.C. (1990) 'Editorial peer review in US medical journals', *JAMA: The Journal of the American Medical Association*, vol. 263, no. 10, pp. 1344–1347, https://doi:10.1001/jama.1990.03440100048007

Wellington, J., & Nixon, J. (2005) 'Shaping the field: The role of academic journal editors in the construction of education as a field of study', *British Journal of Sociology of Education*, vol. 26, no. 5, pp. 643–655, https://doi.org/10.1080/01425690500293835

Wendler, D., & Miller, F. (2014) 'The ethics of peer review in bioethics', *Journal of Medical Ethics*, vol. 40, no. 10, pp. 697–701, https://doi:10.1136/medethics-2013-101364

Wicherts, J.M. (2016) 'Peer review quality and transparency of the peer-review process in open access and subscription journals', *PLoS ONE*, vol. 11, no. 1, pp. 1–19, https://doi.org/10.1371/journal.pone.0147913

Yan, W. (2020) 'Coronavirus tests science's need for speed limits', *The New York Times*, 21 April.

Yankauer, A. (1990) 'Who are the peer reviewers and how much do they review?', *JAMA: The Journal of the American Medical Association*, vol. 263, no. 10, pp. 1338–1340, https://doi:10.1001/jama.1990.03440100042005

Yates, B.J. (2017) 'The "new realities" of peer review', *Journal of Neurophysiology*, vol. 117, pp. 869–871, https://doi.org/10.1152/jn.00058.2017

Yore, L.D., Hand, B.M., & Florence, M.K. (2004) 'Scientists' views of science, models of writing, and science writing practices', *Journal of Research in Science Teaching*, vol. 41, no. 4, pp. 338–369, https://doi.org/10.1002/tea.20008

8

EDITORIAL PEER REVIEW AT BASIC AND APPLIED SCIENCES JOURNALS IN THE SEMIPERIPHERY CONTEXT OF TAIWAN

Cheryl L. Sheridan

Introduction

Investigations into the conditions and practices of scientists around the world, who use English as a second (L2) or foreign language (EFL) for academic writing have grown in the last decade (Curry and Lillis, 2018, 2019). Due to globalization influences on higher education, especially with emphasis on global academic rankings, institutions have pushed researchers to publish in prestigious 'internationally' indexed journals (IIJs), which are overwhelmingly English-medium publications (Mongeon and Paul-Hus, 2016). Discursive and non-discursive challenges create an additional burden for these researchers when writing in English compared to their L1 (Flowerdew, 2019; Hanauer et al., 2019). However, despite these pressures, they weigh commitments and publish in their first academic languages and/or submit manuscripts to 'national' publications (Curry and Lillis, 2004; Duszak and Lewkowicz, 2008).

Much scholarly publishing research has focused on researchers in semi-peripheral contexts, particularly in Europe (Bennet, 2014). The concept of the semi-periphery derives from world systems theory (Wallerstein, 1991) when describing nation states that exhibit characteristics of both the global center (control of economic power and production) and the periphery, which is dependent on the center for progress as production providers. According to Bennet, in global academic work, the United States is the center with its economic might forging the greatest resources into science. Consequently, the majority of scientific publications in the highest ranked journals originate there (Nature, 2020). In contrast, the periphery 'off-networked' researchers (Canagarajah, 2003), typically in the global south (Salager-Meyer, 2008), lack resources to compete globally. Between these, according to Bennet, semi-periphery countries simultaneously depend on the center for economic and scientific development, but also support periphery nations by taking on a relative center role. Thus, researchers in the semi-periphery work under political, economic, and social interests at the international and local levels. As a result, many conduct and communicate scientific work for these two audiences.

While some research on 'national' journals presents periphery publications wrangling for submissions and fighting for survival (Salager-Meyer, 2008, 2015) or operating as 'article mills'

DOI: 10.4324/9781003043782-10

(Kuzhabekova, 2018), other accounts report steady growth (Lundin et al., 2010; Marušić and Marušić, 2014; Sheridan, 2015). Far from just a 'back-up plan', non-center journals can bring fresh knowledge and innovative perspectives to global publishing (Curry and Lillis, 2010) while providing 'intellectual infrastructure for developing and harnessing local knowledge and local knowledge making' (Lillis, 2012, p. 697), which I argue is especially possible with a peer review system that supports authors.

Editorial peer review

Editorial peer review (EPR) provides a mechanism whereby editors call on subject experts to evaluate manuscripts and provide feedback to contributors to improve their work so that journal standards are maintained (Belcher, 2007; Paltridge, 2013). Double-blind review (authors' and reviewers' identities blinded to each other) has been the norm in humanities and social sciences (HSS), but in many natural science disciplines, single-blind review (only reviewers' identities blinded to authors) is common. To mitigate possible bias, some science journals have been opting for double-blind review (Darling, 2015). For more on EPR development, see Mehlenbacher and Mehlenbacher (Chapter 7 in this volume), and for associated issues, see Hyland (Chapter 1 in this volume, pp. 18–19). This study includes journals from science disciplines in Taiwan; therefore, while various review processes may be identified, all will be referred to as EPR.

According to Paltridge (2017), local conditions and discourse community norms dictate EPR, but this has raised criticisms over inconsistent quality standards among journals. Therefore, journals published beyond the Anglophone center, weighing internationalization pressure, may adopt center practices to satisfy membership requirements of citation indexes (Lundin et al., 2010; Sheridan, 2018). Even so, the judgment associated with peer review anywhere leads authors to see EPR as primarily a 'gatekeeping' exercise (McKay, 2003). While Paltridge notes EPR is perceived as a guarantee for the best possible dissemination of knowledge, even among major journals, editors approach it differently based on the roles they adopt. For example, Starfield and Paltridge (2019, p. 254) prefer editors to function as 'custodians' caring for the well-being of journals rather than as gatekeepers policing their borders. From their perspective, the role of editors 'is one of safeguarding and enhancing the journal's reputation, of finding the best reviewers we can for the articles we choose to send out for review, and of shepherding and supporting worthy research articles through the review process' (Starfield and Paltridge, 2019, p. 254).

Some journals beyond the Anglophone center have explicitly trained reviewers (Adamson and Muller, 2012; Adamson and Fujimoto-Adamson, 2016) or have supported authors through mentoring (Lillis et al., 2013), while Marušić et al. (2004) and Marušić and Marušić (2014) advocate an 'author-helpful policy' beginning with a 'pre-peer review' process. It appears that explicit and targeted interventions (Hanauer and Englander, 2013, pp. 134–159) needed to alleviate the burden on multilingual English science writers can take place before, during, or following the peer review process.

Taiwan's scientific publishing context

Taiwanese universities have been impacted by national policies seeking to raise institutions' global rankings (Song and Tai, 2007). Therefore, they have implemented policies that privilege IIJs (Sheridan, 2017b). Concurrently, institutions fund journal production, while the Ministry of Science and Technology (MOST) has endeavored to raise the quality of local HSS journals through establishing citation indexes. An active scientific journal publishing environment is evidenced by Airiti Library, a commercial database of academic journals published primarily in Taiwan listing over 1,600 journals in a comprehensive range of disciplines. Chinese is the language of publication for most publications, but English is the most common foreign language. While the 'quality' varies, most journals have published regularly, even over decades, and many have been admitted to national and international indexes and databases.

Regarding Taiwan's scientists, Sheridan (2017a) found nearly 45% of HSS (humanities and social sciences) and STEM (science, technology, engineering, and mathematics) researchers surveyed had only published in English over the last five years, mostly to reach specialist audiences or fulfill department requirements. The STEM researchers were most interested in publishing in English, especially in IIJs, but they also used Chinese for books and book chapters. Overall, 35% of Sheridan's (2017a) respondents published in both English and Chinese and 12% only in Chinese. Furthermore, they submitted papers to Taiwan-based journals, including indexed and nonindexed publications in both languages. These findings contradict the 'lore' (Curry and Lillis, 2019) suggesting that scientists are only focused on English-medium internationally indexed journals (IIJs). In a complex academic environment, Taiwan-based HSS researchers turn to 'national' journals, even though they may bring fewer points, for several reasons, but often following difficult IIJ experiences because they are seen as easier with faster turnaround (Sheridan, 2017b).

Overall, multilingual scientists in semiperiphery contexts need to publish in English, which constitutes an additional burden (Hanauer et al., 2019). IIJs are more likely to use EPR to reject manuscripts, but 'national' journals are more likely to need manuscripts (Donovan, 2011). Therefore, it may be possible for the latter to use EPR as a learning opportunity for novice authors, especially those on the tenure clock, while simultaneously motivating more and better submissions. However, 'small' journals already face serious challenges. Before knowing what is needed in a context and how editors might be facilitators rather than gatekeepers, it is important to understand their attitudes and circumstances. Therefore, this study seeks editorial committee members' attitudes regarding EPR at basic and applied sciences (BAS) journals published in Taiwan. In particular, an online survey asked respondents why they believe manuscripts are rejected, to identify shortcomings in revisions, and to consider what authors learn from the submission and revision experience.

The study: survey of editorial committee members

The online survey, 'Peer Review Practices at Taiwan-Based Basic and Applied Sciences Journals,' is based on a survey of HSS editorial committee members (Sheridan, 2020). The survey included 37 quantitative and five qualitative items. Those who indicated they are 'not at all involved' in EPR were excluded. Besides basic information about the journals, their EPR processes, and demographics, the survey covered:

- reasons papers are commonly rejected at the in-house and peer review stages,
- shortcomings in authors' manuscript revisions and revision reports,
- ways the journal can help facilitate authors' success,
- ways EPR can benefit authors besides manuscript acceptance, and
- the degree to which authors learn research and publication skills through EPR.

This chapter focuses on the perspectives of editorial committee members of the 38 English-medium BAS journals published between 2014 and 2019 by entities in Taiwan. From Airiti

Library, 169 unique titles published between 2014 and 2019, of which 38 published all articles in English, were found. Because journal practices and structures vary, 'editorial committee members' include co-/chief/section/associate/senior/area/managing editors and editorial board members. Entities include universities, associations, and public research centers. As a result, a total database of 384 individual potential survey respondents was compiled as the sample.

The online survey was distributed to the entire sample of 384 editorial committee members. The broad recruitment criteria aimed to compensate for an anticipated low response rate while allowing as many perspectives as possible. However, this may have led to an uneven distribution of respondents among journals, rendering findings likely unevenly distributed among journals.

Data analysis

With quantitative and qualitative data, a mixed methods analysis was possible. Descriptive statistics provided frequency counts and valid percentages. Cumulative responses from rating questions were totaled and examined holistically. Responses from open-ended questions were coded and analyzed through qualitative methods (Miles et al., 2014). Finally, cross-examined quantitative and qualitative data provided a triangulated view of editorial committee members' perceptions.

Editorial committee members' perspectives on EPR

Of the 384 invited to participate, 73 from earth science and geography, atmospheric science, general science, information science, mathematics, oceanography, and physics responded. Of the 48 who completed most items, 48% identified as editors, 67% as editorial board members, and 10% as reviewers. Six 'not at all involved with peer review' were excluded. Most of the respondents were male (91%) and three-fourths between the ages of 46 and 65 (76%), with over half from research-oriented public universities (58%). For most, their terminal degree came from the United States (44%) or Taiwan (49%). The majority use Chinese as a first language (86%) and English as a second (86%).

EPR process

Participants reported that most journals use single-blind review, which is typical in STEM and contrasts with HSS journals in Taiwan (Sheridan, 2020). Also, manuscripts go through, at most, three rounds of review with 65% reporting that two are typical. Over 67% indicated that the editor makes the final acceptance/rejection decision.

Respondents were asked to list the greatest challenges of EPR. Relevant responses from 39 respondents were classified into four categories. The two most frequently mentioned issues concerned finding appropriate and willing reviewers, a challenge intensified with increasing numbers of journals publishing around the world (Hyland, Chapter 1 of this volume). However, having to negotiate with reviewers to get good-quality reports returned by deadlines was also an issue in eight responses. Finally, four showed just getting enough submissions was a challenge, a fundamental problem of some 'small' journals.

EPR and rejection

After a manuscript is submitted, typically the editor or editorial committee decides whether it is suitable for the journal and has potential to be published. If so, the search for peer reviewers commences; otherwise, it will be 'desk-rejected,' referred to as 'in-house' rejection in Taiwan.

Of 20 respondents who identified as editors, nine believed that only up to 20% of rejected manuscripts are rejected in-house, while five believed 21% to 40% are. This low percentage may indicate that they are willing to give researchers a chance because journals need manuscripts or that journals cover a range of research topics beyond editors' specialties, making it necessary to get outside evaluation: both reasons are typical of smaller national journals.

Respondents were asked to rank their perceptions of the three most common reasons for desk rejection and peer review rejection, which were based on Thrower (September 12, 2012). Based on cumulative ratings of reasons for desk rejection, Table 8.1 shows that respondents considered 'unsuitable topic' and 'lacking originality' as most common, followed by 'major methodological flaws' and 'plagiarism'. This indicates editors view these problems as too serious to rectify with available resources. Unfortunately, about 10% were attributed to poor language use, revealing that writing in English can be an extra burden (Hanauer et al., 2019). Of least concern for desk rejection were poor or insufficient data and inadequate data presentation or interpretation, likely because editors would leave these issues for subject experts, i.e. peer reviewers.

By looking further into the highest three responses in each rank level, some subtlety can be ascertained. Table 8.2 shows the three most frequently mentioned reasons for desk rejection

Table 8.1 Reasons for in-house/desk reject (*n* = 48)

Reason	CR	%
Unsuitable topic	58	21.32
Lacking originality	43	15.81
Major methodological flaws	32	11.76
Plagiarism	32	11.76
Poor language use	29	10.66
Does not meet format and style requirements	23	8.46
Needs additional research	19	6.99
Poor or insufficient data	18	6.62
Inadequate data presentation or interpretation	18	6.62

Note: CR = cumulative ratings from 48 editorial committee members.

Table 8.2 Top three of three most common reasons for in-house/ desk rejection

Variable (n = 48)	Frequency	%
Rank 1: First most common		
Unsuitable topic	16	33.33
Lacking originality	9	18.75
Major methodological flaws	5	10.42
Rank 2: Second most common		
Poor language use	10	20.83
Lacking originality	6	12.50
Major methodological flaws	6	12.50
Rank 3: Third most common		
Plagiarism	9	18.75
Poor language use	6	12.50
Major methodological flaws	5	10.42

among three frequency ranks. It shows one-third of respondents identified 'unsuitable topic' as the most common reason for desk rejection, further emphasizing that picking the appropriate journal is critical for authors' success. 'Lacking originality', the second most mentioned overall (Table 8.1) is still present in second place in Rank 1 and 2, but perhaps not as critical as language and plagiarism, which are in the top position in the second and third ranking respectively. This means language issues and plagiarism may be more serious concerns than originality at this stage. However, all four are still perceived as unrectifiable through EPR processes.

Table 8.3 lists the cumulative total ratings of the nine reasons for rejection during peer review from highest to lowest. It shows over 24% of respondents considered 'major methodological flaws' of primary concern, likely because, similar to being an unsuitable topic, methodology problems are difficult to rectify. Perhaps because editors have desk-rejected the most obviously unsuitable manuscripts, reviewers can focus on criteria related to the specific research quality of their specialties. In contrast, reviewers seem less willing to reject manuscripts because of nonresearch issues such as format and language. This may indicate that even though multilingual authors can experience an extra burden when writing in English, language is not necessarily what should worry them most if their manuscript is not immediately rejected.

Of ranking categories shown in Table 8.4, the top three reasons for rejection during peer review were: 'major methodological flaws', 'lacking originality', and 'poor or insufficient data'. Similar to the cumulative ratings (Table 8.3), the top two here were the same, and the third is related to data quality rather than to presentation or interpretation. This shows these three are consistent concerns, especially as they were so evenly distributed among respondents, each with 13. However, it is not clear whether these issues are disqualifiers during a first or later round of review.

In Table 8.4, 'lacking originality' appears in Rank 3, indicating it is not only an acute problem but a common concern in both desk rejections and peer review, thereby supporting the cumulative ratings. While major 'methodological flaws' is the top Rank 2 reason for peer review rejection, it is also a persistent problem appearing in rank categories over both stages. On the other hand, 'plagiarism' and 'poor language use' show up in the cumulative totals but were not cited in the top three rank categories of peer review rejection. This also appears to indicate that journal editors are screening for and have no tolerance for plagiarism; regarding language, either poorly written manuscripts are rejected in-house, or peer reviewers maintain some tolerance.

Table 8.3 Cumulative ratings of reasons for peer-review rejection

Reason (n = 48)	CR	%
Major methodological flaws	66	24.18
Lacking originality	57	20.88
Inadequate data presentation or interpretation	40	14.65
Unsuitable topic	35	12.82
Poor or insufficient data	30	10.99
Plagiarism	17	6.23
Needs additional research	13	4.76
Does not meet format and style requirements	9	3.30
Poor language use	6	2.20

Note: CR = cumulative ratings from 48 editorial committee members.

Table 8.4 Top three of three most common reasons for peer review rejection

Variable (n = 48)	Frequency	%
Rank 1: First most common		
Lacking originality	13	27.08
Major methodological flaws	12	25.00
Unsuitable topic	11	22.92
Rank 2: Second most common		
Major methodological flaws	13	27.08
Inadequate data presentation or interpretation	11	22.92
Poor or insufficient data	7	14.58
Rank 3: Third most common		
Poor or insufficient data	13	27.08
Inadequate data presentation or interpretation	6	12.50
Lacking originality	6	12.50

Perceptions of authors' performance

Besides ranking reasons for rejection, respondents were asked about their perceptions of authors' performance during peer review. Data showed that of 43 respondents, over 95% ($n = 41$) believe authors respond appropriately to reviewer comments, though only two strongly agreed. Even so, they believe authors face challenges during the peer review process. Responding to open-ended questions, 40 noted various shortcomings in authors' revised manuscripts and revision reports. The greatest number of responses (22 of 60) related to incomplete revision reports, especially authors not addressing reviewers' comments fully or directly, while 16 responses noted that authors may incorrectly respond to reviewer questions because they do not understand the feedback. This can lead to authors arguing or being defensive while going off topic. Regarding manuscript revisions, ten responses indicated problems when authors are unable to fulfill reviewers' requirements due to a lack of knowledge or data, leading to inappropriate or incomplete manuscript revisions. Regarding manuscripts, 12 responses expressed problems with organization, format, and language, with six specifically mentioning English language issues. Otherwise, ten noted that authors' revisions can be inappropriate or incomplete.

It seems reasonable that incomplete revisions and reports could lead to rejection based on inadequate data presentation or interpretation and/or poor or insufficient data. It is fairly obvious that manuscript issues would be related to not meeting format and style requirements; furthermore, language is a serious issue for six respondents. On the other hand, while EPR is frustrating, it can be a learning process (Sheridan, 2018), and those who persevere are likely to eventually succeed (Braine, 2003; Belcher, 2007).

What authors learn

Respondents agreed that EPR implicitly teaches authors how to avoid rejection, especially how to meet format and style requirements (Table 8.5). However, based on the reasons manuscripts are rejected in-house and through peer review just discussed, this is one of the most uncommon reasons that manuscripts are rejected. Actually, when combining cumulative totals of both phases, the most common reasons for manuscript rejection are 'lacking originality', 'major methodological flaws', and 'unsuitable topic', respectively. Fortunately, over half of respondents believe that authors can learn how to improve research methodology. Unfortunately, the fewest

Table 8.5 Skills that the manuscript submission and peer review process implicitly teach authors

Variable	n	Agree[a]		Disagree[b]	
		Frequency	%	Frequency	%
Meet format and style requirements	41	28	68.29	2	4.88
Interpret data	40	25	62.50	0	0.00
Present data	40	23	57.50	0	0.00
Avoid methodological flaws	41	23	56.10	0	0.00
Improve English writing ability	40	21	52.50	1	2.50
Avoid plagiarism	40	19	47.50	1	2.50
Choose a suitable journal for the research topic	41	18	43.90	4	9.76
Find an original research topic	41	15	36.59	3	7.32

Note: [a]Somewhat agree, agree, strongly agree; [b]somewhat disagree, disagree, strongly disagree.

respondents (36.6%) believe the process can implicitly teach how to find an original topic, while 7.3% even disagreed that it can. Likewise, choosing a suitable journal is second to the bottom in Table 8.5, meaning that about 44% agree EPR can implicitly teach authors this skill, although about 10% disagreed. On the other hand, it appears respondents believe that authors can learn to improve data presentation and interpretation and English writing, which are aspects of research articles that can be problematic. Plagiarism constitutes a problem for editors; unfortunately, learning to avoid it is not high on the list.

To complement the quantitative data, respondents were asked to list one to five benefits of EPR they have observed for authors. The 38 respondents, who recognized the potential educational value, offered 33 valid responses. These data were sorted into three categories: quality, perspectives, and feedback. Regarding quality, the respondents expressed confidence that authors have the opportunity to improve their research and writing if the manuscript is accepted or not. Comments categorized as perspectives revealed they believe peer review can change or expand authors' thinking through subject experts' different viewpoints. Finally, they also recognized that reviewers can teach authors through their feedback when it includes useful, critical, or constructive opinions. These responses seem to show a long-term view of the development of publishing scientists.

Supporting authors

Based on these findings, it appears that respondents believe authors can implicitly learn some aspects of scholarly publishing and scientific writing during the submission and peer review process. To explore how this occurs more explicitly, respondents were asked to list steps the editor or editorial board take during the peer review process to facilitate the publication of authors' manuscripts. The 51 valid responses from 45 editorial committee members were sorted into five categories: nothing, communication, assistance, review process, and reviewers and editors. Notably, 11 responses indicated that their journals do not do anything in particular to support authors traversing EPR. In contrast, ten in communication recognized the benefit of constructive feedback for giving advice and encouraging authors to revise manuscripts. Regarding assistance, three responses supported helping authors with English and adjusting review standards. However, the review process, with 18 responses, revealed that the most common perspective on how to facilitate publication is expediting EPR but with care by setting deadlines to keep manuscripts moving. In addition, nine comments recognized the importance of editors recruiting suitable reviewers with appropriate disciplinary knowledge.

Editors as 'custodians'

This chapter views scientific publication from the semiperiphery, specifically Taiwan. It recognizes the globalized institutional conditions in which scientists need to publish their research in English, preferably in 'international' journals, and that such conditions increase the burden of their knowledge work. It also recognizes that despite conventional 'lore' (Curry and Lillis, 2019), 'national' journals are viable outlets and that the academic journal publishing environment is very active in Taiwan.

The survey study reported here investigated EPR at BAS journals in Taiwan from the perspective of editorial committee members. In particular, it examined reasons manuscripts are rejected, author performance, and what authors learn through EPR. Though most manuscripts are sent to peer review, it was found that those with serious problems such as unsuitable topic and unoriginal research are especially likely to warrant desk rejection. However, language issues and plagiarism are also recognized as irreconcilable problems. Manuscripts that are sent to peer reviewers typically go through two rounds of single-blind review and are most likely rejected because of methodological flaws, although originality is also important to reviewers. Most editorial committee members believe authors respond appropriately to reviewer feedback; however, some critiqued manuscript revisions and reports for not fully or properly addressing feedback, possibly because they misinterpreted it or did not know how to revise or respond. According to Paltridge (2013), this is not unusual among novice academics, the likely frequent contributors to national journals (Sheridan, 2017b).

Despite these difficulties, it appears respondents believe authors *implicitly* learn skills that increase scientific publication success. Unfortunately, the most likely skills learned, such as format and style or interpreting and presenting data, are not the most critical to getting published. Overall, the respondents believe that getting papers through EPR quickly but carefully is how journals facilitate publication of authors' manuscripts. This indicates that they see the role of editors running EPR in the 'custodian' mode, shepherding and supporting worthy research articles through the review process (Starfield and Paltridge, 2019). This is important because authors have the right to a timely review with substantive feedback (McKay, 2003). However, a few respondents shared that their journals attempt to support authors with constructive feedback, language assistance, and realistic expectations. This resembles Marušić et al.'s (2004) 'author-helpful' EPR mode, albeit in an ad hoc fashion.

Starfield and Paltridge (2019, p. 264) highlighted the important 'mentoring and mediating functions' of being an editor; however, the idea of supporting authors in this way does not seem to be obvious to most BAS editorial committee members who participated in the study. Considering participants revealed that finding enough good reviewers and receiving enough quality submissions are major challenges in running EPR, perhaps adopting some practices to move from a 'gatekeeping' or 'custodian' role toward an 'author-helpful' perspective could mitigate those difficulties.

Finally, some findings in the current study appear similar to what is found in the HSS journals and in other semiperiphery countries. Unsuitable topic and methodological problems are often mentioned as reasons for rejection across disciplines. However, even though semiperiphery countries share similar pressures, local contexts influence local practices (Paltridge, 2017). The biggest difference between STEM and HSS journals published in Taiwan is that there is no Taiwanese citation index for the former. Therefore, it is likely that publishing in the local

STEM journals 'counts' for even less on institutional evaluation point systems. It is also possible that the science journal editors receive fewer submissions, which can initiate the negative spiral toward inconsistent publications and even closure (Salager-Meyer, 2008). On the other hand, based on Airiti Library, it seems Taiwanese science journals are more likely published by established academic and professional organizations, which have published journals for many years, even decades. Therefore, the motivation for producing and publishing in these journals might be different from HSS publications and should be further investigated.

Making the implicit explicit

Hanauer and Englander (2013, pp. 136–139) presented seven principles on which they based recommendations for institutions to support multilingual scholars who need to publish in English. I suggest that there are two for which the necessary structures are likely already present in the submission and review process at most journals that could be enhanced or constructed relatively easily. The first is 'Multilayered understanding of the research article.' Information explaining journal expectations regarding 'structural linguistic features on the micro and macro levels' could be explicated by editors in 'pre-peer review' (Marušić et al., 2004) or even on the journal website. The second is 'Demystification of the structures and the processes of scientific publication.' Journals are perfectly situated to help demystify their practices through clear information on websites. Most journal websites do post 'author guidelines.' However, based on participants' comments of authors' shortcomings, perhaps more can be done to make the information more accessible. In addition, editors can give talks and workshops at conferences, which have the extra benefit of also demystifying editors' personas.

Ultimately, what is the purpose of 'national' journals? There are multiple centripetal forces from centering institutions on the publishing environment in Taiwan (Sheridan, 2015) and likely any semiperipheral context. However, as Lillis (2012) maintains, 'national' journals are in a position to produce knowledge that is unique from the center, which means they can protect against 'lost science' (Gibbs, 1995). Adamson and Muller (2012, p. 255) maintain that editors' and reviewers' actions have real-world implications for the lives of the many researchers who submit their articles to the journals. Therefore, making EPR a constructive pedagogical process – making the implicit explicit – may help increase editors' satisfaction with authors' performance and raise authors' investment in the process. Ultimately, the overall quality of journals in the semiperiphery can improve while raising the profile of research beyond the Anglophone center.

Acknowledgment

This project was supported by the Ministry of Science and Technology, Republic of China (MOST 109–2410-H-004–158 -MY2).

References

Adamson, J., & Fujimoto-Adamson, N. (2016) 'Sustaining Reviewing Quality: Induction, Mentoring, and Community', *English Scholarship Beyond Borders*, vol. 2, no. 1, pp. 29–57.

Adamson, J., & Muller, T. (2012) 'Editorial investigation of roles and responsibilities in academic journal editorial systems', in *Editorial and authorial voices in EFL academic journal publishing*, Asian EFL Journal Press, pp. 123–169

Belcher, D.D. (2007) 'Seeking acceptance in an English-only research world', *Journal of Second Language Writing*, vol. 16, pp. 1–12, https://doi.org/10.1016/j.jslw.2006.12.001

Bennett, K. (2014) 'The political and economic infrastructure of academic practice: The "semiperiphery" as a category for social and linguistic analysis', in K. Bennett (ed.), *The semiperiphery of academic writing: Discourses, communities and practices*, Hampshire: Palgrave Macmillan, pp. 1–9.

Braine, G. (2003) 'Negotiating the gatekeepers: The journey of an academic article', in C.P. Casanave & S. Vandrick (eds.), *Writing for scholarly publication: Behind the scenes in language education*, Mahwah, NJ: Lawrence Erlbaum, pp. 73–90.

Canagarajah, A.S. (2003) 'A somewhat legitimate and very peripheral participation', in C.P. Casanave & S. Vandrick (eds.), *Writing for scholarly publication: Behind the scenes in language education*, Mahwah, NJ: Lawrence Erlbaum, pp. 197–210.

Curry, M.J., & Lillis, T. (2004) 'Multilingual scholars and the imperative to publish in English: Negotiating interests, demands, and rewards', *TESOL Quarterly*, vol. 38, no. 4, pp. 663–688, https://doi.org/10.2307/3588284

Curry, M.J., & Lillis, T. (2010) 'Academic research networks: Accessing resources for English-medium publishing'. *English for Specific Purposes*, vol. 29, no. 4, pp. 281–295, doi: 10.1016/j.esp.2010.06.002

Curry, M.J., & Lillis, T. (2018) 'Problematizing English as the privileged language of global academic publishing', in M.J. Curry & T. Lillis (eds.), *Global academic publishing: Policies, perspectives, and pedagogies*, Clevedon: Multilingual Matters, pp. 1–20.

Curry, M.J., & Lillis, T. (2019) 'Unpacking the Lore on Multilingual Scholars Publishing in English: A Discussion Paper', *Publications*, vol. 7, p. 27, doi: 10.3390/publications7020027

Darling, E.S. (2015) 'Use of double-blind peer review to increase author diversity', *Conservation Biology*, 29, pp. 297–299, doi:10.1111/cobi.12333

Donovan, S.K. (2011) 'Big journals, small journals, and the two peer reviews', *Journal of Scholarly Publishing*, vol. 42, no. 4, pp. 534–538, http://search.ebscohost.com/login.aspx?direct=trueanddb=a9handAN=61309749andsite=ehost-live; https://doi.org/10.3138/jsp.42.4.534

Duszak, A., & Lewkowicz, J. (2008) 'Publishing academic texts in English: A Polish perspective', *Journal of English for Academic Purposes*, vol. 7, no. 2, pp. 108–120, doi: 10.1016/j.jeap.2008.03.001

Flowerdew, J. (2019) 'The linguistic disadvantage of scholars who write in English as an additional language: Myth or reality', *Language Teaching*, vol. 52, no. 2, pp. 249–260, doi: 10.1017/S0261444819000041

Gibbs, W.W. (1995) 'Lost science in the Third World', *Scientific American, vol. 273*, no. 2, p. 92, doi: 10.1038/scientificamerican0895-92

Hanauer, D.I., & Englander, K. (2013) *Scientific writing in a second language*, Anderson, SC: Parlor Press.

Hanauer, D.I., Sheridan, C.L., & Englander, K. (2019) 'Linguistic Injustice in the Writing of Research Articles in English as a Second Language: Data from Taiwanese and Mexican Researchers', *Written Communication*, vol. 36, no. 1, pp. 136–154.

Kuzhabekova, A. (2018) 'The reaction of scholarly journals to Impact-Factor publication requirements in Kazakhstan', in M.J. Curry & T. Lillis (eds.), *Global academic publishing: Policies, practices, and pedagogies*, Multilingual Matters, Clevedon, UK, pp. 136–150

Lillis, T. (2012) 'Economies of signs in writing for academic publication: The case of English medium 'national' journals', *Journal of Advanced Composition*, vol. 32, no. 3–4, pp. 695–722

Lillis, T., Magyar, A., & Robinson-Pant, A. (2013) 'Putting 'wordface' work at the centre of academic text production: working with an international journal to develop an authors' mentoring programme, in V. Matarese (ed.) *Supporting Research Writing*. Chandos Publishing, Oxford, pp. 237–255, doi: 10.1016/B978-1-84334-666-1.50015-1

Lundin, R.A., Jönsson, S., Kreiner, K., & Tienari, J. (2010) 'The changing face of academic publishing: On the past, present and future of the Scandinavian Journal of Management', *Scandinavian Journal of Management*, vol. 26, no. 3, pp. 309–317, www.sciencedirect.com/science/article/pii/S0956522110000576 (Accessed 22 August 2021)

Marušić, M., & Marušić, A. (2014) 'The Croatian Medical Journal: Success and consequences', in K. Bennett (ed.) *The semiperiphery of academic writing: discourses, communities and practices*. Palgrave Macmillan, Hampshire, UK, pp. 210–220.

Marušić, M., Misak, A., Kljakovic-Gaspic, M., Fister, K., Hren, D., & Marušić, A. (2004) 'Producing a scientific journal in a small scientific community: An author-helpful policy', *International Microbiology*, vol. 7, pp. 143–147

McKay, S.L. (2003) 'Reflections on being a gatekeeper', in C.P. Casanave & S. Vandrick (eds.), *Writing for scholarly publication: Behind the scenes in language education*, Mahwah, NJ: Lawrence Erlbaum, pp. 91–102.

Miles, M.B., Huberman, A.M., & Saldana, J. (2014) *Qualitative data analysis: A methods sourcebook*, Thousand Oaks, CA: Sage, 3rd edition.

Mongeon, P., & Paul-Hus, A. (2016) 'The journal coverage of Web of Science and Scopus: a comparative analysis', *Scientometrics*, vol. 106, no. 1, pp. 213–228, https://doi.org/10.1007/s11192-015-1765-5

Nature (2020)'2020 Tables: Countries/territories', *Nature Index 2020 Annual Tables*, vol. 580, no. 7805, www.natureindex.com/annual-tables/2020/country/all (Accessed 18 May 2021)

Paltridge, B. (2013) 'Referees' comments on submissions to peer-reviewed journals: when is a suggestion not a suggestion?', *Studies in Higher Education*, vol. 40 no. 1, pp. 106–122, doi: 10.1080/03075079.2013.818641

Paltridge, B. (2017) *Discourse of Peer Review: Reviewing Submissions to Academic Journals*, Palgrave Macmillan UK

Salager-Meyer, F. (2008) 'Scientific publishing in developing countries: Challenges for the future', *Journal of English for Academic Purposes*, vol. 7, no. 2, pp. 121–132, DOI: 10.1016/j.jeap.2008.03.00

Salager-Meyer, F. (2015) 'Peripheral scholarly journals: From locality to globality', *Ibérica*, vol. 30, pp. 15–36

Sheridan, C.L. (2015) 'National journals and centering institutions: A historiography of an English language teaching journal in Taiwan', *English for Specific Purposes*, vol. 38, pp. 70–84, doi: 10.1016/j.esp.2014.12.001

Sheridan, C.L. (2017a) 'The burden of writing research articles in a second language: Data from Taiwanese scholars', (MOST 105-2410-H-004-160), Republic of China Ministry of Science and Technology, Taipei, Taiwan, www.grb.gov.tw/search/planDetail?id=11898308 (Accessed 1 November 2020)

Sheridan, C.L. (2017b) *English Medium 'National' Journals Beyond the Anglophone Center: A Qualitative Study of Multilingual Scholars and their Publishing Decisions in Taiwan'*, PhD dissertation, ProQuest Indiana University of Pennsylvania, Indiana, PA

Sheridan, C.L. (2018) 'Blind peer review at an English language teaching journal in Taiwan: Glocalized practices within globalization of higher education', in M.J. Curry and T. Lillis (eds.), *Global academic publishing: Policies, practices, and pedagogies*, Multilingual Matters, Clevedon, UK, pp. 136–150.

Sheridan, C.L. (2020) 'Knowledge Production and Participation: Peer Review at National Journals as an Educational Intervention to Enhance Multilingual Scholars' Publishing Expertise and Promote Taiwanese Research', (MOST- 107–2410-H-004–117-MY2), Republic of China Ministry of Science and Technology, Taipei, Taiwan, www.grb.gov.tw/search/planDetail?id=12664276 (Accessed 1 November 2021)

Song, M-M., & Tai, H-H. (2007) 'Taiwan's responses to globalisation: Internationalisation and questing for world class universities', *Asia Pacific Journal of Education*, vol. 27, no. 3, pp. 323–340, doi: 10.1080/02188790701594067

Starfield, S., & Paltridge, B. (2019) 'Journal Editors: Gatekeepers or custodians', in P. Habibie and K. Hyland (eds.) *Novice Writers and Scholarly Publication: Authors, mentors, gatekeepers*, Palgrave Macmillan, Cham, Switzerland, pp. 253–270.

Thrower, P. (2012) ' "Eight reasons I rejected your article": A journal editor reveals the top reasons so many manuscripts don't make it to the peer review process', *Elsevier Connect*, 12 September.

Wallerstein, I. (1991) *Geopolitics and geoculture: Essays on the changing world-system*, Cambridge: Cambridge University Press.

9

WHY I SHOULD CITE YOU? THE EVOLVING ROLE OF DOCUMENTATION AND CITATION IN SCHOLARLY COMMUNICATION

Tomás Koch Ewertz

Introduction: a theoretical overview of citation practices

Have you ever tried asking Google why you should cite references? If you have, your search results would have led you to many websites with a variety of arguments for doing so. Among the most common reasons are avoiding plagiarism (and the sanctions associated with this kind of misconduct), acknowledging the work of others, demonstrating thoroughness in bibliographic research, and enabling other researchers to trace sources. All of these are good reasons, although they seem to point in at least two different directions. While the first two – avoiding plagiarism and acknowledging the work of others – are closely linked and are aimed at correctly recognizing others' work, the last two are intended to influence readers either by persuading them of the thoroughness of the work carried out or by leading them to other texts.

Several researchers have studied the variety of author motivations for including references in scholarly papers, as well as the functions of these references in the texts (e.g. Erikson and Eridson, 2014; Tahamtan and Bornmann, 2018; Hassan and Serenko, 2019; Case and Higgins, 2000). These studies have found evidence of the complex motivations for using citations, ranging from recognizing the influence of other people's ideas, methods, or data and framing the contribution within a certain field to reinforcing this by citing authoritative scholars and/ or papers. In other words, each time a work is referenced, a particular meaning and intention are attached to the citation, referring either to the paper's content or to the positioning of the journal, institution, or author cited within the prestige structure of science.

The differences in the importance attributed to these two kinds of motivation for using citations – the paper's ideas or its positioning as an authoritative source – have led to two different representations of citations, with this practice being understood either as an instrument for recognition or as a tool for persuasion (see Nicolaisen, 2007; Davis, 2009). While the first approach has its roots in the normative model of science developed since the works of Robert K. Merton (1973) onward, understanding citations as tool for persuasion is based on the socioconstructivist representation of scientific endeavor, with the early work of Nigel Gilbert

DOI: 10.4324/9781003043782-11

(1977) usually recognized as being seminal. Given the centrality of the nature of citations to this chapter's aim, we will now briefly outline these two approaches.

Kaplan's 1965 paper is usually said to have been seminal to identifying citations with science's reward system. This work – as well as subsequent ones using similar approaches (e.g. Merton, 1973; Cole and Cole, 1973) – conceived citations as a formal way of recognizing the scholarly influence of ideas. In other words, they all conceive citations as a mechanism for recognizing the 'intellectual debt' to previous generations of scientists, echoing the twelfth-century expression famously quoted by Newton (1675): 'if I have seen further, it is by standing on the shoulders of giants.' This concept of citations is embedded in a normative representation of science, which stresses the relevance of precepts or values to binding together scientists and orienting their actions.

The influential set of scientific norms identified by Merton in 1942 – communism, universalism, disinterestedness, and organized skepticism (CUDOS[1]) – provides a shared basis for this approach to citations. In this scheme of things, citations represent a sort of coinage that 'repays' the debt with the author by recognizing that they have priority. This recognition is especially important given the communist norm of science that stipulates that published ideas are common property. Moreover, this representation of citations as a way of 'linking ideas' is also derived from the norm of universalism, which secures the relevance of ideas irrespective of their author's characteristics. In this way, citations provide the foundation for the prestige structuring of science based on ideas instead of authors' traits. This conception can be seen as being behind some of the reasons for citing references found in Google searches, such as acknowledging the work of others or avoiding sanctions for contravening this normative imperative (plagiarism).

From a different standpoint, some scholars have argued that the main purpose of citations is not to recognize influence, but to persuade readers of the contribution's worth and to shield it from criticism. This representation of citations rests upon a description of science that emphasizes the actor's practices instead of normative precepts, mostly based on the variety of ways in which the cited material is used, and especially on the so-called 'perfunctory citation' (see Gilbert, 1977; Latour, 1987). These authors question the idea that citations aim to fulfill a normative imperative, indicating that they are primarily a strategy for authors to make an argument more convincing to their readers.

According to this view, citing authoritative papers (that is, ones recognized by a particular scholarly community as valid) circumscribes the paper into a specific community or tradition (Hyland, 2004), while attracting the interest of readers and allowing the paper 'to shine in their [the cited papers'] reflected glory even if they do not seem closely related to the substantive content of the report' (Gilbert, 1977, p. 116). In this light, citations are not an immediate link between the content of publications (ideas) as assumed by the recognition model but are a tool that refers mostly to scientists' rhetoric-strategies, which take into account the authors' position within scientific structuring. Some of the reasons for citations just listed – such as demonstrating the thoroughness of bibliographic research or enabling other researchers to trace sources – seem to fit into this concept better.

Several attempts have been made to empirically test these citation theories (see Tahamtan and Bornmann, 2018; Aksnes et al., 2019). The results, however, are not conclusive. In fact, it seems that both citation functions – rewarding credit and persuasion – are entwined in authors' actual practices (Cozzens, 1989). The following example illustrates this. Imagine a standard paper's introduction following the CARS (Creating a Research Space) model template, as proposed by Swales (1990, 2004). Let us consider the references in the two first moves: establishing a territory and a niche. In the first move, authors typically reference previous papers that provide both the contribution's background and establish the field the paper aims to contribute to.

In the second move, references are commonly used to highlight the gap in knowledge, counter-claiming or continuing a tradition. In any of these uses of citations, do they mainly acknowledge the work of previous authors, or are they merely argumentative devices? Do they exclusively refer to the paper's content or also to social recognition within the field? Or, in more general terms, are they normatively or pragmatically oriented?

In all likelihood, no one can provide a clear answer to any of these questions, even for references in one's own papers. Moreover, the average of references per paper tends to increase over time (Sanchez-Gil et al., 2018). As there are often no limits on the number of references to be included, many different motives can easily exist side by side. In this light, rather than looking for a clear demarcation between these two orientations (or proposing a third one), it seems to be more useful to admit citations' apparent entwined motivations and look at how they actually function in the current world of science (e.g. Wouters, 1999). Small (1978) noted that using references has the symbolic function of indicating both ideas and texts. While the link between ideas depends on the different uses of the cited document (e.g. to reject, comment on, or interpret it), the link between texts is expressed in the references themselves. These links create a network of papers connected through citations, currently illustrated by the variety of metrics, maps, and hyperlinks that digital technologies permit. International scholarly databases and indexes (e.g. the Web of Science or SCOPUS) are especially important for (shaping and) depicting these networks, with their role in scholarship going beyond that of a mere technical instrument. In fact, despite the systematic use of references in scholarly texts since the nineteenth century, the first theorizing efforts on citations came in the 1960s, after the launch of the Science Citation Index (SCI) in 1961, a scholarly index that revolutionized scientific communication by bringing the links between papers (citations) to the fore. In the following section, our attention will turn to some of the effects that this and other broad transformations within the world of science have had on referencing practices, looking at the relevant changes in authorship and readership over the last couple of centuries.

Referencing from a historical perspective. Changes in scholarly authorship and readership

One of the main characteristics of academic knowledge is that it requires that new claims should be put into context with previous ones. In Vanderstraeten's words, 'Each publication is expected to interact with preceding ones, by incorporating into its own line of reasoning arguments developed in other publications; and each new publication, due to the claims it makes to new knowledge, invites reactions and hence further publications' (Vanderstraeten, 2010, p. 561). In this sense, referencing has a history that can be traced back several centuries (see Grafton, 1997). Modern referencing as we know it now, however, is a much more recent phenomenon. As Edelstein et al. (2013) noted, as late as the eighteenth century, referencing was not necessarily a shared practice, among other reasons, due to the still present habit of publishing anonymous texts. In fact, the evolving nature of referencing – as Grafton (1997) noted – makes it hard and rather arbitrary to establish a date for the rise in the importance of modern referencing. However, it can be safely noted that this practice has evolved in parallel with important transformations in scholarly communication, such as the increasing importance of the scholarly journal and the consolidation of modern authorship (see Foucault, 1979; Csiszar, 2018).

These two transformations impacted scholarly communication on several levels. On the one hand, the rise of the scholarly journal as the 'ultimate' form of scientific communication changed the way scientists communicated their findings, going from publishing comprehensive works with detailed explanations of their implications – characteristic of traditional book publishing – to a rapid succession of short and focused communications typical of a paper format. The impact that this transformation had on scholarship was far from trivial and, unsurprisingly, encountered strong resistance (see Csiszar, 2018). Charles Darwin's decision to publish his influential thesis on biological evolution in the traditional book format instead of in the paper format is a good example of the competing publishing formats in the nineteenth century. While this decision was criticized within part of the scholarly community, the new publishing speed imposed by the paper format even made him risk this work's primacy (see Merton, 1973, pp. 306–307). However, this change in format not only affected authors but also influenced reader expectations and reading practices. As opposed to books – which usually focus on a main topic, claim, or perspective – each issue of a scholarly journal is expected to present a picture of the state-of-the-art of an entire field of inquiry, with readers not knowing in advance what they are going to encounter (Vanderstraeten, 2010).

On the other hand, as Foucault (1979) remarked, the rise of the modern author has had far-reaching consequences beyond the mere register of who wrote the text and also attributes authorship to discourses. In this light, the claim of some citation theorists that referencing does not necessarily point to the content of the text but to a school of thought, tradition, or the author's reputation becomes particularly relevant. Authorship is traditionally associated with three functions (Birnholtz, 2006):

1   *The attribution of credit over the primacy of discoveries*: This is particularly relevant in the context of the rapid succession of publications in journal format. The current relevance of this function can be seen in practices such as publishing the date articles were received or including the year published in all available referencing styles.
2   *Ownership*: The previously mentioned communist ethos of science – which states the common ownership of the ideas published – gives this a specific interpretation, responsible for the data published and claims put forward. This function is particularly important in light of the always present possibility of data fabrication or other forms of scientific fraud.
3   *Reputation*: This refers to authorship as the backbone of science's reward system and its prestige structuring, as the cornerstone for representing citations as a reward mechanism. This structuring function has become central for many scholars worldwide, especially after several universities and funding agencies worldwide adopted publications (and citations) as an accountability mechanism.

During the eighteenth and nineteenth centuries, disciplinary communities emerged around the world, with several national networks starting to publish journals and papers in their own languages instead of Latin, science's international lingua franca. Despite the diversifying effect this trend had on networks (and thus on referencing) during the first half of the twentieth century, journals started to standardize both referencing and authors' identification on the basis of recently created style manuals (e.g. *Chicago Manual of Style* in 1906 or *APA Manual Style* in 1929). Clearly identifying authorship became mandatory across disciplines, while rigorous documentation – reflected in proper referencing – was identified as a hallmark of good science (e.g. Koch et al., 2021). Standardization gained momentum after World War II, with the United States' renewed influence facilitating both English becoming the new international language for science and the worldwide influence of a new type of scholarly indexes developed largely within this country (de Swaan, 2001; Archambault and Larivière, 2009).

In 1955, Eugene Garfield – who later went on to become the founder of the Institute for Scientific Information (ISI) – wrote an article in *Science* proposing the novel idea of a multidisciplinary scholarly index based on citations instead of authors, themes, or disciplines. This project came about in 1961 with the launch of the Science Citation Index (SCI). From then on, a series of international, regional, and national scholarly databases and indexes have followed in its footsteps (e.g. SCOPUS, Redalyc, SciELO, the Russian Citation Index, and the African Citation Index). These databases constitute an ecology of indexes (Koch et al., 2021) that have gained salience in the day-to-day practices of several authors, readers, and publishers around the world. While the powerful image of science that these instruments provide has been widely adopted as a shared measure for scholarly outputs, the technical possibilities they have enabled actually came to revolutionize scholarly communication at the end of the twentieth century.

These databases and indexes helped change collaboration patterns, bibliographic research, and publishing practices, either by reinforcing the transformations triggered by the rise of so-called big science (see Price, 1963) or by improving the visibility of and accessibility to (certain) scholarship on the basis of the new possibilities provided by digital technologies. In other words, they contributed to the transformation of traditional representations of both authorship and readership. As far as authorship is concerned, coauthorship trends have increased across disciplines since the 1960s (O'Brien, 2012) and, for some disciplines, multi- or hyperauthorship has become the norm, with papers authored by more than 1,000 people regularly appearing in journals (see Birnholtz, 2006). While these practices question the traditional functions attributed to authorship, transformations in bibliographic research – from reading issues of a journal to searching for papers in digital databases and repositories – have transformed traditional readerships (see Koch and Vanderstraeten, 2021). The network of hyperlinks based on citations that several scholarly databases now provide has allowed researchers to easily surf through related papers by following the links created by citations (either to the cited or citing papers). Moreover, these changes currently challenge the traditional role of journal publishing, in which the traditional 'issue' has lost its relevance in favor of the increasingly common practice of publishing articles directly in journals' websites, with the DOI (digital object identifier) replacing the old coordinates used to locate an article (volume, issue, and page numbers).

The powerful representation of scholarly networks that these databases provide has renewed citation's importance. Furthermore, in our current information-saturated world, databases and indexes have gained ground as mechanisms for certifying the quality and reliability of sources, thus increasingly occupying a position as guardians of academic rigor. In spite of these recognized roles for academic development, other aspects of these databases – such as their coverage bias or performative effects on scholarship – have also received intense criticism. Next, we will reflect on some of the possibilities, challenges, and threats emerging from this new and powerful representation of the world of science, based on citation networks.

The coming of the indexes era: the possibilities and pitfalls of this representation of science

The first multidisciplinary index based on citations (SCI) only covered natural science disciplines. However, in the following decades, the Institute for Scientific Investigation (ISI) launched two new instruments to index social sciences and the humanities: the Social Sciences Citation Index (SSCI) and the Art and Humanities Citation Index (A&HCI). Despite their influence, these instruments were also widely criticized over the next few decades for their coverage bias toward Anglo-Saxon journals (see Archambault et al., 2006). Reflecting the increasing importance of these tools – and as a way to counteract these coverage biases – several regional indexes were launched from the end of the century onward, for example the Latin American SciELO

(Scientific Electronic Library Online) in 1998 (see Packer, 2009). While new regional databases and indexes continued to emerge during the first decade of the twenty-first century, another index with global aspirations (viz SCOPUS) also came into being in 2004.[2] These databases and indexes came to constitute an ecology of indexes occupying a central position in the world of scientific communication and currently accomplish functions that go far beyond the intentions of the indexes' founders (Cronin, 2013).

According to Garfield (1998), the main motivation for creating the Science Citation index was to aid bibliographic research, given the enormous number of published texts. This makes sense today more than ever. Along with the huge amount of nonspecialized information available on the Internet, the number of scholarly journals has rocketed since the 1950s, with conservative estimations calculating that more than 14,000 outlets were regularly published by the beginning of the 2000s (Mabe, 2003), a trend that is likely to have increased during the first few decades of this century. In this context, it makes sense to have a mechanism to discriminate between sources when doing bibliographic research. Scholarly databases have proved very important for this purpose.

The differentiation of science as an autonomous sphere rests largely on a series of mechanisms that allow for its inclusion and exclusion criteria to be delimited. Along with the diversity of specific methodological and theoretical disciplinary standards, certain shared practices gained ground as a hallmark of academic work over the twentieth century. Writing and referencing styles, ethical procedures, and mandatory peer reviews now constitute shared mechanisms to assess the suitability of contributions for scholarly publishing (see Bazerman, 1988; Baldwin, 2015). The incorporation of these elements as criteria for including journals within scholarly databases and indexes – as well as further periodic overseeing – have helped to guarantee these standards and then to constitute these instruments as a valuable source of documentation. Furthermore, the set of statistical usage measures these databases usually provide – such as numbers or citations, readers or mentions in social media – also help assess the interest the articles generate.

Scholarly citation is likely the most important of these measures nowadays. While this indicator provides us with a measure of a paper's 'impact', the possibility the databases give us to 'follow' citations through a network of hyperlinks makes it possible not only to find the references cited by the paper (i.e. the prior literature it claims to be connected to) but also to look at other papers claiming to be connected with the paper (in any of the ways already described in the first section of this chapter). This possibility brings the relevance of proper citation to the fore, with what Garfield (1998) called 'citation consciousness' (i.e. a consciousness of the crucial scholarly practice of explicitly referencing the works relevant to the topic) gaining importance.

Besides the interesting possibilities for bibliographic research, the emphasis of these databases and indexes on citations also allows for the use of indicators whose calculations are based on them (scientometrics) as a proxy for the research's relevance (see Aksnes et al., 2019). Although the lack of a general theory of citation haunted these attempts for several decades – and according to some critics is still a major issue – most of the studies in the field seem to conceive citations as a recognition of influence or simply and pragmatically choose to focus on the connections they actually describe rather than the authors' motivations for using references (e.g. Wouters, 1999). In any case, the number of citations a paper receives is associated with its utility or impact, thus providing a measure for the performance of papers, journals, institutions, or scholars.

Among the measures currently available, perhaps the two most well-known are the Journal Impact Factor, which establishes a journal's importance based on the ratio between published papers and the number of citations received in a period of time (usually two or five years), as well as the h-index that establishes a relationship between the number of publications and the number of citations at an institutional or individual level. These and other measures have helped to depict some important characteristics of science, such as the existence of a research front (Price, 1965) or the core–periphery structure that scholarly systems seem to produce (e.g. Hwang, 2008). However, both the most recognized and controversial employment of these instruments comes from their use as evaluation mechanisms, especially in the context of so-called New Public Management policies (Olssen, 2015). These measures have been widely used by policy makers and science administrators as key evaluation criteria for scholars, and they currently play a major role in universities' international rankings.

Using scientometric indicators as an accountability mechanism for scholars or their use by funding agencies for evaluation purposes is still a very controversial topic. Their defenders have argued that their proper use could provide a shared mechanism for academic accountability while at the same time allowing evaluators to avoid biases such as those produced by the existence of 'old boys' networks (e.g. Alstete et al., 2018). Critics, however, have denounced the reductionist and homogenizing traits of these measures, which are not suitable for depicting the variety of practices and outputs that academia involves. Controversy aside though, most of their defenders and critics seem to agree – for different reasons – on rejecting the naïve use of indicators based on citations as a measure of the quality of scholarship and, in fact, call for them to be used responsibly (e.g. Hicks et al., 2015).

From the standpoint of the scholarly practice of referencing, it is important to note that these instruments have so-called performative effects (Callon, 2010). Their acritical use has the latent capacity of reinforcing underlying unequal structures within the world of science (e.g. due to databases selection biases), influencing the journal's editorial decisions, affecting the research topics and approaches used (Wang et al., 2017), or even encouraging malpractices such as 'salami publications', the disproportionate use of self-citations (Seeber et al., 2019), or the emergence of 'citation cartels' (Fister et al., 2016). Most of these effects are closely connected with the citation consciousness described by Garfield, which seems to be in a process of transformation, influenced by academic databases and indexes. In this sense, a new citation consciousness – highly orientated by these new tools and the indicators calculated as a result – seems to be emerging (Koch and Vanderstraeten, 2019).

Conclusion

Covering all the aspects involved in the scholarly practices of documentation and citation was beyond the aim and possibilities of this chapter. On the contrary, the brief reflections presented here on the evolving importance of these elements aim to provide both background for those initiating on scholarly writing and a guide to some relevant topics and literature for those interested in looking further into these themes. As (I hope) can be gathered from this text, the referencing practice has gained importance in academic writing and is currently a hallmark of this genre. Moreover, the importance that citations have gained for bibliographic research – as well as their current role in the assessment of journals, scholars, and institutions worldwide – reveals that this practice has far more profound implications than initially thought.

As illustrated, there are several motives for referencing texts. All of them, however, have the effect of linking papers to previous literature, to some extent 'addressing' the contribution by establishing the neighborhood of papers to which the text is (or claims to be) connected. This effect is especially important in today's information-saturated world. Citations help readers put the contribution into context (one of the main issues for other kinds of sources, such as

Wikipedia), while also allowing citing authors to play their part in the hierarchizing of authors, ideas, and papers.

The current omnipresence of international databases and indexes and of scientometric indicators in academic communication have both a positive and a negative side. While they help us to identify reliable sources and to navigate through citation networks, their use also tends to marginalize any relevant sources not included in these databases. This effect is especially important to social sciences and the humanities and/or to themes with mostly local relevance. Moreover, their use as an evaluation mechanism tends to orient scholars toward a certain kind of output – internationally oriented papers published in (hopefully top-ranked) indexed journals. This effect could be harmful to academic communities, especially those located in the so-called (semi)periphery of the world of science. When scholars and journals are (mostly) evaluated by their contributions and by citations from within the global networks described in these databases and indexes, they are compelled to resign or diminish their engagement beyond academia and/or to research themes with (mostly) local relevance.

All things considered, proper documentation and citation is a crucial scholarly practice, with citation consciousness gaining a renewed importance within scholarly ethos.[3] Although it is debatable whether citations can represent the whole scientific endeavor – as some policy makers and science administrators seem to believe – it is certainly a very important part of it, possibly today more than ever. Understanding the variety of implications of using citations is nowadays not only important for scholars researching these themes but also for all those who want to participate in the academic communication system. In the current scientometrics-saturated academic world, proper citation consciousness should involve a commitment to thorough documentation that goes beyond top-ranked papers and sources, while also adequately 'addressing' our paper within the citation networks by properly referencing all the sources relevant to the paper.

Notes

1 Communism (or communalism) stands for the mandatory public dissemination of produced knowledge; universalism for the impersonal criteria used to evaluate claims; disinterestedness for the exclusion of personal interests from proper procedures and organized skepticism for allowing and encouraging criticism (Merton, 1973, pp. 270–278).
2 Probably influenced by the emergence of new competitors, the Web of Science databases (SCI, SSCI, and A&HCI) also introduced editorial changes during this decade to increase their global coverage (see Collazo-Reyes, 2014) and included several regional databases (SciELO, the Russian Citation Index, and the Korean Citation Index) as part of the Web of Science collection.
3 Besides the important consequences highlighted, it is important to remark that the new network of papers and the improved visibility of highly cited papers and authors risk also reinforcing false data dissemination through inadequate citation. When citation becomes an important rhetoric mechanism, and citation information is available, the temptation of referencing highly cited papers is high (sometimes without even reading them), even though it can help to reproduce highly cited errors (see Rekdal, 2014).

References

Aksnes, D., Langfeldt, L., & Wouters, P. (2019) 'Citations, citation indicators, and research quality: An overview of basic concepts and theories', *Sage Open*, vol. 9, no. 1, pp. 1–17, https://doi.org/10.1177/2158244019829575.
Alstete, J., Beutell, N., & Meyer, J. (2018) *Evaluating scholarship and research impact (great debates in higher education)*, Bingley: Emerald Publishing Limited.

Archambault, É., & Larivière, V. (2009) 'History of the journal impact factor: Contingencies and consequences', *Scientometrics*, no. 79, pp. 635–649, https://doi.org/10.1007/s11192-007-2036-x

Archambault, É., Vignola-Gagné, É., Côté, G., Larivière, V., & Gingras, Y. (2006) 'Benchmarking scientific output in the social sciences and humanities: The limits of existing databases', *Scientometrics*, vol. 68, no. 3, pp. 329–342, https://doi.org/10.1007/s11192-006-0115-z

Baldwin, M. (2015) *Making "Nature": The History of a Scientific Journal*. University of Chicago Press, Chicago

Bazerman, C. (1988) *Shaping written knowledge: The genre and activity of the experimental article in science*, Madison, WI: University of Wisconsin Press.

Birnholtz, J. (2006) 'What Does It Mean to Be an Author? The Intersection of Credit, Contribution, and Collaboration in Science', *Journal of the American Society for Information Science and Technology*, vol. 57, no. 13, pp. 1758–1770, https://doi.org/10.1002/asi.20380

Callon, M. (2010) 'Performativity, misfires and politics', *Journal of Cultural Economy*, vol. 3, no. 2, pp. 163–169, https://doi.org/10.1080/17530350.2010.494119

Case, D., & Higgins, G. (2000) 'How can we investigate citation behavior? A study of reasons for citing literature in communication', *Journal of the American Society for Information Science*, vol. 51, no. 7, pp. 635–645, https://doi.org/10.1002/(SICI)1097-4571(2000)51:7<635::AID-ASI6>3.0.CO;2-H

Cole, J., & Cole, S. (1973) *Social stratification in science*, Chicago: University of Chicago Press.

Collazo-Reyes, F. (2014) 'Growth of the number of indexed journals of Latin America and the Caribbean: The effect on the impact of each country', *Scientometrics*, vol. 98, no. 1, pp 197–209.

Cozzens, S. (1989) 'What do citations count? The rhetoric-first model', *Scientometrics*, vol. 15, nos. 5–6, pp. 437–447, https://doi.org/10.1007/BF02017064

Cronin, B. (2013) 'From signtometrics to scientometrics: A cautionary tale of our times', *Journal of Information Science Theory and Practice*, vol. 1, no. 4, pp. 6–11, https://doi.org/10.1633/JISTaP.2013.1.4.1

Csiszar, A. (2018) *The Scientific Journal. Authorship and the politics of knowledge in the nineteenth century*, University of Chicago Press, Chicago

Davis, P. (2009) 'Reward or persuasion? The battle to define the meaning of a citation', *Learned Publishing*, no. 21, pp. 5–11, doi:10.1087/095315108X378712

De Swaan, A. (2001) *Words of the world*, Cambridge: Polity Press.

Fister, I., Fister, I., & Perc, M. (2016) 'Toward the Discovery of Citation Cartels in Citation Networks', *Frontiers in Physics*, vol. 4, no. 1, pp. 4–49, https://doi.org/10.3389/fphy.2016.00049

Foucault, M. (1979) 'Authorship: What is an author?', *Screen*, vol. 20, no. 1, pp. 13–34, https://doi.org/10.1093/screen/20.1.13

Garfield, E. (1955) 'Citation indexes for science', *Science*, vol. 122, no. 3159, pp. 108–111, DOI: 10.1126/science.122.3159.108

Garfield, E. (1998) 'From citation indexes to informetrics: Is the tail now wagging the dog?', *Libri*, vol. 48, no. 2, 67–80, https://doi.org/10.1515/libr.1998.48.2.67

Gilbert, N. (1977) 'Referencing as persuasion', *Social Studies of Science*, vol. 7, no. 1, pp. 113–122, www.jstor.org/stable/284636

Grafton, A. (1997) *The footnote: A curious history*, Cambridge: Harvard University Press.

Edelstein, D., Morrissey, R., & Roe, G. (2013) 'To quote or not to quote: Citation strategies in the "Encyclopédie"', *Journal of the History of Ideas*, vol. 74, no. 2, pp. 213–236

Erikson, M., & Eridson, P. (2014) 'A taxonomy of motives to cite', *Social Studies of Science*, vol. 44, no. 4, pp. 625–637, https://doi.org/10.1177/0306312714522871

Hassan, N., & Serenko, A. (2019) 'Patterns of citations for the growth of knowledge: a Foucauldian perspective', *Journal of Documentation*, vol. 75, no. 3, pp. 593–611, 10.1108/JD-08-2018-0125

Hicks, D., Wouters, P., Waltman, L. de Rijcke, S., & Rafols, I. (2015) 'Bibliometrics: The Leiden Manifesto for research metrics', *Nature*, no. 520, pp. 429–431, https://doi.org/10.1038/520429a

Hwang, K. (2008) 'International collaboration in multi-layered center-periphery in the globalization of science and technology', *Science, Technology & Human Values*, vol. 33, no. 1, pp. 101–133, https://doi.org/10.1177/0162243907306196

Hyland, K. (2004) *Disciplinary discourses: Social interactions in academic writing*, Ann Arbor, MI: University of Michigan Press.

Kaplan, N. (1965) 'The norms of citation behaviour: Prolegomena to the footnote', *American Documentation*, vol. 16, no. 3, pp. 179–184, https://doi.org/10.1002/asi.5090160305

Koch, T., & Vanderstraeten, R. (2019) 'Internationalizing a national scientific community? Changes in publication and citation practices in Chile, 1976–2015', *Current Sociology*, vol. 67, no. 5, pp. 723–741, https://doi.org/10.1177/0011392118807514

Koch, T., & Vanderstraeten, R. (2021) 'Journal editors and journal indexes: Internationalization pressures in the semi-periphery of the world of science', *Learned Publishing*, https://doi.org/10.1002/leap.1390.

Koch, T., Vanderstraeten, R., & Ayala, R. (2021) 'Making science international: Chilean journals and communities in the world of science', *Social Studies of Science*, vol. 51, no. 1, pp. 121–138, https://doi.org/10.1177/0306312720949709

Latour, B. (1987) *Science in action*, Cambridge: Harvard University Press.

Mabe, M. (2003) 'The growth and number of journals', *Serials*, vol. 16, no. 2, pp. 191–197, https://doi.org/10.1629/16191

Merton, R. (1973) *The sociology of science: Theoretical and empirical investigations*, Chicago: Chicago University Press.

Newton, I. (1675) 'Isaac Newton letter to Robert Hooke', https://digitallibrary.hsp.org/index.php/Detail/objects/9792# (Accessed 1 May 2020)

Nicolaisen, J. (2007) 'Citation Analysis', *Annual Review of Information Science and Technology*, vol. 41, no. 1, pp. 609–641, https://doi.org/10.1002/aris.2007.144041012

O'Brien, T.L. (2012) 'Change in academic co-authorship, 1953–2003', *Science, Technology, and Human Values*, vol. 37, no. 3, pp. 210–234, https://doi.org/10.1177/0162243911406744

Olssen, M. (2015) 'Neoliberal competition in higher education today: Research, accountability and impact', *British Journal of Sociology of Education*, vol. 37, no. 1, pp. 129–148, https://doi.org/10.1080/01425692.2015.1100530

Packer, A. (2009) 'The SciELO open access: A gold way from the south', *Canadian Journal of Higher Education*, vol. 39, no. 3, pp. 111–126, DOI: https://doi.org/10.47678/cjhe.v39i3.479

Price, D. (1963) *Little science, big science*, New York: Columbia University Press.

Price, D. (1965) 'Networks of scientific papers', *Science*, vol. 149, no. 3683, pp. 510–515, DOI: 10.1126/science.149.3683.510

Rekdal, O.B. (2014) 'Academic urban legends', *Social Studies of Science*, vol. 44, no. 4, pp. 638–654, https://doi.org/10.1177/0306312714535679

Sanchez-Gil, S., Gorraiz, J., & Melero-Fuentes, D. (2018) 'Reference density trends in the major disciplines', *Journal of Infometrics*, vol. 12, no. 1, pp. 42–58, DOI: 10.1016/j.joi.2017.11.003

Seeber, M., Cattaneo, M., Meoli, M., & Malighetti, P. (2019) 'Self-citations as strategic response to the use of metrics for career decisions', *Research Policy*, vol. 48, no. 2, pp. 478–491, DOI: 10.1016/j.respol.2017.12.004

Small, H. (1978) 'Cited Documents as Concept Symbols', *Social Studies of Science*, vol. 8, no. 3, pp. 327–340, https://doi.org/10.1177/030631277800800305

Swales, J. (1990) *Genre analysis: English in academic and research settings*, Cambridge: Cambridge University Press.

Swales, J. (2004) *Research genres: Explorations and applications*, Cambridge: Cambridge University Press.

Tahamtan, I., & Bornmann, L. (2018) 'Core elements in the process of citing publications: Conceptual overview of the literature', *Journal of Infometrics*, vol. 12, no. 1, pp. 203–216, DOI: 10.1016/j.joi.2018.01.002

Vanderstraeten, R. (2010) 'Scientific communication: Sociology journals and publication practices', *Sociology*, vol. 44, no. 3, pp. 559–576, https://doi.org/10.1177/0038038510362477

Wang, J., Veugelers, R., & Stephan, P. (2017) 'Bias against novelty in science: A cautionary tale for users of bibliometric indicators', *Research Policy*, vol. 46, no. 8, pp. 1416–1436, DOI 10.3386/w22180

Wouters, P. (1999) 'Beyond the holy grail: From citation theory to indicator theories', *Scientometrics*, vol. 44, no. 3, pp. 561–580, https://doi.org/10.1007/BF02458496

10

MEASURING IMPACT

Glenn Hampson

Why do we try to measure the impact of science? Why do we question whether we are support-ing the right research, spending our time and money wisely, or discovering the "right" things? What kinds of measurements do we use, how do we know if these measurements are correct, and what affect does all this attention have on science itself? Understanding the answers to these questions is key to understanding how modern science works.

Origins

Science is and always has been an almost dictionary-perfect example of impact. Five hun-dred years ago, a radical idea began taking hold in Europe. The ardent believers in this idea – explorers and philosophers who did their best to avoid prison or death for their heresy – thought it was important to discover objective truths about the natural world rather than simply accept all proclamations of church leaders at face value. This radical idea spread, proved its merit, and gradually grew into the Scientific Revolution, freeing society from centuries of darkness and fueling the intellectual and technological revolutions that created our modern world.[1]

Few inventions have been and remain as essential to the challenges of our modern world as science. And none have been as impactful. Science research and discovery are constantly rip-pling out waves upon waves of impact, both internally and across society. Even internally, as a meritocracy, science is always using impact evaluations of some sort to determine who becomes a scholarly society member, who speaks at conferences, and who wins Nobel Prizes.

Perspectives

So, other than launching the Scientific Revolution and creating decades of Nobel Prize–winning discoveries, what else might we mean by science impact? This impact is obviously everywhere, but where exactly is it coming from?

For example, if we think of science as a "gift economy," then impact might mean the degree to which one science study influences another. From this perspective, measuring impact means trying to map the causes, magnitudes, and patterns of these influences – measuring, for exam-ple, how many times researchers cite a particular journal article.

DOI: 10.4324/9781003043782-12

If we see instead that most of the world's research and development (R&D) investment and spending happen in the business sector and are also highly concentrated in the world's largest economies and in the engineering and technology sectors[2] – think giant technology, auto, and pharmaceutical companies, who are not in the habit of "gifting" anything – then we might be more inclined to use R&D totals, tech sector employment, or patents as barometers of science impact.

Or, maybe our perspective is centered on social impacts, noting how science improves opportunity and well-being in society or builds partnerships between academia and industry. In this case, we might focus on social indicators as key impact metrics.

Our current science impact measurement policies include all of these perspectives and more, depending on who is doing the measuring and for what reasons.[3]

Dimensions

The science impact evaluation universe is broad and varied. There are, however, three common dimensions to keep in mind:

1   When we talk about "science," we often actually mean the broad research ecosystem that includes the natural sciences, engineering, technology, math, social sciences, arts and humanities. Considered together (which it normally is), science, technology, engineering and math (STEM) research is by far the largest and most influential part of this ecosystem, so what happens to STEM research impact evaluation has an outsized effect on what happens to all other types of research impact evaluation. Because it is all related, in this chapter, we will focus more on "research" impacts than on "science" impacts.

2   Most policy makers (less so researchers) group research into three main buckets: basic, applied, and experimental development. Most R&D research is experimental development – the process of turning ideas into products – and almost all this activity is performed by business. Higher education performs the largest share of basic research – the search for fundamental science truths, which are significant unto themselves but also provide the fuel for applied and experimental work. Funding for this work comes mostly (about 68%) from governments. Governments also directly perform basic research work that is of national interest (in nuclear energy, for example).

3   We can only guess at a lot of research impact, except for counting the things that are quantifiable like dollars and journal articles (Bornmann, 2012). We especially have no idea how to account for factors like time, magnitude, and causality. Timewise, judging the impact of research that results in a new product or confirms previous studies is easy. More often, though, it takes years to understand impact. In terms of magnitude, evaluators only occasionally see research that has created a paradigm shift in a particular field. More often, magnitude is estimated by proxy – calculating which work has collected the most citations or has led to the most products, patents, or public benefit.

At a more granular level, the key stakeholder groups in the research evaluation universe – researchers, governments, nonprofits, universities, publishers, and businesses – all see different dimensions of impact and use different metrics to measure it.

Researchers

Surveys have repeatedly shown that impact is the most important concern of researchers (see, for example, Taylor and Francis, 2019). To researchers, "impact" means having their work read

Table 10.1 Top concerns of authors when publishing their findings

Top five concerns (90–98% of authors)	The journal has a good reputation in my field.
	The journal is well read by researchers in my field.
	The journal focuses on my specific field or area of research.
	The journal has a high impact factor ranking in my field.
	The journal is read by a broad audience.
Next five concerns (80–89% of authors)	Speed of review
	Cost of publishing
	Journal belongs to an important scholarly society.
	Cost of making articles available via open access
	Journal offers sophisticated tracking of article impact.

Source: Adapted from Taylor and Francis, 2019.

by the right audiences, and making a solid contribution to their field and to society. The main tools researchers have used over the years to accomplish these objectives are academic journals and direct communication (which often takes the form of conferences). Therefore, the main impact measures used by researchers are journal centric, like counting the number of citations collected, journal articles written (or coauthored), and articles published in the most prestigious journals.

Throughout their careers – in funding, promotion, and publishing evaluations – a researcher's impact will also be evaluated by many other quantifiable metrics, like the value of grants received, the number of conference presentations made, society memberships, teaching record, dissertations directed, scientific achievements, books published and patents awarded. More subjective evaluation measures include the relative merit of their research work, the reputation of their institutions, opinions of colleagues, and notoriety (which is not always a plus). Every impact evaluation is different since every evaluator uses a different definition of impact.

Governments

The research evaluation systems we recognize today gradually evolved during the post–World War II because of heightened oversight practices, especially in the United States during the 1970s. By the 1980s and 1990s, these systems had spread globally, not just to governments but to all government and nonprofit programs (Baldwin, 2018).

As these systems evolved, they sprouted a cornucopia of tangible and intangible metrics for trying to measure impact. For example, all major U.S. government research funding agencies today use expert review panels to assess why the research being proposed matters (in terms of scientific significance and broader social impacts), how the research is new, how research will be done, in what context, what is special about the people involved (investigators, organization, people, researchers, personnel, partners, collaborators, staff), what the return on investment will be (impact, value, relevance), how financial resources will be managed, and how success will be determined. Some of these data points are objective and quantifiable; many are not.[4]

Most government funders around the world use similar approaches. The United Kingdom is an outlier, recently launching its Research Excellence Framework (REF) to try to do a better job of measuring broad social impacts. REF requires that impact case studies be included with all new funding proposals – narratives, using both qualitative and quantitative evidence, that show how research has benefited culture, the economy, the environment, health, public policy, quality of life or society (REF, 2021). In contrast to the citation counts used by researchers to demonstrate academic impact, it is expected that REF will provide a more accurate estimate of broad impact (Ravenscroft, 2017), although REF is not without its critics.

To weigh the impact of completed research, governments also collect mountains of economic statistics on the R&D sector, which get fed back into decisions about which areas of research provide a better return on investment for society, at least in economic terms. Government agencies also conduct and evaluate research, which can spur additional impact evaluation and research. For example, government-funded research showing how fine particulate matter in the air leads to an increased incidence of lung cancer might lead to industry research, supported by government tax incentives, to design the machinery needed to reduce air pollution.

It is important to note that all our meticulous evaluation processes aside, government evaluations can also be hijacked by politics. During the 1970s and 1980s, for example, U.S. Senator William Proxmire regularly humiliated researchers and their funders by handing out "Golden Fleece" awards for work that he thought lacked adequate public benefit. More recently, the Trump administration created a policy (which was later reversed) that would have made it harder for the U.S. Environmental Protection Agency to use science that – in the estimation of the agency's politically appointed leadership – was simply too "secretive" to be trusted (OSI, 2018).

Nonprofits

Nonprofits fund only a small portion of the world's research, but they wield a lot of influence when it comes to focus and impact evaluation. Nonprofits focus mostly on research that governments have underfunded, like disease, education, and maternal health. The Gates Foundation is a good example, leading a global war against malaria (among other notable programs). Other nonprofits have tackled everything from Alzheimer's and heart disease to sustainable agriculture and the search for extraterrestrial life. In terms of evaluation, major nonprofit research funders like Gates, the Wellcome Trust, and the Max Planck Institute have all been global leaders in the push for open access and other mechanisms to help improve the impact of research through more robust transparency and reliability. These organizations also usually punch well above their weight when it comes to publicizing the results of their work and translating science research into societal impact.

Universities

Universities conduct most of the world's basic research, with most of their funding for this work coming from governments. University researchers also author around 80% of the articles published in journals; and most of the articles that business-based researchers author are written in collaboration with university researchers (Larivière et al., 2018).

Research impact evaluation is used by these institutions in many ways. For example, almost every university uses some form of publish-or-perish metric where retention, promotion, and tenure (RPT) decisions consider the impact of a researcher's published body of work.[5]

Research also has a direct impact on universities – especially R1 institutions (top-tier research facilities) in terms of their employment, reputation, and budgets. Universities, like businesses, are huge economic engines, providing employment for a great many research and nonresearch personnel. The overhead charges that universities levy on research grants is vital for supporting university infrastructure, administrative costs, and even unfunded research. In addition, school rankings like those published by *US News & World Report* are calculated using research-related inputs and outputs – the number of researchers employed, amount of research funding received, number of journal articles published, and so on. These rankings are important to universities because of the impact they have on school reputations, which can translate into larger endowments, higher enrollment demand, and better recognition, retention, and funding for researchers.

Publishers

Somewhere between 2.5 million and 3.7 million scholarly journal articles are published annually, depending on who is counting what.[6] Academic publishers of all kinds (not just commercial publishers but nonprofit, society, and university publishers as well) measure the impact of these articles to better understand industry trends and customer needs, develop new titles as needed, and adjust pricing and delivery formats.

The best known metric used by publishers is the Journal Impact Factor (JIF),[7] a trademarked statistic owned and operated by Clarivate, which measures the citation performance of journals over a two-year period. Being published in a high-impact journal is highly prized by most researchers because funders and universities put a lot of weight on JIF when measuring research impact. Other commonly used evaluation tools include the h-index, Altmetric score, page rank, journal half-life, and SCImago Journal Rank (SJR).

Publishers also rely on peer review or other similar mechanisms (for example, desk rejections by editors) to determine whether a research paper will be of interest to their readers, whether it merits being distributed for peer review, and whether it is worth the additional investment of time and money to transform into a published article. These are all important decisions because journals are the center of the academic impact evaluation universe (OSI, 2021).[8]

Table 10.2 Major impact assessment tools for published science

Tool	Description
Journal Impact Factor (JIF)	Measures the average number of citations over a two-year period in a particular journal (this is a journal metric, not an article metric). Less than 1% of the world's research is published in the highest JIF "prestige" journals, which influences public perceptions and researcher incentives about which science is and is not "impactful."
Hirsch index (h-index)	Measures the impact of a particular scientist. *H* is calculated by comparing the number of articles published by a researcher to the number of times these articles have been cited. A researcher with an h-index of five, for example, has published five papers that have each been cited at least five times by other researchers.
Altmetric score	Measures the amount of media exposure a research article receives in the news, blogs, tweets, and more. These scores aren't "good" or "bad." A bad study can generate a high Altmetric score, for example, because it gets widely refuted.
SCImago Journal Ranking (SJR)	SJR catalogs citations, downloads, references, and more, by subject area and category, country and region, type of output (journal article, book, conference paper, etc.), and time period. See www.scimagojr.com/.

Businesses

Most applied and experimental research is funded and conducted by the business sector, and the overwhelming majority of patents are awarded to business-based researchers and their companies (Larivière et al., 2018). Assessing the impact of research in this setting is generally different than in governments and universities. Rather than guessing which research is likely to have the most impact or counting articles and citations, businesses are ultimately concerned with whether research will translate into patents or revenue.[9]

This approach might seem incongruous with a university's approach, but it is actually integral. For one, the basic research done primarily by universities is the fuel that powers the experimental research done primarily by businesses. This relationship has become more dependent in recent years because businesses are increasingly outsourcing their basic research work to universities and focusing more on experimental development, since this kind of work results more immediately in the products and profits that shareholders demand and also spreads the risk of sinking too much money into high-tech labs and personnel (PWC, 2020). To help facilitate closer ties with industry, most R1 universities also run technology transfer centers that try to push research out of universities and into the business sector via patents, licensing agreements, and spin-offs. In addition, many university researchers either work closely with industry (across many disciplines), make a leap to industry so they can focus more intently on developing their high-impact ideas to fruition (Dolgin, 2017), or even get hired away from universities entirely, especially in high demand areas like artificial intelligence (Etzioni, 2019).

Impacts

Impact evaluation thinking has burrowed deep into the mind-set of researchers today. What are the impacts? In general, researchers are pressured to create work that is seen as impactful because impact evaluations determine the trajectory of their businesses and careers.

Publishing distortion

Many have argued for years now that widespread use of the JIF in impact evaluation, combined with the pressure authors feel to publish in high JIF journals, has created a publishing economy where prestige is the driving consideration. Rather than aspiring to publish in the soundest available venue, researchers instead are often incentivized to seek out the most impactful venue because this will pay the most dividends in their research impact evaluations. Soundness and impact are not mutually exclusive, of course, and not all researchers are motivated in this way – it is most often the early and mid-career researchers who feel more pressure to seek out impact.

Since JIFs are only calculated for some of the world's journals, the net effect of this dynamic is to bias our impact evaluations in favor of authors who can afford to publish in these more visible journals or whose work is deemed worthy by journal editors. As a result, the rich get richer, and the "poor" get less recognition for their work and generate less career and research impact.[10]

The JIF and other citation-based metrics create other distortions as well, including biasing us toward valuing popular or active disciplines; rewarding bad work by counting all citations as equal when in fact many citations are negative;[11] impeding the freer sharing of research articles through open access (because high-impact factors are most often tied to older, subscription-based journals); and discouraging authors from publishing findings that show no effect – an outcome that is just as important to science but less likely to produce a citation bump for impact evaluations and careers. More generally, our focus on publishing-based metrics has distorted author behaviors across the seniority and discipline spectrum and has fueled a sharp rise in multiauthored and partial result papers (Plume et al., 2014), predatory journals (promising a quick way to pad publishing resumes, merit notwithstanding), fake JIFs (not only blatantly faked numbers but fake measures as well), manipulated JIFs, and even cash awards for publishing in high-impact venues.

Publishers also tangentially affect research impact evaluation by controlling how much it costs to read research, what format research is shared in, deciding which journals are included in major indexes, which data policies to adapt, and more. It is beyond the scope of this chapter to examine these tangents in detail, but suffice it to say that the impact of publishers is at the forefront of many global reform efforts regarding the future of research (Hampson et al., 2021).

Funding and research influences

In the United States, the success rate of grant proposals has plummeted over the last 30 years as funding has grown tighter and the number of proposals submitted has skyrocketed (Berg, 2020). Therefore, how we decide what work gets funded has become an issue of increasing importance in the research community. Our impact evaluation systems are thoughtfully designed, and certainly science has been remarkably successful over the years, but we may want to revisit these systems since they are becoming increasingly important in deciding what research gets funded. This is doubly important since patterns of bias in all manner of evaluations have emerged, including underfunding researchers from less prestigious institutions (Wahls, 2018), reinforcing scientific conformity (Packalen et al., 2020), cultural and gender bias (Weisshaar, 2017), and favoring some fields over others (Reale, 2018). Indeed, the biases of a few individuals can end up affecting not only individual research careers but the direction and location of research in an entire field (Holbrook and Frodeman, 2011).

Our assessment systems as they now stand are also routinely gamed for benefit. Researchers can be tempted to achieve high impact at the expense of scientific rigor and integrity by exaggerating study results for the sake of newspaper headlines, retrofitting "proof" to fit available evidence (Baker, 2015), committing forgery or plagiarism, or participating in "citation rings" to boost citation counts. As mentioned, not publishing negative results is also common – null findings do not get citations, and citations are needed to establish research impact.

Global South issues distortion

Most R&D is done in major industrialized countries, so naturally, most R&D deals with issues that are important to these countries – which crops we will develop, which diseases we will try to eradicate, and which social issues merit our attention. For the rest of the world, meaning most of the world's countries and people, researchers publish a more limited amount, mostly in regional and specialty journals that are more affordable and accessible, often in local languages, and less in the highest-profile, highest-impact journals of the world. As a result, we have fallen well short of our ideals of science knowledge being able to serve everyone equally. Companies, nonprofits, and multilaterals are actively engaged in filling needs and gaps, but for government-sponsored academic research, our impact evaluation policies may be creating a roadblock to learning, sharing, and doing more globally. (See Figure 10.1.)

Business research influence

Business-based research has also been influenced by impact evaluations. As noted earlier, since the impact metric valued most by business has increasingly been profit, this has resulted in

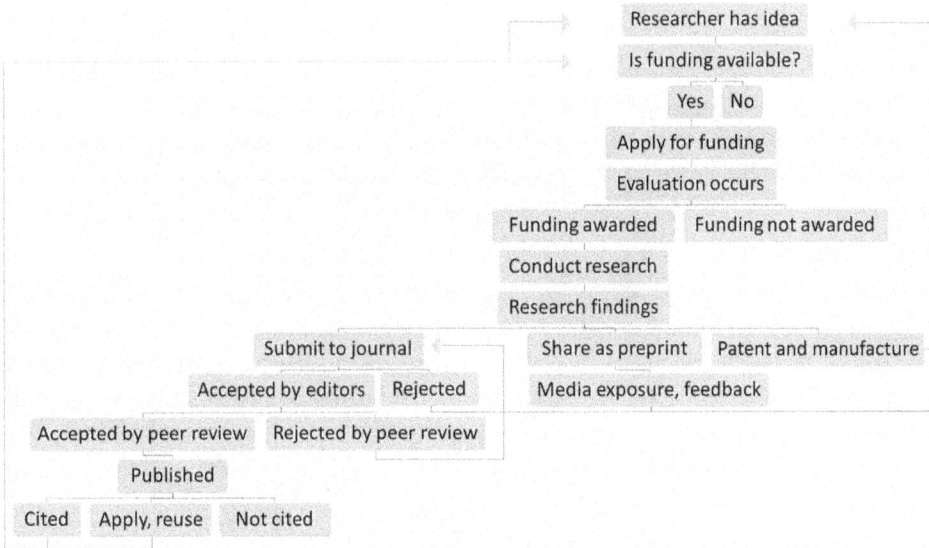

Figure 10.1 The many stages of impact evaluation

businesses spending less on basic research and more on experimental development in recent years, making businesses more reliant on universities for basic research (Wu, 2018). This shift has also meant business-based researchers sharing less with the public through journal articles, favoring instead to lock up their research with more patents, and also tying business-based research and research practices more and more to the practices and evaluation policies used by governments and universities, which is not always a comfortable fit. Motivated by the need to produce business-friendly impacts, there has also been a well established pattern of bias in business-based and business-financed research over the years, as well as sometimes outright deceit (as with research that showed no link between smoking and lung cancer; Fabbri et al., 2018).

Reactions

The fact that our research impact evaluation systems have flaws is not news to researchers, nor is it necessarily a fatal indictment. The research community's reactions to these systems have been mixed over the years (Lok, 2010), but there have not been any large-scale revolts, at least not yet. The questions being asked today are less, "Will we ever stop measuring impact?" and more, "How can we reform our thinking and measurements around impact so we don't end up distorting the very things we are trying to measure?"

For example, the Declaration on Research Assessment (DORA) and the Leiden Manifesto have each collected thousands of signatures from individuals and organizations around the world who are concerned about the misuse of impact measurement practices and are committed to developing better practices. The open movement is also creating change. Many involved in open access, open data, and open science see these efforts as being mostly about improving impact – trying to fix many science impact inequities in one fell swoop, from improving who can access science articles to reducing the influence of gatekeepers who currently decide what is worth sharing and what is not. To facilitate this effort, principles like FAIR are gaining widespread traction – overarching guidance that all research data should be findable, accessible, interoperable, and reusable.

Inside academia, there has also been a trend in recent years to rethink the publish-or-perish metrics used in RPT evaluations (Schimanski et al., 2018), especially the JIF, and to focus more on quality-based metrics instead of prestige- and quantity-based metrics. Along these lines, India recently cracked down on predatory publishers operating inside its borders because these paper mills were being used by Indian scholars to inflate their publishing resumes (Priyadarshini, 2018), and in 2020, China stopped paying cash bonuses to academics who succeeded in publishing in high-impact journals (Mallapaty, 2020).

Different publishing and peer review norms have also been evolving quickly, from the increased use of preprints (research shared "as is," mostly without peer review) to open forms of peer review. We are also witnessing a trend in treating more research outputs as being valuable and impactful – data and code, for example, are increasingly considered "first-class" research objects. Indeed, researchers in business- and government-sponsored research are currently building huge new value chains with these elements (Hampson et al., 2021).

Many other ideas – most of them still in the discussion stage – have been proposed for how to change our impact measurement systems, ranging from simple adjustments to complete overhauls. On the simple end are REF-like ideas that might help us better measure broad impacts. Examples include the Science Impact Framework (Ari et al., 2020), which suggests evaluating impact in terms of dissemination, creating awareness, catalyzing action, effecting change, and shaping the future, and a recent SAGE model (Sage Publishing, 2019) that might be a better fit for humanities and social science work than current STEM-centric models, evaluating research along the dimensions of academic, practitioner, societal, and public impact. (See Figure 10.2.)

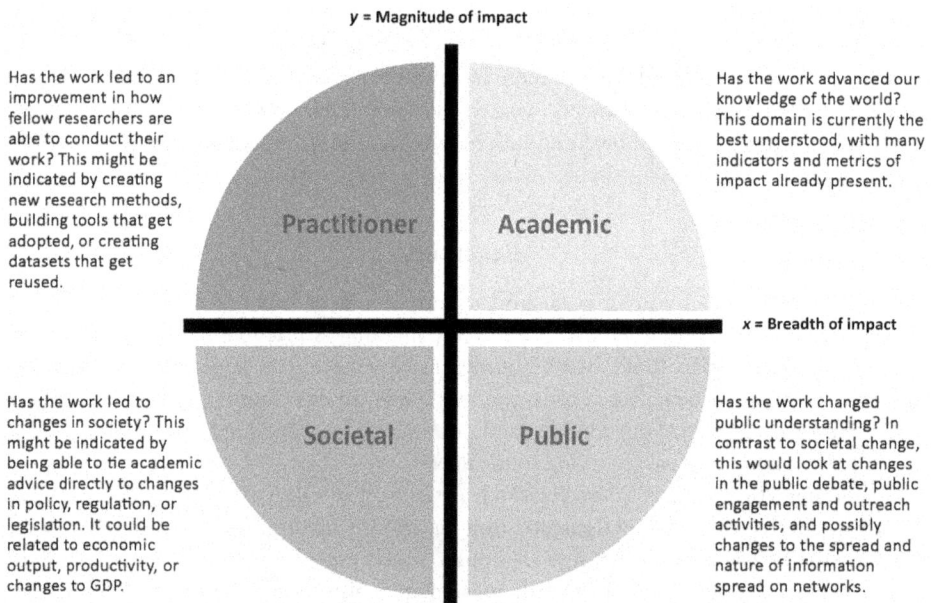

Figure 10.2   Adaptation of the SAGE model of scientific impact

Source: Text from Sage Publishing, 2019. The rest of this model is an adaptation. Axes, axes descriptions, and quadrant locations are not part of the SAGE model, just quadrant descriptions.

In the more modestly ambitious reform range, it has been proposed that we consider introducing "consensus" measures of impact to replace our current metrics, which are too often affected by time, journal, and prestige (Bollen et al., 2009). Metrics that identify high-impact researchers instead of high-impact research have also been invented (Sinatra et al., 2016), as well as systems that use artificial intelligence to look through large bodies of research more objectively (Gordon et al., 2020). Continued developments in the "science of science" field, fueled by big data, are also teasing new (and controversial) insights into the drivers of research impact (Wang et al., 2021).

More ambitious still, some have suggested retrofitting our impact evaluation models to fit the available evidence. Most of our current impact measures just describe program compliance and efficiency – whether research is being properly conducted and disseminated. We can content ourselves with this level of oversight and start gathering more generalizable knowledge about why and how our research investments work so that we can develop better ways to model and predict broad impact (Gugerty et al., 2018).

At the complete overhaul end of the scale, one change that might eventually gain more traction is the science funding lottery system (Fang et al., 2016). Humans are terrible at predicting the future. Distributing funding as broadly as possible by lottery might maximize the likelihood that discoveries will occur while at the same time removing the distortions and perverse incentives caused by the current system. Government research funding in Switzerland and New Zealand is currently being distributed this way (Adam, 2019).

An even more radical idea that some futurists have proposed is to remove government from the equation entirely and use a self-sustaining system instead, where researchers directly buy, sell, and trade research in a marketplace of ideas. In such a model, the best and most useful ideas "win," and no one prejudges where money gets invested or uses publications as a vector or proxy for science impact.

A role for science communication

Can science communication help us improve how we measure research impact? Interpreting and communicating research is, after all, what science communication does for a living. Too often, researchers improvise with their outreach and communication efforts, when engaging science communication professionals would be far more advantageous for their work and for society. Everything from recruiting study participants, to connecting researchers, to writing blog posts and white papers, and on to preparing policy briefs would be greatly enhanced by engaging people who are trained and expert in these tasks.

If we go down this route, then our goal should be to improve not only research impact evaluation but research impact itself. Why? Because one will empower the other: as we do a better job understanding and communicating impact, research will be able to have greater impact. And because we need to do this. The ultimate goal of research, after all, is to make an impact, not to be assigned an impact score.

This effort matters now more than ever. We need to make sure our researchers, policy leaders, and the public that elects these leaders are all singing from the same songbook when it comes to solving urgent problems like climate change, food and water security, and vaccine preparedness. To help this urgent research succeed, we need to make sure that viewing the potential of research through the prism of prestige-based metrics is doing more good than harm. Ideally, our focus, working together, should be to strengthen both research and the systems we use to evaluate research so that our 500-year-old pursuit of truth can continue to serve society, less encumbered and better empowered to succeed at its highest possible level.

Notes

1 See Wootton, 2015 for a good exploration of the invention of science.
2 See UNESCO data tables, SEI, and OSI, 2021, for details.
3 See NRC, 2014, Chapters 4 and 5, for a comprehensive overview of U.S. research impact evaluation processes.
4 Falk-Krzesinski et al., 2015. See also NSF, 2013; NIH, 2013.
5 The "R" in RPT can also stand for "reappointment" or "review."
6 The low-end estimate is from SEI, 2020. The higher-end estimate is from Archambault, 2018, using the 1findr database (which is more comprehensive than databases used for the lower-end estimate).
7 Internally, the key metrics publishers use are revenue and growth, followed by market share and the number of submissions.
8 Internally, these are also important decisions because as a matter of cost control, the number of articles submitted to the prestigious journals is staggering.
9 The basic research units of business do, however, spend a considerable amount of time and energy trying to see the future and predict what society will need. Profit motive is not the only driver.
10 In research, this is known as the Matthew effect.
11 The widely discredited Hirschfeld study purporting to establish a link between vaccines and autism is an extreme example of this dynamic.

References

Adam, D. (2019) 'Science funders gamble on grant lotteries' *Nature news* vol. 575, no. 7784, pp. 574–575 https://go.nature.com/3svELIa. doi: 10.1038/d41586-019-03572-7

Archambault, E. (2018) 'Universalisation of OA scientific dissemination', www.slideshare.net/scielo/2-eric-archambault

Ari, M.D., Iskander, J., Araujo, J., et al. (2020) 'A science impact framework to measure impact beyond journal metrics' *PLoS ONE* vol. 15, no. 12, p. e0244407, doi: 10.1371/journal.pone.0244407

Baker, M. (2015) 'Over half of psychology studies fail reproducibility test', *Nature*, https://doi.org/10.1038/nature.2015.18248

Baldwin, M. (2018) 'Scientific Autonomy, Public Accountability, and the Rise of "Peer Review" in the Cold War United States', *Isis* vol. 109, no. 3, pp. 538–558, doi: 10.1086/700070

Berg, J. (2020) 'Modeling Research Project Grant Success Rates from NIH Appropriation History: Extension to 2020' *bioRxiv*, doi: 10.1101/2020.11.25.398339

Bollen, J., Van de Sompel, H., Hagberg, A., et al. (2009) 'A Principal Component Analysis of 39 Scientific Impact Measures' *PLoS ONE* vol. 4, no. 6, p. e6022, doi: 10.1371/journal.pone.0006022

Bornmann, L. (2012) 'Measuring the societal impact of research: research is less and less assessed on scientific impact alone – we should aim to quantify the increasingly important contributions of science to society', *EMBO Rep*, vol. 13, no. 8, pp. 673–676, doi:10.1038/embor.2012.99

Dolgin, E. (2017) 'Why we left academia: Corporate scientists reveal their motives', *Nature Careers*, http://go.nature.com/3kqt6Y6

Etzioni, O. (2019) 'AI academy under siege', *Inside Higher Education Blog Post*, http://bit.ly/3q3gS91

Fabbri, A., Lai, A., Grundy, Q., et al. (2018) 'The Influence of Industry Sponsorship on the Research Agenda: A Scoping Review', *American Journal of Public Health*, vol. 108, no. 11, pp. e9–e16, doi:10.2105/AJPH.2018.304677

Falk-Krzesinski, H.J., & Tobin, S.C. (2015) 'How Do I Review Thee? Let Me Count the Ways: A Comparison of Research Grant Proposal Review Criteria Across US Federal Funding Agencies' *The Journal of Research Administration*, vol. 46, no. 2, pp. 79–94

Fang, F.C., & Casadevall, A. (2016) 'Research Funding: The Case for a Modified Lottery, *mBio editorial* vol. 7, no. 2, p. e00422–16, doi: 10.1128/mBio.00422-16

Gordon, M., Viganola, D., Bishop, M., et al. (2020) 'Are replication rates the same across academic fields? Community forecasts from the DARPA SCORE programme', *Royal Society Open Science*, vol. 7, no. 7, p. 200566, doi: 10.1098/rsos.200566

Gugerty, M.K., & Karlan, D. (2018) 'Ten reasons not to measure impact – and what to do instead', *Stanford Social Innovation Review*, https://ssir.org/articles/entry/ten_reasons_not_to_measure_impact_and_what_to_do_instead#

Hampson, G., DeSart, M., Kamerlin, L., et al. (2021) 'OSI Policy Perspective 4: Open Solutions: Unifying the meaning of open and designing a new global open solutions policy framework', *Open Scholarship Initiative*, January 2021 edition, doi: 10.13021/osi2020.2930

Holbrook, J.B., & Frodeman, R. (2011) 'Peer review and the ex ante assessment of societal impacts', *Research Evaluation* vol. 20, no. 3, pp. 239–246, doi: 10.3152/095820211X12941371876788

Larivière, V., Macaluso, B., Mongeon, P., et al. (2018) 'Vanishing industries and the rising monopoly of universities in published research' *PLoS ONE* vol. 13, no. 8, p. e0202120, doi: 10.1371/journal.pone.0202120

Lok, C. (2010) 'Science funding: Science for the masses', *Nature* vol. 465, pp. 416–418, doi:10.1038/465416a

Mallapaty, S. (2020) 'China bans cash rewards for publishing papers', *Nature*, vol. 579, p. 18 https://go.nature.com/3qPbB66

National Institutes of Health (NIH). (2013) 'Additional scoring guidance for research applications', https://bit.ly/3uJpsO3

National Research Council (NRC) (2014) *Furthering America's Research Enterprise*, Washington, DC: The National Academies Press, doi: 10.17226/18804.

National Science Foundation (NSF). (2013) 'Chapter II – proposal preparation instructions', http://bit.ly/2O0C2Yx

OSI. (2018) 'Comment on proposed rule, "Strengthening Transparency in Regulatory Science," EPA-HQ-OA-2018-0259', *Open Scholarship Initiative*, https://bit.ly/3bHgN5O

OSI. (2021) 'How do researchers decide where to publish? OSI infographic 3', *Open Scholarship Initiative*, http://bit.ly/37VVvk1

Packalen, M., & Bhattacharya, J. (2020) 'NIH funding and the pursuit of edge science', *Proceedings of the National Academy of Sciences May 2020*, 201910160, vol. 117, no. 22, pp. 12011–12016, doi:10.1073/pnas.1910160117

Plume, A., & van Weijen, D. (2014) 'Publish or perish? The rise of the fractional author', *Research Trends* vol. 38, no. 3, pp. 16–18

PricewaterhouseCoopers (PWC). (2020) 'Pharma 2020: Challenging business models', *White Paper*, http://pwc.to/38codwT

Priyadarshini, S. (2018) 'India targets universities in predatory-journal crackdown', *Nature* vol. 560, pp. 537–538, doi: 10.1038/d41586-018-06048-2

Ravenscroft, J., Liakata, M., Clare, A., et al. (2017) 'Measuring scientific impact beyond academia: An assessment of existing impact metrics and proposed improvements', *PLoS ONE* vol. 12, no. 3, p. e0173152, doi: 10.1371/journal.pone.0173152

Reale, E., Avramov, D., Canhial, K. et al. (2018) 'A review of literature on evaluating the scientific, social and political impact of social sciences and humanities research', *Research Evaluation*, vol. 27, no. 4, pp. 298–308.

Research Excellence Framework (REF). (2021) www.ref.ac.uk

Sage Publishing (2019) *The latest thinking about metrics for research impact in the social sciences (White paper)* Thousand Oaks, CA: Author. doi: 10.4135/wp190522

Schimanski, L.A., & Alperin, J.P (2018) 'The evaluation of scholarship in academic promotion and tenure processes: Past, present, and future', *F1000Res* vol. 7, p. 1605, doi:10.12688/f1000research.16493.1

Science & Engineering Indicators (SEI). (2020) 'US national science board', https://www.nsf.gov/statistics/seind/

Sinatra, R., Wang, D., Deville, P., et al. (2016) 'Quantifying the evolution of individual scientific impact', *Science*, vol. 354, no. 6312, p. aaf5239, doi: 10.1126/science.aaf5239

Taylor & Francis. (2019) 'Taylor & Francis researcher survey', https://bit.ly/3koHgrX

UNESCO. (2021) 'UNESCO institute for statistics (UIS) dataset', http://data.uis.unesco.org

Wahls, W., (2018) 'High cost of bias: Diminishing marginal returns on NIH grant funding to institutions', *bioRxiv* 367847, doi: 10.1101/367847

Wang, D., & Barabási, A.L. (2021) *The science of science*, Cambridge: Cambridge University Press.

Weisshaar, K (2017) 'Publish *and* Perish? An Assessment of Gender Gaps in Promotion to Tenure in Academia', *Social Forces* vol. 96, no. 2, pp. 529–560, doi: 10.1093/sf/sox052

Wootton, D. (2015) *The invention of science: A new history of the scientific revolution*, New York: HarperCollins.

Wu, J. (2018) 'Why U.S. business R&D is not as strong as it appears: Information technology & innovation foundation', *White Paper*, http://www2.itif.org/2018-us-business-rd.pdf

11

COLLABORATIVE MODELS FOR SCIENTIFIC WRITING

Lisa DeTora and Sabrina Sobel

In "Synchronized Editing: The Future of Collaborative Writing," Jeffrey M. Perkel (2020) describes a growing trend toward open-source, publicly visible scientific writing collaboration, which he presents in contradistinction to what might seems to be a general rule: "[d]raft scientific manuscripts are typically confidential" (p. 154). Citing the rise of interactive software, like Manubot, as a means of opening the door to global collaborations that can ameliorate crises like the COVID-19 global public health emergency, Perkel concludes with a call for more scientists to engage with this trend. Yet science, especially the type of biomedical science (and writing) needed to address the current pandemic and potentially avert others, has already long been a highly – and globally – collaborative endeavor. What may be different in Perkel's account is that the process of collaboration in these contexts is becoming more transparent and therefore increasingly subject to rhetorical analysis or intervention by those outside the immediate biomedical research community.

Rhetorical studies of science – scholarship on the function and scope of communications about scientific information – often address published manuscripts, presentations, letters, or reports, situating these genres within specific fields of discursive action. Forms like science journalism, blogs, and science textbooks also receive attention from scholars interested in their language and function. For instance, Halliday and Martin's foundational and influential analysis of scientific style and grammar in *Writing Science: Literacy and Discursive Power* (2015, originally published in 1993) hinges on readings of popular science and high school textbooks. Alan Gross, who walked readers through the scientific literature in books like *Starring the Text* (2006), ended his career with a volume on 'the scientific sublime' in popular science writing (2018), also signaling interest in forms adapted heavily from the scientific literature. Yet little attention has been paid to other forms of scientific writing, like regulatory documentation, guidance documents, or lab manuals, that often underpin the activities of interest to scholars of rhetoric, writing studies, and technical communication. The tension between a commonsense notion of scientific writing by individuals versus that by large multinational teams, as in regulatory documentation, may be of interest for rhetorical or technical communication study.

We define biomedical writing, following general professional practice in groups like the American and European Medical Writers Associations, the Regulatory Affairs Professionals Society, and the International Society of Medical Publication Professionals (see Winchester, 2017; Benau, 2020; DeTora, 2020b), as a broad field of endeavor that includes the regulatory

DOI: 10.4324/9781003043782-13

documentation and resulting publications of studies conducted to investigate various medicinal products, as well as the guidance documents developed to guide such work. Biomedical inquiry includes studies needed to identify potential medical interventions, to evaluate their suitability for investigation in animals or persons, and to manufacture and distribute such products, as well as further studies and reviews that situate the outcomes of studies relative to existing scientific literature, medical practice, and future aims (see www.ich.org). This grouping of research disciplines parallels that cited by the International Committee of Medical Journal Editors (2019), roughly corresponding to the works indexed in PubMed, which include some primary science publications as well as medical journals, ethics journals, and specialty resources in fields like nursing, sports science, or pharmacology. Collaboration is an inherent aspect of all stages of biomedical research projects (see Benau, 2020; DeTora, 2020b; Hamilton, 2014; Montgomery, 2017).

Next, we outline some features of biomedical writing, focusing on regulatory writing, then situate its practices relative to commonsense notions of scientific authorship and existing rhetorical and technical communications scholarship.

Many features of scientific discourse communities are invisible

Since all scientific endeavor is collaborative, scientific reports, publications, and presentations represent the ending of a very long process of lived, cooperative material practice played out across disciplinary and often cultural contexts (see Clemow et al., 2018a; Hamilton, 2014). Writing and authorship in these contexts is also collaborative, even if the traces of such collaboration are hidden from most readers. We discuss scientific collaboration during the regulated research and development of medicinal products like drugs and vaccines. Regulatory documentation represents a vast body of work that informs many scientific publications and additional textual genres, such as lay summaries, that are freely available to researchers and the general public (see Benau, 2020; Bernhardt, 2003; DeTora, 2018, 2020a, 2020b; DeTora and Klein, 2019; Hamilton, 2014). These studies are often also published (see Battisti et al., 2015; DeTora, 2020b; Montgomery, 2017; Winchester, 2017). We will explain some of the particulars of scientific collaboration as it intersects with regulatory writing practice by situating this context against more commonsense notions of scientific writing like those Perkel (2020) presumes.

Science represents many different disciplines and practices that intersect in various ways, even in a specific, narrow context like an individual regulatory document. David B. Clemow and colleagues (2018a, 2018b) provide an "analytical report" that details how competency in medical writing – a central profession within regulatory documentation – intersects specifically with the ability to manage multidisciplinary team environments. These collaborative teams define basic skills, such as scientific and mathematical literacy, project management, and linguistic literacy, as well as the more specialized knowledge of regulations, clinical research methods, statistical analysis, and health authority expectations. Thus medical writing competency relies on deep and specialized knowledge as well as on more basic forms of literacy and relationship building. These forms, like a clinical trial protocol, study report, and common technical document frameworks (Table 11.1), also parallel genres like the lab manual, lab report, and review paper.

One way to understand how collaborations within biomedicine represent individual communities of scholarship is through what linguist John Swales (1990) identified as "discourse communities" (p. 9), or "sociorhetorical networks that form in order to work toward common sets of goals" (p. 9). Swales identifies writing genres as having "features of stability, name recognition and so on" (p. 9) that are self-determined by a specific discourse community. The

Table 11.1 Examples of common regulatory documents and their functions

Regulatory document	Function	Sample guideline
Clinical protocol	Specifies the rationale and planned materials and methods used for a study of a medicinal product in humans	ICH E6 Good Clinical Practice www.ich.org/page/efficacy-guidelines
Clinical study report	Reviews the actual conduct and presents results of a clinical study, highlighting differences between the original plan and actual practice	ICH E3 Structure and Content of Clinical Study Reports www.ich.org/page/efficacy-guidelines
Common technical document	Serves as a dossier of all completed and ongoing studies in all scientific areas associated with a specific medicinal product to aid in health authority decision making	ICH M4 Common Technical Document www.ich.org/page/ctd

Note: ICH = International Committee on Harmonisation of Technical Requirements for Pharmaceuticals for Human Use.

structured research format of introduction, materials and methods, results, and discussion is common to many scientific discourse communities, which means that the presence of these specific features is not always helpful in sorting out the identities of sociorhetorical networks. Regulatory documentation, similarly, comprises specific genres, such as the clinical study report (see Table 11.1), that appear across all types of clinical studies. Hence, subsequent discussions of genre, like Anis Bawarshi and Mary J. Reiff's (2010) notion that genre might represent communities of scholarship that exist in direct conversation are also important. For instance, Julia Phillippi, Frances Likis, and Ellen Tilden's work on authorship grids (2018) suggests that understanding research contributions and relationships can be facilitated by understanding certain shared writing practices that are developed within communities.

Recent studies of biomedical publishing, like Jesse Fagan and colleagues' "Assessing Research Collaboration Through Co-Authorship Network Analysis" (2018), highlight how scientists seek to understand their communities of exchange and to enhance research collaboration. Tracing coauthoring and citation practices using social network analysis, Fagan and coauthors find that academic research networks tend to be strongly defined by current and former departmental relationships. This pattern differs strongly from that of global biomedical research studies, which require broad networks of international collaborators (see Battisti et al., 2015; www.ich. org; Winchester, 2017): thus, understanding one type of scientific discourse community may not be enough to elucidate the subtleties of all scientific writing collaborations.

The stakes of understanding the boundaries and practices of discourse communities extend into material practices of advocacy and educational access. An ethnographic analysis by Randy K. Yerrick and Andrew Gilbert (2011), for example, links problems in entering scientific discourse communities with the perpetuation of existing social and educational inequities. Underrepresented students in STEM fields are marginalized in large part because of the opacity of scientific discourse and the unyielding nature of its discourse communities. Yerrick and Gilbert find that programs intended to transition underrepresented students into STEM majors generally fail to inculcate these students with the mental habits of scientific discourse communities. Worse, still, the programs themselves mobilize discourses that marginalize the students still further, reinforcing rather than mitigating identified inequities. Similar problems may extend to any nonmember of a scientific field, suggesting a need for a better understanding of how these communities self-constitute not only to gain access to scientific careers but also to foster rhetorical analyses.

The individual apprenticeship model of scientific writing

Most professional scientific writing practitioners, including biomedical writers, learn through an apprenticeship model. In *The Chicago Guide to Communicating Science*, Scott L. Montgomery (2017) describes this as a sort of on-the-job "osmosis" (p. 5), where graduate students and post-doctoral fellows either "pick up" (p. 5) writing skills or are tutored individually by mentors in the context of a specific project. These apprenticeship models depend on individual working relationships with an existing member of a discourse community. Collaborations with novice or apprentice writers at the undergraduate level, similarly, require careful oversight by an expert who divides tasks and aligns the text into a cohesive whole. Our coauthor (SS), a PhD chemist who has published dozens of collaborative peer-reviewed papers, describes this process as follows:

> Undergraduate students have limited exposure and little to no experience in writing research papers for publication. These students are unfamiliar with the extreme condensation of information that occurs in such papers, especially the synthesis and analysis information. My role as a mentoring collaborator is to guide the student by assigning tasks, modeling effective writing by completing sections myself, and editing and rewriting in an iterative and collaborative manner.
>
> If the paper presents experimental work, then the formal lab report format can serve as a starting point. For an advanced student, the lab report structure is familiar. Often, I task the undergraduate research student with writing the experimental methods and results. However, there is an accepted method of presentation in a research paper: information is organized and written using unique coding that the experienced reader knows how to interpret. Undergraduate lab reports do not use this coding since there is no page or word count limitation, and professors want to evaluate the data and student conclusions. In published papers, results are often presented in tables, graphs and figures to best condense and clearly represent the information. Creating these graphics is an art requiring strong organizational skills and familiarity with the uses of different formats. For the introduction and conclusions to a manuscript, I work with the students to identify relevant prior research and help them cast their research into a broader context. We research and review relevant prior publications over several meetings, engaging in a dialogue about the previous work and its relevance to this work. This work also requires critical reading skills.

Here, writing skills are imparted via an intensive series of interpersonal interactions, tailored to a specific project and adjusted in the context of individual working relationships, generally for a specific target audience and venue. This mode of working requires a significant investment of time and energy into specific projects and particular relationships. Furthermore, the lead researcher is responsible for organizing and dividing up tasks, as well as reviewing and editing a paper for the intended publication venue, which limits the possibility for true peer interactions.

Our chemist coauthor also relies on an informal site of collaboration in the context of a personal relationship with a family member.

> I have always felt viscerally that writing a paper is a daunting task. To lessen anxiety and bring order to my thinking, I use an outline method that my sister explained to

me years ago (while working on my dissertation). It's easy to outline sections: introduction, discussion, conclusions. From there, I think about what I want to include in each section. I gradually fill in the outline until I am writing practically full sentences. The key to this process is to start with scientific content and terminology, filling in details only as they become necessary. This method prevents rambling and lessens the need for revision and proofreading. This process helps me to break up the paper into digestible chunks in a structured, systematic manner so that each "chunk" is a smaller, more approachable piece that is integrated into the whole (based on a personal communication).

This outline model has been adapted to various contexts for novice writers. Our colleague adjusts this helpful mode of authorship within individual apprentice authoring groups, reinventing the process as needed to suit each new project and collaborator. The methods outlined here can be leveraged to inculcate undergraduate and graduate STEM students with the habits of mind necessary to gain success in a professional environment. Unfortunately, this highly individualized method is not scalable to large biomedical research programs. As research projects increase in size and complexity, collaboration models must be adapted and often made more "lean" in order to optimize efficiency (see Bhardwaj et al., 2017). Breaking up the parts of desired product into achievable chunks for completion by the appropriate expert in the collaboration is one common way of organizing the collaborative effort. Next we discuss biomedical regulatory writing in a setting of international collaboration in the context in which Clemow and colleagues describe medical writing competency.

Biomedical writing competency

Medical writing competency, as described by Clemow and coauthors (2018a), revolves around several major functional areas:

- scientific strategy, or the reasons why a project is being completed
- document preparation, review, and approval, or the means by which a document is compiled, any necessary text is written, and the resulting work is subjected to reviews for scientific accuracy and alignment with regulations and the scientific strategy
- project management, or the manner in which meetings, teams, and relationships should be organized
- developing and evaluating standard operating procedures and authoring tools and templates, or creating an infrastructure within which to establish procedures and retain specific control of document drafts
- relationship building with employees, clients, and vendors, or best practices for managing projects conducted by different institutions and also for outsourcing certain activities
- professional development, or the means by which medical writers develop skills and knowledge and then exchange them

The model of document development Clemow and coauthors describe is fundamentally different from that detailed by our chemist colleague or commented on by Montgomery. Medical writing competency hinges on the knowledge of outside writing experts managing documentation on behalf of the people who designed and conducted a study and then collected data. The major medical writing competency areas are difficult to learn, even before one considers the scientific subject matter, because these collaborations must balance discourse skills separately from content mastery. The feat of acquiring competency in regulatory writing settings is made even more difficult by the fact that most medical writers, as well as their expert scientific and medical collaborators, initially learn to write using the apprenticeship models just outlined.

It may seem that the medical writing function comes in only at the end of a project. However, in biomedical research, every stage of research is regulated, which means it must be documented and reviewed. The regulatory constraints and documentation requirements for research and development form a series of discourses, each of which was consciously developed through collaboration and invention. One of the features of these discourse communities is that their readers, writers, and reviewers are presumed to understand regulatory requirements, which often remain unstated in the final documentation (see Benau, 2020; Clemow et al., 2018a, 2018b; Winchester, 2017). This tension between spoken and presumed knowledge informs all biomedical collaboration, hence its writing practices and documentation. These global collaborations require connections through guidance documents, education, and other forms of governance. For example, the International Committee for Harmonisation of Technical Requirements for Pharmaceuticals for Human Use (ICH; www.ich.org) has been working for decades to provide guidelines regarding the design of studies, statistical analysis, safety reporting, and regulatory documentation. Each of their guidelines is proposed, then developed by a multinational team of experts who provide multiple opportunities for public commentary – including public disclosure of comments – before a final guideline is issued. In fact, public commentary has been a feature of ICH guidelines since the 1990s, well before online interactive software was available. These documents are publicly available at www.ich.org.

Models of regulatory writing collaboration

Since regulatory writing takes place in communities of individuals who learned to write using the apprenticeship model and under the constraints of regulatory practice, their collaborations include cooperation in a commonsense way (i.e. working together to produce a text), as well as more structured activities like review, oversight, and editing. For instance, Sam Hamilton (2014) describes a common regulatory document, the clinical study report, as requiring a great deal of organization in order to coordinate the contributions of dozens of experts who work at geographically distant sites. It is worthy of note that, unlike the work of our expert chemist coauthor, the functional identities of the various contributors – clinicians, statisticians, medical writers, bench scientists, physicians – predefine the nature and scope of their contributions to regulatory documentation. In other words, these contributors all work as independent experts, not as a group of apprentices under the direction of a single expert. Thus relationship management is essential. Clemow and colleagues, in fact, separately detail the knowledge, skills, abilities, and behaviors (2018b) necessary to foster effective collaborations as well as the specific functions, tasks, and activities (2018a) of these positions.

In the second edition of *Regulatory Writing: An Overview* (2020), Danny Benau presents a description of the history of collaborative practices associated with regulatory documentation. These include team building, interpretation of government regulations, and building an understanding of medical writing as a professional field. Much additional advice is available to aid this work. In "Review Process in Regulatory Writing" (2015), Gregory Morley details specific meeting types to "kick off" (p. 97) a regulatory document and to adjudicate comments with authors and the specific needs of different readers. For instance, Morley comments on the difficulties experienced (and caused) by department supervisors and area heads when faced with commenting on the minutiae of projects being completed by their employees and enjoins medical writers to keep such audiences in mind. His paper centers on "fostering goodwill" (p. 97) as an essential element of successful teamwork and hence effective documentation.

Additional work, like Sonia Costa's "Embracing a New Friendship" (2019), links the function of medical writers to the increasing automation of regulatory documentation. However, unlike Perkel (2020), who sees electronic tools in the context of traditional authorship, Costa cites automation as a means by which contributors no longer function as textual composers but instead oversee the assembly of documents by machine intelligence. Benau (2020) similarly comments that machine intelligence and the increasing use of algorithms to pull data from electronic databases into text-based formats are a natural outgrowth of the medical writing field. Tables and figures have long been automated, and the more rigidly structured documents that present research results, as opposed to interpretations, offer another opportunity for computer assistance. If a specific regulatory document or section of a document is intended as a formulaic list of short paragraphs built through an algorithm that adds prescribed information into premade templates, then a lean, computer-based approach increases the goodwill of teams that Morley (2015) encourages by limiting their need to double-check data, spelling, formatting, and style.

Taken together, the advice of experts in regulatory documentation, then, presents several models of collaboration including medical writers:

- A medical writer coordinates a team of collaborators in composing documentation by accepting and editing contributions from expert colleagues, fostering goodwill, and achieving consensus through real-time meetings to ensure that all functional areas are adequately represented (see Clemow et al., 2018a, 2018b; Hamilton, 2014; Morley, 2015).
- A medical writer receives specifications from a group of experts and data owners in order to reflect their ideas effectively in writing (see Winchester, 2017).
- A medical writer conveys specifications (developed by an expert group) to a programmer to obtain documents or parts of documents composed by machine intelligence using algorithms, then oversees a review by content experts (see Benau, 2020; Costa, 2019).
- A medical writer composes hybrid documents containing elements penned by colleagues as well as components generated by computer algorithms, then oversees a review and approval process (that may be adjusted to accommodate time zones and schedules) to ensure that the resulting document meets audience needs (see Benau, 2020).

What each of these models has in common is the central function of a medical writer who combines competencies in functional areas, writing, and interpersonal skills. Like our coauthor leading a group of apprentice writers, the medical writer must help provide a framework for success, helping teams identify the appropriate process steps and compiling documents. Because guidance documents like ICH E3 (see Table 11.1) prescribe the outlines and structures of specific regulatory documents, the medical writer may have to advise contributors on format requirements as well as matching contributions to their appropriate places in a final dossier. Thus the knowledge competency identified by experts like Benau, Clemow and coauthors, Morley, or Hamilton means that medical writers possess important independent expertise.

Unlike our chemist coauthor, however, the medical writer is not the ultimate owner of the document or the data it contains, and the focus has shifted from the lead communicator as lead expert to the lead communicator as only one of a group of equally (if not more) expert colleagues. These relationships are generally bounded by standard procedures and guidance documents. In fact, as Winchester (2017) emphasizes, the named authors or leaders of the work are responsible for providing leadership and guidance to the writer. In these scenarios, it is the presence of the medical writer as a central coordinator with content knowledge, discourse expertise, and project management expertise that allows huge multiperson teams located in disparate geographies to collaborate successfully and meet regulatory audience expectations. This single focus is necessary to accommodate the scale of such projects by eliminating the need for the type of highly personalized adjustment our chemist colleague describes as necessary when working with student coauthors.

Rhetorical and technical communication interventions

Several papers have examined the discursive and rhetorical characteristics of regulatory documentation for audiences in rhetoric, writing studies, or technical communication. Prior studies by Stephen Bernhardt, Gregory Cuppan, and various coauthors (Bernhardt, 2003; Bernick et al., 2008), for instance, address how regulatory document reviewers can avoid abstract notions about "writing" and instead think about the direct content needs of end users and audiences. This work considers the role of regulatory medical writers (like the authors previously cited) working in complex matrices of collaborators and reviewers. In "Principles of Technical Communication and Design Can Enrich Practice in Regulated Contexts" (2018), DeTora later situated this work by recourse to Wilson and Wolford's description of "[postmodern] discourse workers" (2016), who, they suggest, might leverage their knowledge and influence by understanding the underlying power relations that inform the discourses within which they participate. The medical writing competency model suggested by Clemow and coauthors (2018a, 2018b) and Morley's description of building goodwill (2015) certainly seem consistent with such an understanding.

Additional work on regulatory settings suggests further opportunities for intervention in rhetoric or technical communication. For instance, plain language summaries of clinical trial results, as DeTora notes (2018), are intended to be made freely available to a global audience and therefore subject to further analysis for audience appropriateness. A subsequent entry by DeTora and Michael J. Klein (2019) leverages "invention questions" as framed by technical communication experts Bowdon and Scott (2003) as a means of identifying the relationship between technical communicators, content development, and audiences. Clemow and colleagues (2018a), like Bowdon and Scott (2003), emphasize collaboration and audience as essential factors for writing competency. However, the technical intricacies evident in work by and for medical and regulatory writing experts may mask underlying needs that receive greater attention by technical communicators.

Another need is to understand the breakdown of discourse communities within the biomedical research field. In "Competing Mentalities: Situating Scientific Content Literacy Within Technical Communication Pedagogy," DeTora (2020a) describes a means of accessing the unspoken yet shared practices of the biomedical research community as a means of rational self-constitution, in contradistinction to what Carolyn Miller (1979) considered to be the "contextless logic" (p. 612) often attributed to scientific forms. Guidelines for writing and peer-reviewing biomedical documentation, like those published by the Food and Drug Administration or the ICH, are a powerful metadiscourse, or form of writing about writing, developed through collaboration in order to organize and enhance further collaborations. As described by various experts (see Benau, 2020; Clemow et al., 2018a, 2018b; Hamilton, 2014; Morley, 2015; Winchester, 2017), these works lay out specific types of relationships necessary to produce certain types of documentation. Further study is needed to examine the operation of these works as discourse communities and the ways that such understanding can be applied in pedagogical and professional settings.

Next steps

Each element of the work of biomedicine contributes to an overall collaborative portfolio of writing that communicates scientific information to regulatory authorities, peer audiences, and the public. In general, such work is completed by teams of individuals in various specialty areas including bench scientists, clinicians, trial coordinators, data managers, ethicists, health authorities, as well as medical writers. Documentation begins before any study or trial; for example, a clinical study cannot begin before a study protocol, trial registry, and informed consent documents are written (see ICH E6 at www.ich.org). During a trial, safety reports may be made, a statistical analysis plan is completed, and any changes to the study – for instance, the need to change in-person visits to phone contact during the recent pandemic – are also documented (see Benau, 2020; DeTora, 2020b; Hamilton, 2014). After the trial is completed, a statistical analysis is completed, a report written, and results posted to the trial results registry. Publications and lay summaries of study reports and publications may also appear. Multiple guidelines exist describing the best ways to produce these different documents (See www.ich.org; Benau, 2020; Bernhardt, 2003; Clemow et al., 2018a; DeTora 2020b; Hamilton, 2014). Together, these circumstances represent a rich field for inquiry. Despite the publications by researchers like those just cited, many additional areas of feedback and collaboration remain unexamined by technical communication and rhetorical scholarship.

The shift between a lead scientist as lead writer/data owner/expert to a medical writer as a coordinator capable of foregrounding the work of subject matter experts like statisticians and physicians may appear disheartening to those who are interested in writing studies or technical communication. As Clemow and contributors note (2018a, 2018b), medical writing competency seemingly requires advanced qualifications in the sciences and many superadded areas of training, such as regulatory requirements and document formats. In fact, the endeavor of biomedical research would be slowed if not for the presence of specialists who function as what Wilson and Wolford (2016) might call "discourse workers." Significantly, Morley (2015) and Hamilton (2014) comment on the ability of regulatory writers to navigate relationships with various colleagues, hinting at the notion that understanding power relations is essential to success in this milieu.

Where these circumstances leave readers in writing studies, technical communication, and rhetoric is somewhat unclear. Certainly, it is useful to understand that the dominant model of scientific expert as lead writer used by our chemist coauthor to train undergraduate students does not extend into all areas of scientific writing and also to recognize that writing expertise on its own does not confer mastery over scientific contents (or discourses). Most crucially, the collaboration models in regulatory documentation, which situate a "discourse worker" (the medical writer) as the central coordinator of a group of experts emphasizes the importance of fields like writing studies, technical communication, and rhetorics of science, health, and medicine. Within the bastions of biomedical science, as studies become larger, it becomes necessary for certain professionals to concentrate on writing as an expert practice and to break away from the commonsense models that propel undergraduate research and education. Considering the potential for writing as parts of large-scale projects can be an exciting area of research in rhetoric and communication studies as well as undergraduate science and writing education.

Acknowledgment

The authors wish to thank Dr. Laura Janda, Professor of Russian at UiT–The Arctic University of Norway, for early conversations and insightful comments on early drafts of this paper. We also thank Risa Gorelick for the introduction.

Bibliography

Battisti, W.P., Wager, E., Baltzer, L., et al. (2015) 'Good publication practice for communicating company-sponsored medical research: GPP3', *Annals of Internal Medicine*, vol. 163, pp. 461–464; doi:10.7326/M15-0288

Bawarshi, A.S., & Reiff, M.J. (2010) *Genre: An introduction to history, theory, research, and pedagogy*, West Lafayette, IN: Parlor Press.

Benau, D. (2020) 'An overview of medical and regulatory writing', in L. DeTora (ed.), *Regulatory writing: An overview*, Baltimore: Regulatory Affairs Professionals Society, 2nd edition, pp. 1–14.

Bernhardt, S.A. (2003) 'Improving document review practices in pharmaceutical companies', *Journal of Business and Technical Communication*, vol. 4, no. 17, pp. 439–473, https://doi.org/10.1177/1050651903255345

Bernick, P., Bernhardt, S.A., & Cuppan, G.P. (2008) 'The genre of the clinical study report in drug development', in B. Heifferon, & S. Brown (eds.), *Rhetoric of healthcare: Essays toward a new disciplinary inquiry*, New York: Hampton Press, pp. 115–132.

Bhardwaj, P., Sinha, S., & Yadav, R.K. (2017) 'Medical and scientific writing: Time to go lean and mean', *Perspectives in clinical research*, vol. 3, no. 8, pp. 113–117. https://doi.org/10.4103/picr.PICR_11_17

Bowdon, M., & Scott, J.B. (2003) *Service-learning in technical and professional communication*, New York: Longman.

Clemow, D.B., Wagner, B., Marshallsay, C., Benau, D., L'Heureux, D., Brown, D.H., Dasgupta, D.G., Girten, E., Hubbard, F., Gawrylewski, H.M., Ebina, H., Stoltenborg, J., York, J.P., Green, K., Wood, L.F., Toth, L., Mihm, M., Katz, N.R., Vasconcelos, N.M., Sakiyama, N., Whitsell, R., Gopalakrishnan, S., Bairnsfather, S., Wanderer, T., Schindler, T.M., Mikyas, Y., Aoyama, Y. (2018a) 'Medical writing competency model – section 1: Functions, tasks, and activities', *Therapeutic innovation & regulatory science*, vol. 1, no. 52, pp. 70–77, https://doi.org/10.1177/2168479017721585

Clemow, D.B., Wagner, B., Marshallsay, C., Benau, D., L'Heureux, D., Brown, D.H., Dasgupta, D.G., Girten, E., Hubbard, F., Gawrylewski, H. M., Ebina, H., Stoltenborg, J., York, J.P., Green, K., Wood, L.F., Toth, L., Mihm, M., Katz, N.R., Vasconcelos, N.M., Sakiyama, N., Whitsell, R., Gopalakrishnan, S., Bairnsfather, S., Wanderer, T., Schindler, T.M., Mikyas, Y., Aoyama, Y. (2018b) 'Medical writing competency model – section 2: Knowledge, skills, abilities, and behaviors', *Therapeutic innovation & regulatory science*, vol. 1, no. 52, pp. 78–88, https://doi.org/10.1177/2168479017723680.

Costa, S. (2019) 'Embracing a new friendship: Artificial intelligence and medical writers', *Medical Writing*, vol. 4, no. 28, pp. 14–15.

DeTora, Lisa. (2018) 'Principles of technical communication and design can enrich writing practice in regulated contexts', *Communication Design Quarterly*, vol, 1, no. 6, pp. 11–19, https://doi.org/10.1145/3230970.3230974

DeTora, L. (2020a) 'Competing mentalities: Situating scientific content literacy within technical communication pedagogy', in M. J. Klein (ed.), *Effective teaching of technical communication*, West Lafayette, IN: University of Colorado Press/WAC Clearinghouse.

DeTora, L. (ed.) (2020b) *Regulatory writing: An overview*, Baltimore: Regulatory Affairs Professionals Society, 2nd edition.

DeTora, L., & Klein, M.J. (2019) 'Invention questions for intercultural understanding: Situating regulatory medical narratives as narrative forms', *Journal of Technical Writing and Communication*, vol. 50, no. 2, pp. 167–186, doi: 10.1177/0047281620906134

Fagan, J., Eddens, K.S., Dolly, J., et al. (2018) 'Assessing research collaboration through co-authorship network analysis', *Journal of Research Administration*, vol. 1, no. 49, pp. 76–99

Gross, A.G. (2006) *Starring the text: The place of rhetoric in science studies*, Carbondale: Southern Illinois University Press.

Gross, A.G. (2018) *The scientific sublime: Popular science unravels the mysteries of the universe*, Oxford: Oxford University Press.

Halliday, M.A.K., & Martin, J.R. (2015) *Writing science: Literacy and discursive power*, Bristol and London: The Falmer Press.

Hamilton, S. (2014) 'Effective authoring of clinical study reports: A companion guide', *Medical Writing*, vol. 2, no. 23, pp. 86–92, DOI: 10.1179/2047480614Z.000000000211

ICH. (1995) 'ICH E3 Content and format of clinical study reports', www.ich.org/page/efficacy-guidelines

ICH. (2000) 'The common technical document', www.ich.org/page/ctd

ICH. (2016) 'ICH E6 (R2) good clinical practice', www.ich.org/page/efficacy-guidelines

International Committee of Medical Journal Editors (ICMJE) (2019) 'Recommendations for the conduct, reporting, editing, and publication of scholarly work in medical journals', www.icmje.org/icmje-rec ommendations.pdf (Accessed 30 May 2021).

International Committee on Harmonisation of Technical Requirements for Pharmaceuticals for Human Use (ICH). Web site. www.ich.org

Miller, C.R. (1979) 'A humanistic rationale for technical writing', *College English*, vol. 6, no. 40, pp. 610–17, DOI:10.2307/375964

Montgomery, S.L. (2017) *The Chicago guide to communicating science*, Chicago: University of Chicago Press.

Morley, G. (2015) 'Regulatory writing: Review process in regulatory writing', *Medical Writing* vol. 2, no. 24, pp. 97–98

Perkel, J.M. (2020) 'Synchronized editing: the future of collaborative writing'. *Nature*, vol. 580, pp. 154–155, doi: 10.1038/d41586-020-00916-6

Phillippi, J.C., Likis, F.E., & Tilden, E.L. (2018) 'Authorship grids: Practical tools to facilitate collaboration and ethical publication', *Research in Nursing and Health*, vol. 2, no. 41, pp. 195–208, DOI: 10.1002/nur.21856

Swales, J. (1990) *Genre analysis: English in academic and research settings*, Cambridge: Cambridge University Press.

Wilson, G., & Wolford, R. (2016) 'The technical communicator as (post-postmodern) discourse worker', *Journal of Business and Technical Communication*, vol. 1, no. 31, pp. 3–29, https://doi.org/10.1177/1050 651916667531

Winchester, C. (2017) 'AMWA-EMWA-ISMPP joint position statement on the role of professional medical writers', *Medical Writing*, vol. 1, no. 26, pp. 7–8

Yerrick, R.K., & Gilbert, A. (2011) 'Constraining the discourse community: How science discourse perpetuates marginalization of underrepresented students', *Journal of Multicultural Discourses*, vol. 1, no. 6, pp. 67–91, https://doi.org/10.1080/17447143.2010.510909

12

EXPANDING EXPERTISE IN COMMUNICATING SCIENCE THROUGH PUBLIC AND CITIZEN SCIENCE

Jennifer C. Mallette

At the beginning of 2020, Australia's raging wildfires dominated the news. In addition to a focus on the fires as a product of global climate change and a symptom of a climate crisis, news outlets also pointed to a source of knowledge: the Australian Aboriginal people's techniques for managing the land with fire to prevent out-of-control wildfires. As a *New York Times* article noted, "Aboriginal people who have been through very difficult times are seeing their language, customs and traditional knowledge being reinvigorated and celebrated using Western science" (Fuller, 2020, para. 7). In her opinion piece, Alexis Wright (2020), a member of the Waanyi nation and a University of Melbourne professor, argues that not only do "leaders need to recognize the depth and value of Aboriginal knowledge and incorporate our skills in hazard management" (para. 2) but also that "listening to Indigenous knowledge would help curb catastrophic fires of the future" (para. 15). Essentially, what these news stories share is the belief that scientific knowledge – in this case, land management and fire prevention – is not just the purview of academically trained scientists and professionals. Furthermore, nonexperts can and do possess invaluable perspectives that should inform scientific practice. This idea is at the heart of citizen science and public science.

In scientific communication, practitioners should understand the interactions between different stakeholders and the wide variety of scientific knowledge and audiences. For instance, scientific communicators can incorporate the knowledge handed down in Indigenous societies for how to adapt to and work with the land. They can also consider how efforts to engage citizen scientists to collect data affect science and the communication of that science, such as recruiting citizens to scan images of the skies to search for brown dwarfs (Palca, 2017) or other forms of public participation (Bonney et al., 2009). Understanding the potentials for scientific communication with the public can also mean that scientists pursue research funding not from federal agencies or traditional sources but instead from interested citizens through crowdfunding campaigns, such as the one that uBiome ran on Indiegogo in 2013 (Palca, 2013). These campaigns require the use of trans-scientific genres to appeal to both scientific peers and potential public funders (Mehlenbacher, 2019). The complexity of these interactions and the range of sources of scientific knowledge mean that scientific communicators must consider how these efforts impact their communication with both expert and nonexpert audiences. Thus this chapter examines scientific communication through the lens of public science and citizen science,

 DOI: 10.4324/9781003043782-14

examining the intersections between scientific communication and the public. After briefly describing how science and the public interact, this chapter explores nonscientists' roles in both the creation and communication of scientific knowledge and ends by describing integrated models that exemplify the inseparable or intertwined nature of who is a scientist and who is a nonscientist in scientific communication.

Science and the public

Scientific communication is often divided into two audience groups: the expert and the non-expert or public audiences. This division often assumes the scientific reader to be expert and the nonscientist to be nonexpert, with assumptions about who is a scientist and who is not. For U.S. audiences, scientists are generally those who work in scientific fields and possess advanced education in the sciences, whereas the public is characterized as individuals who may not have a high level of scientific literacy, though they may be interested in science as it impacts their daily lives. In *Science in Public*, Gregory and Miller (1998) begin by highlighting the chasm between scientists and the general public, describing the "Movement for Public Understanding of Science" that seeks to address the public's lack of scientific literacy. These movements often position scientists as possessing scientific knowledge, with most of the public lacking the ability to read or understand science, creating a unidirectional understanding of the communication of scientific information. Furthermore, the movement for understanding science is often conflated with a push for the public to appreciate science. This appreciation means that the public is not expected or encouraged to critically examine the impacts of scientific endeavors on individuals and societies (Gregory and Miller, 1998; Perrault, 2013). This tension between misunderstanding the public's potential to understand with a focus on appreciation means that communicators focus on deficit rather than empowerment and that they overlook the ways science communication can engage a range of individuals in scientific knowledge production (see Natasha N. Jones's Chapter 5 in this volume, who uses Black Feminist Thought to examine this tendency to overlook the contributions and perspectives of marginalized individuals and groups, arguing that it presents a form of silencing that perpetrates white supremacy).

Thus this dichotomy between scientists and the public is neither as straightforward as it might seem nor beneficial for scientific communication. As a widely published writer for both scholarly and popular presses, Steven Jay Gould (2002) argues that "the conceptual depth of technical and general writing should not differ, lest we disrespect the interest and intelligence of potential readers who lack advanced technical training in science, but who remain just as fascinated as any professional, and just as well aware of the importance of science to our human and earthly existence" (p. 2). And as Perrault (2013) emphasizes throughout her book, scientists themselves can be nonexperts in fields outside their specific expertise. Justin Mando, in Chapter 21 of this volume, further complicates the split between scientists and the general public by arguing that scientists can and do also act as citizens and thus members of the public. In addition, nonscientific audiences can become sufficiently motivated to explore and understand scientific concepts, particularly if those concepts have some relevance to their lives. Thus the dichotomy between scientist and nonscientist, expert and nonexpert, can blur and be affected by background, interest, relevance, and motivation.

These understandings of *scientist* and *nonscientist* – and disruptions of that binary – are visible in two models of public scientific communication: Public Appreciation of Science and Technology (PAST) and Critical Understanding of Science in Public (CUSP). The PAST and CUSP frameworks help communication scholars articulate how science communication happens between communicators and their audiences. According to Perrault (2013), PAST (or a booster approach) frames communication as unidirectional, as passing from experts to nonscientific publics. It also frames the public audiences as in awe of scientific discovery, only appreciating advancements and developments rather than understanding them more deeply. Furthermore, this model emphasizes a hierarchy built on the binary between scientists as knowing experts and the public as unknowing nonexperts, who are often configured as passive recipients of information. However, the CUSP framework (or the critical approach) complicates this model by demonstrating the ways that public audiences can read or engage with scientific work critically, showing a nuanced understanding of science and offering critiques of that work or pushing against problematic configurations of scientific discourse. The CUSP framework most effectively presents a model of scientific communication that is not built on a simple unidirectional movement of scientific knowledge but accounts for the range of audiences, motivations, goals, and purposes that communicators and audiences bring to scientific texts.

As Perrault (2013) argues, a CUSP framework also unlocks the democratic potential of science. In her examination of risk and the public's ability to understand, she points to several studies that show the ability of nonexperts to quickly understand information that impacts their lives. Perrault notes, "In contrast to the deficit model belief that people are not interested in or able to understand scientific issues, a democratic view respects that Americans 'believe they have a right to be involved – experts alone should not choose what benefits should be sought and what risks the research might acceptably pose'" (pp. 145–146). Here, Perrault distinguishes between technocratic and democratic models to demonstrate the limitations of the technocratic approaches based in PAST frameworks that do not address assumptions. Meanwhile "CUSP model texts also examine facets of a risk situation that a technocratic approach considers neutral" such as addressing "limitations of numeric risk assessment" as well as "emphasizing interpretation, and considering questions of power, control, and other social factors that will affect who experiences the risk and how" (p. 147). In her view, models that minimize the critical understanding of nonscientists disempower individuals, while models that account for nonscientists' motivations, interests, and abilities allow these nonscientists to use scientific knowledge to assess risk and take action to benefit themselves and their communities. And ultimately scientists benefit too by bringing these stakeholders into the development of experiments and through the communication that engages them. The democratic potential emerges when science is aligned with social aims, or "the power to balance scientific optimism with socially informed caution" (p. 159). Thus science is viewed not as separate from the society in which it exists but informed and guided by this "new social contract" (p. 158).

Perrault (2013) asserts that communicators "need to stop chasing the elusive grail of pure science and technocratic certainty, and accept realities about how science works and about the contingent nature of the knowledge it produces" (p. 165). This movement would promote a new social contract that "focuses on science-in-society and values socially robust knowledge" (p. 165). Her aim is to show how popular science writing can invite nonscientists not to be admirers of science but instead to be critically engaged citizens who are able to use their understanding to participate in democracy. Furthermore, she claims:

> I have argued that popular science writing can help foster CUSP because it influences what people know about science and what they know about science-in-society; how people view science and how they think science should be viewed; how people view themselves in relation to science, including what kinds of involvement they consider appropriate; how people think about science's responsibilities toward society; and how people think about society's responsibilities toward science.
>
> *(p. 167)*

In other words, the ways scientists engage with a variety of readers and publics are essential for moving from a unidirectional understanding of science's role in society. Scientists cannot communicate as disinterested and separate entities because they are informed by the cultures and societies they participate in, and their work has impacts on – and is responsible for – the society in which they work and communicate.

In fact, scientists acknowledge that they need to be able to communicate with a range of audiences and that being able to communicate with those outside their immediate discipline is beneficial for conveying information to all audiences they may encounter, from funding agencies to the general public. For instance, Stony Brook University created the Alan Alda Center for Communicating Science, housed in its journalism school. Several graduate programs in the sciences require their students to take classes, and Alda offers improv workshops to train Stony Brook faculty to communicate their findings more clearly (Chang, 2015). As Alda comments, "Now more than ever scientists, researchers and health professionals need to communicate clearly and vividly with the public, with each other, with funders and with policy makers" (Alda Center, 2020, para. 3). This example highlights that the communication expected of scientists is not easily divided into expert and nonexpert audiences. This dualism between science and society is reductionist and does not allow participation from and engagement with those who can offer invaluable perspectives (Perrault, 2013).

Furthermore, political influences not only undermine the scientific enterprise but also how information is communicated and acted on. In the United States, citizens and scientists are facing fallout from the denial of climate change from political leaders, controversies about medical treatments for those affected by the coronavirus, and the widespread distribution of misinformation about supposedly scientific work. In addition, increasing attention to antiracist activism means that long-standing critiques of science that emerge from racism, sexism, and other biases are gaining new attention. As Yu and Northcut (2018) comment, "Indeed, in today's political climate, the ability to invite, engage, and persuade public audiences may be the best way to preserve science" (p. 3). In essence, the communication situation is not knowledgeable scientists versus an unknowledgeable public. Instead, science professionals, communicators, and a broader public must work together to slow the spread of misinformation, to generate action based on scientific knowledge (such as the urgency of addressing climate change), and to ensure that all audiences read and critically examine scientific findings for bias.

Expanding definitions of scientific communication and knowledge makers

Communicating scientific knowledge is often characterized as a binary between scientist communicators and lay audiences, or between the makers of scientific knowledge and the consumers of that knowledge after it has been "translated" for them. Several factors complicate or problematize this oversimplification, among them the ways that individuals participate in citizen science (explored in the next section) and in the scientific knowledge that emerges from the lived experiences of individuals or is passed down through communities. By breaking down the binaries between scientists as knowledge producers and nonscientists as knowledge consumers, science communication can open the door for broader participation that can inform scientific policy and lead to new discoveries with broad-ranging impacts. For one, entire schools of thoughts or perspectives may be silenced in these models, or the impacts of white supremacy and racism can go unchecked, as Jones addresses in her chapter in this volume. Additionally, as Perrault notes, "Drawing on multiple kinds of knowledge supports the knowledge-making mission of science" (p. 94). Furthermore, in understanding scientific communication as more complicated than unidirectional transmissions from experts to lay publics, we can also understand how knowledge may be shared in venues outside of published research articles in scientific journals, often with greater impacts on decision makers and stakeholders. Scientific knowledge can be passed down through practices and stories from one generation to the next or through the approaches developed through lived experiences that are shared as specialist knowledge. As Jones argues in her chapter, ignoring the knowledge and experiences of marginalized groups delegitimizes their communication as scientific and thus perpetuates white supremacy. Understanding communication beyond the narrow focus on published scientific work allows us to look for scientific communication in a range of spaces, from sources as varied as folklore, cultural traditions, the methods of those who engage with natural phenomena daily, the lived experiences of individuals and groups, among other means.

One example is how many Indigenous communities across the world communicate scientific knowledge and practices. Indigenous practices and methods of communicating that knowledge are gaining increased attention from non-Indigenous scientists and land management agencies, particularly around climate change, as Indigenous activists around the world highlight the need to address the impact of Western, industrialized activity and capitalism on biodiversity, waterways, and the environment as a whole (e.g. Lakota Law, 2019; Indigenous Climate Action, 2020; United Frontline Table, 2020). For instance, while Indigenous peoples make up less than 5% of the world's population, they protect 80% of the world's biodiversity (Raygorodetsky, 2018). In these communities, scientific knowledge is communicated through each group's cultural traditions, values, and engagement with their surrounding environment and offers perspectives that should be part of scientific communication and research. To that end, Indigenous groups advocate that they should be partners in scientific research, stressing that policy should move beyond "utilizing their knowledge in policy and science" but instead to "begin in a place of equity, respect, cooperation, and acknowledgment from the scientific community regarding the value of Tribal knowledge, culture, and customs" (Kawerak, 2020, para. 1). Thus this knowledge may emerge from non-Western scientific epistemologies but is no less valid or valuable.

These epistemologies are communicated in ways that may seem nonscientific but are worth attention as scientific communicative practices. For example, as Robin Wall Kimmerer (2013) highlights in *Braiding Sweetgrass*, while many believe that humans are at odds with and harmful to the environment, Native teachings focus on humanity's relationship to the land as positive and beneficial, "with ethical prescriptions for respectful hunting, family life, ceremonies that made sense for their world" (p. 7). In addition to foregrounding a positive relationship between human beings and their environment, Native teachings also present approaches to understanding – and communicating – natural phenomena in ways only recently being recognized by Western science. Kimmerer (2013) shares the story about how her grandfather understood, through Native teachings, the biological mechanisms of how pecan trees determine when to fruit. These teachings acknowledge that trees communicate with one another, a concept dismissed by science until recently, when scientists began to identify how trees share knowledge with one another. Here she demonstrates just one example of a Native teaching that is initially dismissed because it doesn't align with Western science but is later found to be valid. This story also presents a way of communicating scientific information that can be overlooked because it does not fall into prescriptions of acceptable scientific genres.

Thus scientific communication must engage with other ways of transmitting scientific knowledge and examine a broader range of sources. One example is the communication about using fire to manage land. When colonization of the western United States violently forced out Native American tribes from their ancestral homes, cultural burning – or the Indigenous practice of burning land in a controlled way to promote new growth and restore the land – was actively prevented (Cagle, 2019; Sommer, 2020). To be able to intentionally burn in a way that is not only controlled but supports land renewal, practitioners need to understand how wind and terrain affect how the fire moves, as well as the impacts burning would have on the local fauna. This complex knowledge has traditionally been "passed down through ceremony and practice. But until recently, it has been mostly dismissed as unscientific" (Cowan, 2020, para. 4). With the growing understanding that these cultural practices are crucial to land management in light of climate change, tribal leaders are partnering with scientists and fire agencies to communicate this knowledge as a way to preserve the land and prevent destructive, uncontrolled fires each year (Cagle, 2019; Cowan, 2020; Fimrite, 2020; Fuller, 2020; Kusmer, 2020; Sommer, 2020).

By recognizing that scientific knowledge can come from sources other than science professionals trained in Western traditions, scientific communicators should account for knowledge developed through cultural tradition and lived experiences. Perrault (2013) outlines four types of expertise: contributory, or the ability to contribute to a field of study; meta-expertise, or the ability to judge work outside of one's area; experiential, or expertise developed through experience; and interactional, or the ability to communicate across fields (pp. 83–84). Perrault's goal is to disrupt the simple binary of scientist versus nonscientist by highlighting how expertise can take on many forms. In her analysis of the stance on expertise in scientific writing, Perrault gives examples of how "community members – people with local, practical knowledge of an environment – can contribute data that institutional scientists don't have the time or resources to gather" (p. 89). These examples include an avid hiker's observations of local fauna, comprehensive Indigenous knowledge gained from long-standing interactions with the land, or fishers' practices that inform interactions with ecologists. Thus by recognizing broader sources of scientific knowledge and the methods used to communicate it, science communication can disrupt the binary between expert and nonexpert and acknowledge a range of scientific practices.

Scientific participation and communication through citizen science

Another space in which to understand how scientific communication increases CUSP engagement and democratic possibilities is citizen science projects. While citizen science projects have been formally recognized for at least the past 25 years, public participation has long been part of scientific practice in collecting data about bird strikes, bird counts, weather observations, and astronomical observations (Bonney et al., 2009). The Center for Advancement of Informal Science Education outlines three models for citizen science or public participation in scientific research: contributory projects, collaborative projects, and cocreated projects (Bonney et al., 2009). Each model represents a different level of engagement, from citizens contributing data to a scientist-designed project, to projects where citizens are engaged in refining projects, analyzing data, and/or disseminating findings of scientist-designed projects, to projects where citizens and scientists collaborate at most levels of the process from design to dissemination (p. 17). These models present approaches that not only complicate the unidirectional transfer of scientific knowledge but also ideas about who can create that knowledge. Through these projects, nonscientist participants essentially learn to be scientists, gaining knowledge about the scientific process and experience with scientific inquiry, as well as increased scientific literacy (Bonney et al., 2009, 2016).

In various conversations about citizen science, the focus often veers toward engagement and nonexperts' learning, which would appear to fall under a PAST approach. For instance, in Reid's (2019) study of scientific communication with scientists engaged in citizen science projects, she notes that scientists used the strategy of meaning expansion "to foster identification and cultivate wonder in mixed audiences" (p. 72), echoing a PAST focus on marveling at scientific knowledge. This framework is hard to avoid when experts engage with nonexperts; however, citizen science engagement moves beyond the passive unidirectional enjoyment and wonder in science toward active participation. This active participation, according to Bonney et al. (2009), serves a democratic function in that some projects allow participants to acquire "knowledge of community structure and environmental regulation" (p. 44) or in some cases, participate in the development of projects that emerge from community-based concerns. These cocreated projects that are "initiated to meet specific community needs can draw concerned citizens into the scientific process who might not otherwise be involved in science-related activities" and impact how information is communicated, to whom, and for what purpose (Bonney et al., 2009, p. 46). While contributory projects seem to recruit enthusiastic hobbyists, such as birdwatchers and amateur astronomers, cocreated projects offer the most democratic potential. As Bonney et al. (2009) note, co-created scientific research requires "building trust and mutual respect between scientists and participants, having an ethical commitment to protect participants from harmful consequences of participating in research, and creating transparency in the entire decision-making process at every step of the research project" (p. 48). While Bonney et al. (2016) stress that not all citizen science projects are created for socially just ends or to democratize science, the projects do often offer that potential, particularly the ones designed to "empower individuals to become active members of the decision-making process" (p. 14). Thus citizens are engaged in both knowledge creation and communication to support their communities. Furthermore, these participants also become critical thinkers of science as they communicate data and contribute to the development of project proposals and outputs, as should be the goal of public-facing science work and communication.

In fact, Reid (2019) found that Clay, her participant, highlighted the ways citizen science and public engagement positioned nonexperts as *doers* of science, focusing on the process over the results. Through this process, citizens would be able to better understand "the provisionality of scientific meaning making," which might help them realize that "results might be falsified or refined by other scientists down the road" (p. 89). Citizen scientists may also experience the reality that there may be a lot that scientists do not know about the natural world. Furthermore, Reid argues that citizen scientists could be prompted to reflect critically "on their assumptions about what it means to 'do science,' including who can participate in doing science" (p. 89). In this way, citizen science not only provides space for reflection and critical thinking on the part of nonexperts but also draws "reflexive attention to scientific meaning making itself" on the part of scientists (p. 90). In other words, citizen science does not solely benefit the individuals who choose to engage and participate, but it also enables scientists in their communication to reflect, effectively present information multimodally to expert and nonexpert audiences, and consider implications beyond the specific project.

In addition to increased participation and democratic potential, citizen science also affects how scientists engage with audiences as well as the genres through which they communicate. In her book *Science Communication Online*, Mehlenbacher (2019) examines the written products of citizen science through the lens of genre theory. She terms these genres as "trans-scientific" and argues that they "provide grounds where we can bring scientific knowledge together with moral and ethical, policy-driven, and social discourse" (p. 2). While she acknowledges that some online scientific communication is aligned with a popularization (or PAST) approach, "trans-scientific genres describe those forms of science communication that exist within both professional and public spheres of discourse" (p. 9). Her aim is not to examine how scientists use new media to communicate with a general public (that doesn't exist) through popularization but to show how the public is a complicated set of individuals with a range of motivations, expertise, and knowledge who also use new media to engage with scientists and scientific knowledge. Reid (2019) also points out that citizen science projects lead to a variety of multimodal genres that require a range of strategies and stances to account for "human actors and material contexts" (p. 90) and that can lead to understanding scientific communication as including "multidimensionality and complex relationships with other forms of scientific communication in the construction of scientific knowledge" (p. 93). Furthermore, Mehlenbacher (2019) calls attention to the concept of *para-scientific genres*, which "borrow some features from the internal discourse of science without the whole complex of features upon which the epistemic authority depends" (p. 37). These para-scientific genres also do not make use of rhetorical accommodations found in popularizations under the PAST framework, where the emphasis is on wonder or novelty rather than on critical conversations and reflections. Trans-scientific genres, in contrast, "operate alongside conventional genres," and they "borrow values from science" and "borrow values from the larger cultures within which the genres operate" (p. 39). Ultimately, both para-scientific and trans-scientific genres are made possible under citizen science; they demonstrate the ways these interactions between scientists and nonscientists striving for a shared goal can impact communication in addition to eliciting new genres.

Citizen science projects offer one form of public engagement, where nonexperts are asked to become scientists and contribute – or, in some cases, cocreate scientific projects and products and engage in trans-scientific genres. Another form of public engagement is crowdfunding, where scientists go to the public to get funding for projects, such as UBiome's 2013 Indiegogo campaign (Palca, 2013). Mehlenbacher (2019) examines crowdfunding campaigns through the lens of trans-scientific genres because of the ways they "inhabit a deliberative space comprised of an audience of peers and publics" (p. 81). Universities have also sponsored crowdfunding, such as a PonyUp campaign at Boise State University that was used by an astronomy professor to organize scientific education events around the 2017 total solar eclipse (PonyUp, 2020). Unlike applying for grants from national organizations or foundations, crowdfunding requires scientists to communicate for vastly different audiences through social media, engaging in creatively incentivizing participation in funding scientific work. For instance, for the eclipse crowdfunding campaign, the organizing astronomy professor, Brian Jackson, provided his expertise as a high-dollar incentive in the form of an "astronomer house-call" to stargaze with a group. Eclipse glasses were offered at the lowest level of participation, which was lucrative since the total eclipse would be best experienced throughout Idaho than in other parts of the country. Because of the rise of crowdfunded science, platforms specific to raising funds for science have emerged, instead of relying on existing platforms such as Kickstarter (Mehlenbacher, 2019). While some platforms provide physical rewards for funders, Experiment.com rewards backers with the knowledge that they are contributing to scientific advancement, providing them inside glimpses into ongoing scientific work (Mehlenbacher, 2019). In fact, Mehlenbacher notes that crowdfunded science is not meant to replace large-scale, publicly funded science grants but instead to "provide small-scale support for projects that cannot secure funding elsewhere," such as "citizen and otherwise independent or grassroots scientists as well as less experienced or junior researchers" (p. 78). Crowdfunding thus provides an additional avenue to explore how scientists and nonscientists interact, the democratic potentials of these approaches, and how these trends alter scientific communication.

Thus, given the increase in citizen science projects and the trend of crowdfunded science, scientific communication should account for these spaces where scientists and nonscientists interact. In her Chapter 29 in this volume, Reid argues that scientists must attend to how their work will be read by the public to support a holistic scientific literacy, using threshold concepts to facilitate those interactions. Moreover, these interactions are not unidirectional but rather allow scientists and nonscientists to influence and engage each other, and they lead to the emergence of new genres. In scientific communication, models that account for a wider recognition of scientific knowledge and civic participation are critical to the success of science more broadly. In her study of a citizen science project, Reid (2018) argues that "*citizen science recontextualizes science for science*," by which she means "that citizen science practices encourage the recontextualization of scientific activity and science-related communication occurring in public contexts for scientific discourse" (pp. 27–28, emphasis in original). Instead of seeing the path of scientific communication as unidirectional, she presents evidence of a more complicated, multidirectional model, where citizen scientists (or the public) influence how science is communicated or even the development of scientific knowledge itself.

Conclusion

Ultimately, in understanding how public science impacts scientific communication, communication researchers can continue to disrupt the reductive binaries and simplistic dichotomies that often characterize conversations about communicating with public audiences. In reconfiguring the relationships between scientists and the public, communicators can use this complexity – such as the ways Indigenous communities communicate scientific knowledge – to draw attention to a range of scientific epistemologies, reveal the interconnectedness between citizen science and scientific discourse, and offer a richer picture of how the public engages with science. This picture means that the division between the scientist and nonscientist is neither clearcut nor easy to define, given that even scientists can become nonexperts in other disciplines and that nonscientists can own expertise in an area of scientific focus. Thus science communication can reveal the circulation of scientific ideas as connected and even emerging from sources outside traditional science. Furthermore, it can reveal a path to better communicating scientific knowledge to support democratic goals, work toward justice, and engage more fully with a range of audiences to meet common goals.

References

Alan Alda Center for Communicating Science. (2020) 'Alda Center to launch programming to foster online science communication', https://web.archive.org/web/20201205112005/www.aldacenter.org/news/announcements/alda-center-launch-programming-foster-online-science-communication (Accessed 13 November 2020)

Bonney, R., Ballard, H., Jordan, R., McCallie, E., Phillips, T., Shirk, J., & Wilderman, C.C. (2009) *Public participation in scientific research: Defining the field and assessing its potential for informal science education*, A CASIE Inquiry Group Report, Washington, DC: Center for Advancement of Informal Science (CASIE).

Bonney, R., Phillips, T.B., Ballard, H.L., & Enck, J.W. (2016) 'Can citizen science enhance public understanding of science?', *Public Understanding of Science*, vol. 25, no. 1, pp. 2–16, DOI: 10.1177/0963662515607406

Cagle, S. (2019) '"Fire is medicine": The tribes burning California forest to save them', *The Guardian*, 21 November, www.theguardian.com/us-news/2019/nov/21/wildfire-prescribed-burns-california-native-americans (Accessed 13 November 2020).

Chang, K. (2015) 'Attention, all scientists: Do improv, with Alan Alda's help', *The New York Times*, 2 March, www.nytimes.com/2015/03/03/science/attention-all-scientists-do-improv-with-alan-aldas-help.html (Accessed 13 November 2020)

Cowan, J. (2020) 'Alarmed by scope of wildfires, officials turn to Native Americans for help', *The New York Times*, 7 October, www.nytimes.com/2020/10/07/us/native-american-burning-practices-california.html (Accessed 13 November 2020)

Fimrite, P. (2020) 'California Native American tribe has been burning forests for 10,000 years: What can we learn from them?', 14 October, www.sfchronicle.com/california-wildfires/article/California-tribe-offered-solution-to-wildfire-15638137.php (Accessed 13 November 2020)

Fuller, T. (2020) 'Reducing fire, and cutting carbon emissions, the aboriginal way', *The New York Times*, 16 January, www.nytimes.com/2020/01/16/world/australia/aboriginal-fire-management.html (Accessed 13 November 2020)

Gregory, J., & Miller, S. (1998) *Science in public*, Cambridge, MA: Basic Books.

Gould, S.J. (2002) *I have landed: The end of a beginning in natural history*, New York: Harmony Books.

Indigenous Climate Action. (2020) www.indigenousclimateaction.com/ (Accessed 16 November 2020)

Kawerak, Inc. (2020) 'Knowledge sovereignty and the Indigenization of knowledge', 17 April, https://kawerak.org/knowledge-sovereignty-and-the-indigenization-of-knowledge-2/# (Accessed 13 May 2021)

Kimmerer, R.W. (2013) *Braiding sweetgrass*, Minneapolis, MN: Milkweed Editions.

Kusmer, A. (2020) 'California and Australia look to Indigenous land management for fire help', *The World*, 1 September, www.pri.org/stories/2020-09-01/california-and-australia-look-indigenous-land-management-fire-help (Accessed 13 November 2020)

Lakota People's Law Project (2019) 'Spotlight on Indigenous activists,' 10 December, https://lakotalaw.org/news/2019-12-10/spotlight-on-indigenous-activists (Accessed 16 November 2020)

Mehlenbacher, A.R. (2019) *Science communication online: Engaging experts and publics on the internet*, Columbus, OH: The Ohio State University Press.

Palca, J. (2013) 'Scientists pass the hat for research funding', *NPR*, 14 February, www.npr.org/sections/health-shots/2013/02/13/171920261/scientists-pass-the-hat-for-research-funding (Accessed 13 November 2020)

Palca, J. (2017) 'Citizen scientists comb images to find an "overexcited planet"', *NPR*, 1 June, www.npr.org/transcripts/530766624 (Accessed 13 November 2020)

Perrault, S.T. (2013) *Communicating popular science: From deficit to democracy*, New York: Palgrave Macmillan.

PonyUp. (2020) '2017 Idaho eclipse', https://ponyup.boisestate.edu/project/5060 (accessed 2 December 2020)

Raygorodetsky, G. (2018) 'Indigenous peoples defend Earth's biodiversity – but they're in danger', *National Geographic*, 16 November, www.nationalgeographic.com/environment/2018/11/can-indigenous-land-stewardship-protect-biodiversity-/#close (Accessed 13 November 2020)

Reid, G. (2018) 'Shifting networks of science: Citizen science and scientific genre change', in H. Yu & K. Northcut (eds.), *Scientific communication: Practices, theories, and pedagogies*, London: Routledge.

Reid, G. (2019) 'Compressing, expanding, and attending to scientific meaning: Writing the semiotic hybrid of science for professional and citizen scientists', *Written Communication*, vol. 36, no. 1, pp. 68–98, DOI:10.1177/0741088318809361

Sommer, L. (2020) 'To manage wildfire, California looks to what tribes have known all along', *NPR*, 24 August, www.npr.org/2020/08/24/899422710/to-manage-wildfire-california-looks-to-what-tribes-have-known-all-along (Accessed 13 November 2020)

United Frontline Table. (2020) 'A people's orientation to a regenerative economy', *Indigenous Environmental Network*, June, www.ienearth.org/wp-content/uploads/2020/06/ProtectRepairInvesBuildTransformDRAFT02.pdf (Accessed 16 November 2020)

Wright, A. (2020) 'Want to stop Australia's fires? Listen to Aboriginal people', *The New York Times*, 15 January, www.nytimes.com/2020/01/15/opinion/australia-fires-aboriginal-people.html (Accessed 13 November 2020)

Yu, H., & Northcut, K.M. (2018) 'High stakes and great responsibility: An introduction to scientific communication', in H. Yu & K. Northcut (eds.), *Scientific communication: Practices, theories, and pedagogies*, London: Routledge.

13

SCIENTIFIC COMMUNICATION IN ENGLISH AS A SECOND LANGUAGE

Maurizio Gotti

Introduction

In recent years, there has been a great acceleration of the moves toward the globalization of sociocultural and communicative practices. The phenomenon of globalization has strongly favored English, which has become the preferred medium for international communication in many contexts. This spread of English as a second language (ESL) has had relevant implications in the field of English used for scientific communication, where the need for a common language is particularly felt for the development of specialized communication at a global level.

This spread of English has not only been considered a great advantage in terms of better global communication but has also aroused criticism as it has often been seen as a factor of marginalization or even obliteration of important existing differences among non-English-speaking communities, with the possible risk of a 'colonization' process preventing the attainment of authentic intercultural discourse (Scollon and Wong Scollon, 1995; Canagarajah, 1999). As globalizing trends commonly rely on covert strategies meant to reduce participants' specificities, they are likely to hybridize local identities in favor of Anglocentric textual models. Globalization thus offers a topical illustration of the interaction between linguistic and cultural factors in the construction of discourse, both within specialized domains and in wider contexts (Candlin and Gotti 2004, 2007). As language is strictly linked to the setting in which it is used, cultural elements operate as key contextual constraints, influencing both the level of discursive organization and its range of realizations (Pérez-Llantada, 2012).

In this chapter, I investigate the present globalizing trends in a specific field of scientific communication, i.e. in the academic world, highlighting both homogenizing and localizing trends. Indeed, in spite of the homogenizing trends deriving from globalization, scientific communication is not at all uniform but varies according to a host of factors, such as language competence, disciplinary field, community membership, professional expertise, and generic conventions, as well as some factors which clearly reflect aspects of the local tradition and culture. The survey of a few case studies discussed here is mainly based on a research project on *Identity and Culture in Academic Discourse* carried out by CERLIS (Centro di Ricerca sui Linguaggi Specialistici), the research center on specialized discourse based at the University of Bergamo, Italy. This research project has been led by me both as principal investigator and head of the research center. The

 DOI: 10.4324/9781003043782-15

first two case studies are part of the research project described in this chapter; although the third case study has not used the same corpus as the other two, it shares the same aim with them and has been reviewed by me as editor of the Linguistic Insights series published by Peter Lang. The three case studies have been selected as paradigmatic of some of the adaptation phenomena that occur when non-native speakers in other languages adopt the English language in the dissemination of the results of their research.

ESL in the research field

The adoption of English as a second language in the globalization of scientific communication has certainly provided a solution of great practical value but has also aroused fears and complaints in many non-English-speaking academics. The strict English-medium policies adopted by many scientific publications and book series have heightened non-English-speakers' awareness that the increasing use of this language in publishing and higher education might greatly reduce the role of national languages for academic purposes. Indeed, as there is a tendency of scholars to publish what they consider to be their best work in English so as to reach a wider audience, non-English-medium publications are often relegated to the status of local scholarly products providing only a marginal contribution to the mainstream because they are unable to disseminate knowledge through a global language.

These hegemonic tendencies of English are known to have relevant ideological and ethical implications in the marginalization, mitigation, or even obliteration of existing differences among 'colonized' communities. The complex interaction that opposes and often merges globalizing/localizing trends contains evidence of hybrid forms of discourse which are as unstable and provisional as the sociocultural identities they encode (Robertson, 1992; Wright, 2000) and which result in the simplification of discourse strategies, the recontextualization of actor–space–time relations, the enactment of processes of deterritorialization and reterritorialization, and the rise of cultural hybridity (Fairclough, 2006). Indeed, the new, contaminated system generally adopts the norms and features of the language/culture that is dominant in the wider discourse community, but it retains key traits of its users' native languages and cultures. At the same time, as English is the language dominant in scientific exchanges, it has a backwash effect that contaminates and hybridizes native systems. The gradual globalization or hybridization of discursive practices that first appeared in English-speaking environments now significantly affects also smaller languages (Cortese and Riley, 2002), which are subject to standardizing pressures in their semantic, textual, sociopragmatic, and even lexicogrammatical construction. Moreover, the influence of English has greatly conditioned the evolution of local specialized discourses (cf. Scarpa, 2007 for the spread of the nominal style and the related progressive depersonalization in Italian scientific prose).

Hegemonic tendencies have clearly been identified in scientific English, especially in the language policies commonly adopted by major international publications employing English as 'the world's academic lingua franca' (Oakes, 2005; Bennett, 2007). Non-native academics are thus expected to have good English literacy skills so as to be able to present their papers in that language at conferences and publish them in peer-reviewed journals and volumes. This expectation has greatly influenced scientists, with the result that the last decades have seen a massive conversion of journals from other languages to English, thus determining "a real loss in professional registers in many national cultures with long scholarly traditions" (Swales, 2000, p. 67). The story of the Egyptian marine biologist reported by Swales (1990, p. 204) shows that, in order to have her dissertation accepted, she had to rewrite it several times, modifying the original style typical of the Arabic way of writing and adopting the rhetorical conventions commonly shared by the American scientific community.

These trends have a number of serious consequences. The first is the concentration of immense power in the hands of a restricted group of academic gatekeepers, located in very few countries in the world. These countries have attained the right to enforce norms and to certify the scientific recognition of research carried out all over the world. Their academic power in certain disciplines is so strong that it can decide the careers of scholars who need to publish in leading international journals to validate and disseminate their research findings (Lillis and Curry, 2010; Hyland, 2015). There is therefore a risk of linguistic monopoly, scholarly chauvinism, and cultural imperialism. The exclusive use of English disfavors non-native writers who have "the triple disadvantage of having to read, do research and write in another language" (Van Dijk, 1994, p. 276). It may thus give rise to unintentional – or even intentional – discrimination against non-native speakers on the part of the editors of specialized publications (Canagarajah, 2002). The demands associated with writing and publishing in English are usually very strict and can be used by scientific publications to filter foreign contributions. Moreover, since only the British or American varieties are favored, a failure to comply with the journal's linguistic standards is usually penalized with rejection (Anderson, 2010).

Scholarly chauvinism and cultural imperialism may be detrimental to the growth of specialized knowledge itself. There is a risk that 'periphery' perspectives (Canagarajah, 1996) in the various disciplines may have no influence on the trends developed in intellectual centers located in a small number of monopolizing academies. The periphery, instead, may play a healthy role by questioning views prevailing in the center and by providing alternative perspectives. In recent years, there has been a heightened awareness in the academic world of the valuable contribution of non-Anglophone scholars working within dominant research paradigms and agendas. However, this increased awareness has rarely "translated into a recognition that the discipline[s are] also 'owned' nowadays (to use the new management-speak) by a very large number of people for whom English is neither a first, nor a second language" (Kayman, 2003, p. 52). In some cases, 'periphery' publications have changed their language or even title to suggest a more international collocation. For example, in 2006, the *Italian Heart Journal* (which was already published in English) changed its name to the *Journal of Cardiovascular Medicine*. As local journals are regarded as second-class research tools by the Italian medical community and since medical literature is regarded as being more competitive if published in the UK or the United States, the scientific board of the *Italian Heart Journal* decided to conceal the peripheral provenance of the journal by assigning it to an American publisher, while maintaining an Italian editor.

The CERLIS project

Being associated with communities linked to local as well as international conventions, scientific discourse has provided fertile ground for the analysis of intercultural variation, both at a textual level and in the communicative strategies embedded in its textualizations. Several research projects have investigated identity-forming features linked to 'local' or disciplinary cultures, as communicated through English in various academic domains by native and non-native speakers. A recent project on this issue is the *Identity and Culture in Academic Discourse* Project, drawn up by CERLIS. In this project, special attention has been given to the relationship between socioculturally oriented identity factors and textual variation in English academic discourse, focusing in particular on the detection of identity traits typical of different branches of learning (Gotti, 2012). Within such domains, the project investigated to what extent the cultural allegiance of (native or non-native) Anglophone discourse communities to their linguistic, professional, social, or national reference groups is affected by the use of English as a second language. To identify textual variants arising from the use of English as a native language or as the lingua franca of science, a corpus was compiled, formed by English texts for academic communication. The corpus also comprises some Italian texts for comparative purposes. Besides including two different languages, CADIS represents four separate disciplinary areas: law, economics, applied linguistics, and medicine. For each disciplinary area, various textual genres have been considered: abstracts, articles, book reviews, editorials, posters. The structural complexity of CADIS reflects its contrastive orientation: it is designed to be internally comparable, so its texts can be analyzed not only by disciplinary area, genre, language, and culture, but also historically. This is possible because the corpus covers a time frame of over 30 years, from 1980 to 2011. Including all language groups – native speakers and non-native speakers of English, and native speakers of Italian – a total of 2,738 texts (from 635 to 739 per disciplinary area) – have been inserted in the corpus. The corpus includes over 12 million words.

Case study 1: intercultural variations in medical editorials

The CERLIS research project has dealt with identity traits across languages and cultures, as the use of a given language affects the writing of a scholar, especially when it is not his or her native language. This is particularly evident in the case of English, whose recurrent use by non-native speakers requires a degree of adaptation of their thought patterns and expressive habits. This issue has been dealt with by various members of the CERLIS team. Giannoni (2012), for example, has investigated local versus global identities in medical editorials. His analysis of Anglo-American journals, English-medium Italian journals, and standard Italian journals suggests a considerable extent of intradisciplinary variation, both within and across languages/ cultures. The data investigated allow for the observation of the writing behavior of three different kinds of scholars: native-speaker English (NEng), non-native (i.e. Italian) English (ItEng), and native-speaker Italian (NIt). Since medical editorials (henceforth MEDs) are signed by only one or two authors, native-speaker status is relatively easy to determine, based on the author's name and affiliation.

A quantitative overview of the material (Table 13.1) shows interesting differences among the three sections in terms of average length, with NEng texts less than half the size of their NIt counterparts and ItEng somewhere in between. Discoursal complexity, as measured by average paragraph length, is instead greatest in NEng (44% higher than NIt). These figures suggest that while Italian MEDs are lengthier than their native English counterparts, they organize the discourse into far shorter units. For both parameters, the ItEng group occupies the middle ground between the two.

Table 13.1 Average size of texts by section

	Length (tokens)	Range	Tokens/paragraph
NEng	1,046	619–1,809	125
ItEng	1,882	609–3,637	113
NIt	2,429	1,154–5,222	87

Source: Giannoni, 2012, p. 65.

Giannoni's analysis shows that editorialists employ three types of MED, whose prominence and microlinguistic traits vary across the corpus:

1   *Advice editorials* are authoritative reviews of medical issues providing guidance for practitioners.
2   *Comment editorials* are opinionated interpretations of developments affecting the medical community, with recommendations for action.
3   *Message editorials* reinforce the journal's relationship with its readers, keeping them informed of its initiatives and developments.

While the orientation of the first subgenre is mainly teleological – i.e. driven by the need to shape medical practice – the second is evaluative and the third is phatic. A rough indication of the respective weight of these subgenres across the corpus is given in Table 13.2, which includes a fourth column, due to the presence in NIt of three spurious text types presented as *editoriale* (namely a review article, an essay, and a conference talk). Interestingly, the three subgenres are documented across the corpus, with the sole exception of comment editorials. These are indeed the most variable subgroup, accounting for 80% of texts in NEng but none in ItEng. On the other hand, advice editorials are used far less, proportionally speaking, in NEng (10%) than in the two groups authored by Italians.

These data warrant the hypothesis that Italian editorialists: (a) are less likely to comment on current affairs and issues of a (non)medical nature, whether writing in their first language or in English and (b) understand the 'Editorial' not only as a genre but also (in NIt) as a slot for publishing other genres that deserve editorial sanction.

Moreover, unlike their NEng counterparts, Italian writers are likely to incorporate references to their own work – a self-promotional strategy observed in all the ItEng texts and in 40% of the NIt sample. Italian scholars appear therefore to be freer in their use of the MED genre, with no clear-cut distinction between the role of editorialist (knowledge validation) and that of researcher (knowledge construction). The high rate of self-citations in ItEng indicates that the two functions are particularly blurred when editorialists address an international audience through the medium of English.

The purpose of message editorials is essentially phatic, insofar as they seek to forge/maintain a strong relationship with the readership by keeping it informed of editorial decisions and policies. Consequently, editorialists act here in an institutional as well as in an individual capacity. Altogether, this was the least common MED subgenre observed in the corpus, accounting for only 10% of texts in NEng and NIt. The figure rises to 40% in the ItEng group – which suggests that the effort to engage readers overtly is greatest for English-medium publications originating from the periphery. In ItEng, however, message editorials are always metatexts introducing/promoting the journal's advice editorials. The different use of message editorials across the

Table 13.2 Proportion of MED subgenres across the corpus sample

	Advice	*Comment*	*Message*	*Other*
NEng	10%	80%	10%	–
ItEng	60%	–	40%	–
Nit	50%	10%	10%	30%

Source: Giannoni, 2012, p. 67.

CADIS sample is clearly observable in their macrostructure. The texts in NEng are essentially unstructured narratives bringing to the attention of readers important developments in the journal's life (and/or that of its affiliates). Such MEDs span events in the past, present, and near future, as shown by the following excerpt:

1   The wind of change is in the air again. *The British Journal of Plastic Surgery* is a great and almost a venerable title, but it seems that BJPS can never stand still. . . . Many of our readers are discovering the benefits of Science Direct, which carries the full text of BJPS from the very first issue, available on line and fully searchable through hypertext links. . . . Now, from January 1st 2006, our journal will become JPRAS, *The Journal of Plastic, Reconstructive and Aesthetic Surgery*, and will be published every month. (NEng, MEED498[1])

Giannoni's analysis thus shows that, as a consequence of the composite generic profile of the medical editorials analyzed and of the coexistence of no less than three distinct subgenres (Advice, Comment, Message), editorialists are keen to adapt their voice to the specific communicative purpose text, taking on a different identity and evaluating a different target, as summarized in Table 13.3.

Moreover, the multilingual and multicultural environment in which scholars are working within a globalized context implies the fact that editorialists are faced with the challenge of reconciling two 'small cultures' (their local scientific community and lingua-cultural affiliation) with a 'large culture' (the discipline as a global, translinguistic community). The easy option is to concentrate on the latter, forgetting that it can only emerge through a negotiation process involving the former. Italian scholars appear to be avoiding this risk and draft English editorials that do not merely incorporate elements of the Native English/Native Italian repertoire but do so in innovative and at times creative ways.

Case study 2: intercultural variations in argumentative strategies

Maci (2012) has compared the argumentative strategies employed in medical research articles (RAs) written by NEng with those written by ItEng in order to identify any cross-cultural differences in terms of argumentative devices employed by their authors. Analyzing the Discussion Section of 50 articles from two important journals of cardiology, she has identified several differences between the textual organization of English medical research articles written by native and non-native speakers, which seem to be linked to their authors' linguistic and cultural identity. The main differences are rhetorically realized through hedges and other argumentative strategies, such as the use of connectives. Indeed, NEngs tend to exploit more fully modality expressed by modal auxiliaries (such as *may, would*), verbs (such as *appear, suggest*), and adverbs (such as *likely*). The modal verb *may*, in particular, frequently appears in the NEng

Table 13.3 Generic profile of medical editorials

Subgenre	Voice	Target
Advice	Expert	Disciplinary knowledge
Comment	Journal	World
Message	Editor	Journal and editor

Source: Giannoni, 2012, p. 77.

corpus, to such an extent that it can be regarded as a keyword with high keyness (*may* occupies position 15). This is not the case in the ItEng subcorpus, where *may* occupies position 95. The minimal use of hedges in the ItEng subcorpus seems to be counterbalanced by other grammatical devices: whenever the outcome conforms to the expected results and is thus validated, Italian authors tend to interpret outcomes with the use of the present tense of such boosters as *confirm*, *find*, and *show* rather than using hedging devices. If hedges are used, there is a preference for *might*, which may be perceived by NNSs as carrying a stronger connotation of probability than *may*, or *should*, employed whenever a suggestion about the correct scientific procedures and/ or treatment is made. This occurs especially whenever the results do not confirm the initial hypothesis or whenever there is a gap in the existing literature filled by the present research. In these cases, NNSs of English seem to prefer the use of hedges and modal expressions to indicate probable interpretations or possible implications:

2 In our opinion, aortic plaques are those *the most likely to be responsible* for recurrent cerebral events. Furthermore, aortic atheromatosis *should be considered* as a clinical entity itself and should be related to different vascular districts than the cerebral one. This was demonstrated in a study by Pandian et al. [46], who affirmed that. . . . (ItEng, MERA242[2])

3 Although no complications occurred in any patient implicating the safety of cryoenergy, these results are slightly inferior to what can be expected with RF energy in terms of acute success. In 17 patients (nine AVNRTs, eight APs) out of 126 patients (13%) with acute successful ablation, recurrence of the arrhythmia and/or AP was observed. The percentage of recurrence is therefore higher than that usually reported with RF energy. . . . The high rate of recurrences in this series *may be ascribed* to a possible more limited lesion created by cryoenergy, which *can even further* decrease in dimensions in the early post-ablation phase owing to tissue healing. (ItEng, MERA250)

A further differentiation can be seen in the use of connectives. There is a lower frequency of connectives in RAs written by NNSs of English, which seems to reflect the trend already established by Italian authors as far as the use of hedges is concerned: whenever the claim is confirmed and supported by scientific literature in the field, Italian researchers seem less keen on exploiting argumentative strategies, as, apparently, reference to the literature becomes the objective evidence supporting the author's reasoning. For instance, the concordance list of *also* shows a different distribution of the connective: in the NEng subcorpus, it is mainly used to underline the findings resulting from the analysis, which may confirm the researcher's hypothesis; in the ItEng subcorpus, *also* is found in connection with reference literature supporting the researcher's data:

4 . . . the immediate postoperative period also demonstrated that the combination of clopidogrel and aspirin was more effective than aspirin alone in reducing MES. (NEng, MERA204)

5 Moreover, BNP is a strong predictor of mortality not only due to heart failure progression[35-37] but also to sudden death.[38] (ItEng, MERA228)

The more frequent use of *although*, *furthermore*, *hence*, *in contrast*, and *therefore* in the NSs subcorpus is indicative of the presence of a textual organization in which scientific information is offered in a coherent and convincing way. Here, the problematizing proposition is introduced by *although*, which positions the reader in the correct reasoning path: *although* presupposes the presence of a second part of a sentence which the reader expects to carry the right type of information necessary to decode the semantic value offered by the researcher's analysis:

6   Although sharing a common familial environment may inflate the estimates of heritability, we found low to moderate heritability for BMI, which in turn represents the maximal possible contribution of additive genes. (NEng. MERA209)

In the ItEng subcorpus, the extremely high frequency of such connectives as *on the contrary* and *on the other hand* seems to suggest a preference for a type of argumentation in which the author plays with a twist: first there is the introduction of common shared knowledge (and reference literature); then there is a counterclaim, from the author's research, supported by other cited literature. This is further emphasized by a list of evidential elements (and relevant literature), introduced by *first, second, third*, etc. which support the results of the researcher's study, as in the following example:

7   First, with respect to infero-posterior AMI, where sympathetic activation may follow transient signs of vagal hyperactivity,[20,21] anterior AMI is constantly followed by strong and stable signs of enhanced adrenergic tone;[20] thus, we avoided any potential flaw in the interpretation of the changes in vagal and sympathetic effects. In addition, the effects of cardiac rehabilitation have been extensively studied in patients with anterior myocardial infarction and reduced ejection fraction in whom concern for adverse ventricular remodeling has been expressed.[22,23] (ItEng, MERA234)

Italian authors seem therefore to prefer the use of an ipse dixit strategy: whenever a claim finds confirmation in the existing literature, they tend to adopt rhetorical strategies less frequently because the established knowledge is deemed to be sufficient to confirm their hypothesis.

Case study 3: linguistic variations in non-native texts in English

The complexity of the choices made by non-native English speakers depends on the fact that they participate in at least two different communities: the English-speaking scientific community and the global discourse community of their own discipline. To belong to the former community, they have to show that they are able to use English and master its norms of use, including grammar rules, word choice, idiomatic expressions, and technical aspects such as punctuation and spelling. Moreover, in order to be accepted by the English-speaking scientific community, scholars need to be aware of the practices commonly used in expository academic prose, as reflected in the guidelines provided by books on academic communication and by the notes to contributors published in international scientific journals. The following examples (from Noguchi, 2006, p. 57) clearly illustrate some of the expectations of the English-speaking community pointed out by the reviewers of submitted papers:

8   Thus, for colorectal adenocarcinoma, it is more useful to investigate the expression of X as well as that of Y for predicting tumor invasion and metastasis than examining Y only.

Revised version: *Thus, to predict tumor invasion and metastasis in colorectal adenocarcinoma, not only expression of Y but also that of X needs to be examined.*

Comment: The aim of the study can be more quickly grasped if the phrase dealing with the purpose comes earlier in the sentence.

9   However, the number of markers is still insufficient. From this standpoint, the present con-
     tig must be reexamined using a larger number of landmarks. Recently, RG was developed
     as a method to scan a large number of restriction sites distributed on entire genome. RG
     employs . . .

     Revised version: *However, the number of markers is still insufficient. From this standpoint, the*
     present contig must be reexamined using a larger number of landmarks. One solution to this problem
     is offered by RG, a method developed to scan a large number of restriction sites distributed on an entire
     genome. RG employs . . .

     Comment: Adding the discourse signal 'one solution' tells the reader what to expect.

At the same time, membership of the global discourse community of their discipline depends
on scholars' compliance with expectations concerning the specific academic genre to which
the text they are writing belongs. These include textual and paragraph organization in terms of
information presentation and ordering, as well as the need to consider cross-cultural issues. The
'rules', however, are not always easy to identify or define in clear terms, as is shown by the fact
that reviewers and editors often point to problems in the text without being able to indicate
exactly which rules are being violated or which criteria have not been met. Here is an example
of such comments cited by Noguchi (2006, p. 59):

10   Comment: There is a problem with the English throughout the text. It is not a very serious
     one, but it certainly detracts from the message and makes some important statements not
     immediately intelligible. Among the many examples I could quote, I will select these:

     "The clinicopathologic importance of the biologic aggressiveness has been well docu-
         mented in many reports." (First sentence of Discussion, page 8) What does this sen-
         tence mean? I think the authors are trying to say that some clinical and pathologic
         parameters of thyroid carcinomas have been found to correlate with the tumor aggres-
         siveness, but it sure takes a while to decipher the message.
     "The classification by Sakamoto et al. defined both papillary and follicular carcinomas
         as poorly differentiated carcinomas." I assume they are trying to say that Sakamoto's
         poorly differentiated carcinomas include tumors in both the papillary and the fol-
         licular category. (Anonymous reviewer for *The American Journal of Surgical Pathology*,
         December 1997)

Indeed, stylistic/rhetorical structures may differ from culture to culture; for example, Japanese
writers prefer a specific-to-general pattern in contrast to the general-to-specific pattern favored
by American writers (Kobayashi, 1984). Another well-known case is the one visually expressed
by Kaplan (1966) referring to the difference between linear (English) and circular (Oriental)
patterns in the rhetorical structuring of an argumentative paper. Since intercultural differences
are bound to influence the comprehension of events by people belonging to different
cultures, research in the field of contrastive rhetoric (Connor, 1998) has greatly helped the
identification of textual aspects which may be attributed to culturally determined schemata
reproducing a 'worldview' typical of a given culture. It has been shown that the non-native,
when communicating in English, is confronted with a psychocognitive situation where her
L1 linguistic and cultural schemata conflict with the schemata dominant in international
professional communities and is thus forced to negotiate and redefine her cultural identity in
order to successfully communicate in international intercultural settings. The importance of

compliance with such conventions (not only linguistic but also cultural ones) for the acceptance of an academic contribution has been aptly pointed out by Mauranen (1993, p. 263):

> The option of not conforming to the norms of the target linguistic culture is not available with respect to grammatical and lexical use, and, as it seems, at least some textual rules must be included in the same category, possibly more than we are accustomed to thinking at present. Breaking grammatical rules has different consequences from breaking textual or rhetorical rules originating in a national culture: by breaking grammatical and lexical rules, a writer conveys the impression of not knowing the language, which may in mild cases be forgiven and in serious cases cause breakdown of comprehension; by breaking rules of a text-linguistic type, a writer may appear incoherent or illogical; finally, by breaking culture-specific rhetorical rules a writer may seem exotic and command low credibility.

Conclusion

As shown by the present analysis, the use of English as a second language has determined important consequences on the status of scientific discourse. Indeed, research into the writing of non-native speakers in other languages suggests that textual differences should not merely be interpreted as unconscious 'errors' but are clearly part of the strategic creative choices made by authors in view of specific rhetorical objectives. Variation in style, language, and textual organization can also be attributed to the influence of the L1 and its culture. This is due to the fact that non-native writers tend to conform to the discursive rules of the target language, as a result of the inflexible models that are often imposed on them. Such constraints interact, especially in the case of English, with the emergence of intercultural identities independent of local traits, thus complicating the overall picture, with a tendency toward discourse-merging and hybridization. The findings reported here reflect the considerable challenges and opportunities that confront scholars seeking to achieve a delicate balance between their willingness to adhere to the mother tongue norms and conventions and their own individual competences and identity traits. Such factors have been found to interact, producing complex realities giving rise to textual realizations characterized by hybridizing forms deriving from interlinguistic and intercultural clashes.

Notes

1 The acronym MEED stands for 'Medical English Editorials'.
2 The acronym MERA stands for 'Medical English Research Articles'.

References

Anderson, L.J. (2010) 'Standards of acceptability in English as an academic Lingua Franca: Evidence from a corpus of peer-reviewed working papers by international scholars', in R. Cagliero & J. Jenkins (eds.), *Discourses, communities and global Englishes*, Bern: Peter Lang, pp. 115-144.

Bennett, K. (2007) 'Epistemicide! The tale of a predatory discourse', *The Translator*, vol. 13, no. pp. 1-19, https://doi.org/10.1080/13556509.2007.10799236

Canagarajah, S. (1996) 'Nondiscursive requirements in academic publishing, Material resources of periphery scholars, and the politics of knowledge production', *Written Communication*, vol. 13, no. 4, pp. 435-472, https://doi.org/10.1177/0741088396013004001

Canagarajah, S. (1999) *Resisting linguistic imperialism in English teaching*, Oxford: Oxford University Press.

Canagarajah, S. (2002) *A geopolitics of academic writing*, Pittsburgh: University of Pittsburgh Press.

Candlin, C., & Gotti, M. (eds.) (2004) 'Intercultural discourse in domain-specific English', special issue of *Textus*, vol. 17, no. 1. Genoa, Tilgher.

Candlin, C., & Gotti, M. (eds.) (2007) *Intercultural aspects of specialized communication*, Bern: Peter Lang, 2nd edition.

Connor, U. (1998) *Contrastive rhetoric: Cross-cultural aspects of second-language writing*, Cambridge: Cambridge University Press.

Cortese, G., & Riley, P. (eds.) (2002) *Domain-specific English: Textual practices across communities and classrooms*, Bern: Peter Lang.

Fairclough, N. (2006) *Language and globalization*, London: Routledge.

Giannoni, D. (2012) 'Local/global identities and the medical editorial genre', in M. Gotti (ed.), *Academic identity traits: A corpus-based investigation*, Bern: Peter Lang, pp. 59-78.

Gotti, M. (ed.) (2012) *Academic identity traits: A corpus-based investigation*, Bern: Peter Lang.

Hyland, K. (2015) *Academic publishing: Issues and challenges in the construction of knowledge*, Oxford: Oxford University Press.

Kaplan, R. (1966) 'Cultural thought patterns in intercultural education', *Language Learning*, no. 16, pp. 1-20

Kayman, M. (2003) 'The foreignness of English', *The European English Messenger*, vol. 12, no. 2, pp. 52-53, DOI: 10.1285/i22390359v24p47

Kobayashi, H. (1984) 'Rhetorical patterns in English and Japanese', unpublished PhD thesis, Columbia University, New York.

Lillis, T., & Curry, M.J. (2010) *Academic writing in a global context: The politics and practices of publishing in English*, London: Routledge.

Maci, S.M. (2012) 'The discussion section of medical research articles: A cross-cultural perspective', in M. Gotti (ed.), *Academic identity traits: A corpus-based investigation*, Bern: Peter Lang, pp. 95-120.

Mauranen, A. (1993) *Cultural differences in academic rhetoric: A text linguistic study*, Frankfurt am Main: Peter Lang.

Noguchi, J. (2006) *The science review article*, Bern: Peter Lang.

Oakes, L. (2005) 'From internationalisation to globalisation', *Language Problems and Language Planning*, vol. 29, no. 2, pp. 151-176, DOI:10.1075/lplp.29.2.04oak

Pérez-Llantada, C. (2012) *Scientific discourse and the rhetoric of globalization: The impact of culture and language*, London: Continuum.

Robertson, R. (1992) *Globalization, social theory and global culture*, London: Sage.

Scarpa, F. (2007) 'Using an Italian diachronic corpus for investigating the "core" patterns of the language of science', in K. Ahmad & M. Rogers (eds.), *Evidence-based LSP*, Bern: Peter Lang, pp. 75-95.

Scollon, R., & Wong Scollon, S. (1995) *Intercultural communication: A discourse approach*, Oxford: Blackwell.

Swales, J. (1990) *Genre analysis: English in academic and research settings*, Cambridge: Cambridge University Press.

Swales, J. (2000) 'Languages for specific purposes', *Annual Review of Applied Linguistics*, no. 20, pp. 59-76, DOI: https://doi.org/10.1017/S0267190500200044

Van Dijk, T. (1994) 'Academic Nationalism', *Discourse and Society*, vol. 5, no. 3, pp, 275-276, www.jstor.org/stable/42887924

Wright, S. (2000) *Community and communication: The role of language in nation state building and European integration*, Clevedon: Multilingual Matters.

PART 2

Scientific communication genres

14

THE SCIENTIFIC ARTICLE

Variation and change in knowledge communication practices

Marina Bondi

Introduction

The research article features prominently in the study of modern scientific discourse and takes pride of place in major works on scientific discourse, academic discourse, and genre analysis (most notably Myers, 1990; Swales, 1990, 2004; Hyland, 2000, 2015). The conventional rhetorical structure of the article has been widely described both by identifying specific units that perform coherent communicative functions within the standard sections of the research report and by exploring the link between communicative functions and language features. Diversity of approaches has largely been considered when looking at disciplinary variation (e.g. Hyland, 2000; Hyland and Bondi, 2006) and, more recently, at the combination of disciplinary variation and methodological choices: see, for example, Gray (2015) on theoretical, quantitative, and qualitative research and Thompson and Hunston (2019) on types of interdisciplinary research. It must be acknowledged, however, that the IMRAD structure – Introduction, Methods, Results and Discussion – has been the focus of great attention and represents a powerful standard across the field of science.

Attention has also been paid to the development of the genre and its communicative tools in historical linguistics (e.g. Pahta and Taavitsainen, 2010) and in rhetoric (Bazerman, 1988; Gross et al., 2002). It is largely felt that scientific discourse as we know it today bears witness to the development of social and intellectual practices that have characterized the world of science over the past five centuries.

Looking at how the scientific article has evolved and is constantly changing can help us understand science better and communicate it more effectively. Rhetorical perspectives (Bazerman, 1988; Gross, 1990; Gross et al., 2002) have been extremely influential in outlining a dynamic view of genre. Bazerman's (1988) seminal work has clearly shown how the emergence of the experimental report was a social reality, a historical creation, and a fully dynamic phenomenon (1988, p. 8). At the same time, Bazermann saw his own essential message as an educational one, where understanding communication can help us communicate better.

Language studies have often looked at writing practices and at their power of defining how knowledge is agreed upon and codified. Systemic functional linguistics (e.g. Halliday and Martin 1993; Martin and Veel, 1998) has shown how important it is to look at scientific language in retrospect in order to understand scientific writing today (e.g. Halliday and Martin,

 DOI: 10.4324/9781003043782-17

1993, Ch. 3). Halliday acknowledges that features of present-day scientific English, such as the use of technical terms and complex nominal groups, date back to the Middle Ages but points at seventeenth-century England (in particular at Newton's *Treatise on Opticks*) as the keystone creating "a discourse of experimentation" (pp. 58–59), where experience was reported by focusing on "things." Nominalization (the use of nouns to refer to processes, e.g. *the separation of rays*) can be seen as "an essential resource for constructing scientific discourse" (p. 61): nouns objectify processes, and verbs establish a relation between nominal elements or represent the practices of arguing and proving (e.g. *those colors argue, this inequality arises*).

This regrammaticalization seems to be in line with a major epistemological shift that took place in the seventeenth century: knowledge was to be found through the systematic observation of the natural world. This compression of complex phenomena into a single entity is instrumental not only to creating new terminology but also to creating textual sequences, where information that has already been presented in clausal form can be repackaged in nominalized form in order to develop orderly textual sequences and logical steps in the argument.

The style and the structure of the argumentation, however, have changed over the centuries, especially in relation to a changing community of readers. As shown by Gross et al. (2002), through the seventeenth- and eighteenth-century articles were written by and for a very small group of people who were members of scientific societies in England and France, whereas the twentieth century saw forms of highly specialized prose written for a large, specialized, international audience. Historical development is not linear, however: features such as detachment versus involvement seem to undergo what Pahta and Taavitsainen (2010, p. 565) call a "wave-like development," with late medieval texts showing more impersonal and detached formats, the style changing from more detached to more involved in the sixteenth and eighteenth centuries and going back to impersonality in the nineteenth and twentieth centuries.

While well aware of the multiplicity of perspectives that could be considered, I restrict my scope here to the development of the scientific article as a genre, mostly in relation to changes in the scholarly community, with only a few hints at selected key language features. I look at the origins of the genre in the seventeenth and eighteenth centuries, its professionalization in the nineteenth, and its proliferation of specialization in the twentieth. I conclude with a brief look at the present challenges provided by the global dimension of online scientific communities and the requirements of open science.

The age of empiricism and the eighteenth century

As observed by Eisenstein (1979), the invention of the printing press contributed to the development not only of modern science but also of modern means of scientific communication by giving access to knowledge to the general public. The process of production, sharing, dissemination, and standardization of knowledge began toward the end of the Renaissance and continued through the Enlightenment. While later work by Eisenstein herself (2005) highlighted evolutionary rather than revolutionary elements in the impact of the printing press, its introduction clearly marked a beginning for the scientific article as a genre. The development of periodical publications in the seventeenth century was a major contributing factor, as they offered an opportunity to provide the scholarly community with information on inventions, discoveries, and work published in various European countries. Scientists soon realized that periodicals had much greater potential for the diffusion of scientific advancements than the informative letters they used to write to one another, as journals offered the opportunity to reach a wider audience and to exchange views with scholars from different backgrounds.

The birth of the scientific research report is widely attributed to the rise of the experimental essay in the Age of Empiricism, especially in two periodicals: *Le Journal des Sçavans* and *Philosophical Transactions*, both founded in 1665 (Bazerman, 1988; Gross, 1990; Gotti, 1996, 2001; Atkinson, 1999; Valle, 1999, 2004; Gross et al., 2002; Moessner, 2006, 2009; Banks, 2008, 2017). Both journals – each in its own way – set out to disseminate scientific information, rapidly acquiring great international prestige. *Le Journal des Sçavans* – founded by Denis de Sallo – aimed to provide reviews of literary and scientific publications, as well as to publish original reports of various aspects of culture. *Philosophical Transactions* – launched in March 1665 by the Royal Society's first Secretary Henry Oldenburg, also acting as publisher and editor – included numerous original contributions on the most relevant scientific discoveries and was intended to create a *public record* of original contributions to knowledge.

What distinguished the *Transactions* from the French journal was that "the Parisian publication followed novelty, while the London journal was helping to validate originality" (Guédon, 2001, p. 11). Rather than focusing on informing readers about scientific advancements, the *Transactions* soon became a "registry" of scientific inventions, intended to promote the question of "intellectual property," mostly for the printer's benefit (Guédon, 2001, p. 11). This difference is reflected in the proportion of the genres present in the two journals: predominantly reviews in *Le Journal des Sçavans*, predominantly articles and letters in the *Transactions* (Banks, 2017, pp. 48–51).

The success of both journals rapidly led to the creation of similar journals, such as *Acta Eruditorum* in Germany or *Bibliothèque universelle et historique* in Holland. In Italy, this meant the birth of *Giornale de' Letterati* in 1668 (Santoro, 2004), reporting news of new publications but also of natural observations, experiments, and curiosities, as well as translations and extracts from *Le Journal des Sçavans* and *Philosophical Transactions*.

The most important journal for the creation of a scientific tradition is usually thought to be *Philosophical Transactions* of the Royal Society of London. The birth of the Society, a national science academy, is closely linked to empiricist philosophy. As still reported on their website: "The Royal Society's motto 'Nullius in verba' is taken to mean 'take nobody's word for it'. It is an expression of the determination of Fellows to withstand the domination of authority and to verify all statements by an appeal to facts determined by experiment" (https://royalsociety.org/about-us/history/). The institution soon became famous for its defense of the experimental method and for its style, banning figurative expressions. In the words of Thomas Sprat, who was commissioned to write the History of the Society, members exhibited "a constant Resolution, to reject all the amplifications, digressions and swellings of style," "a close, naked, natural way of speaking; positive expressions; clear senses; a native easiness: bringing all things as near the Mathematical plainness as they can: and preferring the language of Artizans, Countrymen, and Merchants, before that, of Wits, or Scholars" (Sprat, 1667, p. 113).

Similarly, Robert Boyle, a leading figure of the Royal Society, characterized the language of the experimental essay at its birth (Gotti, 1996, 2001, 2003; Moessner, 2006, 2009). An explicit emphasis on the "experimental essay" in his *Proemial Essay* (Boyle, 1661) was accompanied by an emphasis on sharing the procedures and the results of experiments with the entire learned world. The main features of Boyle's recommendations appear to be clarity and brevity in style, emphasis on maximal informational content, simplicity of form, objectivity and honesty toward other scientists. Boyle's recommendations are reflected in his own style: after a brief outline of the purpose by way of introduction, what dominates his essays is a clear, full narrative of the experiment, markedly characterized by abundance and precision of detail; the narrative often puts the experimenter to the forefront of the recount and makes frequent use of hedging expressions representing the experimenter as a reliable witness (Gotti, 1996, pp. 55–69, 2001, pp. 227–237, 2003, pp. 227–241).

The number of experimental reports in the *Transactions*, however, is very limited at the beginning and shows all the changing patterns that typically characterize emerging genres. Focusing on a selection of issues of the *Transactions*, Bazerman (1988) points at features such as the contextualization of an issue within a wider context and the decisive role of the experiment in solving academic disputes and structuring the argument, so that the narrative gradually becomes one of demonstration rather than discovery. He actually highlights four stages in the early development of the report through the seventeenth and eighteenth centuries, going from early uncontested reports, to later experimental reports arguing over results, discovery accounts of unusual events, and eventually experimental articles with claims supported by experimental proof (Bazerman, 1988, p. 78). He also notices the emergence of a growing tendency to add illustrations and to compare the results of different experiments.

Atkinsons's (1999) study of the *Transactions* across three centuries combines rhetorical analysis and multidimensional analysis (Biber, 1988), looking at the language features that may define the position of texts according to five dimensions of variation (involved versus informational production; narrative versus non-narrative concern; situation-dependent versus elaborated reference; overt expression of persuasion; impersonal versus nonimpersonal style). His work shows a gradual change of the style of the experimental report "from an author-centred to an object-centred norm" (p. 161), with an increasing use of passives and a marked tendency toward a more abstract style in writing, characterized by nominalizations and complex noun groups and constructions, starting from the eighteenth century. Other diachronic studies of scientific writing based on multidimensional analysis (Biber and Finegan, 1997; González-Álvarez and Pérez-Guerra, 1998; Moessner, 2009) confirm that the century either establishes a trend that will continue over the centuries or marks a turning point. Scientific discourse is shown to move toward more informational and less narrative, more elaborated, less overtly persuasive, and more impersonal forms of writing after the seventeenth century.

Studies of a wider range of journals for the same period (e.g. Gross et al., 2002) highlight that, while the emphasis on fact is clearly in line with the dominant philosophy, the focus of the prose style is still largely the personal narrative, based on reliable testimony. Exposition is chiefly impressionistic, with long sentences, no technical vocabulary, few quantitative expressions and casual use of citations (2002, Ch. 2). The argument is largely observational and only partly experimental, and visual representations are used to illustrate equipment or support the argument (2002, Ch. 3).

The values and writing practices of the discourse community can thus be seen as characterized by a strong preference for observation, description, and inductive argumentation over speculation (Valle, 1999, 2004, p. 60). The editorial voice of the *Philosophical Transactions* was taking control over texts in the early stages and gradually finding a balance with authorial voice (1999, p. 168). This was most clearly seen in the mid-eighteenth century, with the construction of dense epistolary networks with almost as many references to personal contacts as to published literature (see Valle, 1999, Ch. 7).

In the eighteenth century, the persistence of personal, epistolary, and narrative norms was noticeable, together with a decrease in personal and emotional elements, and a focus on "measurement, calculation and empirical observation" (Gross et al., 2002, p. 69). From the point of view of language use, this led to increased use of complex noun phrases, headings and captions for visuals, while textual organization featured introductions and conclusions that increasingly tended to place the study in the context of current interests and to base their credibility on the replicability of results rather than personal trust (p. 98).

Developments across the nineteenth century: professionalization

During the nineteenth century, journals became increasingly specialized, in a process that determined the establishment of the scientific journal as such, providing professionals with collections of articles by different authors, ordered on the basis of specific criteria and published regularly. As traditional academias – involving all the areas of knowledge – lost their importance, the scientific community became more clearly defined around different disciplines. Parallel to science becoming a profession, the scientific journal also became more clearly linked to communication within the profession. It should be noted, however, that there was no clear-cut division between professional and popular scientific journals at the time and most journals addressed both professionals and amateurs.

The growth of science periodicals was impressive throughout the nineteenth century. Shuttleworth and Charnley (2016, p. 297) report a rise from around 100 titles worldwide at the beginning of the century to an estimated 10,000 at the end. This stunning growth coincided largely with the fact that the practice of communicating results in written publications was becoming dominant over the tradition of communication by personal letter or lecture at meetings. Journal editors frequently chose articles to meet the interests of other scientists rather than amateurs. The whole process, however, was rapidly opening up new audiences (see Dawson et al., 2020, Part 3), as well as new social contexts and new technologies for papermaking and printing itself.

Competition among journals and the growth of more specialized scientific societies also led to the establishment of the peer review system. The Royal Society, for example, introduced more rigorous and systematic expert peer review in 1832, while further specializing their publications (Atkinson, 1999). The peer review system became the key element in assessing the scientific value of publications, linking knowledge production to its reception within the scholarly community.

The style of writing became less personal and more formal, with sentence length and clausal density dropping to levels that are fairly close to those of today. Some gentlemanly norms were still used, however. In a study of hedging in medical academic writing, for example, Salager-Meyer and Defives (1998) look at the transition from the nineteenth to the twentieth centuries in terms of a shift from "the gentleman's courtesy" to the "expert's caution," showing that nineteenth-century articles still admitted a lot of approximators and emotionally charged expressions that would later make room for shields and passives.

Citation practice, almost nonexistent in the early stages, became the norm, thus establishing the idea that the credibility of the writer also depends on the existence of an established body of literature in the field (Valle, 2004, p. 63). The diachronic evolution of referencing in a medical corpus, for example, reflects a shift to professionalized medicine and to a highly structured scientific community (Salager-Meyer, 1999).

At the same time, the development of the argument was becoming more complex: observations and experiments were used to build theory, and texts and visuals were increasingly integrated into arguments (Gross et al., 2002, Ch. 7). Tables and figures became indispensable elements, exploiting the cognitive possibilities of visuals and their explanatory power (pp. 148–155), while greater attention was paid to methodology and precision (Atkinson, 1999).

With the aim of simplifying communication, scientific journals developed a remarkable unity of form. Textual organization consolidated the use of headings, with recognizable introductions and conclusions. Introductions approached the standards of twentieth-century articles, as defined, for example, by Swales (1990) in terms of mapping a territory for research, identifying a niche, and occupying the niche. Conclusions added insight, suggesting wider significance and adding suggestions for further research, while still lacking detailed summaries and accepting deviations into the personal and polemical (Gross et al., 2002, pp. 134–136).

The proliferation of disciplines with their different interests makes it difficult to trace a uniform path in the nineteenth century, as each discipline was developing its own set of practices. Chemistry was arguably the most "professionalized" (Gross et al., 2002, p. 135), and quantitative methods appeared to qualify physics and chemistry in particular (p. 159), but the general trend was toward measurement, explicitness, experimental methods, and turning observations into scientific facts. If the physical sciences had been experimental from the beginning, the biological sciences only turned from observational to experimental toward the middle of the nineteenth century (Banks, 2008), and medicine gradually moved narrative genres like the medical case report to a secondary position unless with multiple case reports and statistical assessment of probabilities (Taavitsainen, 2011). Most disciplines moved toward the structure of the experimental report.

The twentieth century: specialization and standardization

In the twentieth century, the proliferation of journals and their centrality in the development of research increased the specialization of international networks of authors and readers. Increasing specialization in science inevitably led to popular and professional scientific journals gradually establishing themselves as different forms of publications with their respective audiences. This also increased the standardization process that characterized the textual and organizational features of scientific publications. By the late twentieth century, scientific articles established a well-defined structure, with arguments that are now supported by numerous citations, rigorous descriptions of methodology and results, and a plethora of visuals. Increasing competitiveness heightened the need to foreground the novelty and significance of papers (Hyland, 2015).

Gross et al. (2002, pp. 215–219) identify two main directions in the evolution of scientific prose over the centuries: objectivity and efficiency. On the one hand, language features like use of passives, impersonal forms, and hedges seem to be related to the dominance of the experimental method and its focus on objects rather than people. On the other hand, aiming at efficiency, the reduction of sentence length compensates for the increasing complexity of nominal elements, and the standard arrangement of the article into its dominant IMRAD structure allows readers to direct their attention to specific components of the argument.

The IMRAD format gained increasing popularity around mid-century. First adopted by health science journals in the 1940s (Sollaci and Pereira, 2004) and widely used by physics in the 1950s (Bazerman, 1988), it became a standard for the vast majority of scientific journals in the 1970s, with the publication of the American National Standard for the preparation of scientific papers for written or oral presentation (ANSI Z39.16–1972) (Day, 1989, p. 18). This organization of the textual structure, with its emphasis on inductive reasoning and the presentations of results before their discussion, certainly does not reflect the processes of scientific thought, as already shown by Medawar (1964). Its success lies rather in the need for readers to locate efficiently the information that is most relevant to them, which is ultimately related to the proliferation of the published literature.

The expansion of scientific journals has also consolidated the peer-review system and led authors to tailor their manuscripts accordingly, influenced by a wealth of guidelines, style manuals, and instructions to authors. The standards have extended to the social sciences, starting from the development of writing in experimental psychology which led to the American Psychological Association *Publication Manual* (Bazerman, 1988).

The century also witnessed the rise of abstracts as a part-genre of the research article. The first discipline to introduce them as a regular feature was physics, with *The Physical Review* in 1920, long before many other sciences (Bazerman, 1988). Most biochemical journals introduced them around the early sixties (Berkenkotter and Huckin, 1995), soon followed by economics, the leading social science (Bondi, 1997). The combination of the informative and promotional value of abstracts (e.g. Bondi and Lorés Sanz, 2014) may be related to the quantitative growth of publications produced (Berkenkotter and Huckin, 1995; Swales and Feak, 2009, p. 1) and to the increasingly international nature of discourse communities, with English as the international language of research publications (Hyland, 2015). Their rise can be related to the same principles of efficiency that led to the establishment of the IMRAD standard: facilitating the extraction of information for selective rather than sequential reading patterns (Bazerman, 1988; Berkenkotter and Huckin, 1995). Structured abstracts in particular – i.e. abstracts that include predefined headings (e.g. Background, Purpose, Research Design, Data Analysis, Findings) – increase the ease of locating information and favor comparative analysis.

The consolidation of the organizational structure of the scientific article in the twentieth century has led to a vast literature focusing on the major sections of the research article (see Samraj, 2016 for an overview). It should be noticed, however, that even within the "hard sciences" there is great variation. Mathematics, for example, tends not to include methods and discussion sections (Graves et al., 2013), and physics has gradually moved toward mathematical reasoning, often adopting the paradigm of theoretical rather than empirical research (Gray, 2015).

A recent study by Hyland and Jiang (2019) has explored language changes at the turn of the twenty-first century by carrying out a corpus-based multidimensional analysis of research articles representing three moments: 1965, 1990, and 2015. The study highlights opposing trends in the hard sciences (science and technology) versus the social sciences: if the social sciences seem to be moving toward greater informational focus and a preference for empirical, experimental, and data-informed investigations, the hard sciences are increasing their use of involvement features such as first- and second-person pronouns or modality, interpersonal and evaluative meanings, self-mention, and engagement markers. They somehow move away from their traditional objective style of writing focused on facts and toward "more involved, stance-laden discourses, which emphasize the role of the interpreting researcher" (pp. 227–230). This may be related to the impact of a wider audience for science in recent years.

Current trends: open science and digital publishing

The world of journal publishing is undergoing radical changes today with new business models, open access, digital libraries and an emerging global dimension (Cope and Phillips, 2014). The digital transformation coincides with major changes in scientific communication. The most obvious are the increasing dominance of English as the language for publishing and the burgeoning community of scientist writing in a second language (see Gotti, Chapter 13 of this volume; Corcoran and Englander, Chapter 28 of this volume). The migration to electronic editions has also meant that academic publications are mostly concentrated in the hands of a few publishers. This in turn has shown the need to transfer research to professionals and to the wider public to favor global access to knowledge, promoting a culture of open access (see Sidler, Chapter 3 of this volume) in publishing and education.

The debate on outreach and open science has contributed to variation in the structure of the scientific article. The IMRAD format remains dominant, but the more recent IRDAM format (Introduction, Results, Discussion, and Methods) has also been advocated, especially in areas that involve the need to communicate with practitioners. This is the case of translational research, which aims at an efficient transition of basic science discoveries into clinical applications – often described as "bench-to-bedside." In the IRDAM format, studies begin with a hypothesis to be tested and are guided by the results, rather than a preset series of experiments or method; this implies emphasizing the Results section, placing it immediately after the Introduction and subordinating the Methods section to the Results. The format, used by many high-impact basic research journals, requires a much more articulate results section, with an overview of the research design and methods, as well as some interpretation of the main conclusions (Derish and Annesley, 2010).

On the other hand, almost at the opposite extreme, there is the very recent development of the "registered report," an emerging form of empirical journal article which – following a two-stage model of peer review – "pre-registers" methods and proposed analyses before conducting the research. This is a way of sharing high-quality protocols, independent of data collection and results. The main focus of the first review is on the methodology, with a view to replicability and transparency, whereas the publication of the full article concentrates on communication of results (Mehlenbacher, 2019). The model, developed in neuroscience and psychology, is now rapidly expanding to the life and social sciences (Chambers, 2019).

Digital media, with their different technical affordances, offer new possibilities to engage in networks with other academics, as well as with nonacademics, determining the birth of new communicative formats and radical changes to traditional ones (Kuteeva and Mauranen, 2018). The very notion of a journal issue has changed: articles can now be part of VSI (virtual special issue), an online-only special issue which is built up gradually, as independently produced articles are published online in the regular journal issues. The articles are linked by an introduction that is regularly updated as new articles appear, thus mapping an evolving territory (Mur-Dueñas, 2018).

Digital publishing has certainly confirmed the central role of abstracts and abstracting services acting as "hubs," but it has also favored the use of highlights and graphical abstracts, adding new visual and verbal features to the selective reading process. Highlights offer key propositions, aiming to promote the research product (Hyland, 2015, Ch. 6). Graphical abstracts offer a visual representation of the research process and/or the processes observed (see Buehl, Chapter 24 of this volume). Another important development can be traced in the birth of the video abstract, often combining use of images, audio, video clips, and text (Spicer, 2014, p. 3), presenting the rationale of the study and the procedures followed. Audioslide presentations can also encapsulate the essentials of the study in a slightly more formal register than videos (Pérez-Llantada, 2016, pp. 31–32).

The structure of the article has also been influenced by intertextuality and multimediality. On the one hand, the whole citation system and other cross-references have rapidly exploited the potential of hypertextual links. On the other, lengthy and careful descriptions of processes have been substituted by video components and animations. The potential of video is particularly obvious in the field of surgery, where the sequences and techniques of an operation are best visualized in supplementary video materials that accompany the verbal text.

The digital transformation is still at work, and it is impossible, within the scope of this chapter, to offer but an extremely limited sketch of the changes underway. The whole knowledge system seems to be changing under many influences – technological, economic, distributional – as well as under the influence of a new emphasis on the social role of knowledge (Cope and Kalantzis, 2014).

References

Atkinson, D. (1999) *Scientific discourse in socio-historical context: The philosophical transactions of the royal society of London, 1675–1975*, Mahwah: Lawrence Erlbaum.

Banks, D. (2008) *The development of scientific writing: Linguistic features and historical context*, Oakfield, CT: Equinox Publishing.

Banks, D. (2017) *The Birth of the Academic Article*. Le Journal des Sçavans *and the* Philosophical Transactions, *1665–1700*, Equinox Publishing, Sheffield

Bazerman, C. (1988) *Shaping written knowledge: The genre and activity of the experimental article in science*, Madison: University of Wisconsin Press.

Berkenkotter, C., & Huckin, T. (1995) *Genre knowledge in disciplinary communication: Cognition/culture/power*, Hillsdale, NJ: Lawrence Erlbaum.

Biber, D. (1988) *Variation across speech and writing*, Cambridge: Cambridge University Press.

Biber, D., & Finegan, E. (1997) 'Diachronic relations among speech-based and written registers in English', in T. Nevalainen & L. Kahlas-Tarkka (eds.), *To explain the present: Studies in the changing English language in honour of Matti Rissanen*, Helsinki: Société Néophilologique, pp. 253–276.

Bondi, M. (1997) 'The rise of abstracts: Development of the genre in the discourse of economics', *Textus*, vol. 10, pp. 395–418.

Bondi, M., & Lorés, S.R. (eds.) (2014) *Abstracts in academic discourse*, Bern: Peter Lang.

Boyle, R. (1661) 'A proemial essay, wherein, with some considerations touching experimental essays in general, is interwoven such an introduction to all those written by the author, as is necessary to be perus'd, for the better understanding of them', in R. Boyle, *Certain physiological essays written at distant times, and on sevaral occasions*, London: Henry Herringman, pp. 11–12.

Chambers, C. (2019) 'What's next for registered reports?', *Nature*, vol. 573, pp. 187–189, DOI: 10.1038/d41586-019-02674-6

Cope, B & Phillips, A. (eds.) (2014) *The Future of the Academic Journal* (Second Edition) Chandos Publishing, Oxford

Cope, B., & Kalantzis, M. (2014) 'Changing knowledge ecologies and the transformation of the scholarly journal', in B. Cope & A. Phillips (eds) *The Future of the Academic Journal*, Chandos Publishing, Oxford, pp. 9–84.

Dawson, G., Lightman, B., Shuttleworth, S., & Topham, J.R. (eds.) (2020) *Science periodicals in nineteenth-century Britain: Constructing scientific communities*, Chicago: University of Chicago Press.

Day, R.A. (1989) 'The Origins of the Scientific Paper: The IMRAD Format', *American Medical Writers Association Journal*, vol. 4, no. 2, pp. 16–18

Derish, P.A., & Annesley, T.M. (2010) 'If an IRDAM journal is what you choose, then sequential results are what you use', *Clinical chemistry*, vol. 56, no. 8, pp. 1226–1228, https://doi.org/10.1373/clinchem.2010.150961

Eisenstein, E. (1979) *The printing press as an agent of change: Communications and cultural transformations in early modern Europe*, Cambridge: Cambridge University Press.

Eisenstein, E. (2005) *The printing revolution in early modern Europe*, Cambridge: Cambridge University Press, 2nd edition.

González-Álvarez, D., & Pérez-Guerra, J. (1998) 'Texting the written evidence: On register analysis in late Middle English and early Modern English', *Text*, vol. 18, no. 3, pp. 321–348, https://doi.org/10.1515/text.1.1998.18.3.321

Gotti, M. (1996) *Robert Boyle and the language of science*, Milano: Guerini Scientifica.

Gotti, M. (2001) 'The Experimental Essay in Early Modern English'', *European Journal of English Studies*, vol.5, no. 2, pp. 221–239, https://doi.org/10.1076/ejes.5.2.221.7307

Gotti, M. (2003) *Specialized discourse: Linguistic features and changing conventions*, Bern: Peter Lang.

Graves, H., Moghaddasi, S., & Hashim, A. (2013) 'Mathematics is the method: Exploring the macro-organizational structure of research articles in mathematics', *Discourse Studies*, vol. 5, no. 4, pp. 421–438, https://doi.org/10.1177/1461445613482430

Gray, B. (2015) *Linguistic variation in research articles*, Amsterdam: John Benjamins.

Gross, A. (1990) *The rhetoric of science*, Cambridge, MA: Harvard University Press.

Gross, A., Harmon, J., & Reidy, M. (2002) *Communicating science: The scientific article from the seventeenth century to the present*, Oxford: Oxford University Press.

Guédon, J.C. (2001) 'In Oldenburg's long shadow: Librarians, research scientists, publishers, and the control of scientific publishing', www.arl.org/arl/proceedings/138/guedon.html (Accessed 20 September 2020)

Halliday, M.A.K., & Martin, J.R. (1993) *Writing science: Literacy and discursive power*, London: Falmer.

Hyland, K. (2000) *Disciplinary discourses: Social interactions in academic writing*, London: Longman.

Hyland, K. (2015) *Academic publishing: Issues and challenges in the construction of knowledge*, Oxford: Oxford University Press.

Hyland, K., & Bondi, M. (eds) (2006) *Academic discourse across disciplines*, Bern: Peter Lang.

Hyland, K., & Jiang, K. (2019) *Academic discourse and global publishing*, London: Routledge.

Kuteeva, M., & Mauranen, A. (eds) (2018) *Digital Academic Discourse: Texts and Contexts*, Special issue of *Discourse, Context & Media*, vol. 24, DOI:10.1016/j.dcm.2018.06.001

Martin, J.R., & Veel, R. (1998) *Reading science: Critical and functional perspectives on discourses of science*, London and New York: Routledge.

Medawar, P. (1964) 'Is the scientific report fraudulent? Yes: It misrepresents scientific thought', *Saturday Review*, vol. 47, pp. 42–43, 1 August.

Mehlenbacher, A.R. (2019) 'Registered Reports: Genre Evolution and the Research Article', *Written Communication*, vol. 36, no. 1, pp. 38–67, https://doi.org/10.1177/0741088318804534

Moessner, L. (2006) 'The birth of the experimental essay', in V. Bathia & M. Gotti (eds.), *Explorations in specialized genres*, Bern: Peter Lang.

Moessner, L. (2009) 'The influence of the Royal Society on seventeenth-century scientific writing', *ICAME Journal*, vol. 33, pp. 65–87

Mur-Dueñas, P., (2018) 'Disseminating and constructing academic knowledge in online scholarly journals: An analysis of virtual special issue introductions', *Discourse, Context & Media*, *24*, pp. 43–52, DOI:10.1016/J.DCM.2018.04.010

Myers, G. (1990) *Writing biology: The social construction of popular science*, Madison: University of Wisconsin.

Pahta, P., & Taavitsainen, I. (2010) 'Scientific discourse', in A. Jucker & I. Taavitsainen (eds.), *Historical pragmatics*, Berlin and New York: De Gruyter.

Pérez-Llantada, C. (2016) 'How is the digital medium shaping research genres? Some cross-disciplinary trends', *ESP Today*, vol. 4, no. 1, pp. 22–42

Salager-Meyer, F. (1999) 'Referential behavior in scientific writing: A diachronic study (1810–1995)', *English for Specific Purposes*, vol. 18, no. 3, pp. 279–305, DOI:10.1016/S0889–4906(97)00042–2

Salager-Meyer, F., & Defives, G. (1998) 'From the gentleman's courtesy to the expert's caution: A diachronic analysis of hedges in academic writing (1810–1995)', in I. Fortanet, S. Posteguillo, C. Palmer, & J.F. Coll (eds.), *Genre studies in English for academic purposes*, Castellon: Universitat Jaume I, pp. 133–169.

Samraj, B. (2016) 'Research articles', in K. Hyland & P. Shaw (eds.), *The Routledge handbook of English for academic purposes*, Abingdon and New York: Routledge, pp. 403–415.

Santoro, M. (2004) 'Il sistema periodico: Breve storia delle riviste tra comunicazione scientifica e pratica bibliotecaria', *Bibliotime*, vol. 7, p. 1.

Shuttleworth, S., & Charnley, B. (2016) 'Science periodicals in the nineteenth and twenty-first centuries', *Notes and Records. The Royal Society Journal of the History of Science*, vol.70, pp. 297–304, DOI: https://doi.org/10.1098/rsnr.2016.0026

Sollaci, L.B., & Pereira, M.G., (2004) 'The introduction, methods, results, and discussion (IMRAD) structure: a fifty-year survey', *Journal of the Medical Library Association*, vol. 92, no. 3, p. 364–371

Spicer, S. (2014) 'Exploring video abstracts in science journals: An overview and case study', *Journal of Librarianship and Scholarly Communication*, vol. 2, no. 2, pp. 1–10, DOI: https://doi.org/10.7710/2162-3309.1110

Sprat, T. (1667) *The history of the royal-society of London, for the improving of natural knowledge*, London: T.R. for John Martyn, Printers to the Royal Society.

Swales, J. (1990) *Genre analysis: English in academic and research settings* Cambridge: Cambridge University Press.

Swales, J. (2004) *Research genres*, Cambridge: Cambridge University Press.

Swales, J., & Feak, C. (2009) *Abstracts and the writing of abstracts*, Ann Arbor: The University of Michigan Press.

Taavitsainen, I. (2011) 'Medical case reports and scientific thought-styles', *Revista de Lenguas para Fines Específicos*, vol.17, pp. 75–98, https://ojsspdc.ulpgc.es/ojs/index.php/LFE/article/view/100

Thompson, P., & Hunston, S. (2019) *Interdisciplinary Research Discourse. Corpus Investigations into Environment Journals*, Routledge, London

Valle, E. (1999) *A collective intelligence: The life sciences in the royal society as a scientific discourse community 1665–1965*, Turku: Anglicana Turkuensia.

Valle, E. (2004) '"A nice and accurate philosopher": Interactivity and evaluation in a historical context', in G. Del Lungo Camiciotti & E. Tognini Bonelli (eds.), *Academic discourse: New insights into evaluation*, Bern: Peter Lang, pp. 55–80.

15

REVIEWING THE SCIENTIFIC REVIEW ARTICLE

Judy Noguchi

Scientific communication is rapidly changing as we move from a paradigm of trying to understand phenomena through experimentation, modeling, and simulation to one in which huge amounts of data are sent through data management systems to utilize information in a way that had not been possible before. The rapidity of developments has led to a change in the way that scientific communities construct their knowledge bases. Traditionally, an individual scientist or group would identify a problem to solve or a hypothesis to test, devise experiments, carry them out, and publish the results as a research article stating their claims about what they had discovered. After such studies had been replicated or similar findings had been reported, until the late 1990s, a review article would be written about the status quo of the field, and this would signal the entry of a claim into the knowledge base of the community. Today, with a rapidly increasing number of papers being published, there are fewer comprehensive review articles and an increased number of shorter ones focusing on specific topics.

This chapter will "review" the scientific review article by describing its current place in scientific communication. Next, it will examine the significance of this genre in the scientific enterprise itself, which is considered to have started in the sixteenth century in Europe. Today, more than 450 years later, we find ourselves in the midst of a paradigm shift which is revolutionizing how humans interact with the data resulting from scientific endeavors. Thus the chapter will close with a prediction about how the scientific review article may evolve in this era of postnormal science.

Introducing the scientific review article

What is a review article?

The word "review" has many meanings and thus misconceptions often arise as to what a review article is. For example, some academics think of a book review or of the peer review of a research paper submitted to a journal for publication. The *review article* to be examined in this chapter is a text type, or genre, that presents a timely, curated view of a research field or subfield

DOI: 10.4324/9781003043782-18

to help researchers obtain vetted information that they can use to further their own work. The need for reviews is pointed out on the *Nature* website (2020):

> With an ever-increasing number of primary-research journals to be monitored, and the expanding volume of information to be found on the web, it is becoming increasingly difficult for scientists and clinicians to find the information that they need in a form that is immediately useful.

The origin of the review article, according to Clarke and Chalmers (2018), can be traced to a treatise by James Lind on scurvy that was published in 1753. Lind, a Scottish naval surgeon, considered it important "to remove a great deal of rubbish" on how to best treat scurvy (Clark and Chalmers, 2018, p. 121). The original document can be viewed at The James Lind Library (www.jameslindlibrary.org/lind-j-1753/), a website launched in 1998 by the Library of the Royal College of Physicians of Edinburgh to commemorate the 50th anniversary of the publication by the Medical Research Council (1948) of another important review: an investigation on the use of streptomycin to treat pulmonary tuberculosis.

As research findings can be conflicting or nonreplicable, reviews can offer guidance on which results to use as the basis of judgments and decisions. Siddaway et al. (2019, p. 750) state that literature reviews integrate "a body of studies in order to (*a*) draw robust conclusions about big questions, principles, and issues, and (*b*) explain how and why existing studies fit together and what that means for theory and future research."

Types of review articles

One of the most important types is the *systematic review*. This type of article began to appear in the 1970s and 1980s (Munn et al., 2018). Today, according to Clarke and Chalmers (2018), more than 10,000 systematic reviews are published every year. The systematic review is defined by the respected Cochrane Database of Systematic Reviews (n.d.) as one which:

> attempts to identify, appraise and synthesize all the empirical evidence that meets pre-specified eligibility criteria to answer a specific research question. Researchers conducting systematic reviews use explicit, systematic methods that are selected with a view aimed at minimizing bias, to produce more reliable findings to inform decision-making.

Cochrane Reviews, selecting the best evidence available to inform health care decisions, assesses and examines interventions, diagnostic test accuracy, methodology, qualitative evidence, and prognosis. The protocol used to develop a systematic review is made public to show the question being addressed, the criteria for inclusion in the review, and how the review process was conducted. Sutton et al. (2019), who conducted a comprehensive review of review families in the health sciences, note that transparency and reproducibility are key features of systematic reviews.

A relatively new approach to synthesizing evidence is the *scoping review*, which has a wider focus and can be considered "a precursor to a systematic review" (Munn et al., 2018). Sucharew and Macaluso (2019) explain, "The purpose of a scoping review is to provide an overview of the available research evidence without producing a summary answer to a discrete research question." Armstrong et al. (2011) compare the systematic review with the scoping review, showing that the former is more focused on a narrow research question, including details on data, quantitative synthesis, and assessment of the quality of the studies examined, in order to present a conclusion. On the other hand, a scoping review addresses broad research questions, may or may not present data, and usually does not include quantitative analysis. Tricco et al., 2018 described how the PRISMA (Preferred Reporting Items for Systematic Reviews and Meta-Analyses) statement was extended to develop a checklist for the scoping review, thus paving a systematic path "to map evidence on a topic and identify main concepts, theories, sources, and knowledge gaps" (Tricco et al., 2018, p. 1). This paper in the *Annals of Internal Medicine* has a 27-item checklist of the relevant information, starting from a title indicating that the article is a scoping review, a structured abstract, an introduction explaining the rationale and objectives, the methods used, the results and their synthesis, a discussion of the findings, and details on the funding for the studies included as well as for the review itself.

The *meta-analysis article* is another type of review article that is based on "a research process used to systematically synthesize or merge the findings of single, independent studies, using statistical methods to calculate an overall or 'absolute' effect" (Shorten and Shorten, 2013, p. 3). This genre is particularly important in health care fields where differences in sample size, study approach, and findings need to be carefully considered to help inform health care practice. Another type of review article is the *evidence mapping article*, which Munn et al. (2018) describe as using visual databases or schematics to highlight the gaps in knowledge.

The *narrative review* is reported by Sutton et al. (2019, p. 206) as "a 'conventional' review of the literature" that came under criticism when contrasted with the systematic review just described. In the field of psychology, Siddaway et al. (2019) describe narrative reviews as trying to integrate the findings of quantitative studies based on different approaches for the reinterpretation of a theory. They can also present a historical overview of a research topic or theory. Siddaway et al. (2019) additionally describe *meta-syntheses*, also referred to as meta-ethnographies or qualitative meta-analyses, that integrate qualitative studies to reveal new explanations for phenomena.

The complexity of defining the review article is captured in this title: "Tensions and Paradoxes in Electronic Patient Record Research: A Systematic Literature Review Using the Meta-Narrative Method" (Greenhalgh et al., 2009). The authors state that "The extensive research literature on electronic patient records (EPRs) presents challenges to systematic reviewers because it covers multiple research traditions with different underlying philosophical assumptions and methodological approaches" (Greenhalgh et al., 2009, p. 729). This variation in research approaches led them to employ a meta-narrative method for examining 25 systematic reviews and 94 primary studies to reveal differences in the conceptualization of EPRs. They considered their primary task to be to tease out "the meaning and significance of the literature rather than producing an encyclopedic inventory of every paper published on the topic" (Greenhalgh et al., 2009, p. 731).

Regarding the issue of information retrieval, Sutton et al. (2019) tried to identify and define the types of review in health-related disciplines. They reported finding 48 types which could be classified into seven families: traditional family (including narrative and state-of-the-art reviews), systematic review family (such as the Cochrane reviews), review of review family (such as overviews or umbrella reviews), rapid review family to glean information in a limited amount of time, qualitative or experiential review family, mixed methods review family that combine qualitative and quantitative methods, and purpose-specific reviews (including expert opinion or technology assessment). Booth and colleagues (Booth, 2016; Booth et al., 2016a, 2016b) present a Review Ready Reckoner–Assessment Tool (RRRsAT) (https://eahilcpd.

files.wordpress.com/2016/11/review-ready-reckoner.pdf), which is a handy table presenting the different types of review families, the questions they try to answer, and even how long it would likely take to write such a review. For example, a mapping review is estimated to require 4 to 16 weeks while a systematic review focusing on effects would probably take 52 to 78 weeks to complete.

An interesting development from the *Proceedings of the National Academy of Sciences of the United States of America* (*PNAS*) is the *Perspectives* genre, which is meant to "present a balanced and objective viewpoint on an important area of research, focus on a specific field or subfield within a larger discipline, and discuss current advances and future directions." A Perspective article can be considered to be a type of review article as it can only be submitted upon invitation from the journal editor and is differentiated by *PNAS* (n.d.) from another of its genres, the Opinion article, which is defined as an essay meant to "further the discourse on a topic via a clearly articulated argument that includes novel ideas or proposals."

As can be seen from all this, there is a great diversity in the types of review articles. Unlike the scientific research article, which has a clear structure that includes the introduction, materials and methods, results and conclusion sections, review articles do not have a set structure or even a common name. From the pedagogical viewpoint of science communication, the review article is thus more difficult to teach from both the reading and writing aspects.

The scientific review article and the science community

A pivotal role in the construction of scientific knowledge

The review article, as just described, is written with the aim of helping researchers make sense of a rapidly expanding database of literature by gaining a timely, curated overview of a field of study, grasping what is already known and identifying gaps in the knowledge base. This is important for finding answers to immediate research questions but, on a grander scale, this curation of information is essential for the process of constructing a reliable knowledge base for the field itself.

This idea of constructing a specialist knowledge base is discussed by Knorr-Cetina (1981) in *The Manufacture of Knowledge: An Essay on the Constructivist and Contextual Nature of Science*. She starts her treatise by proposing that scientists are practical reasoners trying to find answers to questions and then proceeds by considering scientists as indexical reasoners and analogical reasoners as they plan a research project. This is followed by positioning scientists as socially situated reasoners and then literary reasoners who have to transform the laboratory findings into a written paper on the research so that it can benefit society. As an example, she details how laboratory notes from a study on potato protein recovery gradually became molded into a research paper for publication. This heralds another transformation: "As they [the products of science] move from the scientist's desk to the office of a politician, they change into a policy argument" (Knorr-Cetina, 1981, p. 132). The final chapter presents the scientist as a symbolic reasoner who can bridge what is perceived as a gap between the natural and social, or human, sciences. What is being described here goes beyond the specialist community of scientists and touches on how the disciplinary knowledge base can even inform public policy.

Playing an important role in constructing such disciplinary knowledge is the review article. Noguchi (2001), using genre analysis techniques and interviews with field specialists, prepared Table 15.1, which presents a view of the pivotal role that the review article plays in the process of constructing knowledge in the sciences. A "fact" comes to be acknowledged by a discourse community only by going through this process of being introduced, replicated, and verified

Table 15.1 Process of a "scientific fact" contributing to knowledge construction in science

Stage	Manifestation
Conception	Idea in the mind of a researcher
Birth	Written manifestation available to public scrutiny
	Notebook (lab notebooks, records, notes)
	Conference abstract
	Conference proceedings
	Research article
	Follow-up research articles
	Citation in other research articles
Coming-of-age	Acceptance as "truth" by scientific community
	Review
	Popular science literature
Middle age	Acceptance as "truth" by general community
	Textbook
	Popular literature
Death	No longer accepted by community or superseded by a new paradigm
or	
sainthood	Acknowledged as a "law of nature"

Source: From Noguchi, 2001, p. 334.

before being accepted. This concept of "scientific facts having a "life" was proposed by Myers (1990b) when he discussed the narrative of split genes.

Communication for the development of science

Considering the important role that the review article plays in constructing the tacit knowledge base in a scientific field, let us examine some key concepts related to "science" itself. In *The Invention of Science*, Wootton (2015) suggests that the Scientific Revolution was a transformation in the very way that humans think. Prior to that, the first significant transformation occurred in the Neolithic Revolution, between 12,000 and 7,000 years ago, when humans domesticated animals, began agriculture, and started using metal tools. After that, it was not until the Scientific Revolution of the sixteenth to eighteenth centuries that humans began to "discover" that they could directly question natural phenomena and try to interpret what was happening rather than continuing to do what scholars until then had been doing – simply trying to interpret what the ancients had written about the world around them, for example, in religious texts.

In *The Origins of Modern Science*, Butterfield (1957) points to the seventeenth-century contributions by William Harvey on blood circulation that were based on "multiple demonstrations" which, in today's terms, we would call "experiments" (Butterfield, 1957, p. 62). In 1605, Francis Bacon, in his book *Of the Proficience and Advancement of Learning, Divine and Human*, stated that he had discovered how to make discoveries by interpreting nature (Wootton, 2015). This led to what we now consider to be the scientific method: identify a research question, use the experimental method to obtain data about it, interpret the findings, and share your claim so that it can be verified by others.

Sharing discoveries, of course, requires language. "No language, no facts" states Gross (1990, p. 203) because "the truths of science . . . are achievements of argument" (Gross, 1990, p. 103).

This idea is strongly advanced by Myers (1990a), who titled his book *Writing Biology: Texts in the Social Construction of Scientific Knowledge* to argue that it is the writing that actually produces the biology. In a book chapter entitled "Making a Discovery: Narrative of Split Genes," Myers (1990b) explains that a scientific discovery does not suddenly appear but is gradually developed into one as it is examined and discussed by the discourse community of the field. He states that:

> if we want to find a discovery, we can't just go back to the original articles or even earlier to the lab notebooks or recorded conversations or autoradiographs and electron micrographs. We need to look at the interpretations of the articles as their stories are retold in news articles, *review articles*, textbooks, and popularizations.
>
> *(Myers, 1990b, pp. 102–103; emphasis mine)*

To better understand how information of new findings becomes part of the knowledge base of science, let us consider the two concepts of "discourse community" and "genre," which have been developed through work in English for specific purposes (ESP). In 1990, Swales published *Genre Analysis: English in Academic and Research Settings*, in which he describes a discourse community as a group of people who are bound by their communication with one another and are not necessarily in physical proximity. For example, researchers trying to develop a treatment for a deadly illness would want to know what other researchers are doing even though they may be scattered across the globe. These researchers form a discourse community that uses a common knowledge base related to the work they are doing. They increase this knowledge base by examining, curating, and verifying the information that they exchange. To facilitate this work, they develop genres, or types of texts. For example, if a research group has discovered the mechanism by which a deadly virus attacks the human body, they would want to share the information with other researchers so that the information can be used to develop drugs to treat or vaccines to prevent the disease. Of course, the claims that they make based on their studies would need to be reviewed and vetted by other researchers before they can be included in the body of knowledge of their field. Being included in a review indicates that the claim has been deemed worthy of consideration for inclusion in the disciplinary knowledge.

Facing the challenges of the evolving paradigms of science

The process of knowledge construction shown in Table 15.1 was proposed based on work conducted in the twentieth century (Noguchi, 2001). While the process in general can still be said to be functional, the amount of data production in science has increased phenomenally in the twenty-first century. This is pointed out by Jim Gray, an American data scientist, who reviewed the paradigms of science, noting that a thousand years ago, science was empirical and was aimed at describing natural phenomena (Hey et al., 2009, p. xviii). Over the last few hundred years, science was theoretical and used models and generalization to understand natural phenomena. From the mid-twentieth century, science became computational with scientists simulating complex phenomena. Today, we have entered an era of data exploration where scientists are trying to unify theory, experimental findings, and simulation outcomes. The data are captured by instruments or generated by simulation, then processed by software. The role of the scientist is that of analyzing the information that has been stored in the computer using data management software and statistics in order to draw conclusions. This is referred to by Hey et al. (2009) as the "fourth paradigm."

Use of the computer for data management became necessary with the explosion of the amount of information. Ginsparg (1997) writes that since the mid-1970s, researchers in high-energy physics began a preprint culture in which laboratories shared their research ideas and results to circumvent the traditional journal publication system which usually took about six months to a year for a paper to actually appear in a journal. By the mid-1980s, with the advancement of computer technology, the community had begun the regular exchange of electronic information. In August 1991, Ginsparg (1997, p. 43) launched an e-print archive "as an experimental means of circumventing recognized inadequacies of research journals but unexpectedly [it] became within a very short period the primary means of communicating ongoing research information in formal areas of high-energy particle theory." As of the date of this writing (2020), the open-access arXiv (n.d.) has 1,753,040 nonpeer-reviewed scholarly articles in physics, mathematics, computer science, quantitative biology, quantitative finance, statistics, electrical engineering and systems science, and economics.

For the first two paradigms of science, the scientific record, which communicates information, builds up communities of collaboration, and validates findings based on reproducibility, functioned "pretty well" (Hey et al., 2009, p. 179). However, with the rapid growth of the scientific enterprise and the amount of data it was generating, Ginsparg predicted "a transformation in the way we process scientific information, much as the availability of interlinked network resources has led to new nonlinear reading strategies, and the availability of networked mobile devices has altered the way we use our short- and long-term memories" (2011, p. 9). In this age of the fourth paradigm of data-intensive science, there are opportunities to further improve how the scientific record is kept and used. Rather than simply reading individual research papers, we can use databases of literature in the following ways:

> Data and text mining, inferencing, integration among structured data collections and papers written in human languages . . . , information retrieval, filtering, and clustering all help to address the problems of the ever-growing scale of the scientific record and the ever-increasing scarcity of human attention.
>
> *(Hey et al., 2009, p. 183)*

Reviewing the review process

Today's deluge of data and information challenges us to find ways "to remove a great deal of rubbish" as Lind tried to do in 1753 (Clark and Chalmers, 2018, p. 121). As Landhuis (2016) points out, "In the biomedical field alone, more than 1 million papers pour into the PubMed database each year – about two papers per minute." Data from the National Science Foundation in the United States show that the publication output of science and engineering articles went from 1,755,850 in 2008 to 2,555,959 in 2018 (White, 2019). At this rate, writing the systematic review article becomes a very challenging task if a single author or even a group were trying to collate all of the information necessary to present a quality overview of a field or even a small subfield.

This chapter has presented what has been the traditional approach to the scientific review article until the twentieth century, where it has played an important role in constructing the knowledge bases that are used to further advance science and technology. However, here I would like to suggest that the twenty-first century requires new ways of collating and curating information. As predicted by Ginsparg (2011) and suggested by Hey et al. (2009), we need to take advantage of data management tools for computer-assisted analysis of text and multimedia data to enable researchers to grasp trends and patterns to inform decisions and actions. Sutton et al. (2019, p. 216), who presented their comprehensive examination of types of review articles, conclude their paper by noting that "a consistent typology" is needed and that information specialists should be consulted for implementing this. Indeed, data management software could even be used to reveal discrepancies in reports uploaded to preprint servers and point to outliers that do not fit into tacit knowledge networks. This approach could reveal flaws in the data as well as point to new avenues of potential research.

Today, we must deal with very complex systems for which there may be no clear answers. We are in an age of postnormal science which:

> is an approach to interface science and policy in cases where decisions need to be made before conclusive scientific evidence is available. Often, a single and conclusive scientific answer will not be available for highly complex systems such as fisheries, climate, society and the human body. In such cases, more research does not necessarily lead to less uncertainty, but can lead to unforeseen complexity. Values are often in dispute when the potential impacts of decisions based on uncertain science have very large consequences.
>
> *(University of Bergen, 2014)*

A *Nature* editorial (2016, p. 7) states, "Theorists have classified fields such as climatology and global-change research as post-normal science, in which socio-economic stakes are high and decisions are pressing." It also notes, "In pure science, as in art, little is urgent. Gravitational waves were discovered – a triumph for curiosity-driven science – thanks to physicists' patience and imaginative power. That they had waited decades is irrelevant. Alas, not all science has the luxury of timelessness" (*Nature*, 2016, p. 7).

From the beginning of 2020, the world has been experiencing the trials of the COVID-19 pandemic. This pandemic serves as an example of postnormal science where policy decisions are being forced on governments around the world without any conclusive scientific evidence to help. Brainard (2020) reports on the plight of a virologist who cannot keep up with the pace of the torrent of papers on COVID-19, which had amounted to more than 4,000 in the previous week. This led the White House Office of Science and Technology Policy to launch the COVID-19 Open Research Dataset (CORD-19) in March, and machine learning groups began to apply natural language processing methods to not only search the large database "but also extract meaningful patterns from findings across papers." By April 2020, Colavizza et al. had published "A scientometric overview of CORD-19" in bioRxiv, the preprint server for biology. Using the VOSviewer (van Eck et al., 2010) and Dimensions software (Herzog et al., 2020), they presented a term map highlighting temporal trends, topic modeling, and citation clusters and also offered Altmetric analyses to reveal the dissemination of information via science blogs and social media. Colavizza et al. (2020, p. 18) concluded their paper with, "This line of work has the potential to provide valuable information to experts and governments during the current and future pandemics."

Conclusion

This chapter has shown that scientific review articles can be found under various names, indicating that they can take various forms but that all have the common goal of providing scientists in their respective discourse communities with a good-quality curated overview of the research in their field so that they can use the information to further their own work to contribute to the disciplinary knowledge base. The origins of science were traced to point out how review articles play a pivotal role in the construction of these knowledge bases by discourse communities that share common goals and means of communication. What is accepted by the scientific community can then be used to inform public policies and actions. The need for this was evident in trying to deal with the COVID-19 pandemic of 2020.

Today, science is such a gargantuan enterprise that we are inundated by a deluge of research papers. This makes it difficult to collate and curate the data and information to glean what is accurate and useful. The twentieth-century approach to producing the scientific review article is likely to be difficult to apply.

I would like to predict that, considering the staggering amount of research being published annually, the scientific review article of the twenty-first century will very likely take the form of a scientometric overview with recommendations from experienced scientists in the field based on what the data management software reveals. The scientific review article will probably continue to play a pivotal role in the construction of field knowledge, but its production will need to entail the wise use of technology by field experts.

References

Armstrong, R., Hall, B.J., Doyle, J., & Waters, E. (2011) 'Cochrane Update 'Scoping the scope' of a Cochrane review', *Journal of Public Health*, vol. 33, no. 1, pp. 147–150, DOI: 10.1093/pubmed/fdr015

arXiv (n.d.) https://arxiv.org/

Booth, A. (2016) 'EVIDENT guidance for reviewing the evidence: A compendium of methodological literature and websites', working paper, February, www.researchgate.net/publication/292991575_EVIDENT_Guidance_for_Reviewing_the_Evidence_a_compendium_of_methodological_literature_and_websites

Booth, A., Noyes, J., Flemming, K., Gerhardus, A., Wahlster, P., Van der Wilt, G.J., Mozygemba, K., Refolo, O., Sacchini, D., Tummers, M., & Rehfuess, E. (2016a) 'Guidance on choosing qualitative evidence synthesis methods for use in health technology assessments of complex interventions', www.integrate-hta.eu/downloads/ and www.integrate-hta.eu/wp-content/uploads/2016/02/Guidance-on-choosing-qualitative-evidence-synthesis-methods-for-use-in-HTA-of-complex-interventions.pdf

Booth, A., Sutton, A., & Papaioannou, D. (2016b) 'EVIDENT guidance for reviewing the evidence: A compendium of methodological literature and websites', working paper, February, www.research gate.net/publication/292991575_EVIDENT_Guidance_for_Reviewing_the_Evidence_a_compendium_of_methodological_literature_and_websites

Brainard, J. (2020) 'Scientists are drowning in COVID-19 papers. Can new tools keep them afloat?', *Science*, May 13, 2020 doi:10.1126/science.abc7839

Butterfield, H. (1957) *The origins of modern science, 1300–1800,* New York: Palgrave Macmillan.

Clarke, M., & Chalmers, I. (2018) 'Reflections on the history of systematic reviews', *BMJ Evidence-Based Medicine*, 2018, vol. 23, pp. 121–122, DOI: 10.1136/bmjebm-2018–110968

Cochrane Database of Systematic Reviews (n.d.) www.cochranelibrary.com/cdsr/about-cdsr

Colavizza, G., Costas, R., Traag, V.A., van Eck, N.J., van Leeuwen, T., & Waltman, L. (2020) 'A scientometric overview of CORD-19', www.biorxiv.org/content/10.1101/2020.04.20.046144v2

Ginsparg, P. (1997) 'First steps toward electronic research communication', in L. Dowler (ed.), *Gateways to knowledge: The role of academic libraries in teaching, learning, and research*, Cambridge, MA: MIT Press, pp. 43–58.

Ginsparg, P. (2011) 'It was twenty years ago today . . .' arXiv:1108.2700 [cs.DL]

Greenhalgh, T., Potts, H.W.W., Wong, G., Bark, P., & Swinglehurst, D. (2009) 'Tensions and paradoxes in electronic patient record research: A systematic literature review using the meta-narrative method', *The Milbank Quarterly*, vol. 87, no. 4, pp. 729–788, DOI: 10.1111/j.1468–0009.2009.00578.x

Gross, A. (1990) *The rhetoric of science*, Cambridge, MA: Harvard University Press.

Herzog, C., Hook, D., & Konkiel, S. (2020) 'Dimensions: Bringing down barriers between scientometricians and data', *Quantitative Science Studies*, vol. 1, no. 1, pp. 387–395, DOI: https://doi.org/10.1162/qss_a_00020

Hey, T., Tansley, S., & Tolle, K. (2009) *The fourth paradigm: Data-intensive scientific discovery*, Redmond, WA: Microsoft Research.

Knorr-Cetina, K.D. (1981) *The manufacture of knowledge: An essay on the constructivist and contextual nature of science*, Oxford: Pergamon Press.

Landhuis, E. (2016) 'How to manage the research-paper deluge? Blogs, colleagues and social media can all help', *Nature*, vol. 535, pp. 457–458, DOI: https://doi.org/10.1038/nj7612-457a

Medical Research Council. (1948) 'Streptomycin treatment of pulmonary tuberculosis', *British Medical Journal*, vol. 2, pp. 769–782.

Munn, Z., Peters, M.D.J., Stern, C., Tufanaru, C., McArthur, A., & Aromataris, E. (2018) 'Systematic review or scoping review? Guidance for authors when choosing between a systematic or scoping review approach', *BMC Med Res Methodol*, vol. 18, no. 1, p. 143, DOI: https://doi.org/10.1186/s12874-018-0611-x

Myers, G. (1990a) *Writing biology: Texts in the social construction of scientific knowledge*, Madison, WI: The University of Wisconsin Press.

Myers, G. (1990b) 'Making a discovery: Narrative of split genes', in C. Nash (ed.), *Narrative in culture: The uses of storytelling in the sciences, philosophy, and literature*, London and New York: Routledge, pp. 102–126.

Nature (2016) Future present: A young global-sustainability platform deserves time to find its feet. *Nature*, vol. 531, p. 7, https://doi.org/10.1038/531007b

Nature (2020) For authors. www.nature.com/nature-research/for-authors/nature-research-journals

Noguchi, J. (2001) 'The science review article: An opportune genre in the construction of science', PhD thesis, University of Birmingham, Birmingham.

Proceedings of the National Academy of Sciences of the United States of America. (n.d.) www.pnas.org/authors/submitting-your-manuscript#article-types

Shorten, A., & Shorten, B. (2013) 'What is meta-analysis?' *Evidence Based Nursing*, vol. 16, no. 1, pp. 3–4, DOI: 10.1136/eb-2012–101118

Siddaway, A.P., Wood, A.M., & Hedges, L.V. (2019) 'How to do a systematic review: a best practice guide for conducting and reporting narrative reviews, meta-analyses, and meta-syntheses', *Annual Review of Psychology*, vol. 70, pp. 747–770, doi: 10.1146/annurev-psych-010418-102803

Sucharew, H., & Macaluso, M. (2019) 'Methods for research evidence synthesis: the scoping review approach', *J. Hosp. Med.* vol. 14, no. 7, pp. 416–418, doi:10.12788/jhm.3248

Sutton, A., Clowes, M., Preston, L., & Booth, A. (2019) Meeting the review family: exploring review types and associated information retrieval requirements. *Health Information & Libraries Journal*, vol. 36, pp. 202–222, DOI: 10.1111/hir.12276

Swales, J. (1990) *Genre analysis: English in academic and research settings,* Cambridge: Cambridge University Press.

Tricco, A.C., Lillie, E., Zarin, W., O'Brien, K.K., Colquhoun, H., Levac, D., Moher, D., Peters, M.D., Horsley, T., Weeks, L., Hempel, S. et al. (2018) 'PRISMA extension for scoping reviews (PRISMA-ScR): checklist and explanation', *Ann Intern Med.* vol. 169, no. 7, pp. 467–473, doi:10.7326/M18-0850.

University of Bergen, Centre for the Study of the Sciences and the Humanities. (2014) 'Open seminar: Science for policy: Post-normal science in practice', www.uib.no/en/svt/81853/science-policy-post-normal-science-practice

van Eck, N.J., & Waltman, L. (2010) 'Software survey: VOSviewer, a computer program for bibliometric mapping', *Scientometrics* (2010) vol. 84, no. 2, pp. 523–538, DOI:10.1007/s11192-009-0146-3

White, K. (2019) *Publications output: U.S. trends and international comparisons*, National Science Foundation, https://ncses.nsf.gov/pubs/nsb20206/publication-output-by-region-country-or-economy

Wootton, D. (2015) *The invention of science: A new history of the scientific revolution*, London: Penguin.

16

GRANT PROPOSALS IN THE ACADEMIC SPHERE

A historical review

Begoña Bellés-Fortuño

Introduction: grant proposals as a research genre

Within the field of genre analysis (Swales, 1990) and more concretely in the study of written academic genres, the research article (RA) has been considered by scholars as the key genre to disseminate scientific knowledge and reach a vast audience (Bellés-Fortuño, 2016); in Montgomery's words, the RA has been called "the master narrative of our time" (1996, p. i). However, knowledge transfer in higher education institutions nowadays strongly depends on collaborative projects of multidisciplinary teams that acutely need funding. From this perspective, grant proposals have a significant impact on the scientific and academic community since getting funding is paramount for our scientific progress and social development. Despite their importance for the academic community, grant proposals have not been broadly studied compared to other written academic research genres such as the RA (Swales, 1990; Montgomery, 1996; Hyland, 2010) or the thesis and dissertation (Paltridge, 1997, 2002). The reasons can be many, but the truth is that accessibility to grant proposals' texts is not easy. Most researchers feel reluctant to make them public or share them with the academic community due to the sensitive information included in the applications. Grant proposals include personal academic information of the main researcher and the research team, as well as updated, original, and competitive research. Moreover, grant proposal applications are strictly related to the funding agency they are submitted to regarding format and structure; therefore, the grant proposal genre seems to be quite versatile and dynamic compared to other academic research genres, making its study more ambitious.

According to previous literature (Bazerman, 1994; Bhatia, 2004), the research grant proposals and the RA are closely related genres within the academic world; they belong to the same genre colony. The former demands funding to carry out impressive research; the latter disseminates the research carried out. Their interrelation is, therefore, marked. It is quite paradoxical to think that the research grant proposal precedes undertaking any research analysis and later publication; the grant proposal application thus seems to be crucial. However, the RA has received more attention in genre analysis (Swales, 1990; Hyon, 1996; Montgomery, 1996; Hyland, 2010), and few studies have focused on grant proposals compared to the amount of work done on RAs. Maybe this fact influenced Swales (1996) to include the grant proposal within what he called academic *occluded genres*, stating that academic occluded genres are "those which support the

DOI: 10.4324/9781003043782-19

research publication process but are not themselves part of the research record" (1996, p. 1). The need and significance of studying this genre is therefore recognized. Grant proposal research has adopted different perspectives and approaches in order to understand how this crucial genre behaves. In the sections that follow, I will review the study of the research academic genre grant proposal from different approaches and methodologies including the rhetorical move approach, the corpus linguistic view, acknowledging the grant proposal's sociocognitive dimensions, and the more recent view toward a dialogic nature of the genre.

Grant proposals from a rhetorical move approach

Previous research on grant proposals has adopted different perspectives depending on the theory followed; thus three different main approaches to the study of grant proposals can be identified. The first group follows genre analysis as proposed by Swales (1981, 1990, 2004) and Bhatia (1993); that is, academic genres are analyzed according to rhetorical moves. Studies such as the ones carried out by Connor and Mauranen (1999), Connor and Wagner (1998), Connor (1998), or Feng (2008) can be included within the rhetorical move analysis. In his masterpiece *Genre Analysis*, Swales (1990, pp. 186–187) already included the grant proposal as a research genre and described its macrostructure including (1) front matter, (2) introduction, (3) background, (4) description of the proposed research, including this section's method, approach, and evaluation instruments, and finally (5) back matter. After this incipient grant proposal structure description, finding a generic rhetorical structure of grant proposals aroused the interest of genre analysis researchers. Early studies followed the traditional move analysis instituted by Swales (1981, 1990), where a move is defined as "a discoursal or rhetorical unit that performs a communicative function in written or spoken discourse" (Swales, 2004, p. 228). Departing from Swales's early structure of grant proposals, Connor (1998) compared a proposal written by a nonprofit organization with a university research proposal, finding that the latter seemed to follow some sort of moves. But it would not be until a later study carried out by Connor and Mauranen (1999) that a comprehensive move structure for grant proposals emerged. In this study, they analyzed 34 grant proposal submissions to the European Union. The findings identified ten rhetorical moves in no specific order: (1) establishing the research *territory*, (2) indicating a *gap* in the territory, (3) stating the *goal* of the current project, (4) specifying the *means* to achieve the goal (methodology), (5) reporting *previous research*, (6) describing anticipated *results* and *achievements*, (7) presenting the *benefits* and *usefulness* of the project, (8) claiming the main researcher and the research team, (9) making an *important claim* for the planned project, and (10) appealing the relevance of the proposed project to the *interests of the funding agency*. These moves proved to be useful for further studies following the same line.

Based on this initial study and the identification of these ten moves, Connor (2000) later approached the variation of these rhetorical moves in a sample of 14 grant proposals written by five researchers from several disciplines including humanities and hard sciences. She discovered that some moves occurred consistently among the corpus sample, including: research territory, indicating a gap, goal of the study, and means to achieve the goal. However, other moves were not so recurrent such as the move presenting benefits or making an important claim. Other inconsistencies found included move length variation; a consistent length pattern could not be identified for each move, although the methodology move (research *means*), seemed to be the longest in the whole sample.

Although the identification of moves in grant proposals proved very useful for the research community, the previously mentioned studies did not emphasize "the role of the writer in the discourse community and the expectations of that community" (Flowerdew and Dudley-Evans,

2002, p. 465). Other studies, however, combined the analysis of grant proposals moves with interviews (Connor and Wagner, 1998) or ethnographic methods of analysis for the writing process (Myers, 1990; Van Nostrand, 1994). Connor and Wagner (1998) interviewed grant writers from six different Latino nonprofit organizations to explore the identification of Latino identities in the texts and the subsequent need for a reviewing process. In the case of Van Nostrand (1994), the goal was to identify negotiation patterns between the writer and the funding agency. The studies reviewed have undoubtedly provided clear evidence of the combination of textual and contextual methods of analysis, offering new approaches to the analysis of grant proposals.

Grant proposals from a corpus linguistic view

Other studies have analyzed grant proposals from a corpus linguistic perspective focusing on a wide variety of linguistic features based on rich corpora that include a certain number of real texts. Among these studies, we find Connor and Upton's (2004) study which analyzed 68 grant proposals from nonprofit organizations. The analysis was based on a multidimensional technique proposed by Biber (1988) where four dimensions are identified: (1) involved versus informational, (2) narrative versus non-narrative, (3) explicit versus situation-dependent, and (4) overt expression of argumentation. Their aim was to see whether the variations of linguistic features related to the four dimensions tend to prevail in a particular move to unveil patterns of use. The study could not find any evidence relating moves to the multidimensional technique but offered detailed data as to linguistic features variation. Feng and Shi (2004) analyzed a corpus of nine grant proposals based on previous structural move analyses. The corpus, taken from the Social Sciences and Humanities Research Council of Canada, included the grant proposal summaries and the full texts. The analysis was supported by interviews with the nine grant proposal writers, which enabled a deeper exploration into the communicative purpose of the genre. This study remarked on the value of the interaction between reader and writer and the influence of social conventions. Feng and Shi's research described a generic structure of grant proposal summaries and main texts showing that the grant proposal genre is a well designed and prepared genre that uses repeated strategies across sections.

In a later study, Feng (2006) carried out a corpus linguistics analysis of research grant proposals abstracts. She thought that the study of grant proposal abstracts was vital for future grant proposal writers. The grant proposal abstract needs to impress the grant committee and predisposes the consequent full-text reading; it should be persuasive but "without seeming to persuade" (Myers, 1990, p. 42); "it is a genre even more rhetorical than the main texts of the proposals" (Feng, 2006, p. 2). Feng compiled a corpus of 37 abstracts of successful Hong Kong competitive Earmarked Research Grant (CERG) proposals from different disciplines; 19 proposal abstracts belonged to humanities and social sciences and 18 to natural science. The purpose was to examine the corpus by conducting a multilevel analysis using corpus linguistics software such as Wordsmith Tools (Scott, 1996). First, she analyzed six moves corresponding to the semantic and functional units obtained from the data (see Table 16.1) in order to identify the semantic/functional units of the texts and the possible subunits within each move.

In a second instance, word frequency counts were analyzed for the whole corpus and each of the moves established to later compare them with the COBUILD general English Corpus (Sinclair, 1991, p. 143). The third and last step analyzed linguistic and pragmatic features such as hedges and boosters in the grant proposal abstract corpus and their incidence in each of the individual moves proposed. Feng (2006) found that hedges occurred more often when establishing a niche, whereas boosters were prominent in expressing achievement and benefit claims.

Table 16.1 Move analysis applied to grant
proposal abstracts

Move 1	Establishing a territory
Move 2	Establishing a niche
Move 3	Outlining research objectives
Move 4	Describing research means
Move 5	Explanation and justification
Move 6	Claiming potential contributors

Source: Based on Feng, 2006.

Moreover, from a general view, she observed that hedges and boosters were more frequent when establishing a niche. Her study contributed largely to the analysis of grant proposals and, more concretely, of grant proposal abstracts from the perspective of a corpus linguistics methodology.

Fundamentally, grant proposal research from a corpus linguistics approach allows the analysis of certain lexicogrammatical and pragmatic features as to frequency, variation, or distribution in real texts compiled. The use of corpus linguistics hand in hand with genre analysis helps to understand the academic language from a more complete perspective considering not only linguistic aspects but also the rhetorical and functional aspects of the genre.

Grant proposals and their sociocognitive dimension

Beyond the analysis of textual features, other dimensions of grant proposals have deserved attention. Some researchers have delved into the social, cultural, and cognitive dimensions of the grant proposal genre (Tardy, 2003; Martin, 2000; Martin and Rose, 2008; Hyland, 1999, 2004). This approach to the study of grant proposals emphasizes genre knowledge. "Genre knowledge of the system . . . extends far beyond the genre knowledge of the grant proposal itself, requires a certain procedural and socio-political knowledge that gains primary importance" (Tardy, 2003, p. 33). Thus this approach toward grant proposals claims the analysis of other functions of the texts such as "positioning, politeness, arguing or alignment work" (Hood, 2004, p. 30), which directly involve the writers, who need to know the conventions of the genre to convey the pursued message for a successful proposal.

The result is an approach to grant proposals that invokes more comprehensive accounts of language, such as the systemic functional linguistics (SFL) based on Halliday's theory (1985), where interpersonal and ideational meanings can be studied apart from textual meanings (Martin and Rose, 2003). This approach conceives of genres as a genre system; for example, grant writing is a chain of genres which include other subgenres such as project summary, a biographical sketch, budget planning, references, etc. (Tardy, 2003, pp. 14–25). In Martin and Rose's words, we should "map cultures as systems of genres" (2008, p. 260) affected by sociocultural factors.

In the same vein, Pascual and Unger (2010) explored expressions of interpersonal meaning following the SFL theory. They analyzed two English examples of successful grant proposals in the disciplines of chemistry and physics written by non-native Argentinian scholars using the Appraisal framework proposed by Martin (2000) and Martin and White (2005), where language users apply evaluative resources "for negotiating our social relationships, by telling our listeners or readers how we feel about things and people" (Martin and Rose, 2003, p. 19). The grant proposal writers' linguistic awareness of the interpersonal resources used was analyzed. The aim of the study was to analyze writers' positioning according to the ENGAGEMENT system in some obligatory moves of the grant proposal genre such as Benefits and Importance Claim. Based on the Appraisal theory that analyses "semantic resources used to negotiate emotions, judgements and valuation, alongside resources for amplifying and engaging with these evaluations" (Martin, 2000, p. 145), Pascual and Unger chose the system of ENGAGEMENT which "deals with sourcing attitudes and the play of voices around opinions in discourse" (2010, p. 266) because it is of relevance for the interactive and persuasive grant proposal genre. Although limited in the amount of corpus analyzed, Pascual and Unger's study shed some light on the variety of strategies used by grant proposal writers to enlarge a multiplicity of voices and engage other members of the community. The results highlight the importance of interpersonal resources to generate stance in academic writing. They conclude that grant proposal writers might benefit from keeping a "stance of openness to other voices" (Pascual and Unger, 2010, p. 277).

Studying the sociocognitive dimension of the grant proposals genre opens new avenues of analysis. The studies reviewed here have focused on the outstanding linguistic features of the genre, exploring the applicant-writer's choices without ignoring the sociocultural context of the language use. However, the grant proposal writing process includes other variables and participants which have been sometimes ignored, namely the proposal reviewers or the funding agency. Including these elements/participants in the process of grant proposal writing lends itself to a new approach for the analysis of the genre. This new view of the genre implies the study of grant proposals from a cognitive-pragmatic perspective, which is the subject of the following section.

Toward the dialogic nature of grant proposals

Recent studies have emphasized the dialogic nature of grant proposals, where members of the grant review committee and reviewers both stand as potential interactants, while the writer's previous knowledge and reviewers' expectations build up a grant proposal frame from a cognitive and pragmatic perspective (Tseng, 2011). This new approach towards the analysis of the grant proposal genre includes relevant cognitive aspects of language use, together with sociocultural factors that influence the production of the text.

In his study, Tseng (2011) addressed the reviewers' and the funding agencies' discourse patterns to later compare them with preestablished rhetorical moves. His aim was to place the genre in *dialogue* following Bakhtin's premises (1981), moving from a more conventional idea of text toward a more communicative perspective, which considers the interactions among participants. To do so, he analyzed a handout of materials given to potential grant proposal applicants from a National Science Council (NSC) in Taiwan. The documents included suggestions for

Table 16.2 Points included in the NCS handout for grant proposal organization

Points
1. Project contents
2. Budget matters
3. Personal statement
4. Publication samples
5. Ethical issues
6. Evaluation matters

Source: Based on Tseng, 2011.

the preparation of the grant proposal and indications about the reviewing procedures. The handout included six main points or recommendations which coincided with the reviewers' expectations of what a grant proposal should contain.

Tseng compared these points and expectations with the grant proposal's moves presented by Connor and Mauranen (1999). The results found convergences but also differences. For example, when trying to compare point 6 referring to evaluation matters, that is, when "the proposal is evaluated in terms of scholarship, significance, and innovativeness of the proposed subject matter" (Tseng, 2011, p. 2258), no correspondence was found with the move analysis. Innovativeness is a characteristic evaluated by the reviewer, but it cannot generate a rhetorical move. Tseng's new approach to grant proposals intends to place genre in dialogue by considering interactions, including (1) dialogue between reviewer's expectation and applicant's production, (2) dialogue between the institution and applicant, and (3) dialogue between the reviewers and the individual applicant (2011, p. 2259). Such interactions are not being considered in the rhetorical move analysis (Connor and Mauranen, 1999).

The idea of genres having a dialogic nature is not entirely new, though. Already in 2004, Swales characterized genres as having "metaphorical endeavor" (2004, p. 61), and he proposed a flow of six metaphors which found a common ground among them, the "expectations" (2004, p. 68). The six metaphors of genre proposed by Swales are:

1   *Frames of social actions*: Based on Bazerman (1997), the social action of genre affects the communication purpose.
2   *Language standards*: Here, linguistic and generic standards are included in terms of "conventional expectations" (p. 68).
3   *Biological species*: The changeable nature of genres is addressed. Genres develop and evolve, and the causes that produce such change may be affected by expectations.
4   *Families and prototypes*: This genre metaphor establishes connections among genres and highlights their similarities.
5   *Institutions*: Based on Todorov's idea of genres as institutions (1990), Swales claims that genres are institutionalized procedures that generate conventions, standards, and potential expectations.
6   *Speech acts*: The genre as a speech acts metaphor emphasizes the function of the genre and how it is performed.

Swales's metaphors of genre help us understand genres not as static procedures but as evolving actions subject to the environment in which they are performed. This view unveils an incipient dialogic nature of genres where expectations play an essential role. In previous rhetorical moves analyses of genres, the genre expectations were not explicit. It is precisely the expectations frame that gives the grant proposal genres a more interactive role. Genres are expected to change; many factors may lead to this change and arise from individuals or institutions. Thus genres and, more concretely, the grant proposal should be addressed from a sociocultural perspective and consider all participants. In Todorov's words: "It is because genres exist as an institution that they function as 'horizons of expectation' for readers and as 'models of writing for authors" (1990, p. 18). In this line of argument, in the case of grant proposals, the addresser's and addressee's expectations should play an essential role in the grant proposal's development and purpose.

It is precisely this view of the grant proposal in a dialogic form which suggests the observation of the producer and recipients' points of view. In Tseng's words, the grant proposal should be addressed from "a dialogic and performative view of genre" (2011, p. 2266), where the addresser's knowledge about the convention and the reviewer's expectations about the proposal conform to a thinking frame. This view is undoubtedly cognitive and pragmatic in nature.

Approaching the analysis of grant proposal from this cognitive-pragmatic view entails complexity in the sense of interaction analysis and surpasses the study of recurrent textual features of the grant proposal genre. Grant proposal applicants interact with the social and institutional conventions, and at the same time the readers' expectations need to be observed. Therefore, the context is not static but in continuous evolution, responding to participants' interactions. The differentiating trait between previous approaches and the study of grant proposals from this dialogic view relies upon the potential interactions between participants in terms of conventions, performance, and expectations rather than paying attention only to textual features and recurrent patterns.

Concluding remarks

The chapter has provided a historical review of the grant proposal academic genre and the way genre analysts and other researchers have approached the genre. Although approached and studied by relevant researchers, the grant proposal academic genre has not been as largely studied as other extended written academic genres, such as the RA. Departing from a genre analysis approach proposed by Swales (1990, 2004) and the classification of the grant proposal as an occluded academic research genre, the different sections in this chapter have highlighted the most outstanding studies of grant proposals from different views and approaches. The identification and classification of rhetorical moves in grant proposal genres have proven to be useful for the academic community (Connor and Mauranen, 1999; Connor and Wagner, 1998; Connor, 1998; Feng, 2008), alongside large corpora studies of grant proposal texts or even grant proposal abstracts which have provided strategic linguistic features usage (Connor and Upton, 2004; Feng, 2006). By broadening the study of grant proposals to consider a sociocognitive dimension, the scope of analysis has further included interpersonal and ideational meanings in addition to the textual ones (Tardy, 2003; Martin, 2000; Martin and Rose, 2008; Hyland, 1999, 2004). More recent studies have argued for the dialogic nature of grant proposals, where, apart from the writers' and the readers' perspectives, other interactants such as the review committee and the reviewers build up the grant proposal's frame for analysis (Tseng, 2011). In this breakthrough approach, the expectations of all participants influence the production and aims of the text. Cognitive aspects along with sociocultural factors reveal a complex interaction process.

The historical review of the grant proposal genre presented here might provide valuable insights to scholars in upcoming effective grant proposal writing. Research on the grant proposal genre benefits the scientific and academic community, unveiling the essentials of a written genre that has become crucial among scholars in the scientific sphere. New insights in the study and development of grant proposal writing include the observation of all participants in the writing process, as well as an emphasis on the expectations of reviewers and funding agencies.

References

Bakhtin, M.M. (1981) *The dialogic imagination: Four essays*, translated by Michael Holquist, Caryl Emerson, Austin and London: University of Texas Press.

Bazerman, C. (1994) 'Systems of genres and the enactment of social intentions', in A. Freedman & P. Medway (eds.), *Genre and the new rhetoric*, London: Taylor and Francis, pp. 79–101.

Bazerman, C. (1997) 'The life of genre, the life in the classroom', in B. Wendy & H. Ostrom (eds.), *Genre and writing: Issues, arguments, alternatives*, Portsmouth, NH: Boynton/Cook, pp. 19–16.

Bellés-Fortuño, B. (2016) 'Popular science articles vs scientific articles: A tool for medical education', in P. Ordoñez-Lopez & N. Edo-Marzá (eds.), *Medical discourse in professional, academic and popular settings*, Bristol: Multilingual Matters, pp. 55–75.

Bhatia, V.K. (1993) *Analysing genre: Language use in professional settings*, London: Longman.

Bhatia, V.K. (2004) *Worlds of written discourse: A genre-based view*, London: Continuum.

Biber, D. (1988) *Variation across speech and writing*, Cambridge: Cambridge University Press.

Connor, U. (1998) 'Comparing research and non-profit grant proposals', in U. Connor (ed.), *Written discourse in philanthropic fund raising Indianapolis: Issues of language and rhetoric*, Indianapolis: Indiana Center on Philanthropy, pp. 45–64.

Connor, U. (2000) Variation in rhetorical moves in grant proposals of US humanists and scientists. *Text*, 20: 1–28.

Connor, U., & Mauranen, A. (1999) Linguistic analysis of grant proposals: European Union research grants. *English for Specific Purposes*, 18: 47–62, DOI: https://doi.org/10.1016/S0889-4906(97)00026-4

Connor, U., & Upton, T.A. (2004) The genre of grant proposals: A corpus linguistics analysis. In U. Connor & T. Upton (eds.), *Discourse in the professions: Perspectives from corpus linguistics*. Amsterdam/ Philadelphia: John Benjamins, pp. 235–256, DOI: https://doi.org/10.1075/scl.16.10con

Connor, U., & Wagner, L. (1998) Language use in grant proposals by non-profits: Spanish and English. *New Directions for Philanthropic Fundraising*, 22: 56–73, DOI:https://doi.org/10.1002/pf.2205

Feng, H. (2006) A corpus-based study of research grant proposal abstracts. Perspectives. *Working papers in English and Communication*, 17(1): 1–24.

Feng, H. (2008) 'A genre-based study of research grant proposals in China', in U. Connor, E. Rozycki, & V. William (eds.), *Contrastive rhetoric*, Amsterdam: John Benjamins, pp. 63–86.

Feng, H., & Shi, L. (2004) Genre analysis of research grant proposals. *LSP & Professional Communication*, 4: 8–32.

Flowerdew, J., & Dudley-Evans, T. (2002) Genre analysis of editorial letters to international journal contributors. *Applied Linguistics*, 23: 463–489. DOI: https://doi.org/10.1093/applin/23.4.463

Halliday, M.A.K. (1985) *An introduction to functional grammar*, London: Edward Arnold.

Hood, S. (2004) *Appraising research: Taking a stance in academic writing*, Sydney: University of Technology.

Hyland, K. (1999) Persuasion in academic articles. *Perspectives*, 11(2): pp. 73–103.

Hyland, K. (2004) *Disciplinary discourses: Social interactions in academic writing*, Ann Arbor: The University of Michigan Press.

Hyland, K. (2010) Constructing proximity: Relating to readers in popular and professional science. *Journal of English for Academic Purposes*, 9:116–127, DOI: https://doi.org/10.1016/j.jeap.2010.02.003

Hyon, S. (1996) Genre in the three traditions; Implications for ESL. *TESOL Quarterly* 30 (4) 693–772, DOI: https://doi.org/10.2307/3587930

Martin, J.R. (2000) 'Beyond exchange: Appraisal systems in English', in S. Hunston & G. Thompson (eds.), *Evaluation in text: Authorial stance and the construction of discourse*, Oxford: Oxford University Press, pp. 142–175.

Martin, J.R., & Rose, D. (2003) *Working with discourse: Meaning beyond the clause*, London and New York: Continuum.

Martin, J.R., & Rose, D. (2008) *Genre relations: Mapping culture*, London: Equinox.

Martin, J.R., & White, P.R. (2005) *Language of evaluation: Appraisal in English*, London: Palgrave Macmillan.

Montgomery, S. (1996) *The scientific voice*, New York: The Guilford Press.

Myers, G. (1990) *Writing biology: Texts in the social construction of scientific knowledge*, Madison, WI: University of Wisconsin Press.

Paltridge, B. (1997) Thesis and dissertation writing: Preparing ESL students for research. *English for Specific Purposes*, 16(1): 61–70, DOI: https://doi.org/10.1016/S0889-4906(96)00028-2

Paltridge, B. (2002) Thesis and dissertation writing: an examination of published advice and actual practice. *English for Specific Purposes*, 21(2): 125–143, DOI: https://doi.org/10.1016/S0889-4906(00)00025-9

Pascual, M., & Unger, L. (2010) Appraisal in the research genres: An analysis of grant proposals by Argentinean researchers. *Signos*, 43(73): 261–280. DOI: 10.4067/S0718–09342010000200004

Scott, M. (1996) *Wordsmith tools*, Oxford: Oxford University Press.

Sinclair, J.M. (1991) *Corpus, concordance, collocation*, Oxford: Oxford University Press.

Swales, J.M. (1981) *Aspects of article introductions*, Birmingham: University of Aston Press.

Swales, J.M. (1990) *Genre analysis*, Cambridge: Cambridge University Press.

Swales, J.M. (1996) 'Occluded genres in the academy: The case of the submission letter', in E. Ventola & A. Mauranen (eds.), *Academic writing: Intercultural and textual issues*, Amsterdam: John Benjamins, pp. 45–58.

Swales, J.M. (2004) *Research genres: Explorations and applications*, Cambridge: Cambridge University Press.

Tardy, C.M. (2003) A genre system view of the funding of academic research. *Written communication*, 20(1): 7–36, DOI: https://doi.org/10.1177/0741088303253569

Todorov, T. (1990) *Genres in discourse*, Cambridge: Cambridge University Press.

Tseng, M. (2011) The genre of research grant proposals: Towards a cognitive-pragmatic analysis. *Journal of Pragmatics*, 43: 2254–2268, DOI: https://doi.org/10.1016/j.pragma.2011.02.015

Van Nostrand, A.D. (1994) 'A genre map of R&D knowledge production for the US department of defense', in A. Freeman & P. Medway (eds.), *Genre and the new rhetoric,* London: Taylor and Francis, pp. 133–145.

17

SEARCHING FOR THE "MAGIC ELIXIR"

Uncertainty in developing specific aims and broader impacts for NIH and NSF grants

Colleen A. Reilly

General guidelines for writing successful scientific grant proposals abound in scholarship, and significant overlap exists across the advice provided. Beyond admonitions to closely follow funding agencies' instructions, reach out to grant officers, and develop a realistic budget, these guidelines often frame writing successful grants as involving a degree of mystery and serendipity. In her brief overview of the "secrets" to writing successful grants, Sohn (2020) highlights the importance of preparation for grant composition and the use of clear writing and a narrative style. She also encourages researchers to answer the "so-what?" question, include personal information or memorable details, and recruit worthy collaborators. Likewise, while Koro-Ljungberg (2014) provides useful advice about developing successful grants based on her experience, such as highlighting what is innovative about the project and selecting credentialed partners, in the end, she also highlights the need to unpack the "hidden complexities and political transformations" that are influential in grant funding processes (p. 211). Strikingly she concludes with "Who knows – maybe next time your proposal will be funded" (Koro-Ljungberg, 2014, p. 211). Such expressions signal the scientific grant proposal as a complex and challenging genre that many researchers struggle to produce effectively and perceive as requiring access to hidden information for consistent success.

For U.S. scientists, large governmental funding agencies, namely the National Institutes of Health (NIH) and the National Science Foundation (NSF), are central sources of significant grant funding. Earning grants from NIH and NSF has become increasingly competitive. In 2019, according to Lauer (2020), NIH awarded "$29.466 billion" to "55,012 new and renewed meritorious extramural grants." The success rate for NIH extramural Research Project Grants in 2019 was 20.1% (Lauer, 2020). NSF's funding is much smaller, providing $7.778 million in FY 2020 for 19,607 projects for research support, education and human resources, and major research equipment to a variety of organizations, including federal agencies, industries, small businesses, and universities, from across all 50 states and U.S. territories (NSF, 2020a, October). NSF (n.d.a) reports a competitive 28% funding rate for 2020 for an unidentified subset of their awards. The dominant influence of NIH and NSF on the direction and viability of grant-funded scientific research is hardly new as Myers (1985) noted this same phenomenon. Because the grant proposals for both NIH and NSF are quite complex and context specific, dependent in part on the institute within NIH and the component of NSF to which they are directed,

 DOI: 10.4324/9781003043782-20

it is difficult to provide generalized writing strategies. However, there is considerable overlap related to the single-page portions of each agency's grant proposals, namely the specific aims section for NIH and the proposal summary that must include the intellectual merit and broader impacts information for NSF. As the following discussion reflects, in the scholarship from across disciplines, scientists signal a significant amount of uncertainty about successfully producing these portions of the grant proposals. In fact, experienced scientists seek to demystify or promote a magic key, recipe, or formula to aid others in writing these sections of their proposals (Mardis et al., 2012; Brownson et al., 2015; Monte and Libby, 2018; Goldstein et al., 2021). The uncertainty around producing these portions of the grant proposals matters as it poses another hurdle in an already competitive grant funding process that is acknowledged to disadvantage scientists from underrepresented groups, including racial minorities and women (Smith et al., 2017; Hall et al., 2018).

This chapter begins with a brief history of milestones relevant to the specific aims and broader impacts sections as requirements in NIH and NSF grant proposals. The remarkably consistent uncertainty expressed in the scholarship from across scientific disciplines regarding how to approach the development of these sections of the grant proposals will then be reviewed. The chapter concludes with a discussion of the importance of creating mentoring and professional development programs to assist students, early career scientists, and scientists from underrepresented minorities to develop these promotional and gatekeeping sections of NIH and NSF grant proposals and highlights several successful efforts.

Brief history of the development of NIH specific aims and NSF broader impacts

In the United States, NIH and NSF figure prominently in terms of major funding for research in academic medicine and the sciences. In addition, specific portions of their proposals, namely the specific aims section for NIH and NSF's proposal summary, which includes the initial discussion of broader impacts, serve a gatekeeping function within their respective funding processes. Advice to scientists about grant writing found in the scholarship and on institutional websites highlights the requirements for these sections of NIH and NSF proposals, which are viewed as determinative for a proposal's success and which prove to be most challenging for grant writers (Florida State University, n.d.; Santen et al., 2017). From a genre perspective, these portions of the grant proposals have overtly promotional purposes (see Chapter 16 of this volume for Fortuño's detailed history of the scientific grant proposal genre). In fact, in her in-depth analysis of broader impacts in NSF grants, Cotos (2019) asserts that if grant proposals are the most rhetorical genres of scientific writing, broader impacts are the most obviously "rhetorically charged" portion (p. 28). This section details the significant milestones in the development of the requirements for these sections of NIH and NSF grant proposals and highlights the central changes most recently introduced.

After studying how to improve the peer review of grant proposals, NIH revised its application review criteria and proposal requirements beginning with proposals submitted in 2010 (NIH, 2009). As a part of that revision, NIH altered the requirements for both the content and length of the specific aims portion of the grant proposals. This change appears to have provoked the most concern among researchers who apply for NIH grants, as evidenced by the focus on specific aims in the scholarship. Prior to that date, it is difficult to find articles devoted to writing specific aims for NIH grant proposals. Previously, the specific aims section was required to include the following: "List the broad, long-term objectives and the goal of the specific research proposed, e.g., to test a stated hypothesis, create a novel design, solve a specific problem, challenge an existing paradigm or clinical practice, address a critical barrier to progress in the field, or develop new technology" (NIH, 2009). In the revision, NIH (2009) maintained this instruction but added the following: "State concisely the goals of the proposed research and summarize the expected outcome(s), including the impact that the results of the proposed research will exert on the research field(s) involved." Additionally, NIH narrowed the page limitations for the specific aims section from recommending one page to limiting this portion of the application to one page. Overall, NIH reduced the number of pages allowed for every section of the grant proposals in the 2010 revision of the requirements.

As Hallinen (2014) explains in her aptly titled dissertation about broader impacts, *The Many Quiet Tensions*, NSF altered its proposal review criteria in 1997 ostensibly to make more explicit the requirement for funded research to make contributions in external contexts through activities including teaching, dissemination of results, enhancing infrastructure, benefitting society, and broadening the participation of underrepresented groups (Watts et al., 2015, p. 399; see also Nagy, 2016). As of 2013, the new criteria required grant writers to include support for two equally important merit review criteria in their project summaries: the intellectual merit and broader impacts of their work (Hallinen, 2014, pp. 18–19; see also Heath et al., 2014, p. 518). Grant writers were then to expand on the merit review criteria in the project description portion of the proposals. While the project summary section contains the abridged version of the broader impacts (often abbreviated BI), the initial representation in this space is particularly significant as the project summary is the most important portion of the grant proposal for reviewers (Florida State University, n.d.) and is the portion that is publicly available for funded grants. Additionally, Heath et al. (2014) argue that with "decreases in available federal funds and an abundance of high-quality proposals, BIs are increasingly used to distinguish proposals during the NSF review process" (p. 518).

The changes in both the specific aims and the project summary portions of the NIH and NSF proposals enhanced their gatekeeping functions in the grant review processes by requiring writers to succinctly demonstrate to reviewers that their projects have merit, can be accomplished, and will have concrete outcomes that, in the case of NSF, constitute benefits for the community and society. As the following discussion indicates, although the guidelines and requirements for the specific aims and summary portions of NIH and NSF grants are detailed by the funding agencies, articles and pedagogical materials discussing them continually highlight their opacity and the difficulties that scientists may face when effectively developing them.

Negotiating the uncertainty in producing the specific aims section for NIH grants

This section highlights how the uncertainty involved in successfully composing the specific aims section is framed in the scholarship from a range of disciplines. NIH and its component institutes provide specific guidelines for the specific aims portion of the grant proposal (National Institute of Allergy and Infectious Diseases, 2020; NIH, 2020). For instance, they point out that the specific aims section should be separated into four subsections: an introduction to the gap or problem in research fulfilled by the work; the presentation of the team and central hypothesis to be addressed; aims to be used, each accompanied by a secondary hypothesis and actions used to test it; and, lastly, the significant outcomes, innovations, and impacts resulting from the work (Santen et al., 2017; Goldstein et al., 2021). Santen et al. (2017) argue that the specific aims section "is the most important component, as it summarizes the scientific premise, gap in current knowledge, hypotheses, methods, and expected results of the project proposed" (p. 1194; see also Goldstein et al., 2021). The purpose of this section is to sell the proposal to reviewers as evidenced by the overlap between the techniques used in this portion of NIH proposals and sales genres, as well as bolstering the credibility of the writer and foregrounding the inspirational and thoughtfully planned nature of the proposed ideas (Santen et al., 2017, p. 1196; Monte and Libby, 2018, p. 1043).

When addressing the perceived uncertainty inherent in developing a successful specific aims section, many scholars couch their discussions in terms of solving a mystery or locating the secret formula for success. For example, Goldstein et al. (2021) propose what they term as an algorithm that other scientists can deploy, including writing out a hypothesis, crafting an opening paragraph, and then creating an impact statement (p. 819), arguing that adhering to the advice in their article and using this algorithm will elicit success in writing a compelling and persuasive specific aims section. Similarly in their presentation posted online entitled "Secrets of superlative specific aims," Griswold and Deardorff (2012) provide advice that involves connecting with and thinking like reviewers and succinctly detailing the main focus, hypotheses, and innovative and inspirational outcomes for the project. Monte and Libby's (2018) "recipe" for the specific aims section involves detailed instructions for composing the required four paragraphs, which basically reiterate the instructions, but their article does include strategic and stylistic recommendations, such as using action verbs, visually relating the aims, and writing accessibly. Finally, Brownson et al.'s (2015) "magic elixir" for this section includes a focus on clear writing and fronting the most significant aspects of the proposal, which also seems like straightforward advice.

To further attempt to demystify the production of a successful specific aims page, experienced scholars often point to the common errors that they perceive others make when developing this section. Most scholars warn against "'aim dependency' whereby one aim cannot be completed if a prior aim fails" (Monte and Libby, 2018, p. 1045; see also Brownson et al., 2015; Santen et al., 2017). Goldstein et al. (2021) call these interdependent aims "domino aims" (p. 818). Other common mistakes are described using vague terms that appear repeatedly in the scholarship. For example, grant writers are advised to avoid making the specific aims portion of the proposal "overly ambitious," meaning that the reviewers will perceive the proposed activities and outcomes to be unachievable by the project team (Santen et al., 2017; Monte and Libby, 2018; Goldstein et al., 2021). Brownson et al. (2015) make the same point but call it "overpromising" (p. 712). Another common admonition is for writers to avoid proposing to engage in a "fishing expedition" (Monte and Libby, 2018, p. 1045; Goldstein et al., 2021, pp. 818–819), a term which appears in much critical commentary about unfounded science inquiry. In this case, avoiding a fishing expedition means that writers should focus on testing the central hypothesis and not stray into tangential inquiries that are not logically planned and based on evidence. Anderson (2015) from the National Institute on Aging also advises avoiding "fishing expeditions" in his blog about writing specific aims (for a critique of the use of the term, see "Fishing expedition" [2007]). Advising grant writers about what to avoid in these relatively consistent ways, none of which present particularly novel or unexpected perspectives, seems designed to reassure them that there exists a discoverable process for producing this section of the NIH proposal.

Negotiating the uncertainty in developing broader impacts for NSF grants

In the case of NSF grants, one aspect of the required information for the project summary, namely the broader impacts subsection, seems to provoke the most anxiety for grant writers, and the advice-oriented scholarship about grant writing reflects this. According to the literature, scientific researchers generally grasp the requirements for demonstrating intellectual merit but are challenged by the need to illustrate and support the broader impacts of their proposed work (Heath et al., 2014, p. 518). NSF (2020b) describes the two merit criteria requirements for the project summary section of the grant proposals and explains how they can be developed. Notably, NSF (n.d.b) previously provided examples of activities that demonstrate broader impacts but removed them in 2013 to avoid influence and prescriptiveness. Despite the detailed advice within the proposal instructions and on NSF's website, numerous articles spanning two decades from a range of science disciplines focus on explicating and strategizing about developing persuasive broader impacts, which reflects the anxiety and uncertainty around writing this portion of the NSF grant application (for example, see Mardis et al., 2012; Hallinen, 2014; Heath et al., 2014; Wattset al., 2015; Cotos, 2019).

A striking amount of the scholarship from across disciplines approaches the discussion of the broader impacts subsection by posing fundamental questions: What are broader impacts? Or what are examples of broader impacts in *X* field? (Mardis et al., 2012; Hallinen, 2014; Cotos, 2019). In contrast to the scholarship around specific aims in which scholars signal uncertainty indirectly through promising to provide magic keys and recipes, the literature around broader impacts unambiguously foregrounds a lack of clarity surrounding this merit requirement – as Watts et al. (2015) explain, scholars openly seek "to clarify the broader impacts concept" (p. 397). Gould et al. (2019) argue that the requirement to address broader impacts forces scientists to ask existential questions: "Why are we doing science? What topics should we study? What should scholars spend their time on? How and by whom should findings be shared?" (p. 1). Therefore, the scholarship acknowledges that this requirement of NSF grant proposals prompts a reconsideration of how and for what purposes science inquiry should be designed (Watts et al., 2015).

Advice in the scholarship about how to develop and articulate successful broader impacts stresses the promotional aspects of this section of the NFS proposal genre. For example, to help scientists develop a writing process to better explain their broader impacts activities in their grant proposals, Cotos (2019) reviewed grants' broader impacts and proposed a model representing the conventional moves common to successful proposals' articulations. This model, called the contextualize-demonstrate-predict (CDP) model, includes three moves: contextualizing potential impacts, demonstrating tangible impacts, and predicting significance (Cotos, 2019, p. 20); professional development materials have been developed using this model to assist scientists to apply the model to their generation of broader impacts.

The scholarship around broader impacts in NSF grant proposals highlights the need for scientists to expand their ideas of what successful science entails in order to effectively address this requirement. Gould et al. (2019) analyzed the broader impacts from 1,451 abstracts from projects funded between 1997 and 2012 related to biodiversity conservation, specifically looking for creative and innovative approaches to broader impacts activities. Innovative activities included community-based research and citizen science, such as partnering with a local zoo to garner support for their "pollination-related citizen science program" (Gould et al., 2019, p. 7). They also found innovative dissemination activities, such as starting a YouTube channel to cover the science or writing a children's book about the research. Gould et al. (2019) include screen captures and color images of these creative approaches to broader impacts in their article, thereby aiding other scientists to envision alternate ways to approach this requirement and expand their work. As Gould et al. (2019) argue in their conclusion, "[O]ur focus on particularly novel impacts foreshadows a suggestion for future directions that intertwines with rejecting the deficit model" (p. 10) of science communication; this model assumes that making a larger amount of accurate information about science available to publics is sufficient to persuade them of its importance and efficacy.

Although discussions in the scholarship covering broader impacts seek to clarify how to articulate them effectively, writers often acknowledge that their colleagues express consternation about this requirement because they perceive it as external to their scientific projects and training as scientists (Watts et al., 2015). Nagy (2016) attributes this in part to the connection between the broader impacts and community engagement and engaged scholarship, which "struggle[s] to achieve legitimacy and support within the academy" (para. 7). As Hallinen (2014) notes, some opinion columns in major publications question the need for broader impacts and the whole impetus for the public outreach behind their adoption. For example, Sarewitz (2011) argues in *Nature* that while the goal of prompting consideration of the contributions of science to society is laudatory, "doing so in the brutal competition for grant money will yield not serious analysis, but hype, cynicism and hypocrisy" (p. 141). Hallinen (2014) argues that the broader impacts have not fundamentally changed the university and the focus of faculty in part because tenure and promotion requirements do not generally reward community engagement (p. 33). Additionally, Watts et al. (2015) highlight the absence of broader impacts in the publicly available abstracts of NSF-funded projects. While the content of the abstracts is not determined by the scientists, PIs submit drafts of their abstracts and "appear to self-censor their BIAs [broader impact activities] when they draft [their abstracts]" (Watts et al., 2015, p. 403). This contributes to a perception that these activities are not central to the science proposed.

Scholarship analyzing broader impacts statements drawn from NSF grant abstracts, funded proposals, reports, and unfunded proposals often documents obvious deficiencies in the broader impacts proposed and reported. From such research, other scientists might surmise that their peers can and do succeed in earning funding despite not addressing broader impacts in a manner equal to the ways in which they propose and document the activities viewed as integral to their science. Mardis et al. (2012) found that the broader impacts from funded proposals in the National Science Digital Library (NSDL) were uneven and included "aspirational or incomplete claims of impact" (p. 1758). Of the 85 funded project abstracts they examined, 50 revealed "the activities and potential impacts were aligned to the goal of reaching a wider audience" while 35 had a "misalignment between activities and proposed impact or a lack of outreach beyond posting on the Web" (Mardis et al., 2012, p. 1768). Watts et al. (2015), who studied broader impacts as articulated in abstracts, funded proposals, and unfunded proposals, discovered that activities related to outreach to underrepresented groups were reported least often. While Watts et al. (2015) found broader impacts proposed more in funded than in unfunded proposals, they note that this does not mean that all funded proposals included them. They also found that broader impacts related to society appeared less often in awarded proposals than in unfunded proposals, which "perhaps reflects reviewer discretion on the ambiguity or perceived importance of this BIA category" (Watts et al., 2015, p. 403). Such analyses of broader impacts in grant proposals reveal how contested this portion of the grant proposal remains. Scientists are uncertain of the weight of well conceived broader impacts on their proposals' success, the ability of reviewers to assess them consistently, and the lasting positive effects, if any, on public perceptions of science.

Furthermore, other analyses of portions of funded grant proposals reveal that the guidelines provided by funding agencies can be at odds with the characteristics of funded proposals. For instance, NSF "is committed to writing new documents in plain language" to help the public to understand the science that was funded (NSF, 2018). Markowitz (2019) completed an analysis of abstracts from 19,569 NSF grant proposals that should include the broader impacts and determined that the abstracts for projects receiving higher levels of funding actually used more specialized words and were longer than average. Additionally, abstracts whose authors exhibited more confidence through "verbal certainty," which often correlated with complex sentence structure, received higher levels of funding (Markowitz, 2019, p. 268). This disconnect reflects the fissure between the stated NSF requirements and the attitude toward and execution of broader impacts as reflected across the scholarship from a range of scientific disciplines. Such gaps also exacerbate the uncertainties scientists experience when planning and writing their grant proposals.

Professional development for developing successful grant proposals

Across scientific disciplines, professional development for graduate students, early career scientists, and anyone seeking to improve their grant writing skills is often targeted to writing the specific aims and broader impacts sections of NIH and NSF grant proposals (see Chapter 31 of this volume for Barel-Ben David and Baram-Tsabari's discussion of other professional development initiatives for practicing scientists). Approaching these contested and weighty sections of the grant proposals within discipline-specific professional development sessions makes sense considering the close connections between grant proposals and disciplinary norms, which Velarde (2018) terms "discursive institutionalization" (p. 103). Ding (2008) concurs and presents a cognitive apprenticeship framework for grant writing for graduate students to enculturate them into the discourse community and genre system of NIH's funding processes. The uncertainty around successfully writing these key portions of NIH and NSF grant proposals matters because funding is very competitive and research reflects inequalities in application rates and awards to scientists of color and female scientists. The uncertainty involved in writing the sections of NIH and NSF grants explored here amplifies these inequalities by discouraging some scientists from even trying to secure this level of funding or declining to resubmit proposals after receiving an initial rejection.

Many professional development programs target the specific aims and broader impacts portions of grant proposals directly. For example, Heath et al. (2014) created a biology graduate course to teach students to plan their broader impacts through focusing on outreach and engagement activities more broadly. Their curriculum included bringing in speakers from local communications, educational, and cultural organizations, such as radio stations and museums, and locating potential partnerships for communicating and participating in science that could be transformed by the graduate students into broader impacts activities for their grant proposals (Heath et al., 2014). The range of options included could give rise to the creative and innovative proposals for broader impacts advocated by Gould et al. (2019).

At a macro level, the National Alliance for Broader Impacts (NABI) was created by a collaborative online community that supported broader impacts and their creative development and expansion (Adetunji and Renoe, 2017). The leaders of NABI earned a grant from NSF to develop the Center for Advancing Research Impact in Society (ARIS), an organization focused on expanding and supporting broader impacts across scientific disciplines. ARIS holds summits, provides training to universities, offers fellowships and awards, and encourages membership. Strikingly, recalling the search for magic formulas often sought in the scholarship about specific aims and broader impacts previously discussed, ARIS provides a Broader Impacts Wizard (Center for Advancing Research Impact in Society and Rutgers, 2020). This online tool guides researchers through a four-step process to develop a broader impacts plan to bolster their NSF grant proposals.

Universities and government funding agencies also have developed programs to assist early stage investigators (ESIs), specifically from underrepresented groups, in applying for federal grants in order to address the racial and gender disparities in the number of applications submitted and awards earned. Hall et al. (2018) highlight the National Research Mentoring Network (NRMN) that was launched by the National Institutes of Health (NIH) and that has "played a key role in professional development and mentoring of participants through grant writing programs" (p. 2). Hall et al. (2018) examined the success of a program allied with NRMN, the Health Equity Learning Collaboratory (EQ-Collaboratory), that recruited cohorts from across the mainland United States and Puerto Rico, enrolled 69% underrepresented minorities, and operated in a virtual environment providing resources and social support for grant development. They found that ESIs who participated in the online collaboratory were more likely to submit a proposal and had a shorter time to submission than those who did not participate (Hall et al., 2018, p. 8).

Other programs specifically target female scientists. Most important for the issues discussed here, namely the uncertainty surrounding elements of the NIH and NSF grant proposals, Biernat et al. (2019) highlight that women make up a smaller portion of those who attempt to apply for first-time R01 NIH grant applications (grants for independent researchers): "Between 2003 and 2007, 34% of first-time R01 applicants were female, and in an analysis of the NIH biomedical database from 2000 to 2006, women submitted fewer applications overall, and Blacks and women who were new investigators were less likely to resubmit a grant application" (Biernat et al., 2019, p. 141; see also Smith et al., 2017). Biernat et al. (2019) discovered that women were less motivated to resubmit their applications after first receiving a rejection, especially if the feedback referenced the "inadequacy of the researcher" (p. 149). To address the gap in funding for female scientists, Smith et al. (2017) designed a grant writing bootcamp program at Montana State University as part of a larger program called ADVANCE Project TRACS (Transformation Through Relatedness, Autonomy, and Competence Support) that was targeted to women. This program is based on self-determination theory (SDT), which "states that when a task, event, or job is experienced as 'self-determined' (as opposed to controlling), the experience is engaging and fulfilling" (Smith et al., 2017, p. 639). The SDT approach to grant writing and submission directly opposes how successful grant writing is discussed in much of the scholarship previously referenced, namely as something that requires magic or a knowledge of a secret key or formula. Smith et al. (2017) found that women who attended the workshop both submitted and earned a greater number of external grants than those who did not. They plan to create another program focused on prompting female scientists to resubmit unfunded proposals.

Conclusions

Pursuing and securing grant funding from NIH and NSF is a time-consuming and high-stakes process. An article in *Nature* captures the frustration felt by some scientists, quoting one as saying, "'Damned if you do, damned if you don't sometimes seems like a theme at NIH'" (Powell, 2017, p. 401). Reductions in funding have exacerbated the competition and made developing these major grant proposals seem more daunting and less desirable to some scientists. Some researchers have even proposed replacing or supplementing the current grant review process with the use of limited lotteries to select among proposals that pass an entry screening or shifting away from the proposal process to an analysis of past performance for established scientists (Gross and Bergstrom, 2019). In a study of the time expenditures involved in writing grants versus the funds earned and the distractions from doing the science itself, Gross and Bergstrom (2019) found that "the effort researchers waste in writing proposals may be comparable to the total scientific value of the research that the funding supports, especially when only a few proposals can be funded" (p. 1). The uncertainty in the scholarship from across scientific disciplines over how to produce the specific aims and broader impacts sections of the NIH and NSF proposals certainly reflects the difficulty in successfully securing funding from these agencies as well as anxieties over the role of medical and scientific research in society and importance of this research to public audiences. This chapter demonstrates the efficacy of creating targeted professional development focused on grant writing for early career scientists, especially individuals from underrepresented groups. Successful interventions should be designed to help these scientists overcome this uncertainty and enable them to feel empowered – not to seek a magic key, secret, or formula for writing successful grant proposals but instead to develop a process based on planning and the ability to articulate a broader purpose and innovative role for their scientific research.

References

Adetunji, O.O., & Renoe, S.D. (2017) 'Assessing broader impacts', *MRS Advances*, vol. 2, pp. 1681–1686, doi: 10.1557/adv.2017.136

Anderson, D. (2015) 'Strengthen your research plan for a better score – dos and don'ts', *National Institute on Aging*, web log post, 28 January, www.nia.nih.gov/research/blog/2015/01/strengthen-your-research-plan-better-score-dos-and-donts (Accessed 5 May 2021)

Biernat, M., Carnes, M., Filut, A., & Kaatz, A. (2019) 'Gender, race, and grant reviews: translating and responding to research feedback', *Personality and Social Psychology Bulletin*, vol. 46, no. 1, pp. 140–154, https://doi.org/10.1177/0146167219845921

Brownson, R.C., Colditz, G.A., Dobbins, M., Emmons, K.M., Kerner, J.F., Padek, M., Proctor, E.K., & Strange, K.C. (2015) 'Concocting that magic elixir: Successful grant application writing in dissemination and implementation research', *Clinical and Translational Science*, vol. 8, pp. 710–716, doi: 10.1111/cts.12356

Center for Advancing Research Impact in Society (ARIS), & Rutgers University. (2020) 'Broader impacts wizard', https://aris.marine.rutgers.edu/wizard/index.php (Accessed 5 May 2021)

Cotos, E. (2019) 'Articulating societal benefits in grant proposals: move analysis of Broader Impacts', *English for Specific Purposes*, vol. 54, pp. 15–34, https://doi.org/10.1016/j.esp.2018.11.002

Ding, H. (2008) 'The use of cognitive and social apprenticeship to teach a disciplinary genre: initiation of graduate students into NIH grant writing', *Written Communication*, vol. 25, no. 1, pp. 3–52, https://doi.org/10.1177/0741088307308660

'Fishing expedition' (2007) *FemaleScienceProfessor*, web log post, 18 October, https://science-professor.blogspot.com/2007/10/fishing-expedition.html (Accessed 5 May 2021)

Florida State University (n.d.) 'Writing the project summary and project description', *NSF Career Toolkit*, https://www.research.fsu.edu/research-offices/ord/nsf-career-toolkit/home/writing-the-proposal/project-summary-and-project-description/ (Accessed 5 May 2021)

Goldstein, A.M., Balaji, S., Ghaferi, A.A., Gosain, A., Maggard-Gibbons, M., Zuckerbraun, B., & Keswani, S.G. (2021) 'An algorithmic approach to an impactful specific aims page', *Surgery*, vol. 169, no. 4, pp. 816–820, https://doi.org/10.1016/j.surg.2020.06.014

Gould, R.K., Coleman, K.J., Krymkowski, D.H., Zafira, I., Gibbs-Plessl, T., & Doty, A. (2019) 'Broader impacts in conservation research', *Conservation Science and Practice*, vol. 1, no. 11, pp. 1–13, https://doi.org/10.1111/csp2.108

Griswold, M., & Deardorff, D. (2012) 'Secrets of superlative specific aims', https://journals.plos.org/plosbiology/article?id=10.1371/journal.pbio.3000065 (Accessed 5 May 2021)

Gross, K., & Bergstrom, C.T. (2019) 'Contest models highlight inherent inefficiencies of scientific funding competitions', *PLoS Biology*, vol. 17, no. 1, pp. 1–15, https://journals.plos.org/plosbiology/article?id=10.1371/journal.pbio.3000065 (Accessed 5 May 2021)

Hall, M., Engler, J., Hemming, J., Alema-Mensah, E., Baez, A., Lawson, K., Quarshie, A., Stiles, J., Pemu, P., Thompson, W., Paulsen, D., Smith, A., & Ofili, E. (2018) 'Using a virtual community (the Health Equity Learning Collaboratory) to support early-stage investigators pursuing grant funding', *The International Journal of Environmental Research and Public Health*, vol. 15, no. 11, pp. 1–12, doi:10.3390/ijerph15112408

Hallinen, J.R. (2014) 'The many quiet tensions: Perceptions of the broader impacts criterion held by NSF career award holders at very high research institutions of higher education', doctoral thesis, University of Pennsylvania, online ProQuest PQDT Open.

Heath, K.D., Bagley, E., Berkey, A.J.M., Birlenbach, D.M., Carr-Markell, M.K., Crawford, J.W., Duennes, M.A., Han, J.O., Haus, M.J., Hellert, S.M., Holmes, C.J., Mommer, B.C., Ossler, J., Peery, R., Powers, L., Scholes, D.R., Silliman, C.A., Stein, L.R., & Wesseln, C.J. (2014) 'Amplify the signal: graduate training in broader impacts of scientific research', *BioScience*, vol. 64, no. 6, pp. 517–523, doi:10.1093/biosci/biu051

Koro-Ljungberg, M. (2014) 'Trickeries of grant work', *Qualitative Inquiry*, vol. 20, no. 2, 203–212, https://doi.org/10.1177/1077800413510881

Lauer, M. (2020) 'Extramural investments in research: FY 2019 by the numbers', *Extramural Nexus, National Institutes of Health Office of Extramural Research*, web log post, 5 May, https://nexus.od.nih.gov/all/2020/05/05/extramural-investments-in-research-fy-2019-by-the-numbers/ (Accessed 5 May 2021)

Mardis, M.A., Hoffman, E.S., & McMartin, F.P. (2012) 'Toward broader impacts: making sense of NSF's merit review criteria in the context of the National Science Digital Library', *Journal of the American Society for Information Science and Technology*, vol. 63, no. 9, pp. 1758–1772, https://doi.org/10.1002/asi.22693

Markowitz, D.M. (2019) 'What words are worth: National Science Foundation grant abstracts indicate award funding', *Journal of Language and Social Psychology*, vol. 38, no. 3, pp. 264–282, https://doi.org/10.1177/0261927X18824859

Monte, A.A., & Libby, A.M. (2018) 'Introduction to the specific aims page of a grant proposal', *Academic Emergency Medicine*, vol. 25, no. 9, pp. 1042–1047, https://doi.org/10.1111/acem.13419

Myers, G. (1985) 'The social construction of two biologists' proposals', *Written Communication*, vol. 2, no. 3, pp. 219–245, https://doi.org/10.1177/0741088385002003001

Nagy, D. (2016) 'Determinants of broader impacts activities: a survey of NSF-funded investigators', *Journal of Research Administration*, vol. 47, no. 1, (online Gale)

National Institute of Allergy and Infectious Diseases. (2020) 'Draft specific aims', 18 September, https://www.niaid.nih.gov/grants-contracts/draft-specific-aims (Accessed 5 May 2021)

National Institutes of Health (NIH). (2009) 'Notice number: NOT-OD-09–149', https://grants.nih.gov/grants/guide/notice-files/NOT-OD-09-149.html (Accessed 5 May 2021)

National Institutes of Health (NIH). (2020) 'General application guide for NIH and other PHS agencies', 16 October, https://grants.nih.gov/grants/how-to-apply-application-guide/forms-f/general/g.100-how-to-use-the-application-instructions.htm#3 (Accessed 5 May 2021)

National Science Foundation (NSF). (2018) 'Plain language', https://www.nsf.gov/policies/nsf_plain_language.jsp (Accessed 5 May 2021)

National Science Foundation (NSF). (2020a) 'Budget internet information system: Award summary by state/institution', October, https://dellweb.bfa.nsf.gov/starth.asp (Accessed 5 May 2021)

National Science Foundation (NSF). (2020b) 'Chapter II: Proposal preparation instructions', *Proposal & Award Policies and Procedures Guide*, 1 June, https://www.nsf.gov/pubs/policydocs/pappg20_1/pappg_2.jsp#IIC2b (Accessed 5 May 2021)

National Science Foundation (NSF). (n.d.a) 'Funding rate by state and organization', https://dellweb.bfa.nsf.gov/awdfr3/default.asp (Accessed 9 May 2021)

National Science Foundation (NSF). (n.d.b) 'Merit review frequently asked questions (FAQs)', https://www.nsf.gov/bfa/dias/policy/merit_review/mrfaqs.jsp#1 (Accessed 5 May 2021)

Powell, K. (2017) 'The best-kept secrets to winning grants', *Nature*, vol. 545, 25 May, pp. 399–402, doi: 10.1038/545399a

Santen, R.J., Barrett, E.J., Siragy, H.M., Farhi, L.S., Fishbein, L., & Carey, R.M. (2017) 'The jewel in the crown: specific aims section of investigator-initiated grant proposals, *Journal of the Endocrine Society*, vol. 1, no. 9, pp. 1194–120, doi: 10.1210/js.2017-00318

Sarewitz, D. (2011) 'The dubious benefits of broader impact', *Nature*, vol. 475, 13 July, p. 141, doi: 10.1038/475141a

Smith, J.L., Stoop, C., Young, M., Belou, R., & Held, S. (2017) 'Grant-writing bootcamp: an intervention to enhance the research capacity of academic women in STEM', *BioScience*, vol. 67, no. 7, pp. 638–645, doi: 10.1093/biosci/bix050

Sohn, E. (2020) 'Secrets to writing a winning grant', *Nature*, vol. 577, 2 January, pp. 133–135, doi: 10.1038/d41586-019-03914-5

Velarde, K.S. (2018) 'The way we ask for money . . . The emergence and institutionalization of grant writing practices in academia', *Minerva*, vol. 56, pp. 85–107, https://doi.org/10.1007/s11024-018-9346-4

Watts, S.M., George, M.D., & Levey, D.J. (2015) 'Achieving broader impacts in the National Science Foundation, Division of Environmental Biology', *BioScience*, vol. 65, no. 4, pp. 397–407, https://doi.org/10.1093/biosci/biv006

18

EXAMINING TENSIONS BETWEEN TECHNICAL COMMUNICATION PRINCIPLES AND SCIENTIFIC PRACTICE

Does jargon belong on scientific posters?[1]

Kalie Leonard

Introduction

Peer-reviewed journal articles are the typical mode of large-scale dissemination of science. However, Rowe (2017) describes posters as 'the most prevalent medium of disseminating information at today's academic/scientific conferences' (p. vii). The scientific poster is a relatively new genre of communicating science, and the conventions of scientific posters have remained stable throughout the genre's short history (Rowe, 2017). Despite the stability of scientific poster conventions over time, new poster designs are currently being tested and implemented. This chapter presents research about scientific communicators' thoughts on conventional and new scientific poster elements. I define scientific communicators as professionals who conduct scientific research and report it through scientific writing/communication – including scientists at universities and companies, as well as science graduate students. Moreover, Hofmann (2017) defines scientific writing as 'technical writing by scientist[s] for other scientists,' as opposed to science writing that is geared toward 'a general audience' (p. 9). This chapter is primarily concerned with scientific posters as a form of scientific writing/communication.

There exists ample advice about designing effective posters, but it is worth exploring whether or not scientific communicators currently think certain poster elements *actually are* effective. Advice about scientific posters is sometimes written by technical communicators rather than by scientists, so there may be tensions between scientific practice and technical communication theory/advice concerning scientific posters. Tensions between technical communication and science arise from different disciplinary perceptions of jargon. Similar to how Wolfe (2009) explains, '[T]echnical communication textbooks offer advice that conflicts with the knowledge practices and professional values' (p. 353) of engineering fields, technical communication advice about posters may conflict with scientific practices. Drawing out these tensions (or the

DOI: 10.4324/9781003043782-21

lack thereof) can lead to productive interdisciplinary discussions about what constitutes effective scientific poster content and design.

In this chapter, I present research about scientific communicators' thoughts on certain scientific poster elements – particularly jargon. More specifically, this chapter aims to answer the following questions. Should scientific posters include jargon? Do scientific communicators' thoughts about using jargon on posters conflict with technical communication principles? Do conventional and/or new scientific poster elements work well? What might scientific posters be like in the future?

Conventional and new scientific posters

Commonalities arise among sources that discuss the conventional elements of scientific posters. Scientific posters typically follow the IMRAD (Introduction, Methods and Materials, Results, and Discussion) format or some slightly modified version of it (Matthews, 1990; Miller, 2007; Speight, 2012; Hofmann, 2016, 2017; Rowe, 2017). Poster creators are encouraged to use large fonts for poster titles, so the title is easily readable from a distance (Block, 1996; Miller, 2007; Penrose and Katz, 2010; Speight, 2012; Hofmann, 2016, 2017; Rowe, 2017). Additionally, posters should have logically arranged sections to make the information easy for audience members to follow (Matthews, 1990; Block, 1996; Erren and Bourne, 2007; Miller, 2007; Penrose and Katz, 2010; Speight, 2012; Hofmann, 2016, 2017; Rowe, 2017). Together, these sources provide helpful insights into conventional elements of scientific posters that have remained historically stable.

Despite the historical stability of poster conventions, some scientists are currently implementing a new poster design at conferences. Greenfieldboyce (2019) describes Mike Morrison's 'better poster' design:[2] 'It looks clean, almost empty. The main research finding is written right in the middle, in plain language and big letters. There's a code underneath you can scan with a cellphone to get a link to the details of the study' (para. 16). This new poster design looks quite different from conventional scientific posters. Morrison's simpler poster may be more effective than conventional posters at attracting attention and helping audience members to retain information, but its visual and informational (in)effectiveness still needs to be researched (Greenfieldboyce, 2019).

Understanding the purpose of scientific posters can aid in evaluating the (in)effectiveness of poster content and design elements. However, little scholarship currently addresses the purpose of scientific posters. Burns et al. (2003) never explicitly state the purpose of scientific posters, but they define science communication 'as the use of appropriate skills, media, activities, and dialogue to produce one or more of the following personal responses to science (the AEIOU vowel analogy): Awareness, Enjoyment, Interest, Opinion-forming, and Understanding' (p. 183). Because posters can be a form of science communication according to Burns et al. (2003), the purpose of a scientific poster may also be to elicit one or more of the 'personal responses' from the audience. Burns et al.'s (2003) definition of science communication is too generalized and it conflicts with Hofmann's (2017) definition that I described earlier, but their definition may still provide insight into the purpose of a scientific poster. Unlike other sources, Rowe (2017) *does* explicitly state that the main objectives of a scientific poster presentation are 'to share information and network with the peer community' (p. 11). My research asks scientific communicators whether or not they agree with Rowe (2017) or Burns et al. (2003) or whether they describe other purposes of scientific posters. A better understanding of the purpose of scientific posters would provide insight into how posters should be designed and what constitutes an effective poster design/content element.

Tensions on the use of jargon on scientific posters

This chapter largely focuses on jargon as an element of posters because very little scholarship discusses the appropriateness of jargon on a scientific poster. Although jargon often has a negative connotation, I use the term 'jargon' according to Hirst's (2003) 'neutral' definition as 'the specialized vocabulary of any organization, profession, trade, science, or even hobby' (p. 203). In general, technical communication advises against jargon because it often interferes with clarity – arguably the most important stylistic element of good technical writing. According to Strunk and White (2009), style is the way an author's words sound that often 'reveals his [or her] identity' (p. 68). In order to maintain a plain and simple style, writers should '[a]void fancy words' (Strunk and White, 2009, p. 76). Similarly, Markel and Selber's (2018) technical communication textbook advises writers to 'Avoid Fancy Words' (p. 237) and to 'Avoid Unnecessary Jargon' (p. 232) in order to communicate more clearly and concisely (pp. 229–237). Jargon has the potential to disrupt clarity, but Penrose and Katz (2010) also explain, 'style has an ethical dimension' (p. 81) because scientific jargon, for instance, may create 'a false *ethos* of objectivity' (p. 82). Therefore, it may be unethical to use jargon in some contexts. Overall, technical communication scholars often dissuade writers from using jargon.

A few scholars explicitly state that jargon should be avoided on scientific posters. Block (1996) explains scientific poster creators should 'Avoid jargon' because posters by themselves should be able to clearly communicate the poster's message (p. 3528). Similarly, in a poster checklist at the back of a chapter, Hofmann (2016) and Hofmann (2017) ask readers to consider, 'Have all jargon and redundancies been omitted?' (pp. 236, 581, respectively). Miller's (2007) article also advises poster presenters to 'Avoid cluttering the poster with too much technical detail or obscuring key findings with excessive jargon' (p. 315). It seems the halls of almost every university science building are covered with inaccessible, jargon-laden posters, so it is a bit surprising that sources seem merely to advise against jargon. Overall, there is a general lack of discussion about the appropriateness of jargon on scientific posters, and the little discussion that does exist seems to contradict current practice. This discrepancy may be due to differences – or, as I mentioned earlier, tensions – between technical communication principles and scientific practice. Highlighting the potential divergences between scientific practice and technical communication theory may lead to productive interdisciplinary discussions about posters with the goal of working toward effective interdisciplinary approaches to scientific poster design.

Methods for collecting survey and interview data

In order to foster more interdisciplinary discussion about scientific posters, I conducted mixed methods research that proceeded in a sequence of two data collection phases: first, a survey, and second, interviews to provide a more in-depth explanation of survey responses. This research was approved by the University of Wyoming's Institutional Review Board as exempt on October 24, 2019. The first phase of data collection involved self-selected participants completing an online survey on Google Forms. The survey was widely disseminated and accessible via a link sent from one of the following avenues: a university-wide listserv e-mail, a science departmental listserv e-mail, an ECOLOG-L listserv e-mail (for members of the Ecological Society of America), a post from my personal Facebook and Twitter accounts or from others sharing any of the previously mentioned e-mails/posts. Before accessing the survey, respondents had to consent to participation. Respondents were informed that participation in the survey was voluntary and that they could discontinue participation at any time. Additionally, survey responses were collected anonymously, unless respondents elected to provide their names and e-mail addresses for potential follow-up interviews. The survey was completed by 408 respondents. I chose to analyze survey responses only from the 359 respondents who marked their role as 'scientist at a university,' 'scientist at an organization or company,' or 'science graduate student.' Other roles such as secondary science teachers and roles outside of science were excluded from my analysis for purposes of scope because this research focuses specifically on scientific communicators' thoughts on scientific posters.

The second phase of data collection involved interviewing five participants who previously completed the Internet survey (see Table 18.1). Five participants were chosen for interviews because they responded to the last optional survey question that asked for their contact information if they were interested in an interview. Two participants were chosen because they seemed skeptical of new poster designs in their survey responses. The other three participants were chosen because they seemed optimistic about new poster designs in their survey responses. The purpose of interviewing was for the selected participants to explain in more detail some of their survey responses. The interviews were semistructured, and the interview questions were open-ended. Before proceeding with the interviews, participants signed a consent form.

Scientific poster audience and purpose

Similar numbers of survey respondents described themselves either as science graduate students (39% of respondents) or as scientists at a university (47% of respondents) (see Figure 18.1). Additionally, 13% of respondents described themselves as scientists at an organization or company. The majority of respondents (76%) described their primary poster audience as experts in their field of study (see Figure 18.2). This information suggests that most of the survey

Table 18.1 Interviewee descriptions

Interviewee	Role in science	Feelings about new poster designs
1	Graduate student	Optimistic
2	Scientist at a university	Optimistic
3	Graduate student	Optimistic
4	Scientist at a university	Skeptical
5	Scientist at a university	Skeptical

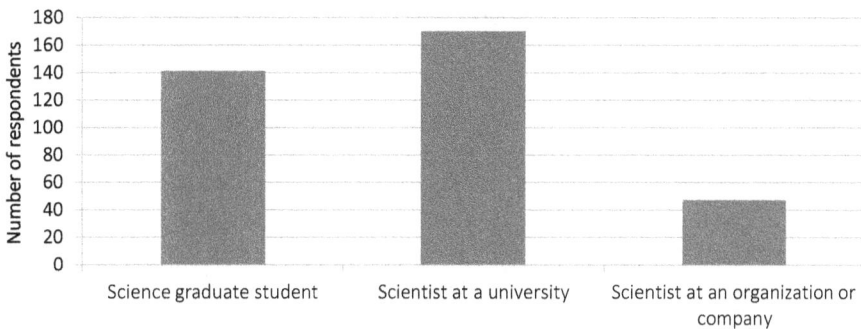

Figure 18.1 Respondent roles in science

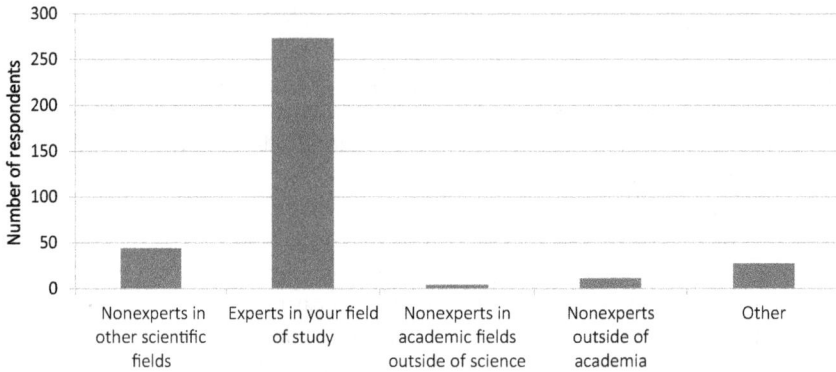

Figure 18.2   Primary poster audience

respondents present posters at conferences in their specialized disciplines. Approximately 16% of respondents described their primary poster audience as nonexperts (in other scientific fields, in fields outside of science, or outside academia), which suggests these respondents may present more at generalized or interdisciplinary academic conferences or at public (rather than academic) conferences. Other respondents may have misunderstood the survey question because some mentioned their primary audience depends on the conference. In sum, most respondents described their primary audience as experts in their field, so it is logical to infer that most participants design their scientific posters for an expert audience.

With a better understanding of the survey respondents' primary poster audiences, I now turn to their thoughts on the purpose(s) of poster presentations. As previously mentioned, Rowe (2017) states that the main objectives of a scientific poster presentation are 'to share information and network with the peer community' (p. 11). The respondents in this survey tend to agree with Rowe. When asked what they believe to be the main purpose of scientific poster presentations, 58% of respondents answered the main purpose is to share information with peers (see Figure 18.3). A notable number of respondents (12%) answered that the main purpose of a scientific poster presentation is to network with their peer communities. Some respondents who answered 'Other' explained that the main purpose of poster presentations is to receive feedback (about their research and/or ideas). From the respondents' answers, it seems poster presentations serve an important social role in science, especially through sharing information and networking.

Defining jargon

Despite the fact that posters serve an important social role among peers and that most survey respondents primarily present to other experts, the survey respondents often seem wary of jargon. Respondents provided their own definitions of jargon, so their responses to questions about jargon are not completely generalizable; however, I still present interpretations of the data. The majority of respondents (66%) defined jargon as field- or discipline-specific language (see Figure 18.4). For example, one respondent defined jargon as 'Terms specific to a certain field.' This respondent's definition and others' closely align with Hirst's (2003) 'neutral' definition of jargon as 'the specialized vocabulary of any organization, profession, trade, science, or even hobby' (p. 203). Almost half of the respondents (48%) also defined jargon as exclusionary to outsiders. For instance, a respondent defined jargon as 'Technical language that

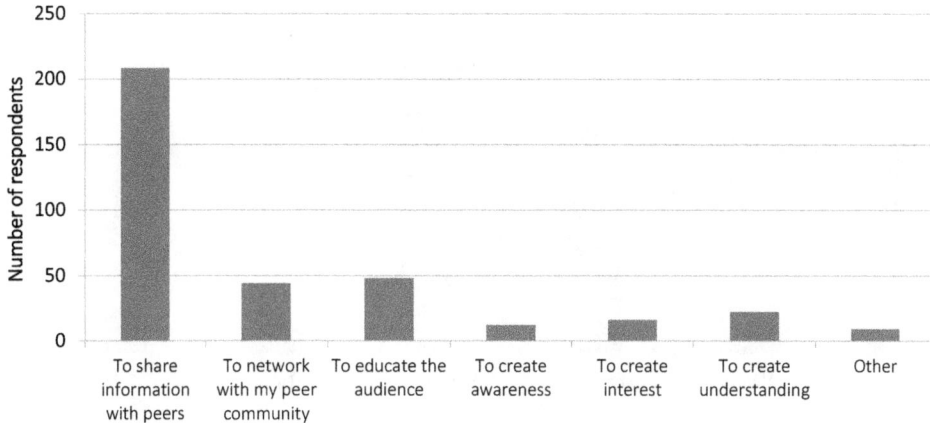

Figure 18.3   Main purpose of a scientific poster presentation

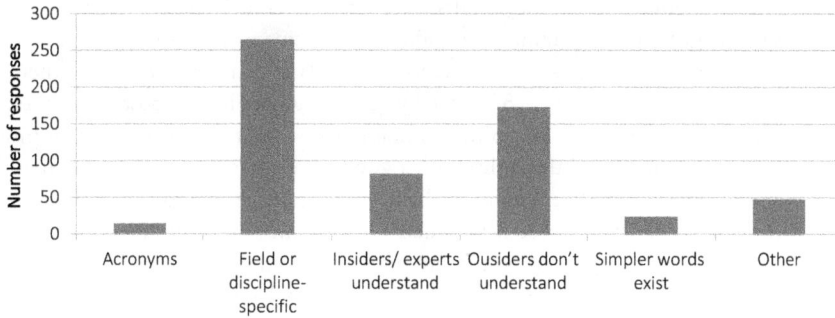

Figure 18.4   Respondent definitions of jargon

is not accessible/understandable outside your field of study.' Definitions that describe jargon as exclusionary to outsiders seem closer to what Hirst (2003) calls 'bad jargon.' Hirst explains, 'Certainly, jargon is bad when ill adapted to its audience' (p. 213). However, Hirst's explanation implies *not* that jargon is bad because it is exclusionary but that jargon is bad when used for an unfamiliar audience. Other answers from respondents included more obviously negative or positive definitions. One respondent described jargon negatively as 'an explanation or statement without getting to the main point; filler information,' and this definition seems to fit what Hirst would describe as a bad or unnecessary use of jargon (p. 213). Another respondent defined jargon more positively as 'specialized words used to concisely convey technical concepts' and this definition aligns with what Hirst calls 'well-used jargon' or jargon that 'achieves its goals of precision, speed, universality, and so on' (p. 218). Although respondents have some differing thoughts about defining jargon, it seems most respondents define jargon neutrally or negatively.

Use of jargon on scientific posters

Respondents' more positive or negative definitions of jargon may have an impact on how often they use jargon on posters. When asked about the frequency of use of jargon on scientific

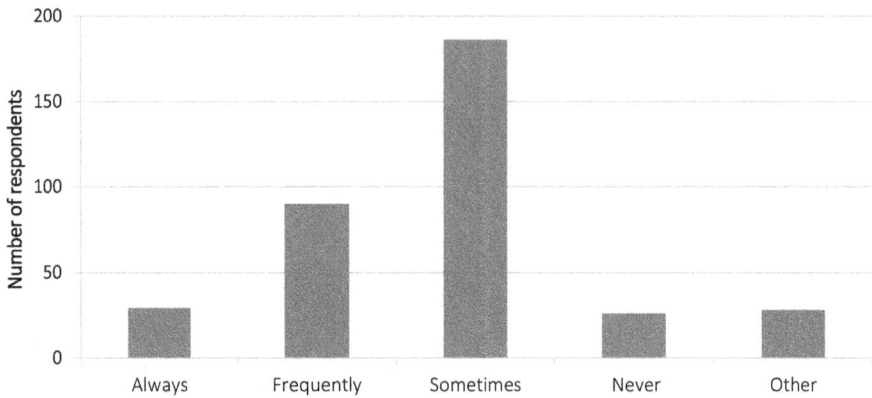

Figure 18.5   How often respondents use jargon on their own posters

posters, most respondents answered that they sometimes (52%) or frequently (25%) use jargon (see Figure 18.5). Respondents who answered 'Other' often explained that they rarely use jargon or that their use of jargon depends upon their audience. It seems many respondents may believe jargon is appropriate in some poster presentation contexts and inappropriate in others since jargon is not always used. Interviewee 5 believed jargon should be left off posters because 'most of our posters have sort of two lives. So there's one that's at the meeting where you probably could get by with jargon – I still don't think it's a good idea – but the other place, then, is we bring them home and we hang them up in the halls, we try to use them to recruit people to our labs – that sort of thing.' Interviewee 5 provided a compelling rationale for not using jargon on posters because, as they mention, posters are often displayed on the walls of science labs or buildings. Thus according to interviewee 5, jargon is not only exclusionary and intimidating to nonexperts, but jargon on posters may also negatively affect lab recruitment. When designing a poster, scientific communicators may want to consider a poster's 'second life' hanging on a wall.

In contrast to interviewee 5, interviewee 2 described a necessity for jargon on posters: 'if you're presenting a scientific poster at a conference on your stuff with the audience being from your field, or field-adjacent, you need jargon. I'm sorry – this anti-jargon, zero tolerance rule, I think is sometimes problematic. . . . The beauty of a technical vocabulary is that it's an established, shared language. And I think that really promotes good communication.' Interviewee 2's description of the 'anti-jargon, zero tolerance rule' seems to directly address disciplinary tensions between science and technical communication. Rather than hindering communication as technical communication theory often posits, interviewee 2 describes jargon as a necessary tool for effective disciplinary communication. From the survey and interview responses, jargon elicits both benefits and drawbacks, which helps explain why most survey respondents only sometimes use jargon on their posters. The range of responses from survey respondents and interviewees highlights the importance of making context-based decisions about whether or not the benefits of jargon outweigh the drawbacks.

Somewhat surprisingly, comparatively low numbers of survey respondents answered that they always (29 participants) or never (26 participants) use jargon (see Figure 18.5). It was surprising that only a small number of respondents feel they always use jargon because the survey respondents often notice jargon on others' posters (see Figure 18.6). The majority of respondents mentioned they frequently (65%) or always (18%) notice jargon on scientific

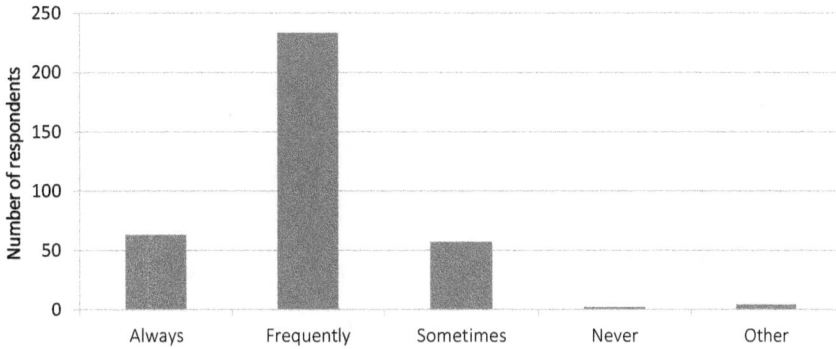

Figure 18.6   How often respondents notice jargon on others' posters

posters (see Figure 18.6). In general, respondents feel they notice more use of jargon on others' posters compared to their own, and this could be interpreted in a few ways. It is possible that the respondents generally use less jargon on posters than is common, and it is also possible that respondents feel they notice jargon more often than they actually do. Further, in examining survey responses to questions about jargon, it is important to keep in mind that each respondent's definition of jargon is subjective and sometimes audience-dependent, so respondents may not agree with what others may or may not classify as jargon on their posters.

In a similar vein, another survey question asked respondents to explain how they make decisions about using (or not using) jargon on scientific posters. A little over half of the respondents (56%) mentioned their decisions about the use of jargon are audience-dependent (see Figure 18.7). For instance, one respondent explained they make decisions about jargon 'Based on who the audience will be.' Similarly, some respondents (9%) mentioned their use of jargon depends on the presentation context. One respondent explained, 'If the poster will only be used at a conference in my field[,] I will use jargon if it is more efficient. At more general conferences or events I will avoid jargon entirely.' Some respondents mentioned they always use jargon (2%), and some mentioned they never use jargon (3%), but it seems most of the respondents use jargon when they deem it audience and/or context appropriate. Overall, most of the scientific communicators in the survey seem to make audience-based decisions about whether or not to use jargon on their posters. Therefore, scientific communicators are often situationally aware of when to use jargon, so their thought processes align well with technical communication theory. Due to seminal works like Miller's (1979) "A Humanistic Rationale for Technical Writing," contemporary technical communication theory is often grounded in rhetoric, and it seems the scientific communicators in my survey make conscious rhetorically based decisions, like using audience-appropriate language.

Elements of conventional vs. well-designed and engaging posters

Multiple-choice questions near the end of the survey asked respondents to select elements of 'conventional' scientific posters and elements of 'well-designed and engaging' scientific posters (see Figure 18.8). The most commonly agreed upon 'conventional' poster elements were visuals/ graphics, large title, logical sequence of sections, IMRAD (Introduction, Methods, Results, and Discussion) sections, concise text, and consistent style. Almost all respondents (96%)

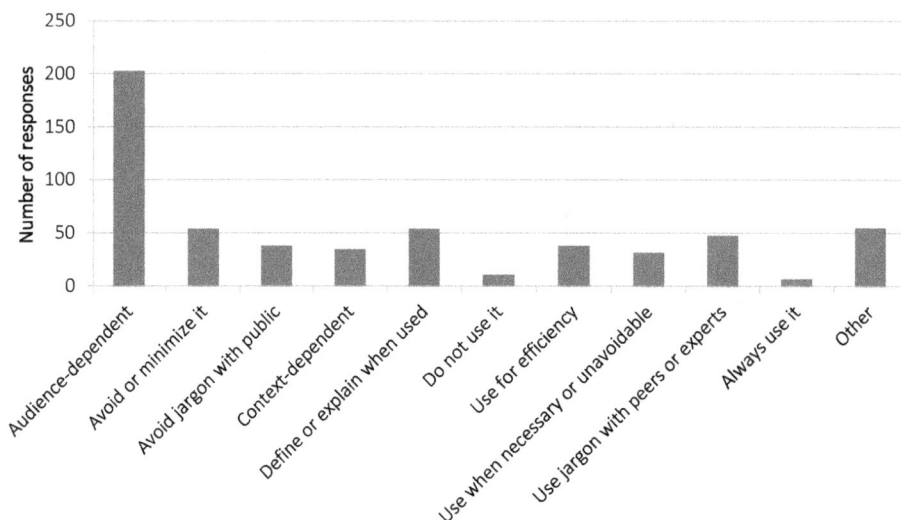

Figure 18.7 How respondents make decisions about using (or not using) jargon on a poster

agreed that conventional scientific posters should include visuals/graphics and that almost all respondents (93%) agreed that conventional scientific posters should include a large title. Because 'visuals/graphics' and 'large title' were the most frequently selected conventional poster elements, it seems scientific communicators recognize the importance of a poster's visual communication function. Corresponding to earlier discussions about jargon, I was surprised to find a small number of survey respondents who answered jargon should be included on scientific posters. Only 36 respondents (10%) selected jargon as an element that should be included on conventional scientific posters, which reiterates the respondents' apparent wariness and rhetorical awareness when using jargon.

A large amount of text was a common element of conventional posters that was not described in the survey but was mentioned by all of the interviewees. For example, as interviewee 3 explained, conventional posters 'tend to be pretty cluttered. In a way, it's kind of an emotional process for a researcher who's been working on a project for multiple years to be able to put it all together on a poster and think, "I'm going to show this amazing work off to everybody." And there's this temptation to put way too much into the poster.' Interviewee 3 provides a seemingly important and overlooked point: conventional posters often include too much text because of the researchers' emotional attachment to their work. Current scholarship about poster design fails to consider emotion as an internal motivation for certain design decisions. However, it is also worth considering whether or not posters are an appropriate outlet for a scientific communicator's emotional attachment to their work – especially if an emotion-driven design results in a poster unsuitable for effective communication. The most frequently selected elements of a 'well designed and engaging' scientific poster were similar to those of 'conventional' posters; these elements include visuals/graphics, large title, logical sequence of sections, concise text, minimal text, and consistent style (see Figure 18.8). Perhaps more unsurprising at this point, only 13 respondents (4%) selected jargon as an element of well designed and engaging posters. The lack of responses for the inclusion of jargon again suggests the respondents' thoughts align with technical communication theory that posits jargon as a hindrance to communication.

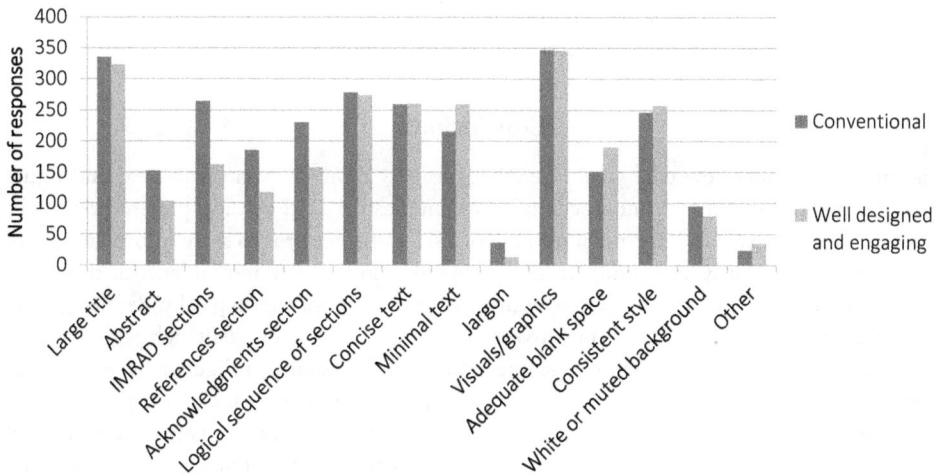

Figure 18.8 Elements of "conventional" and "well designed and engaging" posters

The future of scientific posters

When asked about the future of scientific posters, four out of five of the interviewees described digital or electronic posters as a useful, free, and/or sustainable alternative to printed posters. Interviewee 1 explained, 'Having it on a screen I think is great too, because it's eco-friendly; I've got five posters sitting in my office ready to be recycled. I just feel bad. And it's expensive to print posters, so having it on a screen would be free.' Interviewee 2 was also concerned about the sustainability of printed posters and explained, 'I think about sustainability as one of the things. Do we have to be printing on paper? Do we have to have a new poster every single time? Maybe there's a way that we can do it digitally, like screens and stuff like that that are up.' The future of posters may not be entirely clear, but it seems like digital posters and more sustainable options may become more common.

In addition to more sustainable posters, interviewees 4 and 5 felt future posters should reach an ideal balance of information and simplicity because they felt conventional posters often include too much information and newer designs like Morrison's (Greenfieldboyce, 2019) are too minimalistic. Interviewee 5 explained a 'happy medium' poster design: 'to me, it's sort of the extremes that are the problem. So these new style posters are so minimalistic that other than starting a conversation right then and there, I'm not sure how useful they are. Other posters that have so many words on them – and they can be quite graphically clever, but still too complex to read as you're moving through. . . . But yeah, some sort of a happy medium. The super wordy ones are just as problematic as the super non-wordy ones for me.' Similarly, interviewee 4 described what he calls a 'middle ground' of posters:

> [B]efore this meeting, I was prepping and going to Twitter. . . . And there are some other people that since this [Morrison's] poster has come out, have tried to merge the two ideas a little bit better. So they still have the central theme where here's the main finding and here's the QR code, but then they'd have much more expansion of the information using figures, particularly, to illustrate their points.

Thus according to a couple of the interviewees, future posters may reach an ideal balance between informative conventional posters and minimalistic new posters.

Conclusions

The survey and interview data yield interesting insights about conventional, new, and future scientific posters. In general, most scientific communicators in this study present posters for other experts and agree that the main purpose of a scientific poster is to share information with peers. Despite primarily presenting to other experts, most survey respondents in this study only sometimes use jargon on their scientific posters. Respondents typically defined jargon in neutral or negative terms, and only a very small number of respondents felt jargon should be included on well designed and engaging posters. The respondents' generally negative or wary attitudes about jargon may explain why they only sometimes use it on posters. One potentially overlooked reason why scientific communicators may be hesitant to use jargon is that posters are often hung on walls when they are no longer used for presentations. Thus jargon may not be audience appropriate when the poster hangs on a wall.

Most technical communication scholarship advises poster creators to avoid using jargon because jargon interferes with important stylistic writing elements, like clarity and concision (Strunk and White, 2009, p. 76; Markel and Selber, 2018, pp. 229–237). Therefore, it seems the respondents' often cautious feelings toward jargon are aligned with those in technical communication. However, some respondents and interviewees described jargon as a positive and useful element on a poster, so although it seems science and technical communication are becoming more closely aligned in their negative attitudes toward jargon, disciplinary tensions about jargon still exist. In order to mitigate existing disciplinary tensions about jargon on posters, interdisciplinary discussions should work to distinguish between poster contexts where jargon is appropriate or inappropriate.

A potential limitation of this study is that respondents provided their own definitions of jargon, so the respondents' attitudes toward jargon could vary depending upon what counts as jargon. Furthermore, it seems necessary for science and technical communication fields to agree upon a standard, interdisciplinary definition of jargon in order to avoid confusion about the term.

The future of scientific posters is not entirely clear, but it does seem that sustainable poster options, like digital posters, will become more common. From a couple of interviewee responses, it also seems future poster designs may fall somewhere between conventional text-heavy posters and new minimalistic posters, like Mike Morrison's (Greenfieldboyce, 2019). Whether or not conventional posters will remain in the norm, it seems we are currently experiencing an important reflective period about scientific poster design.

Recommendations for teaching scientific posters

Scientific/technical communication textbooks typically provide good general advice about designing posters, so it may be helpful to use a textbook to establish general guidelines for students designing posters. Participants in my survey and interviews seemed to agree with textbook authors on following certain poster conventions. Based on current literature about posters, as well as my survey results, I recommend scientific posters include the following elements: a large title, visuals/graphics, a logical sequence of sections (potentially IMRAD sections), concise and minimal text, and consistent style. To help students understand the importance of each of the previously mentioned poster elements, consider providing readings, discussions, and examples of each of the elements.

Because Mike Morrison's 'better poster' design is becoming increasingly popular, I recommend introducing students to both Morrison's poster design and more conventional poster designs. Students should be given space to consider the pros and cons of Morrison's design and conventional designs. It would also be worth discussing contexts in which one poster design may be more appropriate than another.

Students may view 'jargon' as a loaded term because, as seen in the survey results, people often equate jargon with exclusionary language. In a classroom setting, it is important for students to examine their own definitions and potential biases about jargon. Teachers should lead a discussion about jargon to help students voice their thoughts about jargon. This discussion may include examples of how jargon can be exclusionary, as well as examples of when jargon can be important for accuracy/precision of language. It would be helpful to have students think of examples of when jargon could be appropriate and inappropriate. Any examples of the (in) appropriateness of jargon should include particular attention to audience/context awareness because the audience is typically the main determining factor of jargon (in)appropriateness.

At the end of the discussion about jargon, I suggest that the class establish an agreed-upon definition of jargon or that the teacher provide Hirst's 'neutral' definition of jargon as a standard definition for the class to use. As a reminder, Hirst defines neutral jargon as 'the specialized vocabulary of any organization, profession, trade, science, or even hobby' (p. 203). Using a standard definition of jargon will help students come from a place of mutual understanding and engage in productive future discussions about jargon.

Rather than providing students with prescriptive rules for when jargon is or is not appropriate, it may be more productive to discuss the appropriateness of jargon on a case-by-case basis with students. These are some helpful questions to help students weigh the trade-offs of using jargon on their posters. What is the purpose of your poster? What is the context in which you will present your poster? Who is your primary poster audience? Will everyone in the audience understand the jargon you want to use on your poster? How could the use of jargon potentially disrupt communication with your audience? If you decide to use jargon on your poster, will there be space on the poster to provide a definition? Is the use of jargon necessary to maintain accuracy/precision of language? Will audience members question your credibility as a researcher/scientist if you choose to omit jargon? This is not a comprehensive list of questions to ask in regard to the appropriateness of jargon in specific contexts, but it is certainly a great starting point.

Recommendations for designing scientific posters

Audience and context are key: they should be the main determining factors for how a poster is designed. Therefore, scientists and technical communicators should keep the audience and context of the poster presentation in mind when designing a poster. For instance, a poster presented at a city council meeting should probably have less jargon and more visuals than a poster presented at a scientific conference. Similarly, it may be more useful to use Morrison's poster design at a large conference, and a more conventional poster design may be more appropriate for a smaller, specialized conference. When deciding whether or not to use jargon on a scientific poster, it will be helpful for scientists and technical communicators to ask the questions listed at the end of the previous section "Recommendations for Teaching Scientific Posters." Also as mentioned in the previous section, I recommend that scientific posters include the following elements: a large title, visuals/graphics, a logical sequence of sections (potentially IMRAD sections), concise and minimal text, and consistent style.

Notes

1 This chapter is based on Kalie Leonard's master's thesis entitled *Examining and Teaching Scientific Poster Design.*
2 Morrison demonstrates his ideas at the following link: https://osf.io/ef53g/.

References

Block, S.M. (1996) 'Do's and don'ts of poster presentation', *Biophysical Journal*, vol. 71, pp. 3527–3529, doi: 10.1016/S0006-3495(96)79549-8
Burns, T.W., O'Connor, D.J., & Stocklmayer, S.M. (2003) 'Science communication: a contemporary definition', *Public Understanding of Science*, vol. 12, no. 2, pp. 183–202, doi: 10.1177/09636625030122004
Erren, T.C., & Bourne, P.E. (2007) 'Ten simple rules for a good poster presentation', *PLoS Computational Biology*, vol. 3, no. 5, pp. 0777–0778, doi: 10.1371/journal.pcbi.0030102
Greenfieldboyce, N. (2019) 'To save the science poster, researchers want to kill it and start over', *National Public Radio*, www.npr.org/sections/health-shots/2019/06/11/729314248/to-save-the-science-poster-researchers-want-to-kill-it-and-start-over
Hirst, R. (2003) 'Scientific jargon, good and bad', *Journal of Technical Writing and Communication*, vol. 33, no. 3, pp. 201–229, doi: 10.2190/J8JJ-4YD0-4R00-G5N0
Hofmann, A.H. (2016) *Writing in the biological sciences: A comprehensive resource for scientific communication*, New York: Oxford University Press, 2nd edition.
Hofmann, A.H. (2017) *Scientific writing and communication: Papers, proposals, and presentations*, New York: Oxford University Press, 3rd edition.
Markel, M., & Selber, S.A. (2018) *Technical communication*, Boston: Bedford, St. Martin's, 12th edition.
Matthews, D.L. (1990) 'The scientific poster: guidelines for effective visual communication', *Technical Communication*, vol. 37, no. 3, pp. 225–232, www.jstor.org/stable/43094875
Miller, C.R. (1979) 'A Humanistic Rationale for Technical Writing', *College English*, vol. 40, no. 6, pp. 610–617, www.jstor.org/stable/375964
Miller, J.E. (2007) 'Preparing and presenting effective research posters', *Health Services Research*, vol. 42, no. 1, pp. 311–328, doi: 10.1111/j.1475-6773.2006.00588.x
Penrose, A.M., & Katz, S.B. (2010) *Writing in the sciences: Exploring conventions of scientific discourse*, New York: Pearson Education, 3rd edition.
Rowe, N. (2017) *Academic & scientific poster presentation: A modern comprehensive guide*, New York: Springer International Publishing.
Speight, J.G. (2012) *Clear and concise communications for scientists and engineers*, Boca Raton: Taylor and Francis.
Strunk, W., & White, E.B. (2009) *The elements of style*, New York: Pearson Education, 50th anniversary edition.
Wolfe, J. (2009) 'How technical communication textbooks fail engineering students', *Technical Communication Quarterly*, vol. 18, no. 4, pp. 351–375, doi: 10.10.1080/10572250903149662

19

THE MEDICAL CASE REPORT AS GENRE AND SOCIAL PRACTICE

Chad Wickman

Introduction: the medical case report

The publication of medical case reports has increased in popularity over the past two decades, and while sometimes characterized as a relatively minor form of research writing, the genre nevertheless continues to draw attention from medical practitioners as well as scholars interested in the social and rhetorical dimensions of biomedical communication. Conventions associated with the genre tend to vary across disciplines and publication contexts; generally speaking, however, the case report exhibits some consistent features: it is typically written from the perspective of a single practitioner or group of practitioners involved in the case; it focuses on an individual patient or a small number of patients; it emphasizes unusual or rare phenomena or, alternatively, common phenomena that present in uncommon ways; and it employs narrative structure and style to render clinical encounters into meaningful textual accounts that can circulate within the medical community and to wider, generalist audiences.

Some scholars have associated narrative aspects of the case report with late twentieth-century developments in case-based "styles of reasoning" (Forrester, 1996) and "processes of ordering" in the human and social sciences (Morgan, 2017, p. 95; see Morgan, 2020). According to Hurwitz (2017), for example, "What John Forrester verbalized as 'thinking in cases', contemporary clinical case reports recount through controlled disclosures of observations and reasoning" (p. 65). Hurwitz specifically emphasizes literary and narrative aspects of the case report and ways in which it tracks, historically, with other genres, including the eighteenth-century novel and nineteenth-century detective fiction (see Montgomery Hunter, 1991; Hurwitz, 2011). Citing Sussex (2010), he suggests that:

> accounts of cases defer telling by withholding information and concealing elements as in a "game of suspense and mystery". The emergence of pattern is orchestrated through a process that stages access to findings by dripfeeding descriptive information in which reasoning and the formation of hypotheses are the intermediary links of narrative beginnings and endings.
>
> *(p. 67)*

Under this description, the case report can be considered performative, and temporally, it constitutes a retrospective and constructive rhetorical activity that functions epistemically by

DOI: 10.4324/9781003043782-22

setting up problems, reporting evidence, and leading readers to some form of resolution (e.g. diagnosis, therapeutic intervention). A growing body of interdisciplinary scholarship continues to highlight the complexity of the genre and its connection to historical and contemporary processes of knowledge-making, circulation, and patient advocacy (see Pomata, 2014; Ankeny, 2017; Böhmer, 2020).

Scholars have shown that the medical case report draws on elements of literary and scientific discourse to ascribe meaning to unusual or rare clinical phenomena. And while relatively stable as a textual form, the genre also continues to evolve due in part to emerging publication models and guidelines that seek to establish the value of case report writing in relation to other forms of biomedical communication. The present chapter engages with some of these developments and aims to complement existing scholarship in three ways: first, by theorizing the genre as a typified but evolving social and rhetorical practice; second, by examining case report writing in light of consensus-based guidelines and reporting standards; and third, by considering how the case report mediates the dynamic between clinical observation, textual documentation, and community expectation. Based on these foci, I will suggest that the medical case report and associated guidelines offer unique opportunities for practitioners to define and negotiate the boundaries of quality reporting, highlight patient experience and agency, and call attention to issues that shape health care in the United States and throughout different parts of the world.

The case report as genre

Researchers in the field of writing and rhetorical studies (WRS) have often examined how "minor" forms of writing contribute to the documentation and reporting of everyday scientific and biomedical activity. While diverse in theoretical and methodological orientation, this body of scholarship often approaches texts – including seemingly mundane ones, like notes and reports – as a form of social action and a means of shaping institutional practices. Drawing on the work of feminist sociologist Dorothy Smith (1974), for instance, Schryer (1993) suggests that:

> documents such as records are the very fabric of organizations, their fact-making mechanisms. They are "the forms that externalize social consciousness in social practices, objectify reasoning, knowledge, memory, decision-making, judgement, evaluation. . . ." Neither sociologists nor researchers into discursive practices can afford to ignore these deeply shared ways of rendering phenomena into language.
>
> *(p. 205)*

Schryer, in part by way of Smith, draws attention to the relationship between organizational texts and the "documentary reality" they both enable and maintain. One way scholars in WRS engage this dynamic is through the concept of *genre*: what Schryer and others have theorized as "stabilized-for-now . . . sites of social and ideological action" (Schryer, 1993, p. 204; see Schryer, 1994) that exhibit recognizable features but are pliable enough to change over time and in response to recurrent situations (Miller, 1984). Scholars in the tradition of rhetorical genre studies tend to define genres not as unvarying textual forms but rather as typified rhetorical actions that emerge and evolve in relation to social processes and user needs (see Bazerman, 1988; Berkenkotter and Huckin, 1995; Berkenkotter, 2008).[1]

Scholarship in WRS has also shown that genres are a common means whereby institutions establish and maintain power over biomedical subjects (see Emmons, 2009). Berkenkotter (2001) suggests, for instance, that practices of note making and report writing in the field of psychiatry "recontextualize" client narratives and draw "individual clients into . . . systems of reimbursement, health care, research, and medical reasoning" (p. 341; see Berkenkotter and Ravotas, 1997). These practices are connected to and shaped by reporting genres that serve a typifying function when it comes to documenting and assigning meaning to clinical encounters. With regard to the case report in particular, the encounter between doctor and patient constitutes a recurrent situation that can be interpreted and represented in different ways, and in the presence of unusual phenomena – what may come to be understood as a "case" – the genre provides "content- and meaning-determining categories" that "exert influence through . . . commonly held expectations about what is appropriate to include" (Winsor, 1999, p. 203). Conventional features associated with case report writing may not ultimately determine what gets included in any particular document; one key function of the genre, however, is to orient practitioners to situations – e.g. clinical encounters – in ways that invite a certain type of response – e.g. narrative accounts that frame patient illness in light of prevailing knowledge and community expectations.

Within the context of this chapter, I specifically discuss two ways the case report functions as a genre in our current sociocultural and technological milieu: (1) it orients practitioners to clinical encounters as a type of recurrent situation, and in doing so, it shapes what is considered document-able and thus know-able when it comes to reporting on the diagnosis and treatment of patients; (2) it mediates between individual action and community expectation with regard to quality reporting, and, in doing so, it shapes processes of disciplinary learning and knowledge-making.[2] The case report in these ways equips practitioners to capture the situational nature of clinical phenomena – e.g. through narrative exposition – while also adhering to community standards for research and publication – e.g. as represented in consensus-based reporting guidelines. I theorize these functions as an affordance of the genre but also as a potential way to "open up" existing rhetorical and epistemological structures in order to highlight diverse stakeholder perspectives.

The case report as social and rhetorical practice

The scholarly literature on case report writing suggests that the genre has evolved throughout different historical periods, sometimes functioning as a common textual practice while other times playing a more marginal role in medical research writing and publication (see Atkinson, 1992; Taavitsainen and Pahta, 2000; Hurwitz, 2006; Pomata, 2011; Taavitsainen, 2011). According to Nissen and Wynn (2012; see also 2014), for example, the case report experienced a downturn in the second half of the twentieth century due in part to the emergence of evidence-based medicine and its "emphasis on large-n quantitative studies" (p. 2). The authors also note, however, that the genre experienced a resurgence in the 1990s, which they associate with multiple factors, including narrative-based medicine, dedicated sections in general medical journals (Bignall and Horton, 1995), publication of evidence-based case reports (Godlee, 1998), and developments in online and open access publishing (pp. 3–4).

Practitioners have often wrestled with the affordances and constraints of the case report compared to other forms of biomedical research and reporting. In a widely referenced editorial, for instance, Vandenbroucke (2001) suggests:

Case reports and series have a high sensitivity for detecting novelty and therefore remain one of the cornerstones of medical progress; they provide many new ideas in medicine. At the same time, good case reporting demands a clear focus to make explicit to the audience why a particular observation is important in the context of existing knowledge.

(p. 330)

The case report from this perspective could be described as heuristic (e.g. orienting the practitioner to novelty and bringing "new ideas to medicine"), as well as sociocultural (e.g. holding practitioners and their writing accountable to the wider community and field-specific knowledge base). These aspects also speak to the epistemic function of the genre: as Rosselli (2002) suggests, "Medical knowledge has been traditionally built case by case. If we acknowledge the limitations of the case report we will not need to neglect its importance in our search for solid evidence" (p. 84).

The relative value of the case report compared to other genres of medical writing remains an ongoing point of discussion. Some practitioners have argued, for instance, that case reports reside at the "bottom of the 'hierarchy' of medical evidence" compared to randomized controlled trials and systematic reviews (Richason et al., 2009, p. 2). Others have described the genre in relational terms, suggesting that case report writing is useful for "generating hypotheses for verification in subsequent longitudinal observational studies and clinical trials," but, when taken in isolation, its "findings are not generalizable, do not address causal inference or explanatory mechanisms, and emphasize low-probability events" (Sun et al., 2013, p. 1065). These apparent constraints aside, practitioners also continue to highlight the value of the genre. As Riley et al. (2017) suggest, "[S]ystematic intervention from the point of care is now possible," and "case reports have the potential to offer evidence . . . that can be useful for clinical research, inform clinical practice guidelines, and improve medical education" (p. 233).

Case-based writing has historically appeared, with greater or lesser frequency, in general medical journals. Over the past two decades, however, an increasing number of venues have emerged that focus on the publication of case reports in particular. Practitioners have approached this development with optimism as well as due caution. For example, Akers (2016) suggests that "mainstream medical journals . . . publish only those case reports describing the most unique and striking clinical situations [while] new journals accept case reports highlighting a wide range of clinical issues" (p. 147). The author adds that while "reputable" journals maintain standards for peer review, "predatory" journals do exist, and they raise questions about the quality of the work being published. Such developments have a role to play in exploring the ongoing evolution of the genre and ways in which it continues to be shaped by new media and emerging publication models; it has also brought renewed attention and ongoing scrutiny to the case report as a valued form of research writing – so much so that practitioners have begun to formalize publication standards in order to mitigate biased and/or potentially misleading information.

One set of consensus-based guidelines, referred to as CARE (CAseREport), were developed circa 2013. Linked to the development of other medical reporting standards (see Moher et al., 2010), the CARE guidelines seek to address some of the exigencies mentioned above: e.g. case reports are becoming more widely published, and as practitioners seek to carve out an appropriate place for them in the context of contemporary medicine, they have articulated a concomitant need to reinforce quality reporting. The guidelines include a "checklist" with multiple descriptive categories (see Table 19.1), and participants in the CARE Group have published articles that continue to elaborate on aspects of guideline development and use (see Gagnier et al., 2013; Riley et al., 2017).

Little research to date has shown how CARE guidelines actually shape individual composing processes. Such guidelines do, however, provide a basis for exploring how practitioners characterize specific features of the case report and what those features may afford for research and writing. The role of narrative in particular is worth emphasizing in this context. Discussing the development of CARE guidelines, for example, Gagnier et al. (2013) suggest:

> A case report is a narrative that describes, for medical, scientific, or educational purposes, a medical problem experienced by one or more patients. Case reports written without guidance from reporting standards are insufficiently rigorous to guide clinical practice or to inform clinical study design.
>
> *(p. 1)*

The emphasis on narrative is of interest here in part because it positions the writer to "tell a story" (p. 3) in a way that allows authorial choice in "focusing the case" and appealing to audiences (Riley et al., 2017, p. 219). At the same time, the authors also suggest that published case reports tend to exhibit inconsistent quality. A tension thus emerges: i.e. narrative is a valuable means of reporting on clinical encounters and sharing patient experience, but because they sometimes lack consistency – e.g. with respect to precision, transparency, and completeness – the practice must be disciplined according to standards defined by practitioners in the field.

These practices extend to other settings as well. For example, some medical journals, like *The New England Journal of Medicine*, offer generalized guidance for authors but limited direction for the writing of case reports in particular.[3] More specialized journals, however, like *BMJ Case Reports (BMJ-CR)*, provide explicit direction in the form of booklets and template-based guidelines. Launched in 2008, the venue expresses a clear interest in the case report as a vehicle for pragmatic communication:

> We want to publish cases with valuable clinical lessons. Common cases that present a diagnostic, ethical or management challenge, or that highlight aspects of mechanisms of injury, pharmacology or histopathology are deemed of particular educational value. It is essential that the learning outcomes of the articles are important and novel.
>
> *(BMJ Case Reports, 2021)*

BMJ-CR has also generated documents that set specific expectations for authors. In 2018, for instance, they published a booklet that provides examples of model case reports, and they have developed related materials that both characterize the "useful and interesting case report" and provide authors with Word-based templates that, similar to the CARE Checklist, define what information should be included in a particular submission and how that information should be composed (see Table 19.1).

Table 19.1 identifies categorical features represented in the CARE Checklist and *BMJ-CR* templates. Some similarities are readily apparent: for instance, both invite information related to context, patient information/case presentation, diagnosis, and intervention/treatment. It is worth noting, too, that they each provide a category related to "patient perspective." The latter is not an insignificant feature. Indeed, while authors are not necessarily required to include patient narratives in their reports, the inclusion of such a category creates space for that possibility, and it reinforces the view that patient perspectives are a welcome and valued part of the research and reporting process.

A recent case report from the *BMJ-CR*, titled "Brief Psychotic Disorder Associated with Quarantine and Mild COVID-19" (Haddad et al., 2020), helps to illustrate how authors respond to these conventions. The piece includes distinctive subject headings that correspond to the

Table 19.1 Case report checklists from CARE Guidelines and *BMJ Case Reports*[4]

CARE Checklist guidelines	BMJ Case Reports guidelines
Title	Title of case
Keywords/abstract	Summary
Introduction	Background
Patient information	Case presentation
Clinical findings, timeline	Investigation (if relevant)
Diagnostic assessment	Differential diagnosis (if relevant)
Therapeutic interventions	Treatment (if relevant)
Follow-up and Outcomes	Outcome and follow-up
Discussion	Discussion, learning points/take home messages, references, figure/video captions
Patient perspective, informed consent	Patient's perspective

BMJ-CR template; it also develops a narrative across different sections. Under "Case Presentation," for example, the authors describe the following scene:

> In May 2020, near the height of the COVID-19 pandemic in Qatar, a 30-year-old man called an ambulance reporting generalised aches and being unable to sleep due to anxiety about his health. He was taken to a major hospital. His symptoms started 4 days earlier after he received a positive COVID-19 test (RT-PCR) that was arranged after a friend had tested positive for COVID-19. Since then he had been searching the internet and social media for COVID-19 information and repeatedly telephoning friends and relatives to seek reassurance about his health. He had no cough or other physical symptoms.
>
> *(p. 1)*

We see in this brief excerpt the foundations of a patient-centered story. The authors provide readers with context; they introduce a central character; they describe action over time and locale; and they present a puzzle or "mystery" to be solved. The remainder of the report is structured according to specific headings – commensurate with the template – that lead readers to diagnosis, treatment, lessons learned, and patient perspective.

The narrative extends across each section, and while it focuses largely on the authors' point of view, it culminates in the patient's own voice. In his words:

> I was seeing excessive Facebook posts where people were saying if you get Corona you will die. Other people around me started to get Corona but when I got it and was sent to quarantine, I took it to heart, could not cope, and felt depressed. I was frightened that this is it and I am going to die. Being away from family was a big factor and I thought I am never going to be able to see them again.
>
> *(p. 4)*

The patient perspective is a vital feature of the report for many reasons. While a complete analysis is beyond the scope of this chapter, I would emphasize two points of interest in the document just cited. First, the "Patient's Perspective" (just under 200 words out of approximately 4,300 in the body of the text) is included at the end of the report rather than used to set up the case; it thus gets folded into an overarching narrative as developed from the authors' point of view. Second, the patient perspective offers a reflection on the patient's experience and provides additional resolution to the case and the care provided (e.g. "[T]hank you so much to the team who looked after me. I am doing well now" [p. 4]). Practitioner narratives and patient perspectives may ultimately bear different epistemological weight in the context of reporting and knowledge-making; yet as Ankeny (2017) suggests, the "inclusion of patient perspectives [does have] the potential to shift understandings of disease away from narrow biological processes to more holistic, socioculturally grounded understandings of wellness and deviations from it" (p. 106).

So how might one rhetorically characterize the relationship between the case report and the type of guidelines that currently inform medical research and publication? Returning to insights from rhetorical genre studies, I would suggest that the CARE checklist and the *BMJ-CR* template act as "metagenres" (Giltrow, 2002) insofar as they offer "pre-emptive feedback" that regulates the "production of a genre, ruling out some kinds of expression, endorsing others" (p. 190). Further study would be needed to determine the precise impact that guidelines have on individual composing and broader publication processes; even so, I believe they offer an important basis for examining how individuals and communities conceptualize, enact, and attempt to reinforce quality reporting – and thus how they view the role of the case report within the broader context of evidence-based research and reporting. In a related sense, these types of metagenres also offer potential sites of intervention both for including multiple voices in case reports and for deliberating about what should be valued in the authoring of medical documents more generally.

Discussion and conclusions

One limitation of the present chapter is that I have been necessarily selective in the scholarship I have reviewed, both from the perspective of breadth – what I have been able to include – and depth – how fully I have been able to contextualize and examine individual studies and commentaries. Indeed, a more systematic review may offer different insights into the case report and its significance for research and clinical practice; a more empirically based study of the discourse would likewise shed light on the variation one may find in individual reports. Yet one thing does seem clear: the case report is a relatively stable but still evolving genre that warrants attention both as a form of medical research writing and as an object of rhetorical study. With that in mind, the remainder of this chapter draws attention to some pathways for further inquiry and possible application.

I would begin by discussing some of the assumptions that appear to inform discourse related to the case report. For one, the development and use of guidelines indicates a general interest in reinforcing the best possible research and best possible good for individual patients and communities. And yet, from a rhetorical perspective, the principles that guide this work – e.g. transparency, completeness – suggest an ontological gap between the clinical encounter and the written report. More specifically: writing is often characterized in these documents as a means of representing cases in more or less faithful ways rather than contributing to a "documentary reality" (Smith, 1974) shaped by material circumstances and community expectations. In saying this, I am not advocating for a relativistic view of reality or report writing (in medicine or elsewhere); I would argue, however, that an alternative model of text production – e.g. where relational and emergent phenomena come to be understood, at least in part, through the act of writing – could provide practitioners with a framework for understanding the social action of the genre and ways in which it gets taken up and used by audiences across time and space.

The role of narrative in case report writing is of relevance in this context as well. Scholars have shown that narrative is an essential means whereby practitioners describe clinical phenomena and establish pattern and consistency in the stories they tell, and elaborations of the CARE guidelines incorporate it as a frame for the checklist (see Gagnier et al., 2013; Riley et al., 2017). In these ways, narrative serves at least two distinctive but related functions: it is a mode of exposition that can be used to detail and share patients' lived experiences, and it is a commonplace from which practitioners can deliberate about, regulate, and assign value to the case report as a genre. The role of narrative thus extends beyond the practice of storytelling; one might say, rather, that it constitutes a "conventionalized way of representing disciplinary knowledge" that demonstrates "tacit and explicit agreements – built communally and negotiated over time – about what constitutes a persuasive story" (Journet, 2012, p. 13). These considerations provide insights into the textual features of case report writing but also the social and rhetorical actions those features have evolved to enact and reinforce.

I believe it would be remiss to discuss the medical case report without mentioning possible applications in our present context. On January 30, 2020, the World Health Organization officially designated the novel coronavirus a Public Health Emergency of International Concern. At the time, the virus was relatively localized, with 82 confirmed cases and no deaths in 18 countries outside China (WHO Situation Report-10). It would not take long, however, for COVID-19 to find its way across the globe and affect nearly all aspects of life as we know it.[5] Research on related infectious diseases has offered some basis for understanding the nature of COVID-19 as an epidemiological phenomenon; yet the social dimensions of the virus – from the way it spreads to the health care systems through which individuals and communities are treated – make it difficult to research and respond to in a systematic way. Indeed, the situation is volatile, and as such, it must be understood as relational and emergent: the novel coronavirus exhibits semipredictable characteristics, but insofar as it is entangled with human action and decision making, the individual case – focused on specific patients situated in specific milieus – becomes a critical locus for engagement and intervention.

I point to the COVID-19 pandemic as a site of investigation, application, and rhetorical possibility because it is one of the most pressing and ongoing matters of concern in the world today. I also do so to emphasize an affordance of the case report as a genre. More precisely: at the time of this writing, long-term studies of the virus and its effects on individuals and populations are in the early stages. We do, however, have case reports that tell us about the people who are currently being affected, and over time, researchers and clinicians will potentially be able to use these documents to develop more broadly generalizable insights in coordination with other types of systematic study (see Ankeny, 2020). It is useful from this perspective to conceptualize the case report not as a marginal genre – e.g. compared to other forms of evidence-based research and reporting – but rather as a social and rhetorical practice that, perhaps more than other genres, may help to foreground patient perspectives and their unique experiences with illness.

We have learned, too, that some individuals and groups are more vulnerable than others and that the pandemic has created conditions which potentially lead to "historically marginalized groups shouldering the greatest burden of disease and disproportionately bearing the social impact" (Laurencin and McClinton, 2020, p. 399).[6] While such findings can and need to be reinforced by long-term, systematic studies, the case report provides one way to highlight the experience of individuals who are suffering in the present. Their voices are critically important not only for what they can tell us about the nature of epidemiological phenomena – but also for what they can tell us about the sociocultural, economic, and racial disparities that shape health care and access to health care in the United States and throughout the world. The case report is not a fix to systemic problems, but through resources like narrative, and through the ongoing development and elaboration of community-based guidelines, it does offer a promising site within which to learn about and engage with them through conscientious rhetorical study.

Notes

1 The scholarship in genre studies is too expansive to cite adequately in the space of this chapter. For a general overview, see Bawarshi and Reiff (2010); for a recent example of rhetorical genre theory developed and applied in the context of scientific communication, see Mehlenbacher (2020).
2 For related work in WRS, see Freedman et al. (1994), Schryer et al. (2003), Schryer and Spoel (2005), and Campbell (2017).
3 The *NEJM* submission portal includes a category for "Clinical Cases" and a subcategory titled "Clinical Problem Solving" rather than "Case Reports" per se. It is also worth noting that the *Lancet* has historically published case reports (see Berman and Horton, 2015), but at the time of this writing (January 2021), the journal is not accepting case report submissions. I mention these examples in part because they reflect the nuanced and evolving nature of case-based writing and publishing.
4 Adapted from Riley et al. (2017) and *BMJ Case Reports* "Standard Case Report Checklist and Template for Authors." I have grouped some categories together for the purpose of comparison.
5 It is important to note that the virus and pandemic have affected and continue to affect different individuals and communities in different ways, especially those in vulnerable and precarious situations (e.g. based on intersections of age, race, ethnicity, income, and sociogeographic location).
6 A deeper discussion of racial and ethnic disparities in health and medicine, in the context of COVID-19 or otherwise, is beyond the scope of this chapter, but I acknowledge it here as a critical and growing area of research and application.

References

Akers, K.G. (2016) 'New journals for publishing medical case reports', *Journal of the Medical Library Association*, vol. 104, no. 2, pp. 146–149, DOI: https://doi.org/10.5195/jmla.2016.62

Ankeny, R.A. (2017) 'The role of patient perspective in clinical case reporting', in R. Bluhm (ed.), *Knowing and acting in medicine*, London: Rowman and Littlefield, pp. 97–112.

Ankeny, R.A. (2020) 'Tracing data journeys through medical case reports: Conceptualizing case reports not as "anecdotes" but productive epistemic constructs, or why zebras can be useful', in S. Leonelli & N. Tempini (eds.), *Data journeys in the sciences*, Berlin: Springer, pp. 59–76.

Atkinson, D. (1992) 'The evolution of medical research writing from 1735–1985: The case of the Edinburgh Medical Journal', *Applied Linguistics*, vol. 13, no. 4, pp. 337–374, https://doi.org/10.1093/applin/13.4.337

Bawarshi, A.S., & Reiff, M.J. (2010) *Genre: An introduction to history, theory, research, and pedagogy*, West Lafayette: Parlor Press.

Bazerman, C. (1988) *Shaping written knowledge: The genre and activity of the experimental article in science*, Madison, WI: University of Wisconsin Press.

Berkenkotter, C. (2001) 'Genre systems at work: DSM-IV and rhetorical recontextualization in psychotherapy paperwork', *Written Communication*, vol. 18, no. 3, pp. 326–349, https://doi.org/10.1177/0741088301018003004

Berkenkotter, C. (2008) 'Genre evolution? The case for a diachronic perspective', in V.K. Bhatia, J. Flowerdew, & R.H. Jones (eds.), *Advances in discourse studies*, New York: Routledge, pp. 178–194.

Berkenkotter, C., & Huckin, T. (1995) *Genre knowledge in disciplinary communication: Cognition/culture/power*, New York: Routledge.

Berkenkotter, C., & Ravotas, D. (1997) 'Genre as a tool in the transmission of practice over time and across professional boundaries', *Mind, Culture, and Activity*, vol. 4, no. 4, pp. 256–274, doi:10.1207/s15327884mca0404_4

Berman, P., & Horton, R. (2015) 'Case Reports', *The Lancet*, vol. 385, p. 1277, DOI:https://doi.org/10.1016/S0140-6736(15)60642-0

Bignall, J., & Horton, R. (1995) 'Learning from stories: The Lancet's case reports', *The Lancet*, vol. 346, p. 1246, doi: 10.1016/s0140-6736(95)91859-0

BMJ Case Reports. (2021) 'Instructions for authors', https://casereports.bmj.com/pages/authors/ (Accessed 15 January 2021)

Böhmer, M. (2020) 'The case as a travelling genre', *History of the Human Sciences*, vol. 33, no. 3–4, pp. 111–128, https://doi.org/10.1177/0952695119897867

Campbell, L. (2017) 'Simulation genres and student uptake: The patient health record in clinical nursing simulations', *Written Communication*, vol. 34, no. 3, pp. 255–279

Emmons, K. (2009) 'Uptake and the biomedical subject', in C. Bazerman, A. Bonini, & D. Figueiredo (eds.), *Genre in a changing world*, Lafayette: Parlor Press, pp. 134–157.

Forrester, J. (1996) 'If p, then what'? Thinking in cases', *History of the Human Sciences*, vol. 9, no. 3, pp. 1–25, https://doi.org/10.1177/095269519600900301

Freedman, A., Adam, C., & Smart, G. (1994) 'Wearing suits to class: Simulating genres and simulations as genre', *Written Communication*, vol. 11, no. 2, pp. 193–226, https://doi.org/10.1177/0741088394011002002

Gagnier, J.J., et al. (2013) 'The CARE guidelines: Consensus-based clinical case reporting guideline development', *Global Advances in Health and Medicine*, vol. 2, no. 5, pp. 38–43, doi: 10.7453/gahmj.2013.008

Giltrow, J. (2002) 'Meta-genre', in R.M. Coe, L. Lingard, & T. Teslenko (eds.), *The rhetoric and ideology of genre: Strategies for stability and change*, Kresskill, NJ: Hampton Press, pp. 187–205.

Godlee, F. (1998) 'Applying research evidence to individual patients: Evidence based case reports will help', *British Medical Journal*, vol. 316, pp. 1621–1622, doi: 10.1136/bmj.316.7145.1621

Haddad, P., et al. (2020) 'A brief psychotic disorder associated with quarantine and mild COVID-19', *BMJ Case Reports*, vol. 13, no. 12, pp. 1–5, http://dx.doi.org/10.1136/bcr-2020-240088

Hurwitz, B. (2006) 'Form and representation in clinical case reports', *Literature and Medicine*, vol. 25, no. 2, pp. 216–240, DOI: 10.1353/lm.2007.0006

Hurwitz, B. (2011) 'Clinical cases and case reports: Boundaries and porosities', in A. Calanchi, G. Catellani, G. Morisco, & G. Turchetti (eds.), *The case and the canon: Anomalies, discontinuities, metaphors between science and literature*, Goettingen: V&R Unipress, pp. 45–58.

Hurwitz, B. (2017) 'Narrative constructs in modern clinical case reporting', *Studies in History and Philosophy of Science Part A*, vol. 62, pp. 65–73, doi: 10.1016/j.shpsa.2017.03.004

Journet, D. (2012), 'Narrative turns in writing studies research', in L. Nickoson & M. Sheridan (eds.), *Writing studies research in practice: Methods and methodologies*, Carbondale, IL: SIU Press, pp. 13–24.

Laurencin, C.T., & McClinton, A. (2020) 'The COVID-19 pandemic: A call to action to identify and address racial and ethnic disparities', *Journal of Racial and Ethnic Health Disparities*, vol. 7, no. 3, pp. 398–402, https://doi.org/10.1007/s40615-020-00756-0

Mehlenbacher, A. (2020) *Science communication online: Engaging experts and publics on the internet*, Columbus, OH: OSU Press.

Miller, C. (1984) 'Genre as social action', *Quarterly Journal of Speech*, vol. 70, no. 2, pp. 151–167, https://doi.org/10.1080/00335638409383686

Moher, D., et al. (2010) 'Guidance for developers of health research reporting guidelines', *PLoS Medicine*, vol. 7, no. 2, pp. 1–9, https://doi.org/10.1371/journal.pmed.1000217

Montgomery Hunter, K. (1991) *Doctor's stories: The narrative structure of medical knowledge*, Princeton, NJ: Princeton University Press.

Morgan, M. (2017) 'Narrative ordering and explanation', *Studies in History and Philosophy of Science Part A*, vol. 62, pp. 86–97, https://doi.org/10.1016/j.shpsa.2017.03.006

Morgan, M. (2020) '"If p? Then what?" Thinking within, with, and from cases', *History of the Human Sciences*, vol. 33, no. 3–4, pp. 198–217, https://doi.org/10.1177/0952695119899349

Nissen, T., & Wynn, R. (2012) 'The recent history of the clinical case report: A narrative review', *Journal of the Royal Society of Medicine Short Reports*, vol. 3, no. 12, p. 2.

Nissen, T., & Wynn, R. (2014) 'The clinical case report: A review of its merits and limitations', *BMC Research Notes*, vol. 7, no. 264, pp. 1–5, doi: 10.1186/1756-0500-7-264

Pomata, G. (2011) 'Observation rising: Birth of an epistemic genre, 1500–1650', in L. Daston & E. Lunbeck (eds.), *Histories of scientific observation*, Chicago: University of Chicago Press, pp. 45–80.

Pomata, G. (2014) 'The medical case narrative: Distant reading of an epistemic genre', *Literature and Medicine*, vol. 32, no. 1, pp. 1–23, doi: 10.1353/lm.2014.0010

Richason, T., et al. (2009) 'Case reports describing treatments in the emergency medicine literature: Missing and misleading information', *BMC Emergency Medicine*, vol. 9, no. 10, pp. 1–7, doi: https://doi.org/10.1186/1471-227X-9-10

Riley, D.S., et al. (2017) 'CARE guidelines for case reports: Explanation and elaboration document', *Journal of Clinical Epidemiology*, vol. 89, pp. 213–235, doi: 10.1016/j.jclinepi.2017.04.026

Rosselli (2002) 'The case report is far from dead', *The Lancet*, vol. 359, no. 9300, p. 84, DOI:https://doi.org/10.1016/S0140-6736(02)07311-7

Schryer, C. (1993) 'Records as genre', *Written Communication*, vol. 10, no. 2, pp. 200–234

Schryer, C. (1994) 'The lab vs. the clinic: Sites of competing genres', in A. Freedman & P. Medway (eds.), *Genre and the new rhetoric*, London: Taylor & Francis, pp. 87–103.

Schryer, C., & Spoel, P. (2005) 'Genre theory, health-care discourse, and professional identity formation, *Journal of Business and Technical Communication*, vol. 19, no. 3, pp. 249–278, ttps://doi.org/10.1177/1050651905275625

Schryer, C. et al. (2003) 'Structure and agency in medical case presentations', in C. Bazerman & D. Russell (eds.), *Writing selves/writing societies: Research from activity perspectives*, Fort Collins, CO: WAC Clearinghouse, pp. 62–96.

Smith, D. (1974) 'The social construction of documentary reality', *Sociological Inquiry*, vol. 44, no. 4, pp. 257–268, https://doi.org/10.1111/j.1475-682X.1974.tb01159.x

Sun, G.H., Oluseyi, A., & Hayward, R.A. (2013) 'Open-access case report journals: The rationale for case report guidelines, *Journal of Clinical Epidemiology*, vol. 66, no. 10, pp. 1065–1070, doi: 10.1016/j.jclinepi.2013.04.001

Sussex, L. (2010) *Women writers and detectives in nineteenth-century crime fiction: The mothers of the mystery genre*, Basingstoke: Palgrave Macmillan.

Taavitsainen, I. (2011) 'Medical case reports and scientific thought-styles', *Revista de Lenguas para Fines Específicos*, vol. 17, pp. 75–98.

Taavitsainen, I, & Pahta, P. (2000) 'Conventions of professional writing: The medical case report in a historical perspective', *Journal of English Linguistics*, vol. 28, no. 1, pp. 60–76

Vandenbroucke, J.P. (2001) 'In defense of case reports and case series', *Annals of Internal Medicine*, vol. 134, pp. 330–334, doi: 10.7326/0003-4819-134-4-200102200-00017

Winsor, D. (1999) 'Genre and activity systems: The role of documentation in maintaining and changing engineering activity systems', *Written Communication*, vol. 16, no. 2, pp. 200–224, https://doi.org/10.1177/0741088399016002003

20

SCIENTIFIC LETTERS AND COMMENTARIES IN THEIR HISTORICAL AND SOCIAL CONTEXTS

Maureen A. Mathison

Introduction

Of the many genres relevant to scientific journals, the subject of this chapter – letters and commentaries – may be one of the least studied in science communication. While practices related to publication have been examined, such as peer review and citations (Berkenkotter and Huckin, 1995), it is the ubiquitous research article that has garnered the most attention (Atkinson, 1998; Bazerman, 1988; Gross et al., 2002). And yet letters and commentaries have played pivotal roles in the practice of science. As opposed to the research article, they are known to be explicitly subjective, with information that is intended for specific audiences, though larger, more public audiences have been taken into consideration when written. They are of particular interest in the practice of science because of their impact, most notably, on the dissemination of novel information and its critique but also in more subtle ways that point toward the social life of science in historical context.

The history of letters in science date back millennia, but their most widely known use in science can be traced back to the seventeenth century when empirical science was emerging as a force in society across various locales and continents. Henry Oldenburg, the secretary of the Royal Society of London established the journal *Philosophical Transactions of the Royal Society* in 1665, creating a central repository for novel information in natural philosophy through the medium of letters. Commentaries have been studied less than letters. What is known has been approached from the perspective of rhetorical analysis of the reception of scientific novelty (Ceccarelli, 2001) or through linguistic analyses of documents (Hyland, 2004; Swales, 1990), but these have tended to focus on critical reviews (see Noguchi, Chapter 15 of this volume). Even less studied are the typified social actions (Miller, 1984) of letters and commentaries and, more importantly, across time as they change to meet the social standards of the periods for which they have been relevant.

In this chapter, I provide a brief account of letters and commentary throughout various critical historical moments and then offer a lengthier account of contemporary uses of letters and commentary. A diachronic account demonstrates that within the social action of a typified genre reside various typified actions; that is, the attributes of the letter and commentary are motivated by the multiple exigencies of place and time and thus are expressed uniquely. On

DOI: 10.4324/9781003043782-23

one level they are generally identifiable as letters and commentary, but on another, their accomplishments are more complex because the actions they respond to are less uniform than previously held, even within shared historical moments. It should be noted that the account here is Western based and that future accounts need to be more inclusive of science in other cultures and periods to provide a fuller account of letters and commentary.

Letters in science

Letters in antiquity and the Middle Ages

Letters as part of the ancient system of scientific communication were written to exchange and share information about the natural world as well as to apprise patrons of novel discoveries and give praise for their support (Taub, 2017). Letters written by natural philosophers were more unidirectional with an instructional purpose. They tended to be more didactic. Rather than the characteristic feel of dialogue, such letters "flowed from the writer to the addressees in the form of teaching or instruction" and, even when addressed to an individual, were written with the intention of being shared with a larger audience (Muir, 2009, p. 9). Senders and receivers of scientific and technical letters would have been keenly aware of status, with the sender being more knowledgeable or, at the very least, an intellectual equal (Langslow, 2007).

Letters to patrons who supported the work of natural philosophers were distinct. They often started with conventional salutations – "To King Ptolemy Eratosthenes sends greetings" (Taub, 2017, p. 57) – but then contextualized the problem and previous attempts to solve it by others before describing a solution. Some even provided a description of its application to other places of merit. Jargon, proofs, and diagrams could be included, and in a rare instance, Taub (2017) found a letter that also incorporated poetry in an epigram – the author's own poetry and the quotation of a known poet. The various "genres within genres" is unusual but demonstrates the uses of rhetoric in not only appealing to a patron ruler acknowledging proper use has been made of the patronage but, again, to a broader audience with whom the letter might be shared.

In 380 CE, when Christianity was established as the official religion of the Roman Empire, the work of natural philosophers was considered pagan; to imagine truth outside of consideration of God was inappropriate. Augustine (354–430 CE), a teacher of rhetoric prior to his conversion to Christianity, reconciled faith and science, expounding the value of classical writers to Christianity. Natural philosophy became the "handmaiden" of Christianity in its service to help explicate but not to overpower the word of God. Religion was paramount to the practice of science in the early Middle Ages, with letters shared among those in positions to utilize the ancients in the service of theology. As science opened up to secular practices, letters were transformed into instruments of dissemination.

Letters in early science

The seventeenth century gave rise to what has been metaphorically termed the "republic of letters," "network," and "invisible college," systems of the learned sharing information and investigations through correspondence (Kronick, 2001, p. 29). Letter-writing across distances brought those with an intellectual curiosity about the natural world together, sometimes loosely, but increasingly in formalized communities, where letters became a mainstay of scientific practice. One of the first scientific societies formed in the early seventeenth century was the Academy of Lincei. Its founder, Federico Ceci, idealized a collection of natural philosophers collaborating and sharing knowledge, locally and at a distance. To facilitate his ideal of scientific practice he encouraged a correspondence network that would transcend national boundaries. The academy was formed for members to gather and converse about natural science, and when they could not be present because of great distance, conversations took place through correspondence (Van Miert, 2009). Importantly, Ceci considered the letters as accounts of science; it was the Lincei that published the letters on sunspots of Galileo, one of their most widely known members (Savoia, 2011).

One of the most researched societies in terms of letters is the Royal Society of London, founded in 1660 under a charter by Charles II. Members were upper-class, representing aristocrats, lawyers, surgeons, scholars, and merchants (Hunter, 1994). Participation in the Society occurred, like the Lincei, in person and also through correspondence. At the center of the communication exchanges was Henry Oldenburg, its first secretary, who established its journal *Philosophical Transactions of the Royal Society* in 1665. Oldenburg was a master of advocating for the society, having created a collection of correspondents prior to his position. In his travels throughout Europe, Oldenburg had made it a point to remain in contact with those he met and maintained a system of correspondents, adding new ones as the community of natural philosophers grew (Boas Hall, 2002). His reach extended from England, to France, to Germany, and beyond as he invited natural philosophers to keep him abreast of their work. Oldenburg personally received and edited letters to be read aloud and discussed at society meetings. Many of the initial reports included in the journal were letters to Oldenburg recounting a firsthand witnessing of nature. Early letters in the journal provide unmediated accounts of wondrous scenes and events, such as birthed monstrous two-headed calves and fireballs streaming across the sky, alongside accounts of an experimental nature.

Epistolary reports were written with a "strong authorial persona:" authors referred to themselves in first person and included their thought processes and feelings (Atkinson, 1998, p. xxiii). Early letters such as that of Newton in 1672 describing light through a prism, employ the phrases "celebrated phaenomena" and "pleasing divertissement" (Montgomery, 2003, p. 11), language that became superfluous as science became a more formalized practice with the object at the center. Gradually, letters written with the "researcher characterized as a full participant in the events" (Atkinson, 1998, p. 142) became less common. In the latter half of the seventeenth century, England became the center for experimentalism (Henry, 1992); increasingly human observation made way for human intervention in the pursuit of the natural world through the manipulation of phenomena and instrumentation.

Many consider the Royal Society to have lost some of its dominance with the 1677 death of Oldenburg, whose mastery of multiple languages and interpersonal skills supported a robust correspondence network of natural philosophers. The receipt of letters, however, continued well throughout the eighteenth century, increasing to the point of needing an institutionalized body, the Committee of Papers, to oversee them in 1752 (Rusnick, 1999, p. 157). Letters continued to be read at meetings and selected for publication in the society's journal, *Philosophical Transactions*, though evaluative judgments about reports were not perceived to be the purview of the society as correspondents were anticipated to be gentlemen and the accounts, therefore, were taken to be honest and true. As in the past, social status and trustworthiness remained a requirement for publication in the journal. Correspondents were men of stature who had affiliations within the network who could vouch for their credibility or act on their behalf, as in the case of Antonie van Leeuwenhoek whose initial letter was sent through a gentleman mediary (Davis, 2020). Letters were written with appropriate etiquette, with knowledge of standing taken into consideration.

The epistolary report persisted in *Philosophical Transactions* as did emotional responses to observations in correspondents' writing. In their analysis of eighteenth-century journals, Gross et al. (2002) found that many of the features of letters in the previous century were maintained. Accounts tended to be personalized, though, as stated earlier, this decreased toward the end of the century. Not until the early nineteenth century did letters become more like the research report today with scientific rhetoric having moved from "cookbook novelties" to more "experimental claims and proofs" (Bazerman, 1988).

Natural philosophers responded to the context of the scientific community as it was becoming more professionalized. Whereas before natural philosophers comprised a group of learned men (almost exclusively) whose observations and accounts contributed to a growing body of information, the nineteenth century heralded a new type of natural philosopher who increasingly was educated and whose expertise came to be seen as a particularized field within science, such as chemistry or geology. They were paid academics whose research targeted others in their same disciplinary community (Fyfe and Geilas, 2020). Rather than merely describing the process of experimentation as they had in letters, natural philosophers became more adept at creating arguments about the claims that could be gleaned from them. Bazerman's analysis of *Philosophical Transactions* tracks the transformation of the letter into the research report as the genre responded to the exigencies of the new, specialized scientific community. The report overtook the letter, which had been a more personalized account of uncovering nature rather than a collection of natural philosophers working with hypotheses, creating detailed accounts of methodology, and arguing for particular claims:

> The emerging form of the experimental report offered a way to harness stories of the smaller world of the laboratory to general claims about the regularities of the larger world of nature.
>
> *(Bazerman, 1988, p. 79)*

Several genre features of the scientific article today have their origins in the nineteenth century. Gross et al. (2002) point out that headings became more common to differentiate the unique sections of the article, and reference to other natural philosophers became a staple, with half of the reports including them, but they did not resemble contemporary citations. In fact, there was no codified use of them; they could be found within a text, at the bottom of the page, with or without page numbers (Gross et al., 2002, p. 85). Generic features of the research report were starting to take shape without consistency. The letter, having been a primary means for disseminating and circulating information, no longer held the same currency for scientific reporting.

Contemporary letters

One of the most common uses of the letter today is not only to disseminate information but, more importantly, to be acknowledged for being the first to discover or create something. The ability to rapidly publish enables a researcher to claim priority when in competition to be recognized, that is, to be credited with the discovery or innovation (Merton, 1957). Going back at least as far as the Royal Society, priority was an issue for natural philosophers. Having a letter read before members of the Society or having a letter logged in the record book afforded someone the right to priority. If a dispute occurred, the decisive proof of priority could be shown through the society's record. With the proliferation of scientists – and many studying the same phenomenon – the need to publish quickly accelerated.

While publishing a report provided a means to assert priority, the lag time in publishing could jeopardize one's novel claim. Baldwin (2014) recounts how at the turn of the nineteenth century, Ernest Rutherford, the Nobel Prize winner known for his discoveries in nuclear physics, initiated using *Nature*'s "Letters to the Editor" column regularly to publish his results after the Curies published findings on radioactivity first, weakening the impact of originality in his work. Thereafter, he sent letters to the editor to announce his innovations. Others followed suit. Letters to the editor were more expedient if one wanted to disseminate information quickly and to also create a name for oneself. Previously, the column's primary purpose had been for scientists to comment on one another's research as well as issues in the field.

Right to priority remains a critical issue in scientific practice. The accolades and perks of someone being "the first" enhances a career. Warren and Marshall, winners of the Nobel Prize in 2005 for their findings on the bacterial cause of ulcers (*H. pylori*), published a letter in the journal *Lancet* in 1983 to ensure that they would receive credit for their discovery. Marshall was sure that others would soon be publishing results similar to theirs and persuaded Warren to first send a letter to a prestigious journal while they worked on the research report, which came out one year later in the same journal (Marshall, 2002).

To counter delays incurred through the peer review and publishing processes, entire journals have been created that utilize letters to disseminate information faster, among them *Biology Letters* and *Physical Review Letters*. The history of *Physical Review Letters* demonstrates the logic behind the establishment of such journals. Blakeslee (1994) explains how the journal evolved out of the "Letters to the Editor" column of the journal *Physical Review*, its parent journal, which reported on research in progress in a timely manner. *Physical Review Letters* was established as an independent journal in 1958 to account for the overabundance of letters due to the rapidity at which the field was advancing. Letters were short, did not need extended peer review, and were published more quickly than research reports.

The journal has served as a main forum for physicists to disseminate their preliminary research quickly. Over time the genre of the letter in *Physical Review* has transformed so that early letters are quite distinguishable from later letters, which look more like research reports. As Blakeslee notes, the genre became blurred as it took on the appearance of a research report. Within decades, the letter no longer followed the genre, though authors recognize that the journal refers to them as such. Today, scientists' letters resemble a truncated research report, with introductions and method sections, but they insist on its belonging to the letter genre. Though earlier and later submissions explicitly refer to themselves as letters with the purpose of presenting novel contributions, earlier letters are shorter with less detail, and it is easier to differentiate them as letters rather than reports.

Physical Review Letters filled a gap in physics journals to disseminate information more quickly than a journal requiring a lengthy review process for research reports. In one sense, the journal brought information to the community of physicists faster and at the same time laid the groundwork for scientists wishing to establish a right to priority.

The social action of letters, always multipurposed (Bazerman, 2000), continues to blur boundaries. Letters do not limit themselves to establishing novelty and priority but extend to respond to published innovation with the goal of improving it and/or elevating one's own related research (Magnet and Carnet, 2006). In an editorial in *JAMA*, editors Winker and Fontanarosa (1999) emphasize the importance of letters, first as "Research Letters" that include short research reports and second as "Letters to the Editors" wherein critiques of original research act as a post–peer review process, discussed in "Commentary in Science."

Summary

Studies of the available letters that have survived through the centuries show that they have fulfilled many purposes for scientists as they have lived in unique contexts with unique exigencies impacting the genre. They have been an instrument addressed for public and private reading and for intellectual exchange among peers and a medium central to the creation and maintenance of networks among natural philosophers and scientific societies. Letters have been critical for the dissemination of knowledge and, lastly, utilized to establish a right to priority.

Commentary in science

The commentary, like the letter, is an ancient genre that has performed manifold actions associated with the historical and social contexts in which they have been situated. But unlike the letter, the commentary is a "text about a text" (Mathison, 1996): whereas letters originate in the writing of a text, commentary originates in the reading of a text in order to comment on it. "Commentaries," says Shuttleworth-Kraus, "are readings" (2002, p. 4). Letters traditionally addressed people, but commentary directly addresses ideas. The exigency for commentary has shifted throughout the ages, with some of its purposes being summary, interpretation, critique, correction, and instruction. In all cases, commentary wields influence over the text and its potential reception.

Commentary in antiquity and the Middle Ages

Commentary played an important role beginning in the third century BCE in the elucidation and understanding of texts. Baltussen (2015) provides an account of early commentary, prior to its establishment as a genre, explaining that initially it was a normal activity of responding to and critiquing the work of other philosophers, "an exegetical activity at the level of normal discourse," but it was not until the end of the seventh century that commentary become more scholarly and formalized (pp. 173–175). Though originally oral, he explains, commentary eventually became a more consistent written genre when philosopher schools were mandated to close, necessitating a means to elucidate ancient texts to their students. Baltussen posits that the closer the teacher was to the school founder and the more knowledgeable about the work, the better able the teacher was to provide insight into the original text. As a pedagogical tool, commentary in the form of lecture notes was read aloud to students as a way to instruct. Issues of vocabulary and interpretation were of interest to commentators, and some even pointed out errors. Students took notes of the lectures and often added their own comments to them (Taub, 2017).

Texts arriving in Western Europe from centers of Islamic culture in the Middle Ages not only were translations of the original ancient texts but also consisted of commentaries on those texts. Transcribers, who were sometimes natural philosophers, interpreted the ancients and also made improvements to their ideas, either correcting or advancing them. The translations ultimately allowed for the transmission of scientific knowledge from one region to another, oftentimes with updated modifications made through commentary.

Commentary in early science

Natural philosophers comprised a group of privileged males where rules of conduct were expected. And though there were disagreements, they were generally expressed appropriately, that is, in a civil manner. Commentary about matters of fact were discussed, but overtly aggressive commentary was generally excluded from written public display. Early, disputes about matters of fact in *Philosophical Transactions* might be presented and adjudicated by members of the Royal Society. A prime example is the debate between Johannes Hevelius and Adrien Auzout, astronomers studying the path of the same comet. When Auzout received a copy of Hevelius's observations, he pointed out an error, which Hevelius denied. The dispute was handed over to particular members of the Royal Society to adjudicate. After many months, they declared that both astronomers were correct: the comet Hevelius had recorded at that location was likely a different, second comet. In this way, neither of the natural philosophers' credibility was challenged and their work, critical to the development of astronomical knowledge, could continue unimpeded by a lack of trust (Shapin, 1994).

Prior to columns dedicated to commentary in journals, natural philosophers found ways to critique one another's science. After Isaac Newton's ascendency to president of the Royal Society in 1705, however, communication in the journal became overtly more critical as certain theoretical approaches were stridently supported over others (Atkinson, 1998). Bazerman's (1988) analysis of *Philosophical Transactions* shows how scientific writing became less focused on natural phenomena, as most letters did, but on the process of discovery with the experimental article. "[The natural philosopher] is not simply reporting the self-evident truth of events, but rather is telling a story that can be mooted," he writes (p. 78).

Bazerman recounts that toward the beginning of the nineteenth century, natural philosophers argued over claims with a focus on method. Correspondents would take issue with others' findings or claims in their letters and even negate them. To avert criticism, natural philosophers became more adept at anticipating such commentary in their writing (Atkinson, 1998). With the growth of science and the diminished use of letters, commentary became a critical component of what became the experimental article and was subsumed within the new genre. Authors were likely to report on previous work and to point out weaknesses that their own work overcame. It was not, however, until the nineteenth century that commentary was explicitly included in scientific journals.

According to Magnet and Carnet (2006), one of the first "Letters to the Editor" to appear in biomedical journals was in the journal *Lancet* in a column titled "Miscellaneous" in 1823, the year the journal was established. Their analysis shows that, at the outset, the purpose of the genre was "mere clarifications aiming to provide further knowledge on a given research topic . . . [but that letters to the editor] gradually became a tool to question previously validated research" (Magnet and Carnet, 2006, p. 176). Between 1996 and 2018, approximately 928,368 such pieces have been published in medical journals alone, or 2% of all medical publications listed in Web of Science (Turki et al., 2018, p. 1285). Other journals have included commentary from early on. The journal *Nature*, for example, founded in 1869, became a central location for scientists to respond to one another's ideas (Baldwin, 2015).

Contemporary commentary

Today, commentary can be found under various headings in scientific journals: "Letters to the Editor," "Commentary," "Correspondence," or some variant. As of 2018, for example, the journal *Nature* has a section called "Matters Arising" (*Nature*, 2018, p. 460). The genre remains a critical venue to further knowledge, either supporting of or contesting some aspect of research, whether it be theory, methods, or claims. Formal commentary in journals, in the words of former *Journal of the American Medical Association* (*JAMA*) editors, serves to "facilitate and document discussion and debate" (Winker and Fontanarosa, 1999, p. 1543). In their editorial, the editors make it clear they seek ways to improve upon research and to not extol the merits of it as the purpose is to move the research forward.

In addition to its role in moving research forward, commentary is now considered a "postreview" genre, a place where scientists can identify weaknesses in research that peer reviewers or editors may have missed (Horton, 2002, p. 2843). The history of peer review shows the process has transformed from singular editors making decisions about publication prior to the twentieth century to more contemporary practices where papers are reviewed by experts in the field (see Mehlenbacher and Mehlenbacher, Chapter 7 in this volume). Peer review, however, has come under critical discussion as the volume of scientific research has expanded at an unprecedented rate. Because of the mass and velocity of publication, peer reviewers face increasing challenges in adjudicating quality. They are flooded with requests in a world where it is increasingly challenging to keep up. The number of postreview commentaries published annually across the spectrum of scientific journals demonstrates the need for closer inspection of published articles. Commentary serves as a means for science to correct that which may have been overlooked in initial peer reviews. Commentary is critical in "keeping journals accountable to the scientific community" (Brown, 1997, p. 792).

A lesser known aspect of commentary is its address of social issues that concern scientists, including the term "scientist" itself. In 1924, common terms to describe those who produced science included "man of science" and "scientist worker," terms Norman R. Campbell and others debated in the journal *Nature*. Campbell wrote a letter to the editor and pleaded with him to have the journal start using the word "scientist" because it was more manageable and more inclusive considering the increasing number of women entering science. Replies to Campbell were lively, with some vehemently deploring the term and others accepting it, though sometimes grudgingly. In response to such diverse responses, the editor, Gregory, proclaimed that the journal would allow authors to use it, but the editorial decision was to have journal staff avoid using it (Baldwin, 2015). Concerns about the relationship between science and society have been published (Kuznick, 1987), as well as commentary regarding conflicts of interest in science (Angell, 2005).

Commentary has also been a powerful tool to draw attention to the social implications of research, especially the inequities that can result. Forums provide a site for scientists to argue about the relationship between science and publics. For example, debates about genetically modified organisms (GMOS) have been represented in the literature by scientists, generally focused on health effects, but, earlier, one scientist, Martha Crouch, drew attention to the ecological devastation of science and its impact on communities (Crouch, 1990). In an analysis of a commentary she wrote to the journal *Plant Cell* and the responses to it, Mathison (2014) found that Crouch, a plant biologist, drew attention to a perspective little shared by other scientists about genetic modifications in plants. In her commentary, Crouch politely questioned anthropomorphic basic research and advocated for a more ecocentric approach to scientific inquiry. Rather than manipulating nature to solve problems, Crouch suggested scientists might work *with* nature, adapting to its patterns. In reflecting on her own career, Crouch (1990) indicates, "Basic plant science in the United States is inextricably linked to technology" (p. 275), which is employed for profit to the detriment of others.

Social justice concerns have arisen as inequities in the health system have come to light. A recent "Letter to the Editor" in *The New England Journal of Medicine* reported on a recent research study that questioned the reliability of the pulse oximeter, an instrument that measures oxygen in the blood. The instrument, not initially tested on African Americans when developed, measures oxygen by shining light through the skin. The authors of the letter, Sjoding et al. (2020), found that the instrument was less reliable for Black than for white patients, leading them to question the negative effects of racial bias in its use and to advocate for more racially astute research in medicine.

Summary

Commentary has been a force for stability and change in science, with natural philosophers adjudicating the merit of work through their support – or lack of support – of particular ideas or processes. Commentary acknowledges the value of a work but also recognizes the importance of making it relevant to particular contexts, social as well as historical. In this way, science can be improved upon as well as updated. Commentary also provides a medium through which debate or controversy can be illuminated. Due to the abundance of journals, contemporary uses of commentary are becoming second-level peer reviews, a solution to the challenge of peer review with increasing research productivity. Finally, commentary is a form of gatekeeping, a means to ensure quality and rigor.

Conclusion

As stated earlier, there is a dearth of scholarship regarding letters and commentaries in science. Though much has been written about the individuals who wrote them, there is little that discusses them as a genre. This chapter has made headway in mapping some of the ways that letters and commentaries have been used by gleaning from such accounts some of the contexts that may have influenced the purposes in writing them. What the chapter shows is that unlike the research report, which has been more thoroughly studied, letters and commentaries have a number of purposes to achieve multiple typified social actions. They remain an uncharted territory for science communication.

References

Angell, M. (2005) 'Is academic medicine for sale?', *New England Journal of Medicine*, vol. 342, no. 20, pp. 1516–1518.

Atkinson, D. (1998) *Scientific discourse in sociohistorical context: The philosophical transactions of the Royal Society of London, 1675–1975*, London: Routledge.

Baldwin, M. (2014) 'Keep in the race': Physics publication speed and national publishing strategies in "Nature", 1895–1939. *The British Journal for the History of Science*, vol. 47, no. 2, pp. 257–279.

Baldwin, M. (2015) *Making "Nature": the history of a scientific journal*. University of Chicago Press, Chicago, IL.

Baltussen, H. (2015) 'Philosophers, exegetes, scholars: The ancient philosophical commentary from Plato to Simplicius', in C.C. Kraus & C. Stray (eds.), *Classical commentaries: Explorations in a scholarly genre*, Oxford: Oxford University Press.

Bazerman, C. (1988) *Shaping written knowledge: The genre of the experimental article in science*, Madison, WI: University of Wisconsin Press.

Bazerman, C. (2000) 'Letters and the social grounding of differentiated genres', in D. Barton & & N. Hall (eds.), *Letter writing as a social practice*, Amsterdam: Benjamins, pp. 15–30.

Berkenkotter, C., & Huckin, T.N. (1995) *Genre knowledge in disciplinary communication – cognition/culture/power*, Hillsdale, NJ: Lawrence Erlbaum.

Blakeslee, A.M. (1994) 'The rhetorical construction of novelty: presenting claims in a letters forum'. *Science, Technology, & Human Values* vol. 19, no.1, pp. 88–100.

Boas Hall, M. (2002) *Henry Oldenburg: Shaping the Royal Society*, Oxford: Oxford University Press.

Brown, C.J. (1997) 'Unvarnished viewpoints and scientific scrutiny: letters to the editor provide a forum for readers and help make a journal accountable to the medical community', *Cmaj*, vol. 157, no. 6, pp. 792–794.

Ceccarelli, L. (2001) *Shaping science with rhetoric: The cases of Dobzhansky, Schrödinger and Wilson,* Chicago: University of Chicago Press.

Crouch, M.L. (1990) 'Debating the responsibilities of plant scientists in the decade of the environment', *The Plant Cell,* vol. *2*, no. 4, p 275.

Davis, I.M. (2020) 'Antoni van Leeuwenhoek and measuring the invisible: The context of 16th and 17th century micrometry', *Studies in History and Philosophy of Science Part A,* vol. 83, pp. 75–85.

Fyfe, A. & Geilas, A. (2020) 'Editorship and the editing of scientific journals, 1750–1950'. *Centaurus,* vol. 62, no. 1, pp. 5–20.

Gross, A.G., Harmon, J.E., & Reidy, M.S. (2002) *Communicating science: The scientific article from the 17th century to the present,* Oxford: Oxford University Press.

Henry, J. (1992) 'The scientific revolution in England', in R. Porter & M. Teich (eds.), *The scientific revolution in national context,* Cambridge: Cambridge University Press.

Horton, R. (2002) 'Post publication criticism and the shaping of clinical knowledge', *JAMA,* vol. 287, no. 21, pp. 2843–2847.

Hunter, M. (1994) *The Royal Society and its fellows 1660–1700: The morphology of an early scientific institution,* London: British Society for the History of Science.

Hyland, K. (2004) *Disciplinary discourses: Social interactions in academic writing,* Ann Arbor, MI: University of Michigan Press.

Kronick, D.A. (2001) 'The commerce of letters: networks and "invisible colleges in seventeenth- and eighteenth- century Europe', *The Library Quarterly,* vol. 71, no. 1, pp. 28–43.

Kuznick, P.J. (1987) *Beyond the laboratory,* Chicago: University of Chicago Press.

Langslow, D.R. (2007) 'The Epistula in ancient scientific and technical literature, with special reference to medicine', in R. Morello & A.D. Morrison (eds.), *Ancient letters: Classical and late antique epistolography,* Oxford: Oxford University Press, pp. 212–234.

Magnet, A., & Carnet, D. (2006) 'Letters to the editor: Still vigorous after all these years?: A presentation of the discursive and linguistic features of the genre', *English for Specific Purposes,* vol. 25, no. 2, pp. 173–199.

Marshall, B. (2002) *Heliobacter pioneers: Firsthand accounts from the scientists who discovered heliobacters, 1892–1982,* Victoria: Blackwell.

Mathison, M.A. (1996) 'Writing the critique, a text about a text', *Written Communication,* vol. 13, no. 3, pp. 314–354.

Mathison, M.A. (2014) 'Controversial texts about controversial issues: An academic arguing for and in the name of science', paper presented at the International Society for the Study of Argumentation Conference, Amsterdam, June.

'Matters arising: A venue for commentary'. (2018). *Nature,* vol. 562, p. 460, https://pubmed.ncbi.nlm.nih.gov/30356198/

Merton, R.K. (1957) 'Priorities in scientific discovery: a chapter in the sociology of science'. *American Sociological Review,* vol. 22, no. 6, pp. 635–659.

Miller, C.R. (1984) 'Genre as social action', *Quarterly Journal of Speech,* vol. 70, no. 2, pp. 151–167.

Montgomery, S.L. (2003) *The Chicago guide to communicating science,* Chicago: University of Chicago Press.

Muir, J. (2009) *Life and letters in the ancient Greek world,* London: Routledge.

Rusnick, A. (1999) 'Correspondence networks and the Royal Society, 1700–1750', *The British Journal for the History of Science,* vol. 32, no. 2, pp. 155–169.

Savoia, A.U. (2011) 'Federico Cesi (1585–1630) and the correspondence network of his Accademi dei Lincei', *Studium,* vol. 4, no. 4, pp. 195–209.

Shapin, S. (1994) *A social history of truth: Civility and science in seventeenth-century England,* Chicago: University of Chicago Press.

Shuttelworth-Kraus, C.S. (2002) 'Introduction: Reading commentaries/commentaries as reading', in R.K. Gibson & C.S. Kraus (eds.), *The classical commentary: Histories, practices, theory,* Boston: Brill, vol. 232.

Sjoding, M.W., Dickson, R.P., Iwashyna, T.J., Gay, S.E. & Valley, T.S. (2020) 'Racial bias in pulse oximetry measurement', *New England Journal of Medicine,* vol. *383* no. 25, pp. 2477–2478.

Swales, J. (1990) *Genre analysis: English in academic and research settings,* Cambridge: Cambridge University Press.

Taub, L. (2017) *Science writing in Greco-Roman antiquity,* Cambridge: Cambridge University Press.

Turki, H., Taieb, M.A.H. & Aouicha, M.B. (2018) 'The value of letters to the editor', *Scientometrics*, vol. 117, no. 2, pp. 1285–1287.

Van Miert, D. (2009) *Humanism in an age of science: The Amsterdam Athenaeum in the golden age, 1632–1704*, London: Brill, vol. 179.

Winker, M.A. & Fontanarosa, P.B. (1999) 'A forum for scientific discourse', *JAMA*, vol. 281, no. 16, p 1543.

21

SCIENTIST CITIZENS

Nontraditional and alternative approaches to scientific communication

Justin Mando

Introduction

A young girl's body is found decomposing in the desert. Three forensic entomologists investigate arthropods found on the body to help determine the post mortem interval, but these scientists reach different conclusions that offer the accused an entomological alibi. In the end, a preponderance of evidence including DNA leads to the defendant being found guilty despite the entomologists' testimony.

Faith Zerbe, a biologist for the Delaware Riverkeeper Association, speaks at a March 6, 2018, Delaware River Basin Commission hearing determining whether or not to allow hydraulic fracturing and its related processes. Zerbe proclaims, "I am a scientist. I do a lot of water quality monitoring as part of my profession. I spend a lot of time up in the Upper Delaware River Basin. We have about 300 volunteer monitors collect data in tributary streams that were initially threatened by hydraulic fracturing. . . ." In one short speech, she speaks as a scientist, a local citizen, and an advocate for the environment. Her speech is as fact driven as it is emotive.

Tens of thousands at over 600 cities around the world shout slogans and carry signs proclaiming "Trust Scientific Facts Not Alternative Facts," "Science Not Silence," and more. Scientists had many reasons for joining the march. Nadia Santini, a plant ecologist at University of New South Wales simply stated, "It's important for scientists to get more involved in what's going on in the world." Max Planck Institute's Moritz Hertel agrees: "Science needs to play a stronger role in policy-making. . . ." (Abbott et al., 2017, pp. 404–405)

In all three of these vignettes, scientists engage the public in ways that are atypical of their work as researchers. In one they apply their expertise in a venue where they are invited but heavily scrutinized by nonscientists. In the next, scientists engage a public-science controversy uninvited and of their own accord. In the last one, scientists speak out and raise awareness. Through these nontraditional communication situations, scientists leave the scientific arena and enter the arena of the public. Their words become public record without peer review, their mistakes can be scrutinized by those who are not trained in the scientist's own discipline, and their emotion can be both a sign of courage and a reason to dismiss them. These are *scientists engaging the public*. This formulation does not necessarily put these two entities ("scientists" and "the public") at odds, but it does make them appear mutually exclusive. This division may have deeper ramifications than we suspect.

 DOI: 10.4324/9781003043782-24

We now commonly refer to the activities of nonscientists as they collect water data in the Chesapeake Monitoring Cooperative or affix radiation detectors to their bicycles in Fukushima as a type of *citizen science*. Such initiatives are most certainly noble efforts, but this framing as well separates scientists from citizens. When citizens engage in scientific activities, they are deemed citizen scientists, but why do we not recognize scientists who engage the public as "scientist citizens"?

This chapter argues, especially in nontraditional and alternative forms of science communication, that scientists regularly assume the role of citizen and that recognizing their efforts as civic service or engagement is important both for scientists and for citizens (inclusively defined). Rather than view scientists as in opposition to citizens, advisors of the public, or servants of the public, we should reframe scientists' efforts to engage public issues as a fulfillment of their duties as citizens. This reframing of scientific participation in public affairs can be accomplished by taking a rhetorical view of citizenship. Discourse-focused interpretations of citizenship or rhetorical citizenship positions the role of citizen not simply as a characteristic of someone born in the boundaries of a nation-state or a title given to members of a society who vote but as a status that is enacted through talk and action. Such a view is important as we encourage engagement that is dialogic instead of deficit based and mutually beneficial instead of one-directional. Scientist citizens may be activists who "alert communities or affected populations about a set of exposures or a disease pattern." They may also make accessible scientific writing and education in the name of equitable development (McCormick, 2009, p. 37). Activism is not a requirement of citizenship, however. As I will show, a rhetorical model of citizenship allows us to recognize enactments of civic participation in the alternative modes of communication in which scientists already engage and even in normal science.

To make this argument, I compare nontraditional and alternative modes of science communication with traditional relationships between scientists and the public. I follow with a specific focus on developments in democratic participation as *rhetorical citizenship* and Irwin's (2014) third-order thinking about scientific communication. Three brief analyses of nontraditional science communication –elaborations of the preceding vignettes– broaden citizenship so that it embraces the work of scientists. This can help us move toward more socially responsive and critically reflective science communication (Irwin, 2014).

What are traditional and nontraditional relationships between scientists and the public?

We tend to view the spheres of science and the public as in contact but separate. A Venn diagram would be an inappropriate depiction as it suggests overlap. Rather than spheres, we may employ the metaphor of arenas wherein scientists may enter and impact proceedings in the public arena, but the public has no access or influence in the scientific arena. These are certainly not "peer group[s]" (Peters, 2013). When we consider the prophetic role scientists at times play in their public interactions, mutual influence seems all the less likely (Walsh, 2013). This division between scientist and citizen is in part rooted in the prevalent metaphor of scientists as frontiersmen, courageous individuals entering the unknown (Ceccarelli, 2013). Such a characterization allows broad appeals to scientific independence, public support, and funding without strong public involvement. In the seventeenth century, scientists' spaces deemed public, private, and experimental were permeable. Robert Boyle and Robert Hooke, along with many other early scientists, conducted experiments in private residences and, following customs, admitted other gentlemen into their experimental quarters, which were "constantly open to the curious" (Shapin, 1988, p. 386). This relationship between the public and scientists was based on direct observation and personal interaction with scientists themselves. Shapin contrasts this with credibility today built on "visible display of the emblems of recognizable expertise and because their claims are vouched for by other experts we do not know" (404). We must take these emblems and assurances by experts as sufficient because the sites of contemporary knowledge-making in the sciences are not public. Yet the access to arenas of science and the public have shifted before and may well shift again.

Today's lack of access of the public in the scientific arena can limit interactions to the deficit model where scientists' role is to inform and educate (Wynne, 1995; Irwin, 2014; Davis et al., 2018; Nisbet, 2014, p. 180). As a "specialist knowledge," this bars entry to many. In a sense, this rests upon the Jeffersonian notion that an informed citizenry governs itself.[1] Scientists are needed to educate the citizenry, even though this is not a role many scientists have been trained (or have time) to fill.[2]

It is in this role as public expert that scientists most frequently enter the space of nontraditional communication situations. Scientists stretch from their traditional role to a nontraditional one when scientific knowledge becomes not an intrinsic good but a foundation for practical application. This process of applying scientific knowledge to practical issues is not a simple task. Scientists must adapt and contextualize research to fit the practical situations before them (Peters, 2014, p. 70). They may fill formal advisory capacities such as in the United Kingdom's Scientific Advisory Group for Emergencies (SAGE) and the European Academies Science Advisory Council (EASAC) or in organized watchdog roles such as Advocates of Science and Technology for the People (Agham) based in the Philippines or the Union for Concerned Scientists in the United States. Scientists like these and others in nontraditional capacities who must contextualize and adapt scientific knowledge for practical application and confront competing discourses from, as Peters notes, "everyday knowledge, special knowledge based on practical experience, or traditional knowledge stemming, for example, from religion, folk wisdom or indigenous culture" (p. 75). It is when scientists must argue for their own value and that value is not itself given that they enter nontraditional spaces. The courtroom, public hearings, protests, social media sites, and white papers are some examples of these.

Nontraditional interaction occurs in some cases when barriers protecting the scientific arena from outside influence are broken and public involvement is expected to shape the course of science. Yuan et al. (2019) surveyed over 1,064 scientists and learned that their top communication objective is "Helping inform the public about scientific issues" along with "Getting people interested or excited about science." In this list of nine separate objectives, the bottom four in descending order are:

6 Showing that scientists share community values
7 Framing research implications so members of the public think about a topic in a way that resonates with their values
8 Hearing what others think about scientific issues
9 Showing the scientific community's expertise (p. 113)

These least preferred communication objectives position scientists not simply as experts who must inform the citizenry but as citizens who should confer with other citizens to find common ground. As stated by the former director of the Leopold Leadership Program, scientists should govern "with" people instead of "over" them (Gold, 2001, p. 43). Despite these survey results, scientists do engage in these activities directly through the various institutions of participatory democracy.

This reflexive engagement begins when scientists recognize citizens as counterparts with potentially overlapping values. Efforts to communicate science to a "general public" misses the mark as the particularities of an audience's values, attitudes, and existing knowledge cannot be effectively addressed at such a scale (Nisbet and Scheufele, 2009, p. 1767). Even though those who engage in science communication can be categorized into reliable audience types (Besley, 2018), considering the particularities of audiences is where reflexive engagement begins. When a scientist engages a public scientific controversy as an uninvited speaker at a public hearing, she must build her ethos as a scientist and as a citizen. This requires rhetorical effort to show she shares community values and holds relevant expertise. Acts of protest function similarly where scientists argue for the benefit of all even if their own interests are foregrounded. Even in the case of forensic entomology, the unusual approach to forensics requires efforts to build ethos and relate scientific knowledge to practical application. These links between scientists' agendas and the needs of those they engage in nontraditional spaces are critical for effective communication (Nisbet and Scheufele, 2009, p. 1774).

By reconciling their own goals with the particularities of the nonscientists they engage, scientists enter into a reflexive relationship where such interaction may shape science itself. We may view the scientific arena not as a walled-off fortress but as a sphere of discursive interaction. This is much like the public sphere that is conceived by Hauser (1999) as reticulate, permeable, and vernacular (71). Even working-class Victorians, prompted by newfound leisure time, access to scientific texts, and social organizations dedicated to science, participated in lay efforts that shaped the public's valuation of the new technologies of the age (McLaughlin-Jenkins, 2003). Twentieth-century Americans, intrigued by model rocketry, astronomy, and other scientific pursuits by the popular magazine *Scientific American*, trod nontraditional paths to science that led new generations to formally study science while celebrating adult, amateur "inveterate tinkerers" (Johnston, 2018). From these, other nontraditional approaches to science communication and citizen science efforts, we can see that boundaries are indeed permeable and that communication is, at times, vernacular. Efforts for scientific communication to become dialogic instead of deficit-based are attempts to distribute power. Yet it is not unusual to consider the power the public holds in the development of science. Bucchi (1996) shows that early dissemination to the public can impact the traditional sequence of scientific publication by providing credence to research with wide support. The following cases show how public engagement can lead to new research trajectories.

What is a rhetorical model of citizenship?

What does it mean to be a citizen, and why are scientists dissociated from citizens? Citizenship is a legal status, but many now consider citizenship to be constituted in practice (Isin and Nielsen, 2008, p. 2). It is in practice, especially in their role as arbiters of information, that scientists are rhetorically separated from citizens. This has an illuminating corollary. The deficit model of science communication resembles the common unidirectional model of citizenship in which citizens merely vote for representatives who act as "channels relaying preferences of their constituencies" (Kock & Villadsen, 2012, p. 3). Instead of the deficit model and unidirectional engagement between scientists and citizens, we may strive for dialogic models that focus not on *what* counts as citizenship but on how citizenship is *enacted* (Asen, 2004; Isin and Nielsen, 2008).

Kock and Villadsen (2017) argue that scholars have long supported perspectives on citizenship that emphasize the discursive and dialogic, whether that is by seeing a citizen as a "co-author" of culture (Boele van Hensbroek, 2010, p. 317) or as needing literacy and reasoning skills related to political institutions and processes (Crick, 1999, p. 348). Rhetorician Robert Asen's (2004) *discourse theory of citizenship* "recognizes the fluid, multimodal, and quotidian enactments of citizenship in a multiple public sphere" (p. 191). This processual view of citizenship can help us close the divide between citizen and scientist, enabling many nontraditional forms of scientific communication to be seen in a different light. This is not to say that simply by considering the advisory role a scientist may play in a deliberative forum as an act of citizenship substantially changes the relationship. What may make a difference is becoming comfortable with the view of science as service toward the common good akin to military service.

We have more to gain when we view citizenship not only as dialogic and discursive but also as *rhetorical* (Kock and Villadsen, 2012, 2017). In essence, a rhetorical approach requires attention to the contextual factors of democratic engagement. It can be seen as "communicative and deliberative practices that in a particular culture and political system allow citizens to enact and embody their citizenship" (Keith and Cossart, 2012, p. 46). Such an approach opens citizenship to scientists through practices that both enrich and are enriched by interaction with diverse stakeholders. This also means that normal science may be considered an act of citizenship when it is directed toward the public good.

This may appear as the dialogue model of science communication, but I argue that the rhetorical approach is more closely aligned with Irwin's (2014) third-order thinking about scientific and risk communication. This is not a new way of viewing communication to be contrasted from the first order (deficit) and second order (dialogue) but one that interrogates the relationship between these modes of engagement (160). The third-order approach enables education from the first order and dialogue from the second but also seeks to take "heterogeneity, conditionality and disagreement as a societal resource" (167). This view of difference as a resource is a hallmark of the rhetorical approach to citizenship. Uniformity and collective identity are more features of antidemocratic nations than of those with free democracy (Benhabib, 1996, p. 68). Stated succinctly, "A core concern in rhetoric is how citizens can live together productively under conditions of *dis*sensus" (Kock and Villadsen, 2017, pp. 573–574). This acceptance of difference and disagreement does not preclude the need for information or in any way undercut the importance of knowledge.

In our current times, especially in the context of the United States, scientific fact is routinely called into question or outright dismissed by vocal segments of the population. The recommendations of experts are ignored. Irwin and those who value rhetorical citizenship would not simply chalk up this failure as a societal resource, nor would they shrug their shoulders in frustration. This approach can encourage scientists and science communicators to "[promote] new frames of reference and cultural voices; [diversify] policy options and technological choices; [invest] in civic capacity for public deliberation" (Nisbet, 2014, p. 179). These efforts to engage across difference represent science as service for the common good.

Exemplars of rhetorical citizenship

I now turn to three exemplars that represent rhetorical citizenship as discourse-based engagement in public issues. Through the lenses of Irwin's (2014) third-order thinking and the concept of rhetorical citizenship, we gain insight into a more inclusive and reflexive scientific communication. This is a framing of scientific communication that welcomes scientists into the public arena not as visitors but as citizens. From this vantage, we explore how scientists enact citizenship (Asen, 2004). Here I feature efforts that celebrate the value of science, advocacy on issues of public concern that relate to one's scientific expertise, and normal science directed at the public good as well as other forms of engagement that constitute scientific citizenship.

March for Science

The April 2017 March for Science provided a platform for tens of thousands of scientists and supporters to perform citizenship through public advocacy. The march has spawned an organization that hosts events worldwide on Earth Day and beyond. Their website boasts success and calls out to scientists and science advocates alike. They inspire audiences through quotations that boost the relationship between science and society ("Science and everyday life cannot and should not be separated" by Rosalind Franklin), through merchandise with unifying slogans (T-shirts printed, "There is no Planet B"), and by finding local groups to focus efforts ("Join the world's largest grassroots network of science advocates as we mobilize advocates across the globe to support science-informed policies") (March for Science, 2020).

Protest is a widely recognized act of citizenship (Isin, 2008; Tyler, 2013; Brändle et al., 2018; Handley, 2019). As acts of rhetorical citizenship, protest is predominantly epideictic rhetoric as it argues for values while bestowing praise and blame. Epideictic discourse, in the context of science, typically features arguments to support scientific advancement. Cutrufello (2015) explains, "The function of epideictic discourse is to highlight underlying values of the particular audiences that must be shared amongst discourse community members for scientific work to continue" (p. 279). Indeed, Cutrufello has it right that the epideictic discourse he has studied in Henry Rowland's public address seeks the assent of discourse community members. The March for Science functions similarly as epideictic rhetoric, but the scale is larger. In this case, epideictic rhetoric is not aimed at the narrower collective of a discourse community; it is meant for citizens. March for Science protesters and supporters of the persistent organization enact citizenship in place-based, situated contexts of public advocacy. The context of this advocacy is important as cities and regions allow for focused communication that connects global issues like climate change to local contexts (Nisbet, 2014, p. 182). Such a connection to local contexts makes more likely the meaningful relationship between science and public concerns that is a hallmark of third-order thinking (Irwin, 2014).

Public testimony on hydraulic fracturing and place

Across the world, energy companies have been seeking to develop shale gas reserves with unconventional hydraulic fracturing techniques. To determine risks and possible impacts, scientists take traditional routes through research articles (McKenzie et al., 2012; Vengosh et al., 2014) and, less typically, advocacy documents like the report by the Physicians for Social Responsibility's *Compendium of Scientific, Medical, and Media Findings Demonstrating Risks and Harms of Fracking* (2019). As a situation that calls for deliberative rhetoric, federal-, and state-level hearings gather expert testimony alongside the comments of local citizens. Where hydraulic fracturing is under consideration, many local governments hold public hearings to determine whether to allow this industry into their communities. Scientists participate in these hearings of their own volition alongside citizens of all stripes. In the voluntary public hearing, scientists are not addressed by title or given any special consideration. They receive a typical three minutes to speak just like everyone else.

A critical part of providing testimony in these hearings is to establish *ethos*, an argument for one's relevance to the topic. Scientists and others with technical or political expertise use their status, but this is not all. A common feature of public hearings on hydraulic fracturing are place-based appeals. These arguments draw on one's experience in place to establish *ethos* and to situate arguments in sites of common concern. Not only members of the public at large but scientists as well situate their arguments in shared experience and values related to place (Mando, 2016, 2021; Scarff, 2021). In the vignette that opened this chapter, Zerbe claims her status as a scientist and also explains that she "spends a lot of time in the Upper Delaware River Basin." This history-in-place is used by hydraulic fracturing supporters as well. Consider noted booster and geophysical scientist, Tony Engelder, in a Point Counterpoint article in *Nature*: "I grew up with the sights, sounds and smells of the Bradford oil fields in New York State" (Howarth et al., 2011). Such place-based arguments can be seen as acts of citizenship by deliberating in public and situating common good in experience (Handley, 2019). Especially in the nontraditional venue of the public hearing, scientists enact citizenship in ways we may consider third-order thinking through "reflection-informed practice" (Irwin, 2014, p. 167) enabled by their lack of status and the rhetorical effort required to regain it.

Forensic entomologists and expert testimony

The courtroom is the venue of forensic rhetoric, yet we see the expert witness serving an epideictic role as educator. In Robert D. Hall's (2010) chapter on the role of forensic entomologists as expert witnesses, he reiterates that their role is to "teach the jury and instruct them why his or her theory of the case is correct and should be believed" (470). Despite being retained by the prosecution or defense, the scientist as expert witness should never be swayed into bias as, "The truth only comes out one way, regardless of who pays the bill!" (Cohen, 2016, p. 38). By virtue of being truth tellers who use their expertise to shed light into issues of scientific value, such as the post mortem interval determined by the life stage of black blow flies on a corpse and the conditions it takes for the fly to progress to that stage, the scientist as expert witness clearly serves a first-order or deficit function in Irwin's (2014) model. They educate jurors on issues too technical for their understanding as a one-way, top-down rhetorical situation.

Yet entomologists who take the stand as expert witnesses fill a vital role as citizen not only by providing testimony but by shaping their science in response to "societal concerns and priorities" (Irwin, 2014, p. 167). In the noted trial that started this chapter, forensic entomologists all interpreted the case quite differently, which called into question the credibility of this branch of science as a whole. This led to the formation of the North American Forensic Entomology Association (Wallace, 2020) and the American Board of Forensic Entomologists (Byrd and Castner, 2010, p. 12). Researchers have directed their efforts toward better understanding topics of relevance to investigators (Merrit and Wallace, 2010; Tomberlin et al., 2012). When we see forensic entomologists enacting citizenship by directing their research to societal priorities and engaging in public proceedings, it is confirmation that scientists can be citizens while losing none of their integrity.

Closing remarks

If we accept this argument that scientists' public efforts can be framed as acts of citizenship, what changes? When scientists see themselves as "scientist citizens," does their relationship to other people, to places, to their work transform? This chapter has argued in the affirmative that scientists' involvement in issues of communal concern are more meaningful when we recognize them as rhetorical acts that constitute citizenship. Just as the legal status of "citizen" bestows rights upon and expresses expectations of political participation of immigrants who attain it, referring to scientists' activity as citizenship associates their work with this same kind of participation. Nonscientists may see the work of scientists as service that is worthy of appreciation similar to military service.

Like many immigrants before they attain legal citizenship, scientists are already engaging in behavior that constitutes acts of citizenship; we just tend not to recognize it as such. To associate community-focused research collaboration, public addresses that apply scientific expertise to social issues, expert testimony in trials, and other acts of citizenship is to come closer to Irwin's (2014) third-order thinking about scientific communication. As citizenship, these acts are more readily placed in the wider social context to which they belong. This rhetorical, contextualized view is important to help scientists consider the social implications of their research, and the extent to which their research can aid communities. Such engagement may also help scientists regain the trust of groups who are skeptical of science such as those who deny climate change or evolution and those actively resisting vaccines or genetically modified foods.

The difficulty of determining what is *not* an act of citizenship remains. Considering hydraulic fracturing, should the work of Terry Engelder, geoscientist and consultant for the industry, be considered an act of citizenship? Is he working for the public good? Certainly, he is well compensated for his consulting work. Does monetary compensation preclude one's scientific engagement as an act of citizenship? Expert witnesses such as forensic entomologists are paid for their time, yet I argue here that their work can be seen as a fulfillment of their civic duty, especially when they direct research toward the public good. These gray areas of what constitutes citizenship and what does not may be sharpened by future research that clarifies discourse-based, participatory citizenship. There is reason for hope.

Notes

1 Thomas Jefferson writes in a 1798 letter to Richard Price, "[W]herever the people are well informed they can be trusted with their own government; that whenever things get so far wrong as to attract their notice, they may be relied on to set them to rights."
2 Many programs train scientists to be public communicators like the Earth Leadership Program (formerly Leopold Leadership Program), the Alan Alda Center for Communicating Science, and growing numbers of science communication courses at universities around the world.

References

Abbott, A., Callaway, E., Casassus, B., Phillips, N., Reardon, S., Rodriguez Mega, E., & Witze, A. (2017) 'March for science attracts thousands across the globe', *Nature*, vol. 544, pp. 404–405.

Asen, R. (2004) 'A discourse theory of citizenship', *Quarterly Journal of Speech*, vol. 90, no. 2, pp. 189–211, doi: https://doi.org/10.1080/0033563042000227436

Benhabib, S. (1996) 'Toward a deliberative model of democratic legitimacy', in S. Behabib (ed.), *Democracy and difference: Contesting the boundaries of the political*, Princeton: Princeton University Press.

Besley, J.C. (2018) 'Audiences for science communication in the United States', *Environmental Communication*, vol. 12, no. 8, pp. 1005–1022, doi: https://doi.org/10.1080/17524032.2018.1457067

Boele van Hensbroek, P. (2010) 'Cultural citizenship as a normative notion for activist practices', *Citizenship Studies*, vol. 14, no. 3, pp. 317–330, doi: https://doi.org/10.1080/13621021003731880

Brändle, V.K., Galpin, C., & Trenz, H. (2018) 'Marching for Europe? Enacting European citizenship as justice during Brexit', *Citizenship Studies*, vol. 22, no. 8, pp. 810–828, doi: https://doi.org/10.1080/13621025.2018.1531825

Bucchi, M. (1996) 'When scientists turn to the public: Alternative routes in science communication', *Public Understanding of Science*, vol. 5, no. 4, pp. 375–394, doi: https://doi.org/10.1088/0963-6625/5/4/005

Byrd, J.H., & Castner, J.L. (2010) *Forensic entomology: The utility of arthropods in legal investigations*, Boca Raton: CRC Press.

Ceccarelli, L. (2013) *On the frontier of science: An American rhetoric of exploration and exploitation*, East Lansing: Michigan State University Press.

Cohen, K.S. (2016) *Expert witnessing and scientific testimony: A guidebook*, Boca Raton: CRC Press.

Crick, B. (1999) 'The presuppositions of citizenship education', *Journal of the Philosophy of Education*, vol. 33, no. 3, pp. 337–352, doi: https://doi.org/10.1111/1467-9752.00141

Cutrufello, G. (2015) 'The public address and the rhetoric of science: Henry Rowland, epideictic speech, and nineteenth-century American science', *Rhetoric Review*, vol. 34, no. 3, pp. 275–291, doi: https://doi.org/10.1080/07350198.2015.1040303

Davis, L.; Fahnrich, B.; Claudia Nepote, A.; Riedlinger, M., & Trench, B. (2018) 'Environmental communication and science communication – Conversations, connections, and collaborations', *Environmental Communication*, vol. 12, no. 4, 431–437, doi: https://doi.org/10.1080/17524032.2018.1436082

Gold, B.D. (2001) 'The Aldo Leopold Leadership Program: Training environmental scientists to be civic scientists', *Science Communication*, vol. 23, no. 1, pp. 41–49, doi: https://doi.org/10.1177/1075547001023001004

Hall, R.D. (2010) 'Forensic entomologist as expert witness', in J.H. Byrd & J.L. Castner (eds.), *Forensic entomology: The utility of arthropods in legal investigations*, Boca Raton: CRC Press.

Handley, D.G. (2019) ' "The line drawn": Freedom Corner and rhetorics of place in Pittsburgh, 1960s-200s', *Rhetoric Review*, vol. 38, no. 2, pp. 173–189, doi: https://doi.org/10.1080/07350198.2019.1582239

Hauser, G.A. (1999) *Vernacular voices: The rhetoric of publics and public spheres*, Columbia: The University of South Carolina Press.

Howarth, R.W., Ingraffea, A., & Engelder, T. (2011) 'Should fracking stop?', *Nature*, vol. 477, pp. 271–275, doi: https://doi.org/10.1038/477271a

Irwin, A. (2014) 'Risk, science and public communication: Third-order thinking about scientific culture', in M. Bucchi & B. Trench (eds.), *Routledge handbook of public communication of science and technology*, New York: Routledge.

Isin, E.F. (2008) 'Theorizing acts of citizenship', in E.F. Isin & G.M. Nielsen (eds.), *Acts of Citizenship*, London: Zed Books.

Isin, E.F., & Nielsen, G.M. (2008) *Acts of citizenship*, London: Zed Books.

Jefferson, T. (1798) 'Thomas Jefferson to Richard Price', 8 January, www.loc.gov/exhibits/jefferson/60.html

Johnston, S.F. (2018) 'Vaunting the independent amateur: *Scientific American* and the representation of lay scientists', *Annals of Science*, vol. 75, no. 2, pp. 97–119, doi: https://doi.org/10.1080/00033790.2018.1460691

Keith, W., & Cossart, P. (2012) 'The search for "real" democracy: Rhetorical citizenship and public deliberation in France and the United States', in C. Kock & L.S. Villadsen (eds.), *Rhetorical citizenship and public deliberation*, University Park: Pennsylvania State University Press.

Kock, C., & Villadsen, L.S. (2012) *Rhetorical citizenship and public deliberation*, University Park: Pennsylvania State University Press.

Kock, C., & Villadsen, L.S. (2017) 'Rhetorical citizenship: Studying the discursive crafting and enactment of citizenship, *Citizenship Studies*, vol. 5, pp. 570–586, doi: https://doi.org/10.1080/13621025.2017.1316360

Mando, J. (2016) 'Constructing the vicarious experience of proximity in a Marcellus Shale public hearing', *Environmental Communication*, vol. 10, no. 3, pp. 352–364, doi: https://doi.org/10.1080/17524032.2015.1133438

Mando, J. (2021) *Fracking and the rhetoric of place: How we argue from where we stand*, Lanham: Lexington.

March for Science. (2020) 'Our mission', http://marchforscience.org/our-mission

McCormick, S. (2009) 'From "politico-scientists" to democratizing science movements: The changing climate of citizens and science', *Organization & Environment*, vol. 22, no. 1, pp. 34–51. doi:10.1177/1086026609333419

McKenzie, L.M., Witter, R.Z., Newman, L.S., & Adgate, J.L. (2012) 'Human health risk assessment of air emissions from development of unconventional natural gas resources', *Science of The Total Environment*, vol. 424, pp. 79–87, doi: https://doi.org/10.1016/j.scitotenv.2012.02.018

McLaughlin-Jenkins, E. (2003) 'Walking the low road: The pursuit of scientific knowledge in late Victorian working-class communities', *Public Understanding of Science*, vol. 12, pp. 147–166, doi: https://doi.org/10.1177/09636625030122002

Merrit, R.W., & Wallace, J.R. (2010) 'The role of aquatic insects in forensic investigations', in J.H. Byrd & J.L. Castner (eds.), *Forensic entomology: The utility of arthropods in legal investigations*, Boca Raton: CRC Press.

Nisbet, M.C. (2014) 'Engaging in science policy controversies: Insights from the US climate change debate', in M. Bucchi & B. Trench (eds.), *Routledge handbook of public communication of science and technology*, New York: Routledge.

Nisbet, M.C., & Scheufele, D.A. (2009) 'What's next for science communication?: Promising directions and lingering distractions', *American Journal of Botany*, vol. 96, no. 10, pp. 1767–1778, doi: https://doi.org/10.3732/ajb.0900041

Peters, H.P. (2013) 'Gap between science and media revisited: Scientists as public communicators', *Proceedings of the National Academy of Sciences of the United States of America*, vol. 110, pp. 14102–14109.

Peters, H.P. (2014) 'Scientists as public experts: Expectations and responsibilities', in M. Bucchi & B. Trench (eds.), *Routledge handbook of public communication of science and technology*, New York: Routledge.

Physicians for Social Responsibility. (2019) 'Compendium of scientific, medical, and media findings demonstrating risks and harms of fracking', www.psr.org/wp-content/uploads/2019/06/compendium-6.pdf

Scarff, K. (2021) 'Mapping vicarious proximity: Holistic and dualistic metaphors in a Mountain Valley Pipeline public hearing', *Environmental Communication*, DOI: 10.1080/17524032.2020.1867602

Shapin, S. (1988) 'The house of experiment in seventeenth-century England', *Isis*, vol. 79, no. 3, pp. 373–404, DOI: http://dx.doi.org/10.1086/354773

Tomberlin, J.K., Byrd, J.H., Wallace, J.R., & Benbow, M.E. (2012) 'Assessment of decomposition studies indicates need for standardized and repeatable research methods in forensic entomology', *Journal of Forensic Research*, vol. 3, no. 5, pp. 1–10, doi: DOI: 10.4172/2157–7145.1000147

Tyler, I. (2013) 'Naked protest: The maternal politics of citizenship and revolt', *Citizenship Studies*, vol. 17, no. 2, pp. 211–226.

Vengosh, A., Jackson, R.B., Warner, N., Darrah, T.H., & Kondash, A. (2014) 'A critical review of the risks to water resources from unconventional shale gas development and hydraulic fracturing in the United States', *Environmental Science & Technology*, vol. 48, no. 15, pp. 8334–8348, doi: dx.doi.org/10.1021/es405118y

Wallace, J. (2020) *Personal correspondence*, 10 September.

Walsh, L. (2013) *Scientists as prophets: A rhetorical genealogy*, New York: Oxford University Press.

Wynne, B. (1995) 'Public understanding of science', in S. Jasanoff, G.E. Markle, J.C. Petersen and T. Pinch (eds) *Handbook of Science and Technology Studies*, Thousand Oaks, London and New Delhi: Sage, 361–388

Yuan, S.; Besley, J.C., & Dudo, A. (2019) 'A comparison between scientists' and communication scholars' views about scientists' public engagement activities', *Public Understanding of Science*, vol. 28, no. 1, pp. 101–118, doi: https://doi.org/10.1177/0963662518797002

PART 3

Scientific visuals and multimedia

22

SPEAKING TO THE EYES

A historical overview of data visualization in the sciences

Danielle DeVasto

Introduction

Given the recent explosion of available data and visualization tools, data visualization is often thought of as a contemporary phenomenon. As American data scientist Edward Tufte (2001) noted, "Much of the world these days is observed and assessed quantitatively – and well-designed graphics are far more effective than words in showing such observations" (p. 87). Debates of effectiveness aside, there is no denying that visualization is a powerful – and prevalent – way to understand the world and the "information [that] gently but relentlessly drizzles down on us" (Von Baeyer, 2004, p. 3). From public transportation to company reports to news websites to phone apps, data visualizations are part and parcel in many areas of our lives, perhaps none more so than within the realm of science.

Indeed, scholars have shown that visual displays are prevalent in various technical and public STEM contexts (Allen, 2018; Barrow, 2008; Bucchi and Saracino, 2016; Desnoyers, 2011; Gross and Harmon, 2013; Krause, 2017; Pauwels, 2006; Ruivenkamp and Rip, 2010). Contemporary science communication relies on data visualizations to explore and analyze data, convey findings, provide models of actual and theoretical worlds, etc. For some scientific fields, communicating data through visual means is a requirement; for others, it's a growing need, particularly as the necessity of communicating with wider audiences (e.g. nonexpert stakeholders, cross-disciplinary collaboration) increases. (FEMA, 2015; Frankel and DePace, 2012; Grainger et al., 2016; International Federation of Red Cross, 2011; Kostelnick et al., 2013; Rodríguez Estrada and Davis, 2015).

While recent enthusiasm for and ubiquity of data visualization in science communication is clear, the idea of visualizing data is not new. Going back at least 8,000 years to the earliest mapmaking, the visual display of quantitative information has developed in response to philosophical, methodological, and technological advancements. As this chapter will show, data visualization has long been a powerful means for making sense of scientific, medical, and technological data, advancing arguments, and communicating information to others. After reviewing the roots of data visualization, this chapter will showcase notable examples in the evolution of data visualization. As these examples will show, data visualizations come in a variety of forms. They have not always been constructed specifically to advance finalized visual arguments but also as heuristics to further analyze and interpret the specific issue at hand.

 DOI: 10.4324/9781003043782-26

Finally, this brief review will serve as a historical backdrop for contextualizing current trends in data visualization.

Distinguishing data visualization

Even though the act of visualizing quantitative information is not new, its vocabulary is still evolving. So, before moving forward, at least three intersecting terms need disambiguation: information graphic, data visualization, and scientific visualization. These first two terms, in particular, are often used interchangeably, however, there are notable differences. While all three types of visualizations are relevant to science communication, this chapter must necessarily confine itself to data visualizations. (See Figure 22.1.)

Generally, the objective of an information graphic (better known as an infographic) is to communicate information quickly and clearly using striking, engaging visual design. Displayed as a separate piece of content or embedded within a larger text, an infographic combines various visual and textual elements, such as one or more data visualizations, images, illustrations, facts, quotes, and captions, to show a process or relationship or to tell a story that cannot be automatically discerned from data alone. It is generally static and can be exclusively qualitative. An infographic also tends to be more highly customized, even when utilizing easily accessible templates; it is generally specific to the collected data and the narrative being constructed, requiring significant effort to recreate with different data.

On the other hand, a data visualization is a graphical display of qualitative or quantitative data "in ways that other forms – text and tables – do not allow" (Friendly and Palsky, 2007, p. 208). Compared to infographics, data visualizations tend to be more interactive and data rich, generally dealing with larger and more abstract datasets. Often in the form of various graphs, charts, or maps, the same form (e.g. bar graph) can be reused with different datasets (see Ferdio [n.d.] for a comprehensive archive of types of data visualizations). They can be created or augmented by hand, but data visualizations are primarily generated with computer support or automated methods. Objectivity is sometimes referenced as a distinguishing characteristic, but it should be noted that all data visualizations are inherently persuasive (Kostelnick, 2007). Generally, they serve two main purposes: (1) to make sense of data and (2) to then communicate that to others. In the first case, scientists are increasingly using data visualizations to grapple with complex data – to enhance pattern recognition, to identify anomalies and variations, to increase working memory, to develop hypotheses. In the second case, data visualizations are used to communicate scientific findings beyond the individual or group directly involved in the research. This includes presenting detailed findings and recording data, as in scientific articles or presentations, as well as conveying summarized findings and conceptual understandings, as in overview figures and visual abstracts.

Finally, scientific visualization is concerned with representing data about scientific phenomena in the form of realistic images. Like data visualization, the primary objective is insight. Scientific visualizations are used to do things like visualizing objects that have not been directly observed (e.g. black holes), assisting in medical diagnosis, and modeling atmospheric data. As these examples suggest, they tend to visualize volumes and surfaces in 3-D, often with a dynamic component. But they also might include more 2-D figurative depictions, like medical cutaways or illustrations.

Rooting data visualization

Data visualization has evolved across (and incorporates the knowledge of) various fields, including survey design, applied statistics, psychology, medicine, technical communication,

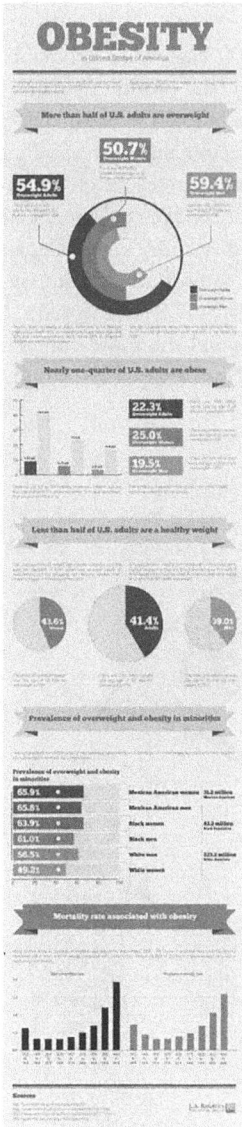

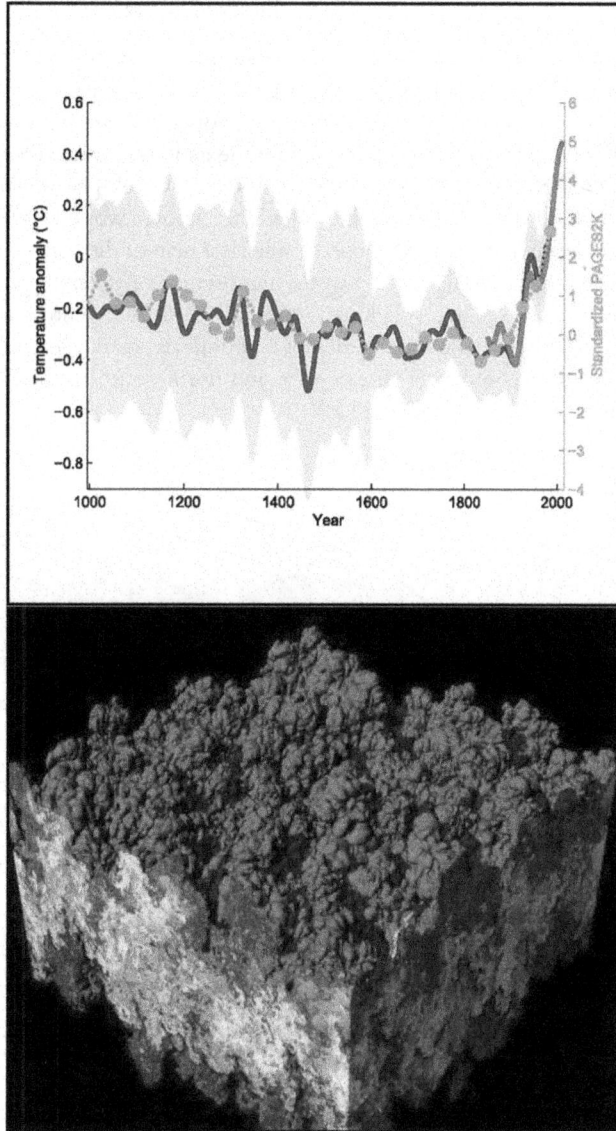

Figure 22.1 Sample visualizations: infographic about obesity (left), data visualization of annual mean temperatures (top right), scientific visualization of a large simulation of a Rayleigh-Taylor instability problem (bottom right)

Sources: "Obesity Infographic from LA-Bariatrics" by Chris Morris CA is licensed under CC BY 2.0; Wikimedia.

and technology and information design. But some of its deepest roots lie in cartography and astronomy (Azzam et al., 2013; Friendly, 2006). This section will highlight several early examples and proto-data visualizations to provide context for later developments. It will also review three catalysts – empiricism, data collection, and visual thinking – that helped data visualization move forward from those early examples.

Early visualizations

Many of the earliest forms of data visualization, dating back to 6200 BCE, were geographic maps. These early visualizations depicted concrete, specific things in the world. As early as 1400 BCE, ancient Egyptians used the idea of coordinates to survey the Nile flood basin, and by 200 BCE, lattice systems similar to latitude and longitude were used to visualize terrestrial and celestial positions (Wainer, 2005). In 550 BCE, Greek philosopher Anaximander developed the first world map, visualizing the scope of the known world beyond what the eye could see. Later that century, Ptolemy developed a spherical map of the earth that used latitude, longitude, and mathematical calculation – not the importance of different places – to proportion countries. His work served as a benchmark until the fourteenth century (Friendly, 2008a). It would take roughly 5,000 years for the idea of a coordinate system to evolve beyond the more concrete, geographic coordinates of east–west and north–south to the more abstract variables *X* and *Y* (Tufte, 1997). (See Figure 22.2.)

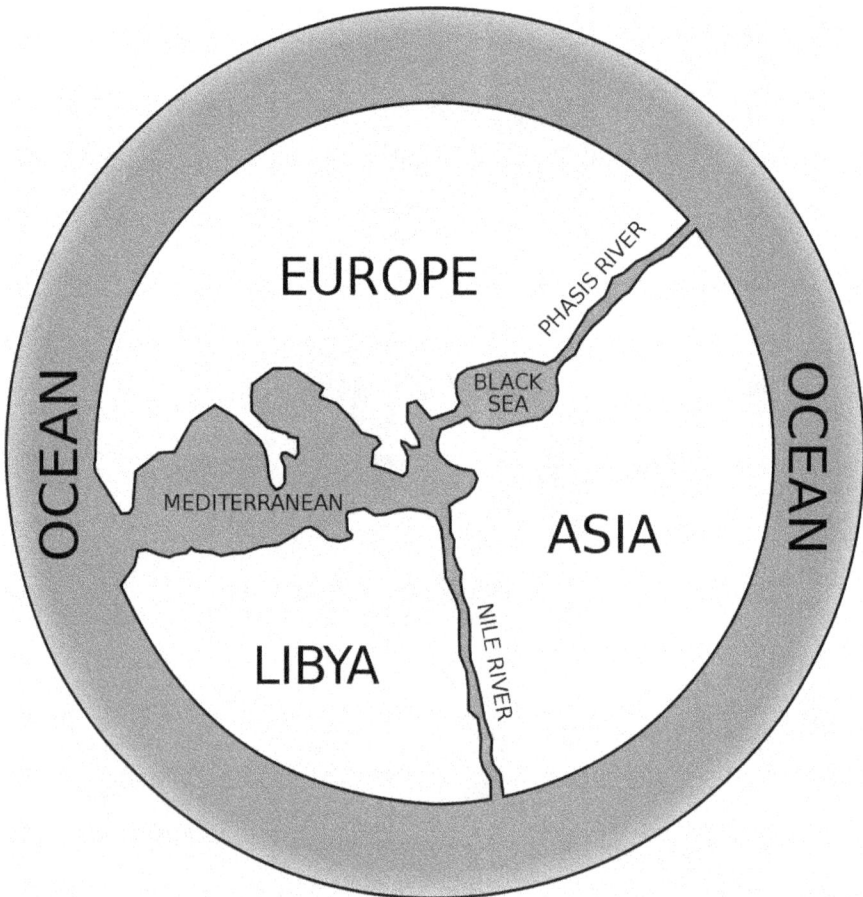

Figure 22.2 Re-creation of Anaximander's world map, sixth century
Source: Wikimedia.

The twelfth century *Yu ji tu* or "Map of the Tracks of Yu," a highly sophisticated map of China's coastline and network of rivers, was the first to use a grid to denote scale.[1] Each of its 5,000 squares representing one Chinese *li* (approximately 500 meters or 1,640 feet). (See Figure 22.3.)

The Middle Ages also saw some the earliest "tree maps," precursors to modern-day network diagrams. Perhaps most notable is Ramon Llull's *Arbor scientiae*, a collection of tree-based maps of science and human knowledge that would, centuries later, influence the classification work of scholars such as Bacon, Descartes, and Darwin (Lima, 2011). (See Figure 22.4.)

While limited to more pictorial, literal visualization, these early maps had the elements and ideas necessary for making more abstract, quantitatively based visuals (Tufte, 1997).

Alongside these more common, concrete visualizations, more abstract, theoretical ones were also beginning to develop. The earliest known attempt to show changing values graphically is

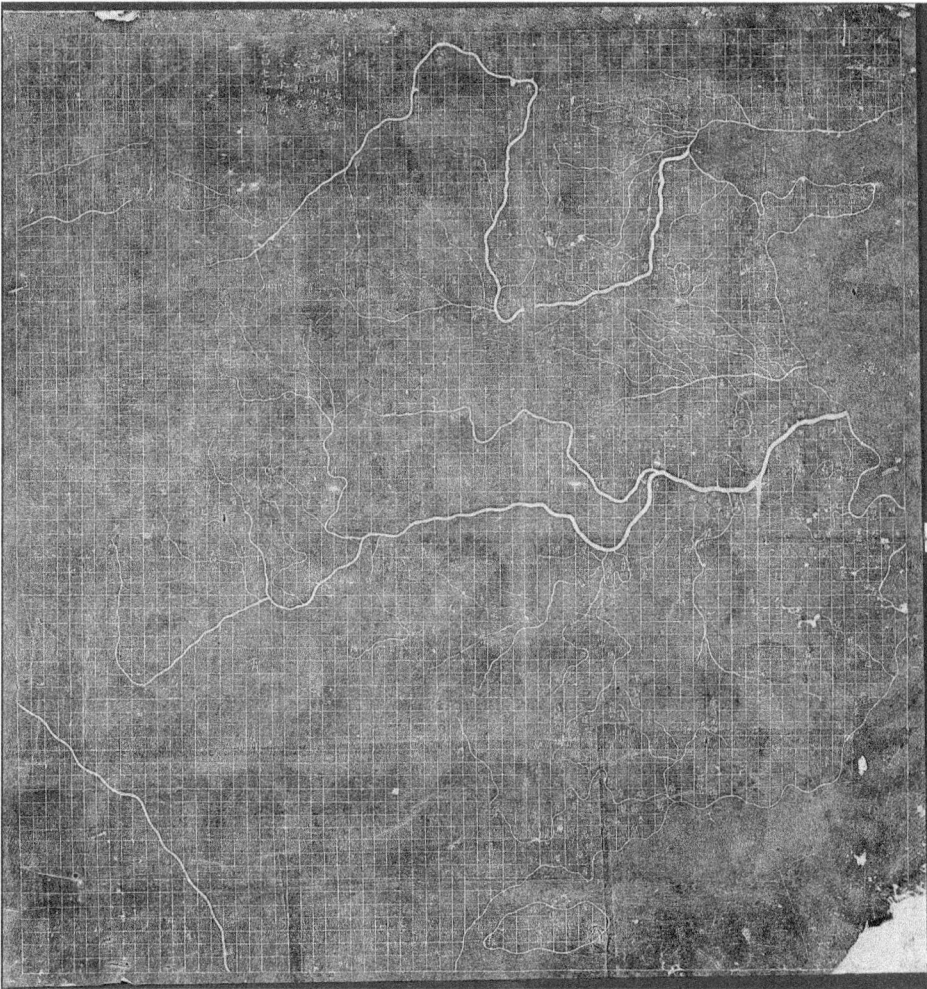

Figure 22.3   Yu ji tu ("Map of the Tracks of Yu"), twelfth century
Source: Library of Congress, Geography and Map Division.

Figure 22.4 *Arbor scientiae* ("Tree of Science"), Ramon Llull, 1515.

Source: Wikimedia.

an anonymous tenth-century time-series of planetary orbits. (See Figure 22.5.) While it has some noted discrepancies, it is distinguished for using a grid as a background for drawing curves. Other examples of time-series would not begin to appear in scientific writing for another 800 years (Tufte, 2001).

Around 1350, Nicole Oresme plotted the velocity of a constantly accelerating object over time. A kind of proto-bar chart, Oresme's work was missing one thing – data. (See Figure 22.6.)

Developing a taste for data

As the previous section shows, many of the core mechanical pieces of data-based visual displays have been in use for a long time. And yet data visualization was relatively rare until the late

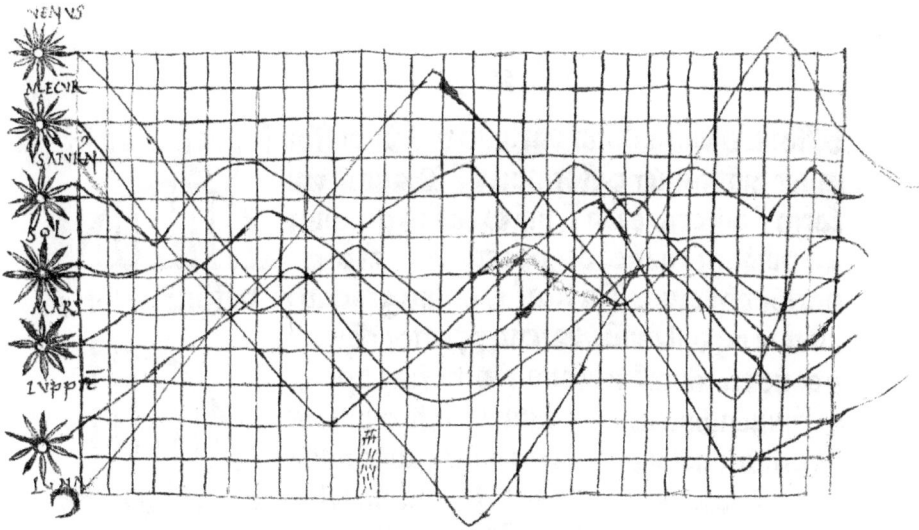

Figure 22.5   Visualization of planetary cycles, tenth century
Source: Wikimedia.

eighteenth century. In addition to the development of tools like coordinate systems and grids, data visualization was furthered by three key shifts in scientific thought and practice: empiricism, data collection, and visual thinking.

Prior to the Scientific Revolution, many discoveries were entirely based on reason and logic. If an idea made sense, then it had to be true. Abstract reasoning dictated science, making science more like philosophy. But with the rise of universities, increased contact with non-Western societies, and the general questioning tone of the Renaissance, empiricism began to take root. In contrast to rationalism, scholars such as Sir Francis Bacon promoted the role of experience and the senses as the source of understanding, which ultimately contributed to the popularization of inductive methods of scientific inquiry, the development of the scientific method, and the uptake of objectivity as an epistemic virtue.[2] As Friendly (2006) notes, the natural and physical sciences were reluctant to embrace empiricism. But following a series of successful empirically based discoveries (e.g. Copernican heliocentrism, Galileo's telescopic observations, Playfair's economic analyses), there was a growing acceptance that many important, scientific questions could be better addressed by a data-driven, empirical approach rather than by logic or religion alone.

Unsurprisingly, under the reign of rationalism, data were rare. But with the rise of empiricism and developments in instrumentation, European scientists began to collect data, particularly data about population, the weather, and economic activity. In 1669, mathematician Christiaan Huygens applied the idea of coordinate systems and probability theory to social data from the bills of mortality to give the first graph of a continuous distribution function (Friendly, 2008a). In 1684, English naturalist Robert Plot recorded the daily barometric pressure in Oxford over six months. He later plotted that data in an early line graph, which he described as a "wandring prickt line" (Kostelnick and Hassett, 2003, p. 121). (See Figure 22.7.)

Another early example comes from Edmond Halley who, after collecting data from ocean navigators and his own experience in the tropics, charted the direction of trade winds and monsoons in what is recognized as the first meteorological map. (See Figure 22.8.)

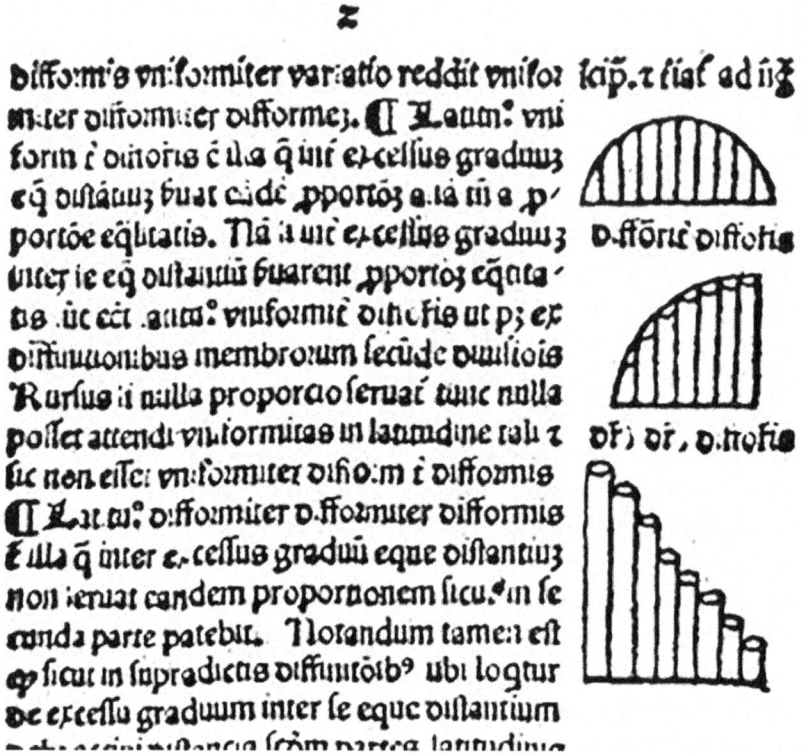

Figure 22.6   Diagrams of the velocity of an accelerating object against time, Nicole Oresme, fourteenth
century

Source: Wikimedia.

As countries began to realize the value of data for planning, managing commerce, and addressing pressing social issues, the collection, publication, and public appetite for data picked up steam, fueling the growth of data visualization (Friendly, 2006).

As empirical, data-driven approaches became more widespread, scientists began to realize the advantages of visual thinking. That is, working with data via visual, graphical methods (rather than through words or tables) often helped them "find both the structure and the surprises hidden in data" (Wainer, 2005, p. 4). One early example of visual thinking was drawn by Flemish astronomer Michael Florent van Langren. Possibly the first visual representation of statistical data, this one-dimensional line graph visualizes different astronomers' and cartographers' estimates of the distance in longitude between Toledo and Rome. (See Figure 22.9.)

Langren could have used a table, but a graphical method sends a clear message about the wide variation in estimates. Casting doubt on the veracity of these other estimates, Langren could then bolster his own method for calculating longitude (Tufte, 1997). Another early example of visual thinking was Christoph Scheiner's use of "small multiples," a series of similar graphs or charts using the same scale and axes, to easily compare the changing configurations of sunspots over time (Tufte, 2001). By the late eighteenth century, the pendulum of preference had

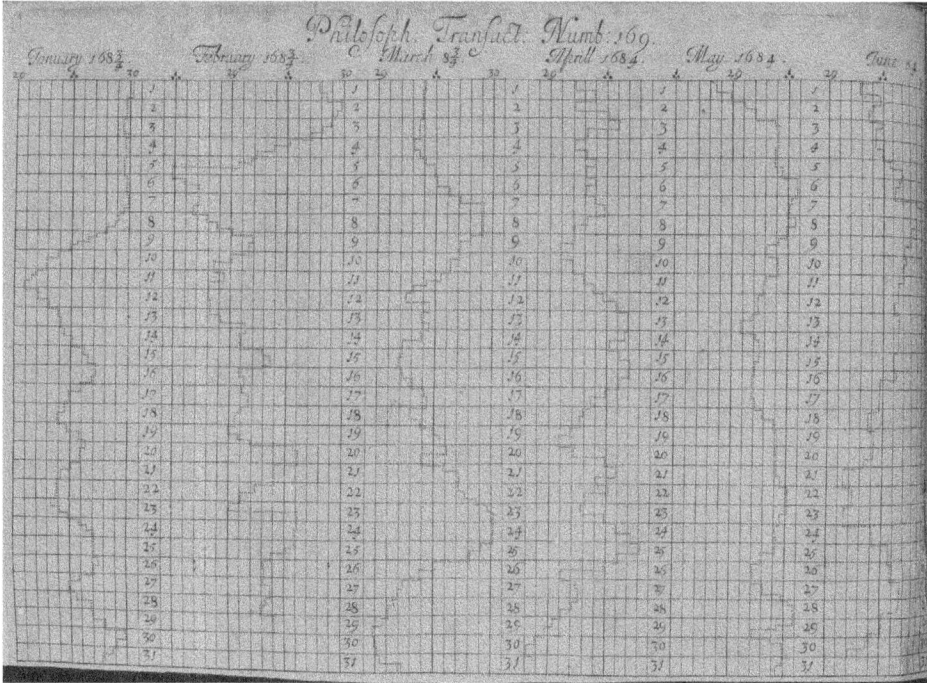

Figure 22.7 "History of Weather," Robert Plot, 1685

Source: Plot, Robert. A Letter from Dr. Robert Plot of Oxford, to Dr. Martin Lister F. of the R. S. concerning the use which may be made of the following History of the Weather, made by him at Oxford throughout the year 1684. *Philosophical Transactions of the Royal Society of London*, vol. 15, no. 169 (March 23, 1685), pp. 930–943 with folding plate.

swung in favor of visual modes of thinking, reflecting the growing scientific ideal of mechanical objectivity and the belief that language, "born before science . . . was often inappropriate to express exact measures or definite relations" (Datson and Gallison, 2007, p. 81).

Another noteworthy example of the role of visual thinking comes from the development of plate tectonics. As Wainer and Friendly (2020) describe, when the entire world was first mapped in the sixteenth century, these visuals caused scientists to comment on the near-perfect fit of the coasts of South America and Africa and wonder how the world came to be. The answer, nearly 400 years later, would also rely on visual thinking, namely, Marie Tharp's plotting of sonar data to map the undersea Mid-Atlantic Ridge (Blakemore, 2016) and Mason and Raff's (1961) "zebra stripes" map of the northeastern Pacific. (See Figure 22.10.)

Visual thinking was a means not only for scientists to persuade using data or to understand data but also to "graphically birth" new questions and lines of inquiry.

Data visualization showcase

From these roots, modern data visualization blossomed. Beginning with the eighteenth century, the history of data visualization is predominantly one of innovation and explosive growth. Freed from direct analogy to the physical world, data visualizations began to relationally represent all kinds of datasets. While there are certainly many more than can be included in this brief

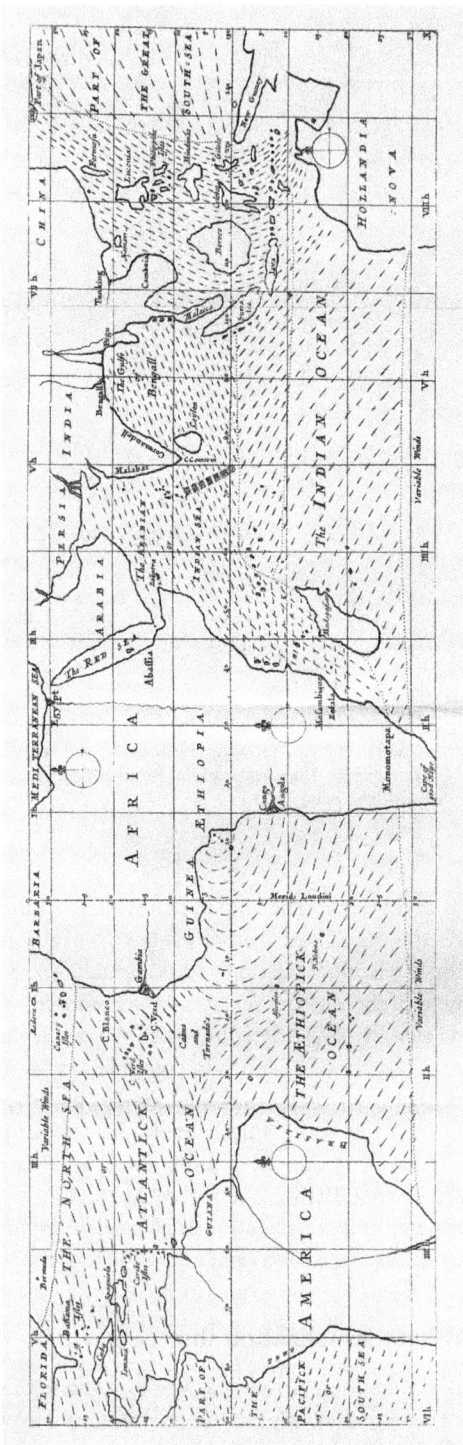

Figure 22.8 Chart of the trade winds and monsoons, Edmond Halley, 1687
Source: Wikimedia.

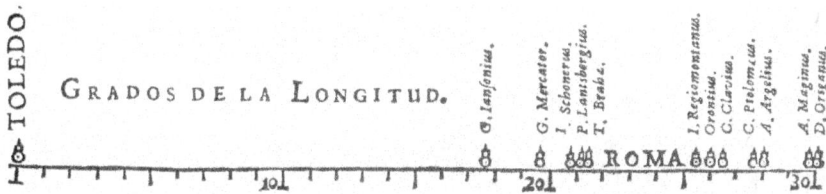

Figure 22.9   Graph of determinations of distance from Toledo to Rome, Michael van Langren, 1644

Source: Tufte, 1997, p. 15.

chapter, the following section presents a sampling of notable science-related data visualizations from the eighteenth to twentieth centuries.[3] These samples all retain evidence of the three key shifts in scientific thought and practice (i.e. empiricism, data collection, and visual thinking) that fostered the development of modern data visualization. In addition to originating from the sciences, these data visualizations were selected because (1) they model novel graphic forms or applications, (2) they supplied visual insights that directly led to important scientific advances or discoveries, or (3) they played a role in catalyzing new areas of scientific inquiry. These visuals were also selected to showcase a variety of voices, both those deemed canonical and those lesser recognized. As these examples show, data visualizations come in a variety of forms. They often summarize and advance arguments, but they can also further analytical and interpretive work.

Mapping the invisible

As scientists warmed to the idea of visual thinking, one major advancement was thematic mapping. Rather than solely visualizing where something was, the thematic maps that began to emerge were more abstract, emphasizing spatial distribution of some select data (Friendly and Palsky, 2007). To help visualize the influx and variety of data being collected, now common conventions, like isolines, contours, and proportional circles, as well as chloropleth techniques, were invented. Uncoupled from geographic or time coordinates, thematic mapping became relevant to all manner of quantitative inquiry, though initially it was most commonly applied to geologic, economic, and medical data.

One notable example is geologist William Smith's geological map of Great Britain. (See Figure 22.11.)

Moving beyond visualizing geographic features alone, Smith's massive 6-by-8.5-foot map visualizes the differences in a particular variable from place to place – in this case, more than 30 years' worth of empirical data on the local sequences of rock layers and fossilized organisms hidden beneath the land. A significant scientific breakthrough for stratigraphic geology, Smith's map advanced contested claims about evolution and the age of the earth (Friendly and Palsky, 2007; Winchester, 2001).

Perhaps a lesser known example is the first modern weather map created by Sir Francis Galton, published in 1875. Born from crowdsourced data organized into a large, published collection of graphs and maps, this visualization uses glyphs on a map to indicate areas of similar air pressure and barometric changes. (See Figure 22.12.) See Friendly (2008b) for a more detailed recounting of the development of the collection of graphs and their critical role in Galton's graphical discovery of the relation between barometric pressure and wind direction that now forms the basis of modern weather maps.

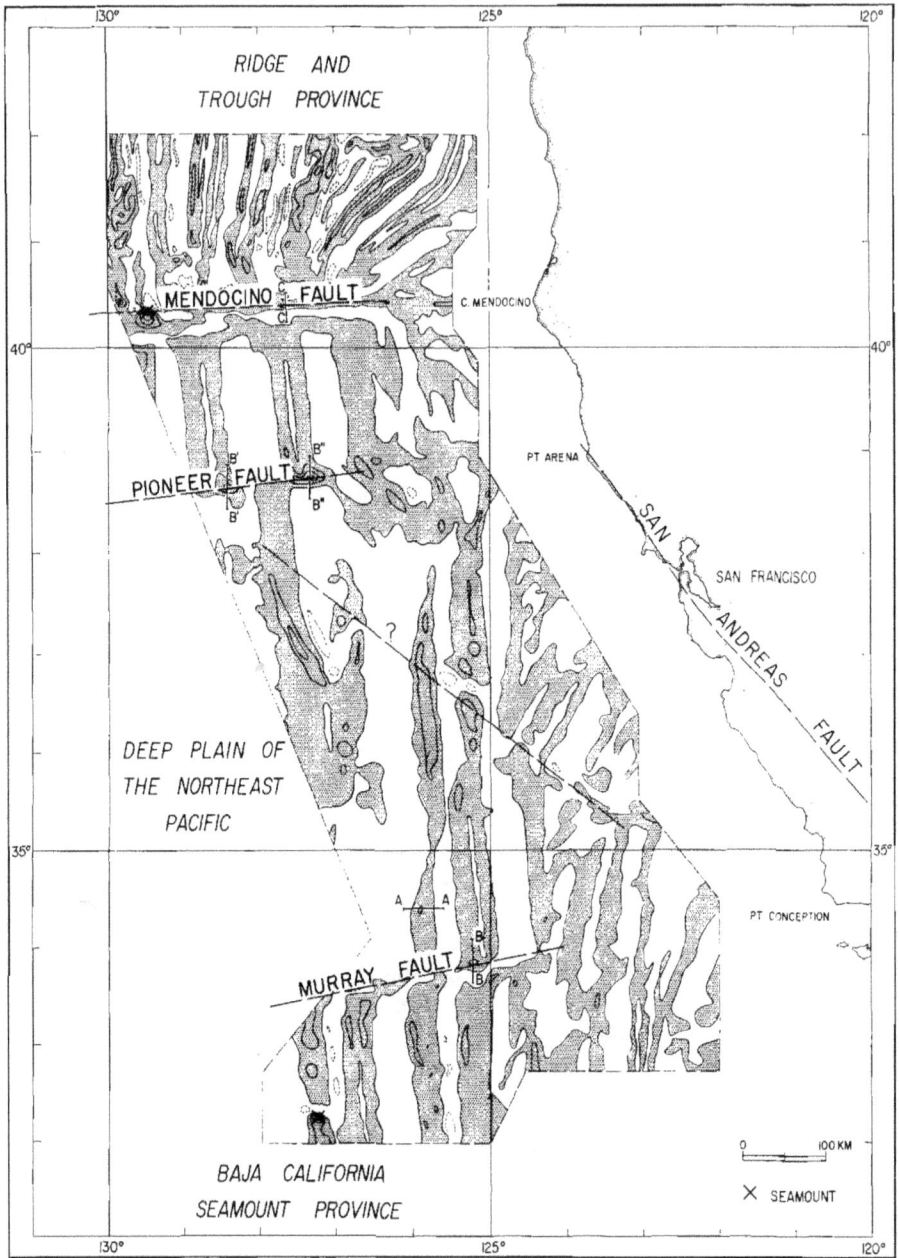

Figure 22.10   Zebra pattern of magnetic stripes on the seafloor, Ronald Mason and Arthur Raff, 1961

Source: Courtesy of the Geological Society of America.

Figure 22.11 Delineation of the strata of England and Wales with part of Scotland, William Smith, 1815
Source: Wikimedia.

WEATHER CHART, MARCH 31, 1875.

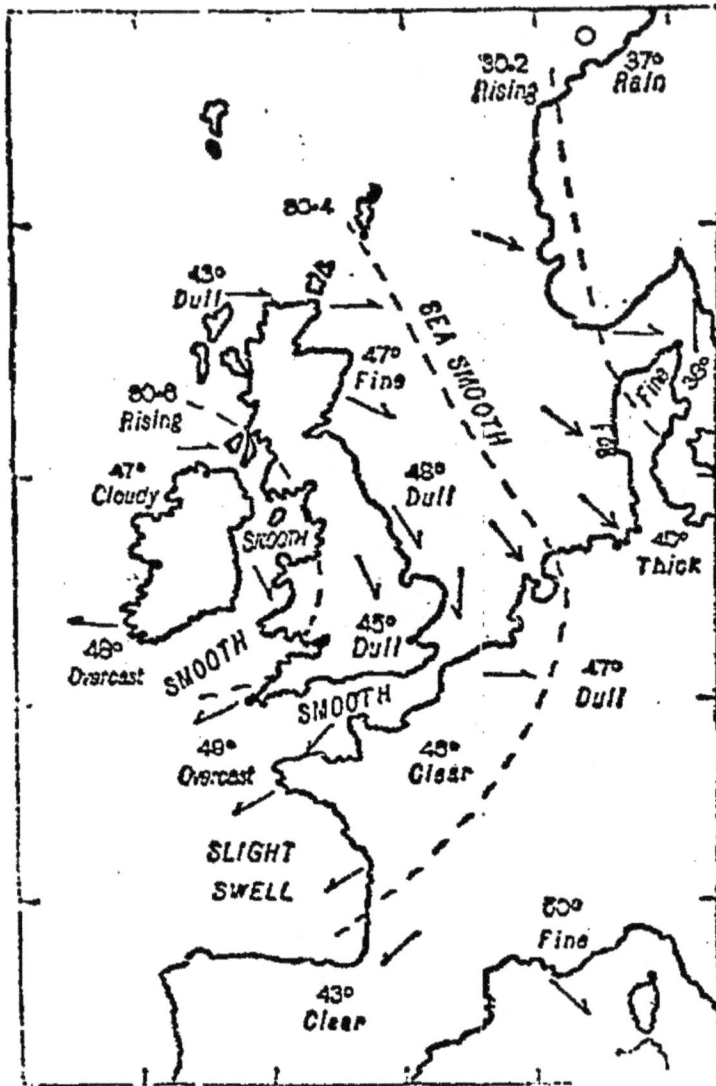

Figure 22.12   Weather chart, Francis Galton, 1875

Source: Galton, F. (1875) 'The Weather', *Times*, p. 12, April 1. *The Times Digital Archive*.

The influencers

Scottish political economist William Playfair (1759–1823) is often credited with being the father of modern statistical graphics. As just shown, data-based visualizations emerged in the sciences and engineering before Playfair, but they were narrowly deployed. Given Playfair's proximity to science via his brother John Playfair and his assistantship to James Watt, he likely was exposed to some of these early visualizations and saw great potential in them. Seeking clarity in a time of political upheaval, Playfair applied these techniques to economic data. He published *The Commercial and Political Atlas* (1786), an at-the-time novel collection of the most common graphical forms (i.e. line graphs and bar charts) in use today, and later *The Statistical Breviary* (1801), containing possibly the first pie chart. (See Figure 22.13.)

Showing their potential for visual thinking, influence, communication, and beauty, these graphical forms proliferated after being deployed by Playfair. In particular, they circulated back into the natural sciences where they continued to be extended and modified. (Kostelnick and Hassett, 2003; Wainer, 2005).

American mathematician John Tukey (1915–2000) is often recognized as the father of exploratory data analysis, the process of performing initial visual investigations on collected data with the help of simple data visualizations. Both his approach to data analysis and his varied contributions (e.g. stem-and-leaf plot, multiple comparisons, ANOVA, interactive and multivariate graphics) are credited with "mak[ing] graphical data analysis both interesting and respectable again" (Friendly, 2006, p. 23). Like Playfair, his work influenced the sciences, perhaps none more so than his adaptation of the box-and-whiskers plot. The basic boxplot form was established in the early 1950s by data visualization pioneer Mary Spear; it was one of the few plot types invented in the twentieth century to be taken up more widely following Tukey's use of it in the late 1970s (Jones, 2019).

Plotting medical data for social and political change

In 1858, the British nurse and statistician Florence Nightingale published the "Diagram of the Causes of Mortality in the Army in the East." This polar-area graph (also known as a rose chart or coxcomb diagram) uses a cyclic model of time. (See Figure 22.14.)

While an earlier example of a polar-area chart exists in meteorology, it became a landmark data visualization in medicine because it visually persuaded the British government of the importance of sanitation and better battlefield care with the help of an empirically driven data collection. The use of color and space clearly communicates that the majority of deaths were not at the hands of the enemy; rather, soldiers were dying from preventable factors, like infectious diseases and lack of hygiene.

Another example that is often mentioned is Dr. John Snow's dot map of the 1854 London cholera outbreak. While neither the pivotal discovery tool nor the first of its kind as it is often depicted, the map was used by Snow after the outbreak to try to persuade the medical establishment of his previously published theory of how cholera spread through water (Johnson, 2006). The map serves as a notable part of the foundation for modern epidemiological mapping, visually situating the spread of infectious disease in space and time in order to argue for public health changes.

Organizing the astronomical and the atomic

In the early twentieth century, data visualizations became increasingly popular in the sciences, leading to a number of new insights, discoveries, and theories in astronomy, physics, biology, and other sciences. Among these is one of the more famous data visualizations in astronomy, the Hertzsprung-Russell diagram. Devised by the independent efforts of two astronomers, Ejnar Hertzsprung and Henry Norris Russell, this scatterplot summarizes the relationship between stars' temperature and luminosity.[4] (See Figure 22.15.)

At the time, astronomers had thought it was just as likely for a star to be hot but dim or hot but bright as it was to be cool and bright. But the Hertzsprung-Russell diagram showed that the properties of stars were not haphazard. The visual relationships shown in the diagram would aid astronomers' thinking about stellar evolution, laying the groundwork for modern astrophysics.

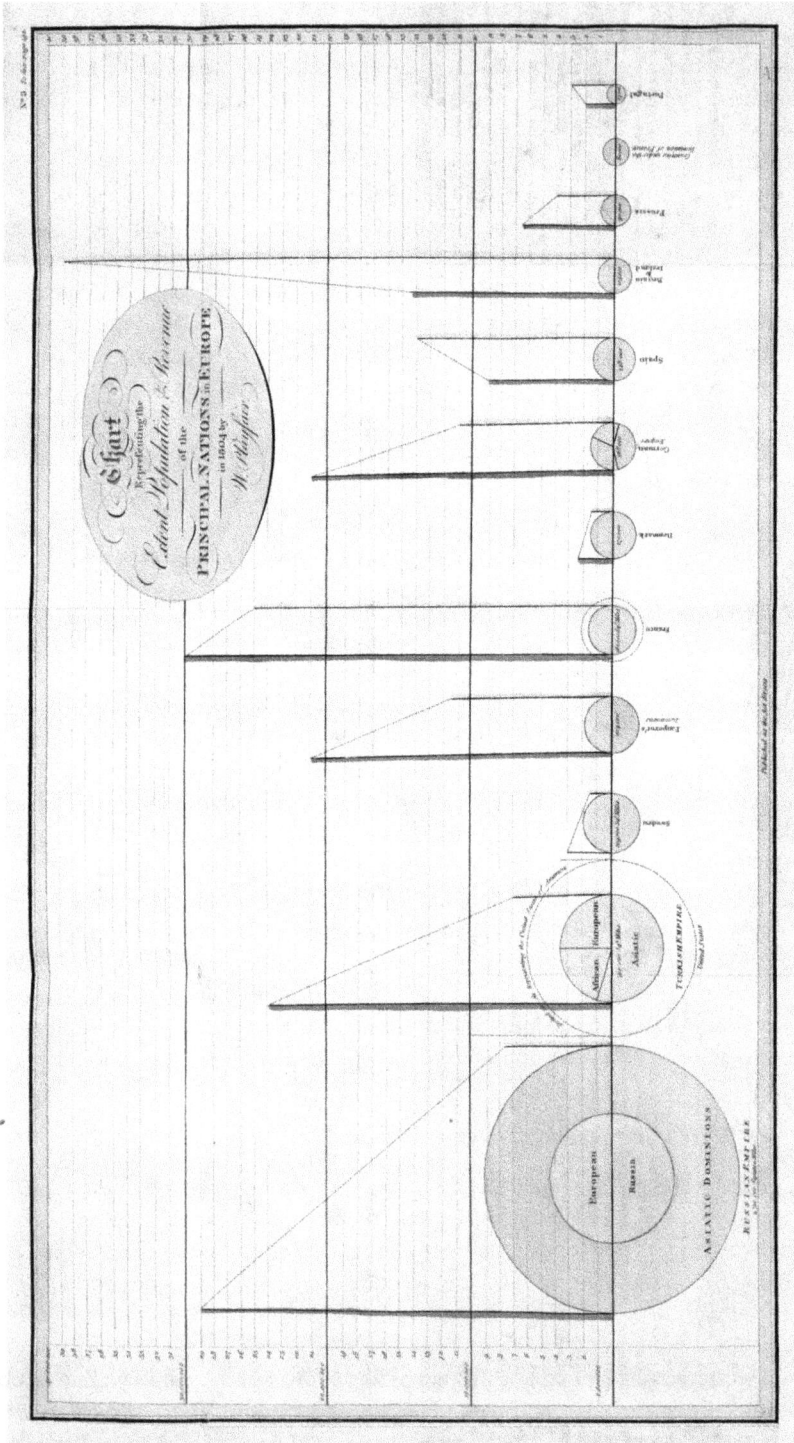

Figure 22.13 "Chart Representing the Extent, Population, and Revenue of the Principle Nations in Europe in 1804," William Playfair, 1805

Source: Playfair, W. (1805) *An Inquiry into the Permanent Causes of the Decline and Fall of Powerful and Wealthy Nations.* Greenland & Norris, London. Courtesy of the Thomas Fisher Rare Book Library, University of Toronto.

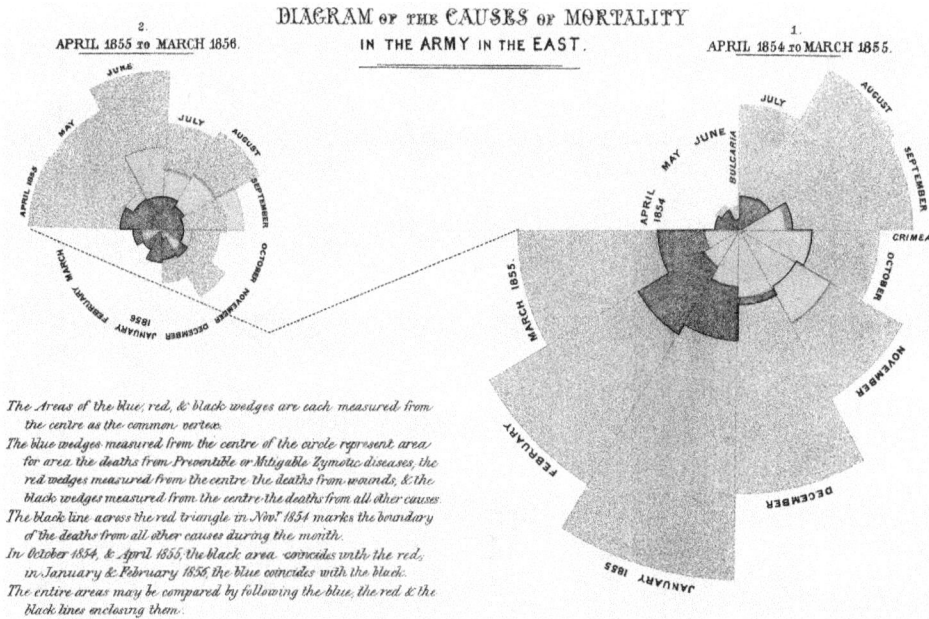

Figure 22.14 "Diagram of the Causes of Mortality in the Army of the East," Florence Nightingale, 1858
Source: Wikimedia.

The concept of the atomic number is another significant graphically influenced discovery. Since Mendeleev, elements in the periodic table had been arranged by atomic mass, but in 1913, British chemist Henry Moseley proposed that it instead be arranged by atomic number. Moseley supported this proposal by graphing his collected X-ray spectroscopy data. This graph showed a linear relationship between atomic number and the measurable property of the nucleus. From the visual gaps on the graph, Moseley also predicted the existence of three elements that would later be discovered.

Current trends and potential future directions

As Friendly (2006) notes, providing an overview of the most recent developments in data visualization can be quite challenging because they are so varied, and they have occurred at an accelerated pace and across a wider range of disciplines. With this disclaimer, I will highlight three themes in the remainder of this chapter:

Big data and technological advances

Perhaps unsurprisingly, the biggest development in the last 30–40 years has been the advent of computing. Initially, computers were the sole preserve of large, well funded research groups studying "big" problems like how stars form, weather forecasting, or simulating nuclear explosions. The invention of personal computers further changed scientific practice and communication, making science more visual and more immediate. Indeed, the ubiquity of computers, the growing programming literacy, the easy creation and modification of visuals,

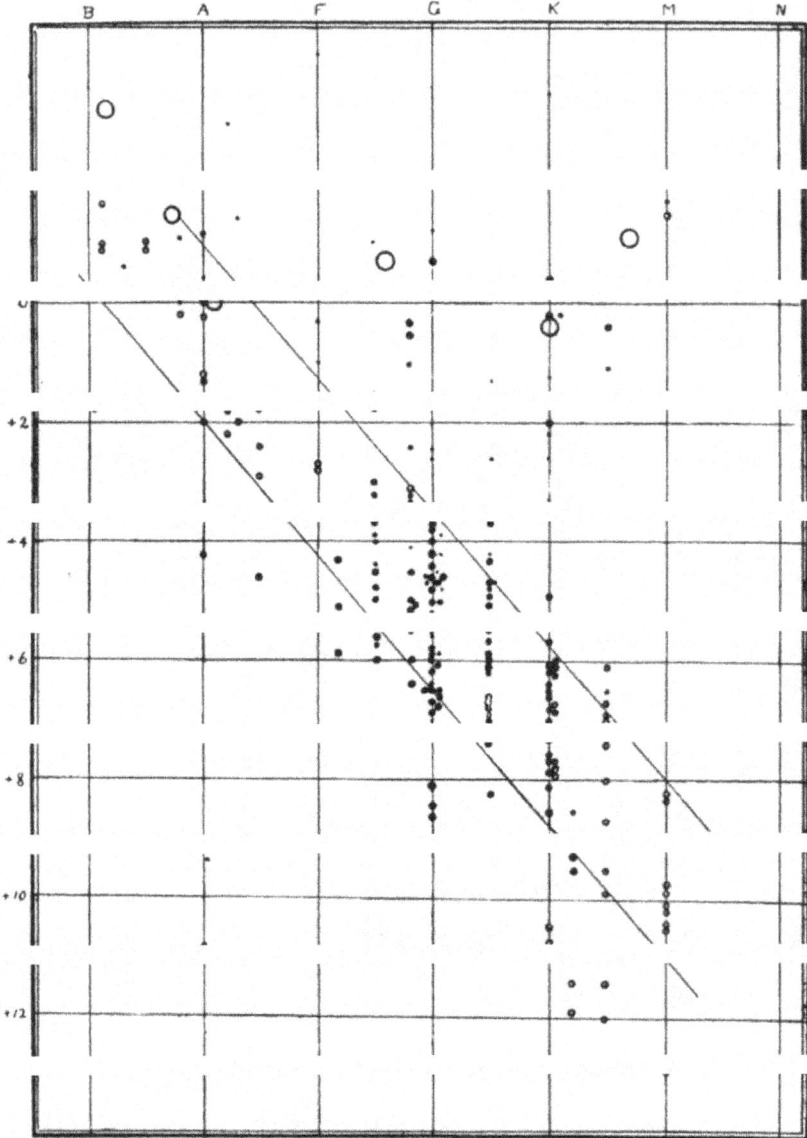

Figure 22.15 One of the original scatterplots for the Hertzsprung-Russell diagram, HN Russell, 1914

Source: Reprinted by permission from Springer Nature. *Nature*. "Relations Between the Spectra and Other Characteristics of the Stars." Russell, HN (1914).

and the wealth of data unleashed at the turn of the twenty-first century have undoubtedly impacted (and, in many ways, democratized) data visualization.

Technological developments offer the opportunity to collect data with increased detail and on a larger scale. Scientists are also working with data in more ways than ever before. For example, seeking to represent more relations between more dimensions of data than is possible with one-dimensional bar charts and two-dimensional scatterplots. Here, too, technological developments for data analysis, like algorithms and increased computing power, are crucial in helping sort through gigabytes or terabytes of data and automate tasks like classification and correlation. In all of this, data visualization has been recognized as key in making sense of big data and making it approachable. As Lima (2011) argues, until recently, visualization has primarily been "a substantiated filter of relevance, disclosing imperceptible patterns and hidden connections . . . [It] will become imperative not solely as a response to the growing surge of data, but also as a supporting mechanism to the various political, economic, cultural, sociological, and technological advances shaping the coming years" (p. 245).

With these advances comes a number of possible opportunities. One area is crowdsourcing and citizen science. While these are not new practices in the history of science, thanks to new technology like wearable devices, smartphones, the Web, and cloud computing platforms, it is easier than ever for individuals to take active roles in their communities by collecting data. Here again, data visualization provides an opportunity to help deal with the flood of data generated. Take, for example, the Safecast map produced with data from roughly 900 volunteers (at the end of 2016). Armed with DIY Geiger counters, these volunteers collected finer-grained radiation contamination measurements than the Japanese government was providing for months following the Fukushima nuclear reactor meltdown. These data were processed by a team of U.S.-based programmers and engineers to create a dynamic, open-access visualization of radiation risk. (See Figure 22.16.)

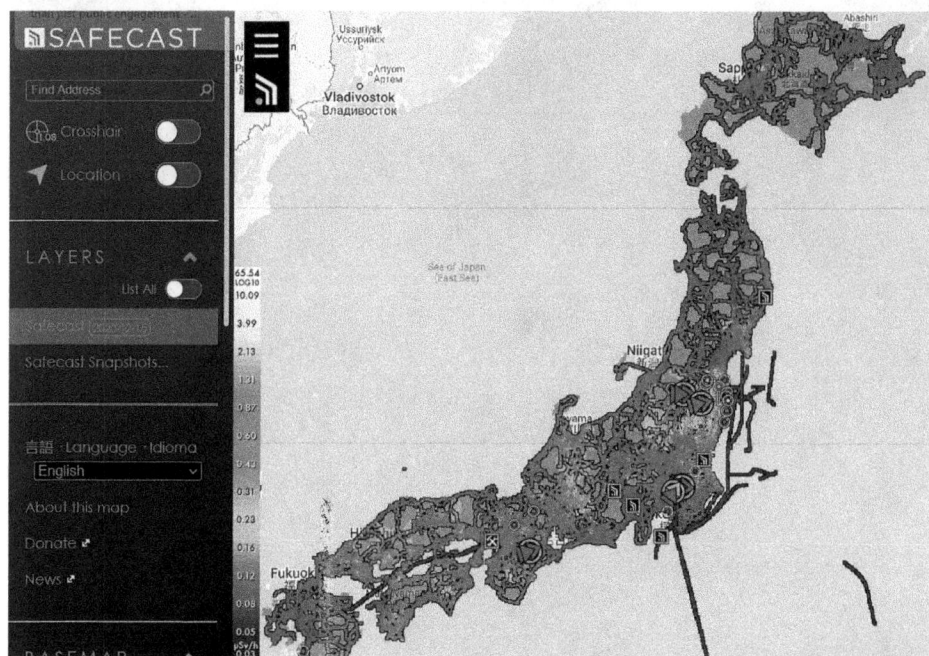

Figure 22.16   Screenshot from Safecast web map, 2020

Source: https://map.safecast.org/, licensed under CC BY-SA 3.0.

Networked and interactive visualization

Driven by developments in computing power, storage, and availability of data, scientists in many fields are now applying methods of analysis, modeling, and simulation, such as network visualization, in order to tackle problems of an increasingly complex and interconnected nature. Over recent years, network visualization has shed light on an incredible array of scientific subjects, including protein interaction, genes, disease spread and correlation, brain function, seismic activity, and ecological networks. Compared to more traditional forms, network visualizations help to convey the depth of big data because they enable multivariate analysis (Lima, 2011). One recent example is a set of three interactive network visualizations of the cosmic web (Center for Complex Network Research, 2016). Using data from 24,000 galaxies, an interdisciplinary group of researchers constructed multiple models of how the galaxies might be connected in order to better understand how the universe might be structured. (See Figure 22.17.)

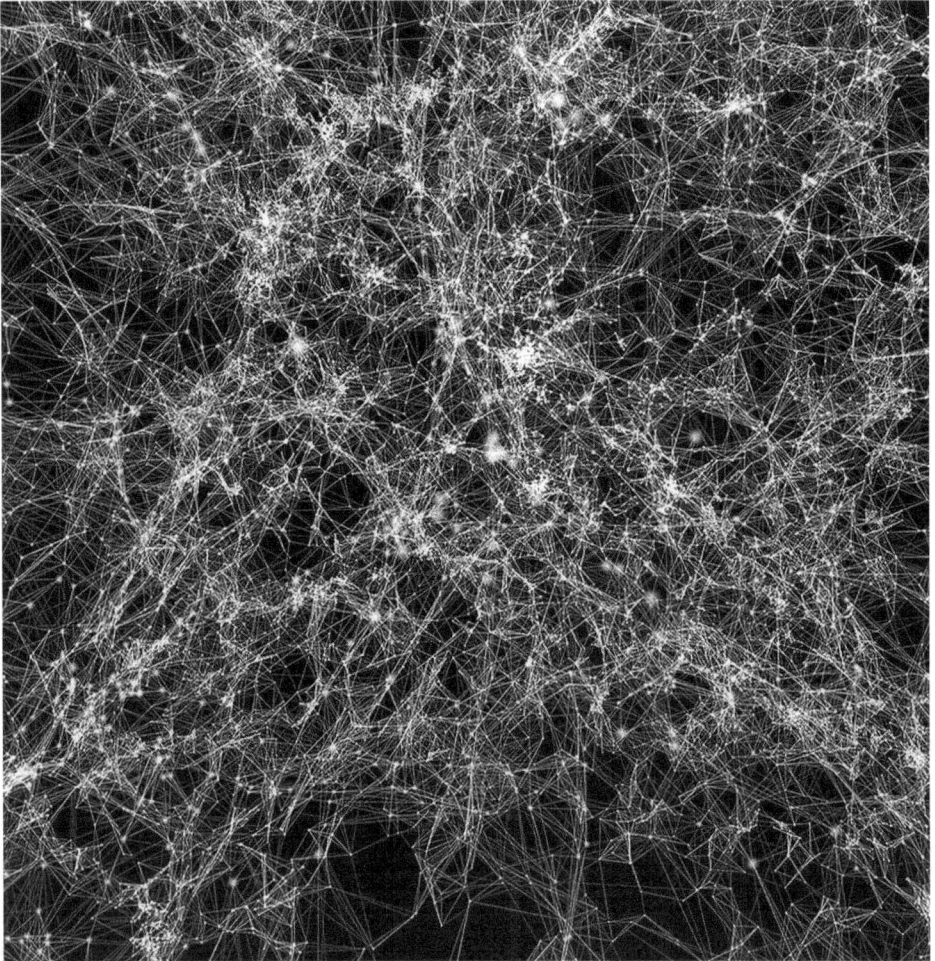

Figure 22.17   Screenshot from Cosmic Web, Kim Albrecht, 2016
Source: www.kimalbrecht.com/vis/.

This example (as well as the enthusiasm for network visualizations more generally) also highlights two additional trends. First is an increasingly collaborative approach to data visualization. The creation of the cosmic web visualization just discussed is credited to no less than eight interdisciplinary collaborators. Second is a more overt emphasis on design and aesthetics (not just visuals that strive for clarity and efficiency) and their entwinement with science (Lima, 2011). Indeed, a number of efforts, like NASA's Scientific Visualization Studio, are already exploring how a greater attention to aesthetics in visualization and mapping can provide new perspectives on critical ecological interactions.

Humanizing data visualization

While there is a lot of excitement around data visualization, along with many new tools for doing it, there are also calls to think more critically about the politics and ethics of representation. As these calls recognize, data visualization is a form of power, a power derived from its generalized, scientific, visually based, seemingly neutral nature. As shown in this chapter, it has been used for good (e.g. to improve health outcomes), but it has also been used to dehumanize, police, and exclude. Take the dominant narratives around data visualization – they are overwhelmingly Western, white, and male. In other words, we tend to imagine that our dashboards show us the whole picture, but they often don't.

Recognizing this fact, one trend has been to ask how to represent more responsibly. One answer is expanding the scope of visual ethics beyond issues of distortion and deception to "develop a genuine sensitivity to the human implications" of data visualization and incorporate human elements into their design (Dragga and Voss, 2001, p. 266; Lupi, 2017). Another response has been to apply feminist approaches to develop a range of alternative visualization practices that emphasize the situated, subjective nature of the claims made with data. For example, one way is to create data visualizations that represent limitations, reference the material economy behind the data (i.e. who gathered and funded it, how and why it was gathered), and facilitate interaction and dissent (D'Ignazio and Klein, 2016). Feminist data visualization has also emphasized the recovery of elided work in the history of data visualization – for example, Almira Hart Lincoln Phelps (1793–1884) and her illustrator Thirza Lee's adaptation of Alexander von Humboldt's *Naturgemälde* for their textbook, *Familiar Lectures on Botany*. (See Figure 22.18.)

While Humboldt's visualization is notable in its own right, Phelps and Lee's work is significant for its innovation in style and method. Namely, their version presents an alternative approach to botanical data that values affective response. Their visualization seeks to synthesize intellectual and emotional connections to the world, bringing material experience and sensation to bear in a way that was different from the approaches of canonical figures, like William Playfair, or the dominant data visualization narratives (Klein, 2014).

Conclusion

We have arrived at the end of our brief venture into the history of data visualization. The objective of this chapter has been to focus on a particular slice of that history as it pertains to the sciences, highlighting specific moments, exigencies, examples, and people in its development. As these examples show, data visualization, in all its varied forms and applications, has long been a powerful means for making sense of scientific, medical, and technological data, for advancing arguments, and for communicating information to others. Its history is entangled with other advances in philosophy, technology, data collection, and visual thinking. This history shows

Figure 22.18 "Top of Chimborazo 21000 Feet from the Level of the Sea," Almira Hart Lincoln Phelps, 1831

Source: Courtesy of the American Antiquarian Society.

that, while data visualization is undeniably malleable and fundamental, at the same time we need to be aware of its practical and ethical complexities. Hopefully, this overview will seed an appreciation for the rich history behind the current enthusiasm for data visualization, informing the work at hand and inspiring us to consider what other choices might be made and how they might be designed differently.

Notes

1 The history of data visualization as it is most often told is heavily western, white, and male-centric. For Joseph Needham, the great historian of Chinese science, the *Yu ji tu* revealed "the extent to which Chinese geography was at that time ahead of the West."

2 See Daston and Gallison (2007) for an extended history of scientific objectivity.

3 The selected data visualizations should not necessarily be considered "milestones," which implies a sense of linear progression or as the first examples of a particular form, application, etc. As the reader will note, the examples in this section are not presented strictly chronologically.

4 While line and bar charts are far more common in journalism and business, the scatterplot dominates science journals. Statistician Edward Tufte once estimated that more than 70% of all charts in scientific publications are scatterplots. It's no surprise, then, that what may be the earliest known scatterplot was created by a scientist, British astronomer Sir John Herschel (Friendly and Denis, 2005).

References

Allen, W.L. (2018) 'Visual brokerage: communicating data and research through visualization', *Public Understanding of Science*, vol. 27, no. 8, pp. 906–922, DOI: https://doi.org/10.1177/0963662518756853

Azzam, T., Evergreen, S., Germuth, A.A., & Kistler, S.J. (2013) 'Data visualization and evolution', in T. Azzam & S. Evergreen (eds.), *Data visualization, part 1*, San Francisco: Jossey-Bass, pp. 7–32.

Barrow, J.D. (2008) *Cosmic imagery: Key images in the history of science*, London: Bodley Head.

Blakemore, E. (2016) 'Seeing is believing: How Marie Tharp changed geology forever', *The Smithsonian Magazine*, www.smithsonianmag.com/history/seeing-believing-how-marie-tharp-changed-geology-forever-180960192/

Bucchi, M., & Saracino, B. (2016) 'Visual science literacy: images of public understanding of science in the digital age', *Science Communication*, vol. 38, no. 6, pp. 812–819, DOI: https://doi.org/10.1177/1075547016677833

Center for Complex Network Research (2016) 'The network behind the cosmic web', http://cosmicweb.barabasilab.com/

Datson, L., & Gallison, P. (2007) *Objectivity*, Cambridge: MIT Press.

Desnoyers, L. (2011) 'Toward a taxonomy of visuals in science communication', *Technical Communication*, vol. 58, no. 2, pp. 119–134.

D'Ignazio, C., & Klein, L.F. (2016) 'Feminist data visualization', IEEE VIS Conference, www.kanarinka.com/wp-content/uploads/2015/07/IEEE_Feminist_Data_Visualization.pdf

Dragga, S., & Voss, D. (2001) 'Cruel pies: The inhumanity of technical illustrations', *Technical Communication*, vol. 48, no. 3, pp. 265–274.

FEMA. (2015) 'Earthquake information for public policy makers and planners', www.fema.gov/earthquake-information-public-policy-makers-and-planners

Ferdio (n.d.) 'The data visualization project', datavizproject.com

Frankel, F., & DePace, A.H. (2012) *Visual strategies: A practical guide to graphics for scientists and engineers,* New Haven: Yale University Press.

Friendly, M. (2006) 'A brief history of data visualization', in C. Chen, W. Hardle, & A. Unwin (eds.), *Handbook of computational statistics,* Berlin: Springer-Verlag.

Friendly, M. (2008a) 'Milestones in the history of thematic cartography, statistical graphics, and data visualization', www.math.yorku.ca/SCS/Gallery/milestone/milestone.pdf

Friendly, M. (2008b) 'The golden age of statistical graphics', *Statistical science*, vol. 23, no. 4, pp. 502–535, DOI: 10.1214/08-STS268

Friendly, M., & Denis, D. (2005) 'The early origins and development of the scatterplot', *Journal of the History of the Behavioral Sciences*, vol. 41, no. 2, pp. 103–130, DOI 10.1002 /jhbs.20078

Friendly, M., & Palsky, G. (2007) 'Thematic maps and diagrams', in J.R. Akerman & R.W. Karrow (eds.), *Maps: finding our place in the world,* Chicago: Chicago University Press, pp. 205–251.

Grainger, S., Mao, F., & Buytaert, W. (2016) 'Environmental data visualization for non-scientific contexts: Literature review and design framework', *Environmental modelling & software*, vol. 85, pp. 299–318, DOI: https://dx.doi.org/10.1016/j.envsoft.2016.09.004

Gross, A.G., & Harmon, J.E. (2013) *Science from sight to insight*, Chicago: University of Chicago Press.

International Federation of Red Cross. (2011) 'Public awareness and public education for disaster risk reduction: A guide', www.ifrc.org/Global/Publications/disasters/reducing_risks/302200-Public-awareness-DDR-guide-EN.pdf

Johnson, S. (2006) *The ghost map*, New York: Riverhead Books.

Jones, B. (2019) 'Credit where credit is due: Mary Eleanor Spear', *Nightingale*, https://medium.com/nightingale/credit-where-credit-is-due-mary-eleanor-spear-6a7a1951b8e6

Klein, L.F. (2014) 'Feminist data visualization [presentation]', HUMlab Umea universitet, http://stream.humlab.umu.se/?streamName=feminist_data

Kostelnick, C. (2007) 'The visual rhetoric of data displays: The conundrum of clarity', *IEEE Transactions on Professional Communication*, vol. 50, no. 4, pp. 280–294, DOI: 10.1109/TPC.2007.908725

Kostelnick, C., & Hassett, M. (2003) *Shaping information: The rhetoric of visual conventions*, Carbondale: Southern Illinois University Press.

Kostelnick, C., McDermott, D., Rowley, R.J., & Bunnyfield, N. (2013) 'A cartographic framework for visualizing risk', *Cartographica*, vol. 48, no. 3, pp. 200–224, DOI: 10.3138/carto.48.3.1531

Krause, K. (2017) 'A framework for visual communication at *Nature*', *Public Understanding of Science*, vol. 26, no. 1, pp. 15–24, DOI: https://doi.org/10.1177/0963662516640966

Lima, M. (2011) *Visual complexity: Mapping patterns of information*, New York: Princeton Architectural Press.

Lupi, G. (2017) 'Data humanism', *Print*, 30 January, www.printmag.com/information-design/data-humanism-future-of-data-visualization/

Mason, R.G., & Raff, A.D. (1961) 'Magnetic survey off the west coast of North America, 32° n. latitude to 42° n. latitude', *Geological society of America bulletin*, vol. 72, pp. 1259–1266, DOI: 10.1130/0016-7606(1961)72[1267:MSOTWC]2.0.CO;2

Pauwels, L. (2006) *Visual cultures of science: rethinking representational practices in knowledge building and science communication*, Hanover: Dartmouth College Press.

Playfair, W. (1786) *The commercial and political atlas*, London: T. Button.

Playfair, W. (1801) *The statistical breviary: Shewing, on a principle entirely new, the resources of every state and kingdom in Europe*, London: T. Bensley.

Rodríguez Estrada, F.C., & Davis, L.S. (2015) 'Improving visual communication of science through the incorporation of graphic design theories and practices into science communication', *Science communication*, vol. 37, no. 1, pp. 140–148, DOI: https://doi.org/10.1177/1075547014562914

Ruivenkamp, M., & Rip, A. (2010) 'Visualizing the invisible nanoscale study', *Science studies*, vol. 23, no. 1, pp. 3–36, DOI: https://doi.org/10.23987/sts.55255

Tufte, E. (1997) *Visual explanations*, Cheshire: Graphics Press, 2nd edition.

Tufte, E. (2001) *The visual display of quantitative information*, Cheshire: Graphics Press, 2nd edition.

Von Baeyer, H.C. (2004) *Information: The new language of science*, Cambridge: Harvard University Press.

Wainer, H. (2005) *Graphic discovery: A trout in the milk and other visual adventures*, Princeton: Princeton University Press.

Wainer, H., & Friendly, M. (2020) 'The graphical birth of plate tectonics', *Chance*, vol. 33, no. 1, pp. 44–47.

Winchester, S. (2001) *The map that changed the world: William Smith and the birth of modern geology*, New York: Harper Perennial.

23

STATISTICS AND DATA VISUALIZATION

Vaclav Brezina and Raffaella Bottini

Introduction

This chapter is about numbers and graphs and their various functions in scientific communication. While numbers help us quantify reality and bring precision to scientific discourse, the primary role of data visualization is abstraction – highlighting main trends in a dataset. Let us look at a simple example from the history of science to illustrate the role of statistics and visualization. In the seventeenth century, a number of new scientific discoveries were made as empirical science emerged (e.g. Ben-Chaim, 2017). These discoveries were enabled by the invention of scientific tools such as the telescope, the microscope, and the thermometer. The frequencies of the terms 'microscope(s)'/'telescope(s)'/'thermometer(s)' in the body of preserved writings from the seventeenth century indicate a clear trend of a dramatic increase from circa the 1650s. With the aid of corpus linguistic techniques, we can measure with precision the mentions of these scientific tools per every million words of preserved text. These measurements are represented by a long line of numbers (frequencies), which stretch over multiple lines:

0, 0, 0, 0, 0, 0, 0, 0, 0, 0, 0, 0, 0, 0, 0, 0, 0, 0, 0, 0, 0, 0, 0, 0, 0, 0, 0, 0, 0, 0, 0, 0, 0, 0, 0, 1, 0, 0, 0, 0, 0, 0, 0, 0, 0, 0, 0, 0.1, 0, 0.1, 0, 0.16, 0.17, 0.4, 0.76, 0.57, 0.33, 0.37, 0.7, 0.77, 0.31, 0.42, 3.89, 1.3, 1.25, 6.49, 6.89, 9.64, 0.34, 5.22, 1.5, 15.66, 10.72, 1.12, 3.36, 5.78, 6.37, 3.23, 6.35, 1.97, 7.87, 0.44, 1.55, 4.52, 1.08, 6.16, 1.53, 5.71, 3.64, 8.48, 0.9, 4.14, 1.55, 19.05, 1.11, 7.76, 1.71, 2.54, 3.1, 6.89, 2.31

The same information can be visualized using a small line graph called the sparkline ( ). The sparkline shows the overall trend in the data with the minimum and the maximum values highlighted. In this particular case, a visual display of the size of a word captures 100 different data points and allows us to make a general sense of the data, something that is crucial to the understanding of the gist of empirical evidence. The sparkline also demonstrates another aspect of successful visualization, which is information-rich presentation: in our example, in a very small space, 100 data points are displayed, making the sparkline a very efficient type of graph.

In contrast to words and numbers, data visualization has one important characteristic, which is iconicity (Stjernfelt, 2000). A graph resembles data showing trends such as increase or decrease, sizes, and proportions. In Figure 23.1, the line (upward slope) shows a steady increase

DOI: 10.4324/9781003043782-27

up to point 3 and then acceleration in the increase, while the height of the bars is indicative of the values they represent.

The iconic character of graphs and other types of visualization has a profound effect on how visualization works with space. It is important to realize that words and numbers, which are much less iconic and more arbitrary in representing reality, can be ordered on a page contiguously, one next to another, potentially infinitely, adding more detail and information. Graphs, on the other hand, inhabit the space of a page not as an unfolding string of symbols but as one visual entity. Consider this: conceptually, it is possible to perceive all of Figure 23.1 at once, while reading about it in this paragraph involves processing multiple words at different stages. This property of visualization is both its blessing and its curse.

On the positive side, effective data visualization allows us to abstract by providing information about the overall pattern in the data. It answers the question of what is the gist? Thus it occupies a central place in scientific communication as a pathway to theory. On the other hand, data visualization is restricted by the space of the page, be it a printed page or a screen. Tufte (2006) calls this space a 'flatland'. Flatland has two dimensions, the x-axis and the y-axis, which are in tension with typically multidimensional data. The art (and science) of data visualization thus lies in successfully resolving this tension and providing the reader with the best access to the data. This chapter outlines some of the solutions of successful visualization starting with a historical overview, followed by general principles and practical techniques.

Visualization across disciplines and a historical overview

Visual representation of statistical information has been used extensively in many disciplines with differences in the preferred 'to go' techniques, reflecting the specific needs of each discipline and different research traditions. Cross-disciplinary comparisons can thus be a useful tool for researchers to learn from one another across the disciplinary spectrum, sometimes finding inspiration in unlikely corners of the scientific practice. In this section, we provide references to the interdisciplinary debate and examples from the history of scientific visualization in different

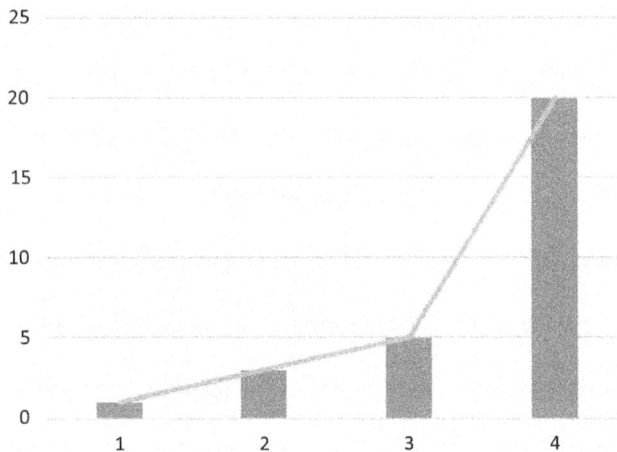

Figure 23.1   Graph demonstrating iconicity

fields. The use and development of graphs have been analyzed across disciplines (e.g. Cleveland, 1984; Wang and Tao, 2017), from environmental science (e.g. Grainger et al., 2016) to psychology (e.g. Best et al., 2001) and sociology (e.g. Healy and Moody, 2014). The principles of data visualization – functions, features, and effectiveness of graphical displays – have also been described in detail (e.g. Heer et al., 2010; Wainer and Velleman, 2001). Specifically, Tufte (2001) and Few (2012) examined human perception of visual displays and ways for graphs and images to communicate scientific results in an effective way. Borkin et al. (2013) investigated how memorable different types of visualizations are depending on their graphic attributes. Recent advances in technologies have also inspired research, e.g. on interactive maps used for risk assessment (e.g. Seipel and Lim, 2017; Welhausen, 2018). Finally, studies have also investigated visual literacy (e.g. Ruchikachorn and Mueller, 2015) and visual discourses that link data visualization with cultural assumptions, values, and ideologies (e.g. Li, 2019).

This debate is, of course, grounded in the rich history of scientific thought and communication, reflecting the milestones, from simple visual representation of numbers and categorizations, to graphic displays of variables and distributions and the visualization of complex relationships between multiple variables (Wainer and Velleman, 2001). Historical development of visual displays has been well documented in the literature (e.g. Friendly, 2008). Here, we provide a brief historical overview of the milestones of data visualization referring to the examples in Table 23.1.

The first phase of data visualization is linked to spatial measures and consists of maps and geometric diagrams. In ancient Greece, visual representations were drawn in sand to demonstrate geometrical postulates and mathematical proofs (Perini, 2005). For example, Figure 23.2 illustrates how mathematical theorems were displayed in Euclid's *Elements*. One of the earliest graphs which visualized two quantitative variables dates back to the tenth century (Friendly, 2008). The graph, presented in Figure 23.3, plots the inclination of planetary orbits (y-axis) as a function of time (x-axis), and its use of a grid suggests an understanding of coordinates before their introduction by Descartes in 1637. From the thirteenth century, graphical displays of hierarchical structures spread in the form of tree diagrams. As Heer et al. (2010) note, tree diagrams would later evolve to display nodes and links between variables, as shown in Figure 23.12. In the fourteenth century, the study of natural phenomena, such as changes in heat and motion, further inspired graphical displays of data. For instance, Figure 23.4 presents a prototype of the bar graph by the French mathematician Nicole Oresme. As Wainer and Velleman (2001) argue, early visualizations were mainly used to display theoretical and a priori relationships since natural science was based on natural philosophy and adopted an approach to data which was rational rather than empirical.

Graphs based on empirical evidence were produced from the seventeenth century. This phase represents "the beginnings of visual thinking" according to Friendly (2008, p. 6). It is characterized by an interest in physical measurements (e.g. time, distance, and space), new tools such as the telescope and the barometer, and a consequent growth in the collection of empirical data in a number of disciplines, such as astronomy, physics, economics, and demography. Innovations include, for example, Descartes's analytic geometry, Napier's invention of logarithms, and Pascal and Fermat's probability theory. Galileo Galilei's display of Jupiter and its four main moons in Figure 23.5 and Harvey's graphical display of blood vessels in Figure 23.6 exemplify how graphs collate evidence and support scientific discoveries. Two key visual techniques were introduced by Christopher Scheiner in the same period: the display of figures enlarged or reduced in scale and the use of 'small multiples' (Tufte, 2001) based on the repetition of the

Table 23.1 Examples of scientific milestones of data visualization

| Figure 23.2 Geometrical graphic in Euclid's *Elements* [300 BC] (1482) | Figure 23.3 First graph that combines two variables [tenth century] [Funkhouser, 1936, p. 261] | Figure 23.4 Prototype of the bar chart by Nicole Oresme [circa 1350] (Funkhouser, 1937, p. 276) |

Figure 23.7 Scheiner's (1630) graph of observed spots on the Sun

Figure 23.10 Example of curve fitting by Lambert (1779)

Figure 23.6 Harvey's (1628) graphical display of blood vessels

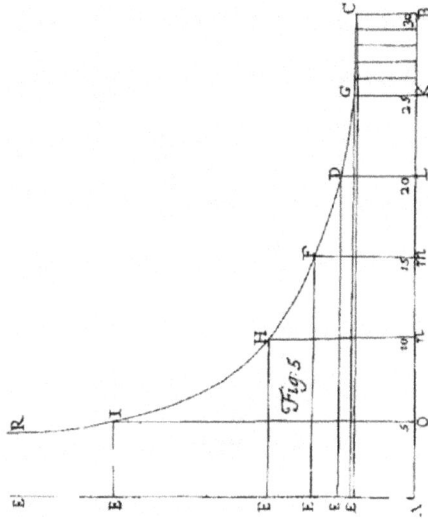

Figure 23.9 Priestley's *Chart of biography* (1764)

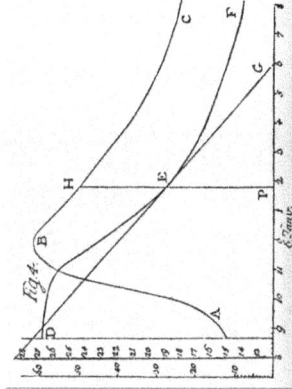

Die decimaquinta , hora noctis tertia in proximè depicta fuerunt habitudine quatuor Stellæ ad Iouem ;

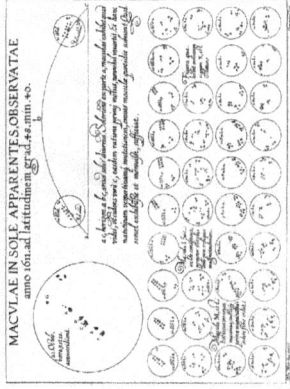

Ori.   O   .   *   *   Occ

Figure 23.5 Graph from Galilei's *Sidereus nuncius* (1610, p. 18)

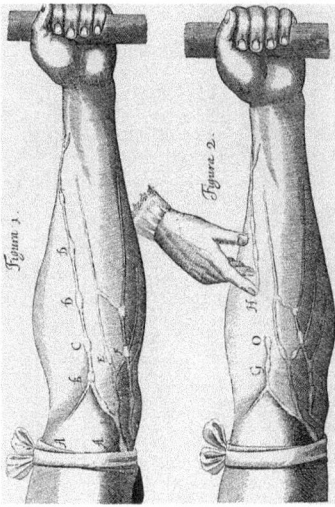

Figure 23.8 Halley's (1686) curvilinear plot of barometric pressure as a function of its distance above sea level

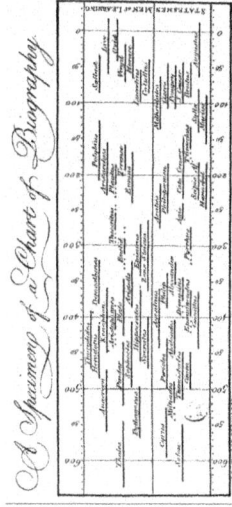

(Continued)

Table 23.1 (Continued)

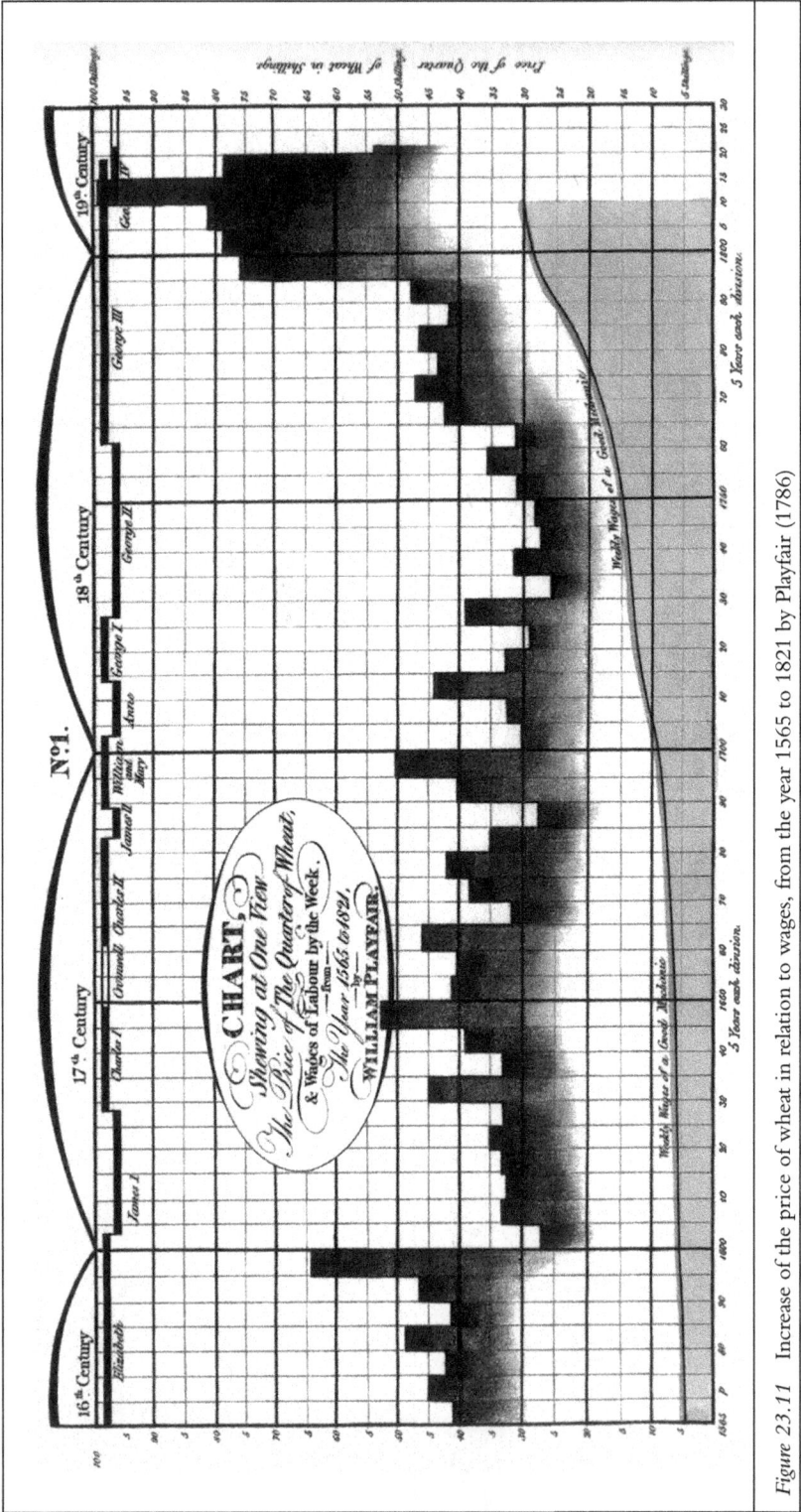

Figure 23.11   Increase of the price of wheat in relation to wages, from the year 1565 to 1821 by Playfair (1786)

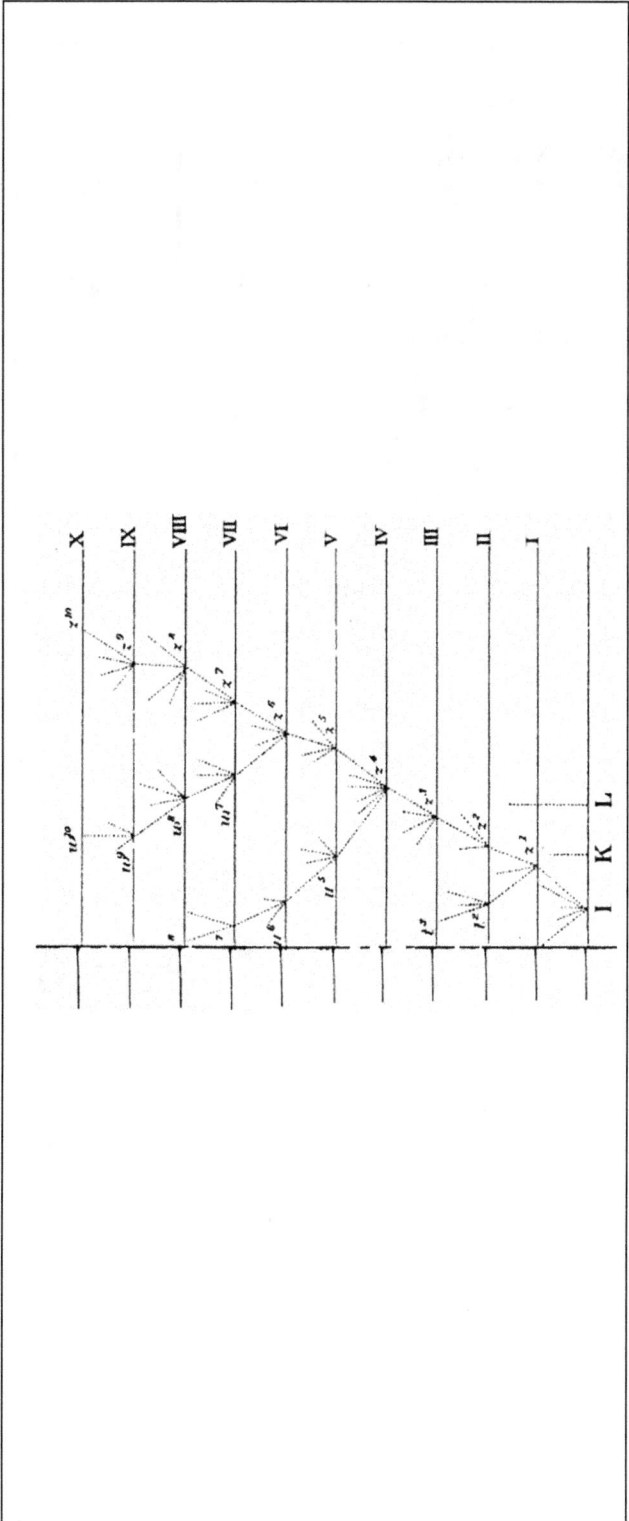

Figure 23.12 Diagram showing the evolution of species in Darwin's *On the Origin of Species* (1859, p. 117)

Table 23.1 (Continued)

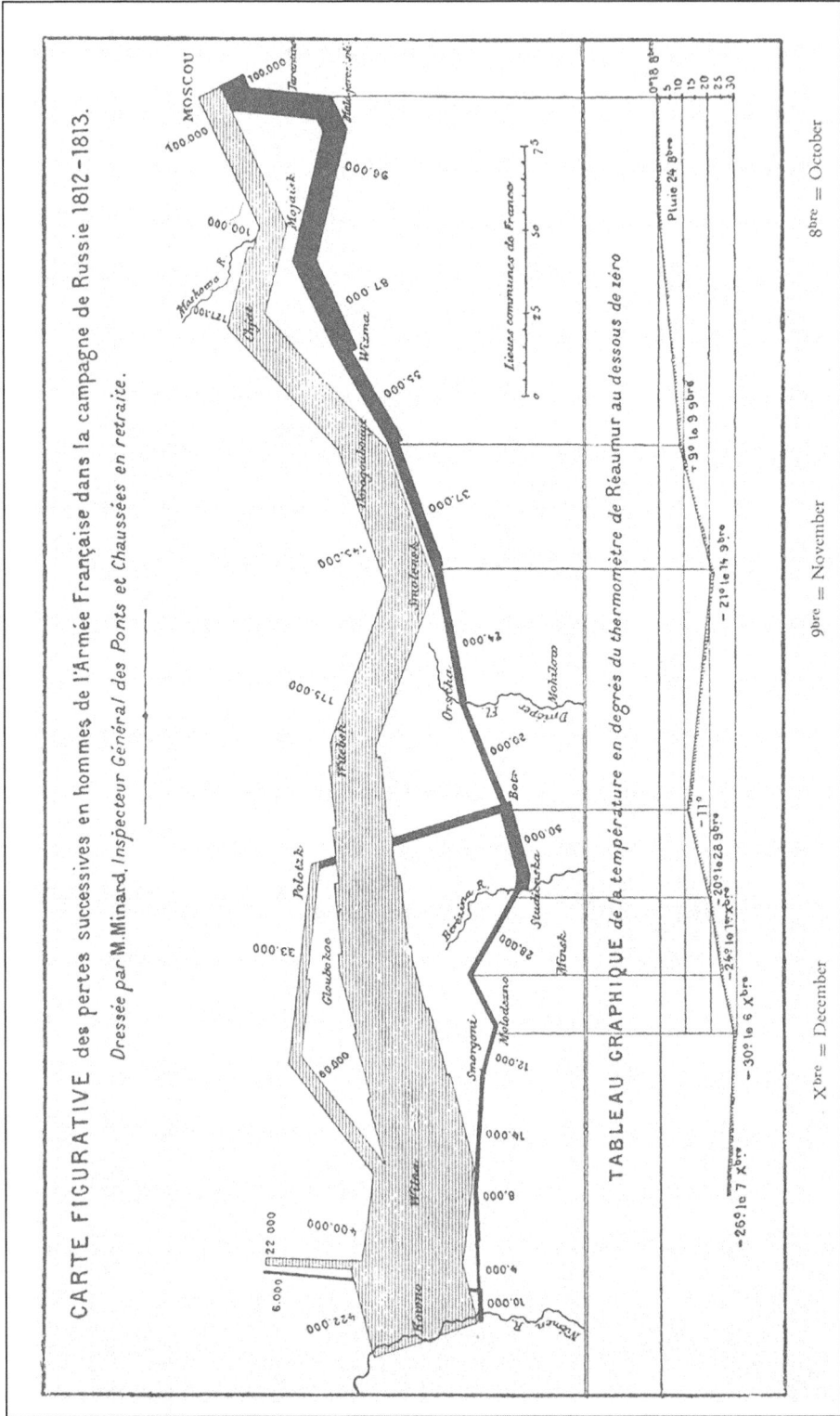

Figure 23.13   Minard's (1861) map showing the losses in men of the French army in the Russian campaign, 1812–1813

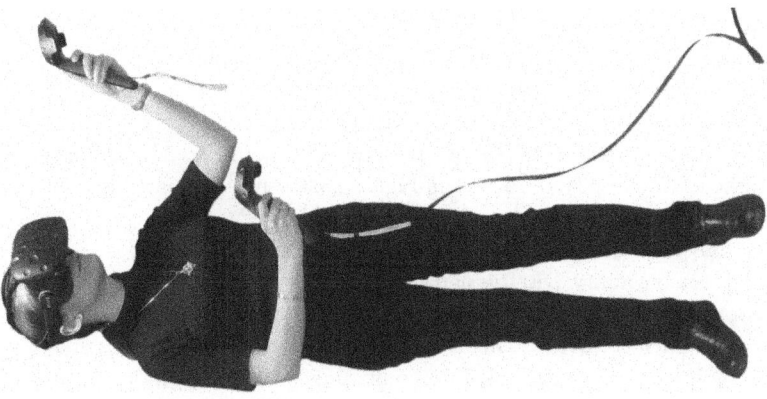

Figure 23.16 Use of virtual reality in data visualization

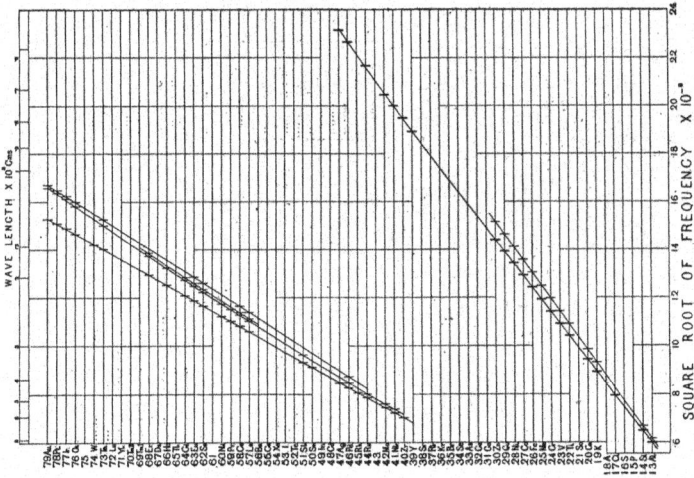

Figure 23.15 High-frequency spectra of the chemical elements (Moseley, 1913, p. 709)

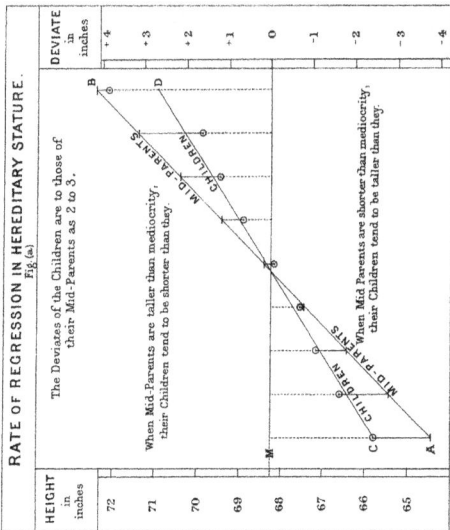

Figure 23.14 Regression in hereditary stature (Galton, 1886, pp. 249–250)

same graphical structure to represent changes of one or more variables across time. For instance, Figure 23.7 presents the surface of the Sun 37 times to show how sunspots change over a month. This facilitates comparisons without displaying all details in one graph, which would clutter individual differences. In the same years, Huygens and Grant visualized continuous variables in life tables that displayed population growth. Graphs were therefore used mainly to visualize single variables and their distribution. Toward the end of the century, relationships between two variables started being displayed, such as in Figure 23.8, which presents the first curvilinear plot showing the relationship between pressure and altitude.

New visual representations and innovations, such as three-color printing and lithography, were introduced in the eighteenth century. Displays of data were encouraged by improvements in scientific communication, such as the first English encyclopedia and the increase in scientific periodicals (Wainer and Velleman, 2001). Thematic maps were used to show trends and patterns of economic, demographic, and medical data, combining quantitative information with their geographical distribution (Friendly and Denis, 2001). Timelines were introduced to annotate historical events and biographies; for example, Priestley visualized time using bars. Figure 23.9 shows his 'Chart of Biography', which contains 2,000 names and their life spans. Lambert was the first to use the idea of curve fitting (Friendly, 2008) to display the relationship between water evaporation and temperature (Figure 23.10). However, most of the graphical innovations of this century were introduced by William Playfair, who designed line, circle, bar, and pie graphs to display multiple variables, therefore being considered by Tufte (2001) the inventor of statistical visualization. Figure 23.11 shows his plot of how the price of wheat changed in relation to wages across the sixteenth and seventeenth centuries in England, combining a bar chart, a line graph, and a timeline of the British monarchs. According to Wainer and Velleman (2001), Playfair applied graphical displays to social sciences such as economics and finance, thus reaching a wide audience and contributing to the popularity of statistical visualizations.

Data visualization continued to rise during the nineteenth century, inspired by Gauss's normal density function and Laplace's Bayesian interpretation of probability. Multivariate plots were extensively used by Minard, who designed the use of flow maps to visualize the changes of a quantitative variable in time and space (Heer et al., 2010). For instance, Figure 23.13 shows Napoleon's campaign in Russia in 1812–1813. The width of the upper line in this graph represents the size of the army at different times and places on the map, while the darker line at the bottom symbolizes the retreat from Moscow showing the reduction in the number of soldiers (Tufte, 2001). Details about the direction of the campaign and temperatures are included in the lower section of the graph. As these examples demonstrate, the first decades of the century helped establish visualizations to communicate information. In the second half of the nineteenth century, Galton introduced key statistical concepts such as central tendency, standard deviation, correlation, and regression analyses. He used graphs to study anthropology, heredity, and psychology; Figure 23.14, for example, visualizes the relationship between a child's and their parents' height. Finally, a rise in statistical studies on commercial, demographic, and social themes inspired the spread of statistical atlases which combined geographical maps with data visualization techniques, such as pie diagrams and flowlines (Li, 2019). In this way, data were made more accessible to nonprofessional audiences.

While the beginning of the 1900s did not produce new forms of data display, statistical graphics continued their spread in different fields (Friendly and Denis, 2001). Key statistical techniques were developed, such as sampling distributions, randomization, likelihood, and the analysis of variance. Graphic displays such as boxplots were introduced to highlight patterns in data, an innovative concept that Tukey (1977) named 'exploratory data analysis'. For example, in 1913 Moseley discovered the concept of the atomic number by observing the plotting of the serial number of the atomic elements in the periodic table and their square root frequencies from X-ray spectra (Figure 23.15). From the second half of the twentieth century, multivariate data started being displayed. Biplots and networks, for instance, visualize both single observations (represented by different data points) and a range of variables (displayed through the position, size, and color of points) in a single graph. Other innovations in data visualization include the use of scatterplot and mosaic matrices to display relationships among multiple variables (Heer et al., 2010).

Recently, computer software has introduced high-resolution, interactive, and multidimensional graphical displays for large datasets (Blundell, 2008). Interactive graphs and maps provide an overview of main patterns, while allowing users to zoom, filter, and click on data points to visualize details and display new graphics. Another recent type of graphic displays is real-time updating visualizations. These are informed by live data feeds, thus adding the fourth dimension of time to traditional static two- or three-dimensional graphs (Hepworth, 2017). Finally, virtual and augmented reality as emerging technologies offer a new type of experience of large amounts of data in an unlimited virtual space, which the data and the analyst occupy together (El Beheiry et al., 2019; see Figure 23.16).

Key principles of data visualization

Having discussed prominent examples that shaped scientific visualization throughout history, we can now turn to key principles, which can be derived from the tradition of visual displays in empirical science.

1 *Focus on the data*: The single most important aspect of scientific visualization is the focus on data because the main purpose is to evaluate empirical evidence. Tufte (2001, p. 93ff.) uses the data-ink ratio to calculate how effective a particular graphic is in this regard. The ratio is calculated as

$$\frac{\text{Data ink}}{\text{Total ink used to print the graphic}}$$

It operates on a scale between 0 (not effective) and 1 (effective). For instance, Figure 23.17 shows two versions of the same visual display for which the data-ink ratio has been calculated, with a much more effective graph on the right (with a 0.93 data-ink ratio).

2 *Provide a bigger picture (abstraction)*: Abstraction allows the focus on the main trends, patterns, and other important features in the data. This is especially important in situations in which there is not enough space for all the details to be included clearly (Strothotte, 1998, pp. 13–15). A good example of abstraction is the computational map, such as Google maps (Google Maps, 2021). At its most abstract, the map shows the globe with country and ocean boundaries and labels only. More geographical information (cities, rivers, peaks, roads, streets, etc.) is added as the user zooms in and the screen space readjusts to the new requirements for the appropriate detail.

3 *Use eight visual variations recognisable by the eye*: (1) horizontal position, (2) vertical position, (3) size, (4) value (lightness or darkness), (5) texture, (6) color, (7) orientation, and (8) shape. Operating within the 'flatland' (Tufte, 2006) created by a sheet of paper or a

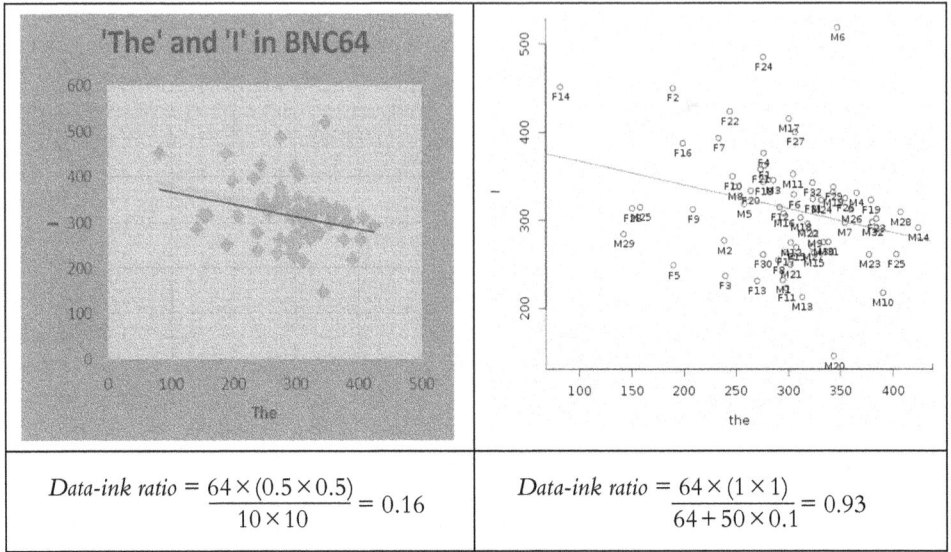

| Data-ink ratio $= \dfrac{64 \times (0.5 \times 0.5)}{10 \times 10} = 0.16$ | Data-ink ratio $= \dfrac{64 \times (1 \times 1)}{64 + 50 \times 0.1} = 0.93$ |

Figure 23.17   Example of a scatterplot with a low (left) and a high (right) data-ink ratio

screen and defined by the x-axis (horizontal position) and the y-axis (vertical position), we have additional six visual variations at our disposal (Bertin, 1981, p. 186). The size of a data point in the display (• ●) can indicate the quantity of the measured variable. Its value (●●●) can be used to communicate a scale variable, while the texture (⊜●◖), color (●●●), orientation (◖ ◑ ◕) and shape (• ▲ ■) can successfully indicate categories.

4   *Use psychological principles to inform the design of visuals*: Cognitive psychology can help us make useful decisions about the design of visual displays (Hegarty, 2011). Working memory and attention are important considerations when thinking about a visual that can communicate information efficiently. Based on the evidence from cognitive psychology, Harold et al. (2016) offer the following four guidelines:

- Direct visual attention.
- Reduce complexity.
- Support inference making.
- Integrate text with graphics.

5   *Make visualizations accessible*: It is always important to think about a range of readers and their particular needs. This includes accessible font size and clear color contrasts, if color is employed. Although multiple redundancy in data visualization is best avoided because it creates clutter, a certain amount of redundancy might be useful to cater for the needs of the audience. For example, if a range of colors is used to indicate categories of a variable, some other (redundant) indication of the categories might be suitable to assists readers with color vision deficiency (CVD; e.g. Kvitle et al., 2016). CVD is common in the population: 8% of men and 0.5% of women have some form

of CVD (Colour Blind Awareness, 2016). For accessible visualizations, use a limited set of colors with a simple contrast that can be easily recognized by readers with different forms of CVD.

Useful statistical and visualization techniques

Data visualization has multiple functions. It is essential to explore and understand the main trends in a dataset, highlighting patterns and behaviors that can guide analyses. It can also summarize results and a large quantity of information through immediate, powerful, and memorable visual communication. As Hudson (2015) notes, the choice of statistical graphs depends on three main elements: (1) the data (type and number of variables), (2) the communicative goal (descriptive or inferential), and (3) the audience. A description of the main statistical graphics is given next, paying particular attention to their characteristics in terms of variables and function. The dataset used for all visual examples in this section (except Figure 23.22) is the Trinity Lancaster Corpus (TLC), which is composed of transcripts of spoken exams of English as a second language (Gablasova et al., 2019).

An initial exploration of data can be carried out using histograms and boxplots. They both have a descriptive function and display the distribution and frequency of variables. Specifically, they highlight whether the distribution of a variable is normal or skewed, therefore guiding the choice of statistical techniques for further analyses. Histograms use bars which indicate a normal distribution when they have a bell shape and are symmetrical about the mean. For example, Figure 23.18 shows the distribution of a linguistic variable – the number of words used by test-takers in a spoken language exam – in the TLC.

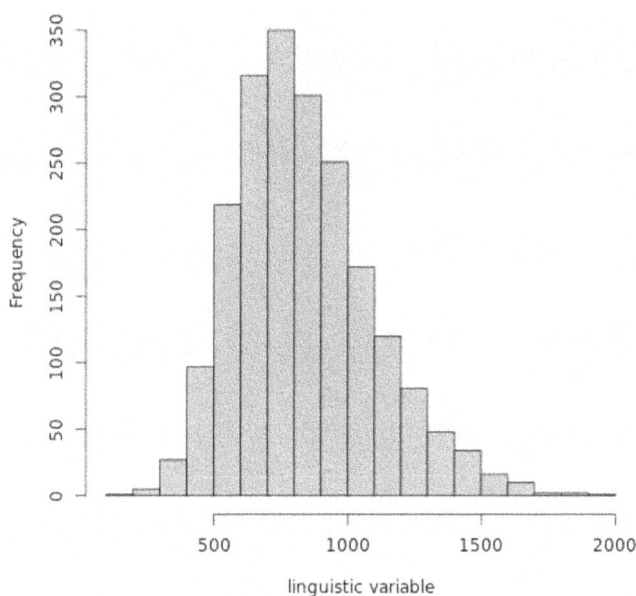

Figure 23.18 Distribution of a linguistic variable – number of words – in the Trinity Lancaster Corpus (N = 2,053)

The histogram shows that the distribution of the data is nonsymmetrical and positively skewed ($M = 839.79$, $SD = 256.511$, median $= 803$). While histograms are a common technique of data visualization, they have some limitations: they only display the distribution of a variable without including other statistical details, and they artificially group the data into bins.

Boxplots can be used to explore the distribution of a quantitative variable, to compare the distributions of different variables or groups, and to visualize the variation of a diachronic variable across time. They include details about minimum and maximum values, lower and upper quartile, median, and mean. In addition to the shape of the distribution, boxplots also display existing outliers in the dataset. For example, Figure 23.19 shows the distribution of the age of test-takers across proficiency levels – preintermediate (B1), intermediate (B2), and advanced (C1/C2) – in the TLC.

The inside of each box shows the interquartile range, which represents 50% of the values. If the interquartile range is small, the distribution of the data is close to the median, as in groups B1 and B2 in Figure 23.19. The box that represents the interquartile range for the age of speakers at C1/C2 level is larger, therefore signaling a wider distribution. The two horizontal lines in each box represent mean and median, providing further details about the shape of the distribution. In this case, the distribution of the speakers' age is positively skewed since the distance of the median is closer to the bottom line of the box (lower quartile) in all groups. Conversely, when the position of the median in a boxplot is exactly in the middle of the box, equidistant from the lower and upper quartile, the data are symmetric (Cleveland, 1993). The maximum and minimum values are signaled by the extremes of the two whiskers above and below each box, while outliers are displayed outside the range of the whiskers.

Error bars can be used to visualize both descriptive and inferential statistical details. Descriptive error bars can display measures of dispersion such as range and standard deviation (Cumming et al., 2007). Inferential error bars are more common and visualize confidence intervals (CI). A 95% CI is computed to provide an estimation of the value of a particular statistical test in the population. Specifically, the confidence interval is identified around the mean of a statistical measure based on our sample. This interval includes the true value of the same measure for

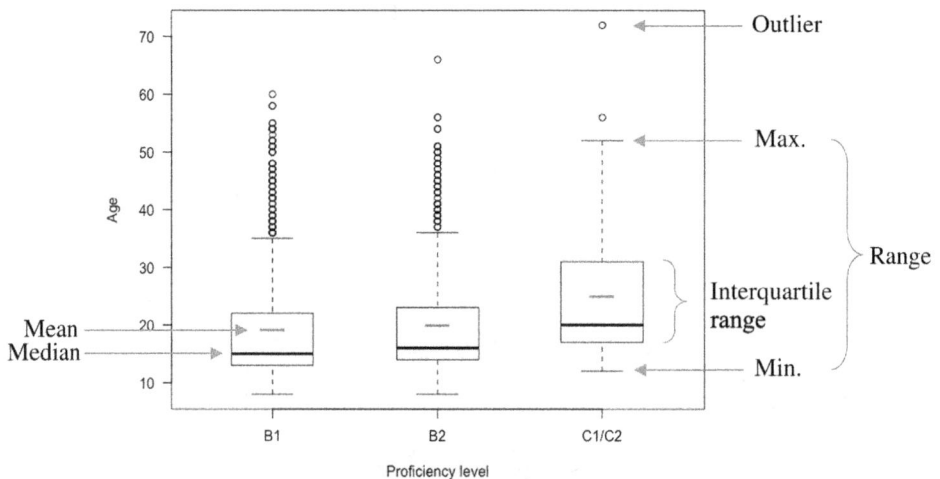

Figure 23.19 Distribution of the age of L2 English speakers at different proficiency levels in the Trinity Lancaster Corpus ($N = 2,053$)

95% of the samples taken from the population (Brezina, 2018). Error bars are useful to compare samples from two or more groups. In this case, when error bars largely overlap, the two samples represent the same population. For example, Figure 23.20 shows the 95% CI around the mean value of a linguistic variable (lexical density) in three groups of test-takers in the TLC. The error bar for the preintermediate group (B1) does not overlap with the other two bars, while the CIs for the intermediate (B2) and advanced (C1/C2) groups partially overlap, demonstrating that the B1 group represents a different population from the two highest proficiency levels for what concerns lexical density.

Line graphs are another common display of statistical data with a descriptive function. They show the change of diachronic values (*y*-axis) across time (*x*-axis) to reveal trends and fluctuations. Values are plotted on the graph, and a line is drawn to connect the data points. This type of graphic is effective for an initial exploration of the progress in time of a variable; however, it does not provide information about the distribution of data. Multiple line graphs can also be compared, providing the same scale is used for the axes (Brezina, 2018, p. 225).

A graphic used to display the relationship between two variables is the scatterplot with an optional line of the best fit (regression line). The scatterplot presents each individual data point using the *x-y* coordinates for the values of the two variables. A regression line can be drawn by finding the best fit for all the data points in the scatterplot to approximate a linear trend. The line of the best fit, which corresponds to the equation of a linear regression analysis, is an inferential statistical technique which shows whether the relationship between the two variables is directly or inversely proportional. Figure 23.21 exemplifies a directly proportional relationship between lexical diversity (measured using the MTLD index) and the number of words in the TLC.

When the relationship between two variables is nonlinear, for example in the case of diachronic data, a curve of the best fit can be used. This is a curve that, like a regression line, finds the best fit to a set of data points on a scatterplot. Its mathematical function can be computed through a nonlinear regression analysis via a generalized additive model (GAM) (Brezina, 2018).

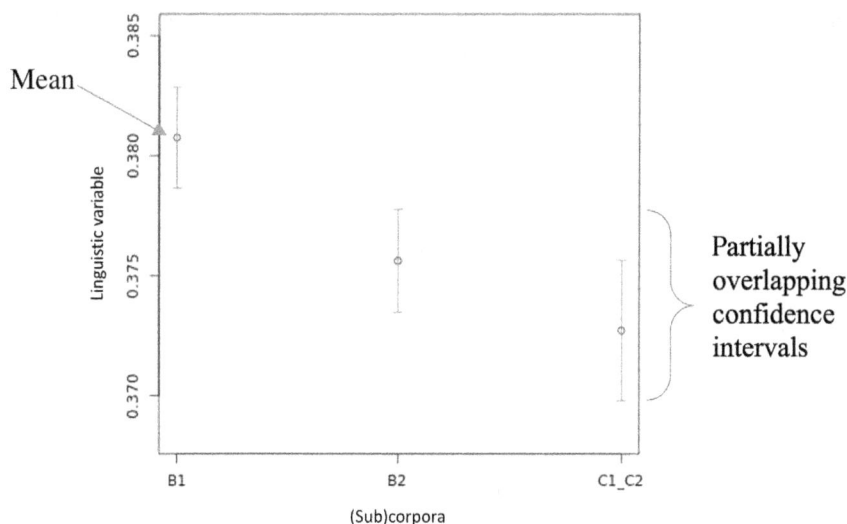

Figure 23.20   Error bars that display 95% confidence intervals for lexical density across different proficiency levels in the Trinity Lancaster Corpus (*N* = 2,053)

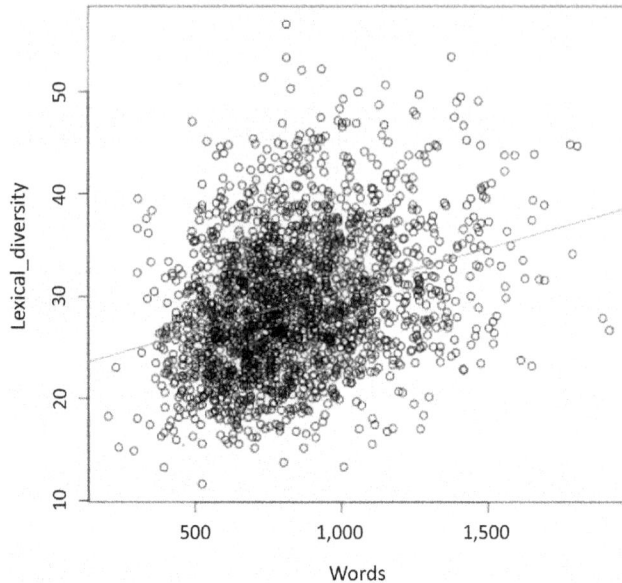

Figure 23.21   Regression line through a scatterplot that displays the relationship between lexical diversity and the number of words in the Trinity Lancaster Corpus ($N = 2{,}053$)

Another example of data visualization is geomapping. Geomaps have a descriptive function and allow the exploration of data to compare their proportions and geographical distribution. Data points are displayed on a map using either a color-coding system or markers such as circles. For example, they are effective in risk communication to visualize the spread of diseases or to formulate hypotheses about their spatial and temporal outbreaks (Welhausen, 2018). An example is provided in Figure 23.22, which shows the number of new cases of COVID-19 in the period December 22–29, 2020 in the ten most populated countries in Europe. Data were retrieved from the WHO website (2020). Circles are displayed on a geographical area with their diameter proportional to their values in the dataset.

Other more complex types of data visualization exist, for example, three-dimensional graphs, which are not presented here for reasons of space. Different tools can be used to produce scientific visualizations. They range from Microsoft Excel to the statistical package R (R core team, 2020), which is arguably the most flexible tool on offer with numerous packages allowing both simple and complex visualizations. The R package is employed in the backend of the user-friendly web interface Lancaster Stats Tools online (Brezina, 2018), used for the creation of the graphs in this section.

Conclusion

Data visualization is as much science as it is art. The science of data visualization combines perspectives from statistics (e.g. Brezina, 2018; Tufte, 2006; Tukey, 1977), computer science (e.g. Strothotte, 1998; Unwin, 2015), and cognitive psychology (e.g. Bertin, 1981; Hegarty, 2011) to create effective, accurate and impactful graphs, diagrams, and schemas. These perspectives have been highlighted in this chapter and can be further explored by following the references listed at the end of the chapter. As a general principle, we suggest following the best practice in a particular field of natural or social

Figure 23.22 Geomap that shows the number of cases of COVID-19 in the period December 22–29, 2020 in the ten most populated countries in Europe

science that the readers are interested in because the conventions and contexts of effective visual displays may vary. At the same time, we encourage readers to look over the fence surrounding their discipline, finding inspiration in the best practices of other disciplines as well. Crucially, however, we would encourage the readers to approach visualizations with a critical eye, considering the key principles outlined in this chapter. As for the art aspect of data visualization, we need to realize that successful data visualization also involves an element of creativity, which goes beyond any rules or guidelines. This is partly why so far there has not been a unified theory of data visualization (cf. Unwin, 2008, p. 61). Data visualization thus opens up a space for different means and modes of scientific communication (cf. Figure 23.16), which is limited only by our imagination.

References

Ben-Chaim, M. (2017) *Experimental philosophy and the birth of empirical science: Boyle, Locke and Newton*, London: Routledge.

Bertin, J. (1981) *Graphics and graphic information processing*, Berlin: Walter de Gruyter.

Best, L.A., Smith, L.D., & Stubbs, A. (2001) 'Graph use in psychology and other sciences', *Behavioural Processes*, vol. 54, pp. 155–165, DOI: 10.1016/s0376–6357(01)00156–5

Blundell, B.G. (2008) *An introduction to computer graphics and creative 3-D environments*, London: Springer.

Borkin, M.A., Vo, A.A., Bylinskii, Z., Isola, P., Sunkavalli, S., Oliva, A. & Pfister, H. (2013) 'What makes a visualization memorable?', *IEEE Transactions on Visualization and Computer Graphics*, vol. 19, no. 12, pp. 2306–2315, DOI: 10.1109/TVCG.2013.234

Brezina, V. (2018) *Statistics in corpus linguistics: A practical guide*, Cambridge: Cambridge University Press.

Cleveland, W.S. (1984) 'Graphs in scientific publications', *American Statistician*, vol. 38, pp. 261–269, DOI: 10.1080/00031305.1984.10483223

Cleveland, W.S. (1993) *Visualizing data*, Summit, NJ: Hobart Press.

Colour Blind Awareness. (2016) www.colourblindawareness.org/ (Accessed 10 July 2021)

Cumming, G., Fidler, F., & Vaux, D.L. (2007) 'Error bars in experimental biology', *Journal of Cell Biology*, vol. 177, no. 1, pp. 7–11, DOI: 10.1083/jcb.200611141

Darwin, C. (1859) *On the origin of species by means of natural selection, or the preservation of favoured races in the struggle for life*, New York: Appleton.

El Beheiry, M., Doutreligne, S., Caporal, C., Ostertag, C., Dahan, M., & Masson, J.B. (2019) 'Virtual reality: beyond visualization', *Journal of Molecular Biology*, vol. 431, no. 7, pp. 1315–1321, DOI: 10.1016/j.jmb.2019.01.033

Euclid. (1482) *Elementa geometriae (Elements of geometry)*, Venice, Italy: Erhard Ratdolt.

Few, S. (2012) *Show me the numbers: Designing tables and graphs to enlighten*, Oakland: Analytics Press.

Friendly, M. (2008) 'A brief history of data visualization', in C Chen, W Härdle & A Unwin (eds.) *Handbook of data visualization*, Springer-Verlag, Berlin, pp. 15–56, DOI:10.1007/978-3-540-33037-0_2

Friendly, M., & Denis, D.J. (2001) 'Milestones in the history of thematic cartography, statistical graphics, and data visualization', www.datavis.ca/milestones/ (Accessed 23 July 2021)

Funkhouser, H.G. (1936) 'A note on a tenth century graph', *Osiris*, vol. 1, pp. 260–262, DOI:10.1086/368425

Funkhouser, H.G. (1937) 'Historical development of the graphical representation of statistical data', *Osiris*, vol. 3, no. 1, pp. 269–404, DOI: https://doi.org/10.1086/368480

Gablasova, D., Brezina, V., & McEnery, T. (2019) 'The Trinity Lancaster Corpus: development, description and application', *International Journal of Learner Corpus Research*, vol. 5, no. 2, pp. 126–158, DOI: https://doi.org/10.1075/ijlcr.19001.gab

Galilei, G. (1610) *Sidereus nuncius (The starry messenger)*, Venice, Italy.

Galton, F. (1886) 'Regression towards mediocrity in hereditary stature', *The Journal of the Anthropological Institute of Great Britain and Ireland*, vol. 15, pp. 246–263, DOI: https://doi.org/10.2307/2841583

Google Maps. (2021) www.google.com/maps (Accessed 14 April 2021)

Grainger, S., Mao, F., & Buytaert, W. (2016) 'Environmental data visualisation for non-scientific contexts: literature review and design framework', *Environmental Modelling & Software*, vol. 85, pp. 299–318, DOI: https://doi.org/10.1016/j.envsoft.2016.09.004

Halley, E. (1686) 'On the height of the mercury in the barometer at different elevations above the surface of the earth, and on the rising and falling of the mercury on the change of weather', *Philosophical Transactions*, pp. 104–115, DOI: https://doi.org/10.2307/2683467

Harold, J., Lorenzoni, I., Shipley, T.F., & Coventry, K.R. (2016) 'Cognitive and psychological science insights to improve climate change data visualization', *Nature Climate Change*, vol. 6, no. 12, pp. 1080–1089, DOI: https://doi.org/10.1038/nclimate3162

Harvey, W. (1628) *Exercitatio anatomica de motu cordis et sanguinis in animalibus (An anatomical exercise on the motion of the heart and blood in living beings)*, Frankfurt am Main: Fitzeri.

Healy, K., & Moody, J. (2014) 'Data visualization in sociology', *Annual Review of Sociology*, vol. 40, pp. 105–128, DOI: https://doi.org/10.1146/annurev-soc-071312-145551

Heer, J., Bostock, M., & Ogievetsky, V. (2010) 'A tour through the visualization zoo', *Communications of the ACM*, vol. 53, no. 6, pp. 59–67, DOI:10.1145/1743546.1743567

Hegarty, M. (2011) 'The cognitive science of visual-spatial displays: implications for design', *Topics in Cognitive Science*, vol. 3, no. 3, pp. 446–474, DOI: https://doi.org/10.1111/j.1756-8765.2011.01150.x

Hepworth, K. (2017) 'Big data visualization: promises & pitfalls', *Communication Design Quarterly*, vol. 4, no. 4, pp. 7–19, DOI: https://doi.org/10.1145/3071088.3071090

Hudson, T. (2015) 'Presenting quantitative data visually', in L. Plonsky (ed.), *Advancing quantitative methods in second language research*, New York: Routledge, pp. 78–105.

Kvitle, A.K., Pedersen, M., & Nussbaum, P. (2016) 'Quality of color coding in maps for color deficient observers', *Electronic Imaging*, no. 20, pp. 1–8, DOI: https://doi.org/10.2352/ISSN.2470-1173.2016.20.COLOR-326

Lambert, J.H. (1779) *Pyrometrie*, Berlin: Haude und Spener.

Li, L. (2019) 'Visualizing Chinese immigrants in the U.S. statistical atlases: a case study in charting and mapping the other(s)', *Technical Communication Quarterly*, vol. 29, no. 1, pp. 1–17, DOI: https://doi.org /10.1080/10572252.2019.1690695

Minard, C.J. (1861) *Des tableaux graphiques et des cartes figuratives (Graphic tables and figurative maps)*, Paris: Thunot et Cie.

Moseley, H. (1913) 'The high-frequency spectra of the elements', *Philosophical Magazine*, vol. 26, pp. 1024–1034, DOI: https://doi.org/10.1080/14786441308635052

Perini, L. (2005) 'The truth in pictures', *Philosophy of Science*, vol. 72, no. 1, pp. 262–285, DOI: https:// doi.org/10.1086/426852

Playfair, W. (1786) *Commercial and political atlas: Representing, by copper-plate charts, the progress of the commerce, revenues, expenditure, and debts of England, during the whole of the eighteenth century*, London.

Priestley, J. (1764) *A description of a chart of biography*, Warrington.

R Core Team. (2020) *R: A language and environment for statistical computing*, Vienna: R Foundation for Statistical Computing, www.R-project.org/ (Accessed 25 July 2020)

Ruchikachorn, P., & Mueller, K. (2015) 'Learning visualizations by analogy: promoting visual literacy through visualization morphing', *IEEE Transactions on Visualization and Computer Graphics*, vol. 21, no. 9, pp. 1028–1044, DOI: 10.1109/TVCG.2015.2413786

Scheiner, C. (1630) *Rosa Ursina sive sol ex admirando facularum & macularum suarum phoenomeno varius (Rosa Ursina or the sun observing its flares and sunspots)*, Bracciano, Italy: Andream Phaeum.

Seipel, S., & Lim, N.J. (2017) 'Color map design for visualization in flood risk assessment', *International Journal of Geographical Information Science: IJGIS*, vol. 31, no. 11, pp. 2286–2309, DOI: https://doi.org /10.1080/13658816.2017.1349318

Stjernfelt, F. (2000) 'Diagrams as centerpiece of a Peircean epistemology', *Transactions of the Charles S. Peirce society*, vol. 36, no. 3, pp. 357–384, doi: www.jstor.org/stable/40320800

Strothotte, T. (1998) *Computational visualization: Graphics, abstraction and interactivity*, Berlin: Springer.

Tufte, E.R. (2001) *The visual display of quantitative information*, Cheshire, CT: Graphics Press.

Tufte, E.R. (2006) *Beautiful evidence*, Cheshire, CT: Graphics Press.

Tukey, J.W. (1977) *Exploratory data analysis*, Reading, MA: Addison-Wesley.

Unwin, A. (2008) 'Good graphics?', in C. Chen, W. Härdle, & A. Unwin (eds.), *Handbook of data visualization*, Berlin: Springer-Verlag, pp. 57–78.

Unwin, A. (2015) *Graphical data analysis with R*, Boca Raton: CRC Press.

Wainer, H., & Velleman, P.F. (2001) 'Statistical graphics: mapping the pathways of science', *Annual Review of Psychology*, vol. 52, no. 1, pp. 305–335, DOI: https://doi.org/10.1146/annurev.psych.52.1.305

Wang, C., & Tao, J. (2017) 'Graphs in scientific visualization: a survey', *Computer Graphics Forum*, vol. 36, no. 1, pp. 263–287, DOI: https://doi.org/10.1111/cgf.12800

Welhausen, C.A. (2018) 'Visualising science: Using grounded theory to critically evaluate data visualizations', in H. Yu & K. Northcut (eds.), *Scientific communication: Practices, theories, and pedagogies*, New York: Routledge, pp. 82–106.

World Health Organisation (WHO). (2020) www.who.int (Accessed 30 December 2020)

24

GRAPHICAL ABSTRACTS

Visually circulating scientific arguments

Jonathan Buehl

Introduction and literature review

Graphical abstracts are important emerging forms of scientific communication. Like textual abstracts, these visuals summarize scientific research articles, but they summarize graphically rather than with text alone. For example, Figure 24.1 summarizes the claim that modifying the biomolecule neurexin negatively affects synaptic activity by disrupting the compound heparan sulfate; thus, the abstract visually supports the article's titular claim: "Heparan Sulfate Organizes Neuronal Synapses Through Neurexin Partnerships."

Publishers deploy graphical abstracts as browsing aids on journal portal pages, and they circulate them through social media platforms. Thus an increasing number of journals now require manuscripts to include graphical abstracts. However, despite graphical abstracts' prevalence and relevance, only a few studies have examined how they function in, between, and beyond scientific discourse communities. Moreover, those studies' insights have not been fully operationalized for practitioners and instructors of scientific writing.

This chapter describes the history and rhetorical purposes of graphical abstracts, reviews how researchers from several fields have approached them, and demonstrates a syncretic procedure for analyzing them. Case studies of *Cell* (a cellular biology journal) and *Applied Surface Science* (a materials science journal) demonstrate strategies for (1) assessing general and discipline-specific rhetorical tactics for creating graphical abstracts and (2) assessing graphical abstracts' online circulation.

Graphical abstracts are not meant to replace textual abstracts, which remain standard parts of research articles; rather, journals use graphical abstracts in ways that reflect how knowledge circulates in the twenty-first century. For example, when *Cell* adopted a new format for online articles in 2010, its editors extolled graphical abstracts on landing pages as new communication assets that "complement the traditional Summary text and promote article browsing" (Marcus, 2010, p. 9).

Although recent interest in graphical abstracts is tied to the digitization of scientific publishing, they are not entirely "born digital" phenomena. As Lane et al. (2015) observed, the German chemistry journal *Angewandte Chemie* began publishing graphical abstracts in 1976. Other print chemistry journals (e.g. *Tetrahedron Letters* [1986], *Chemical Communications* [1994], and *Phytochemistry* [1997]) also adopted the practice; however, these journals' line

DOI: 10.4324/9781003043782-28

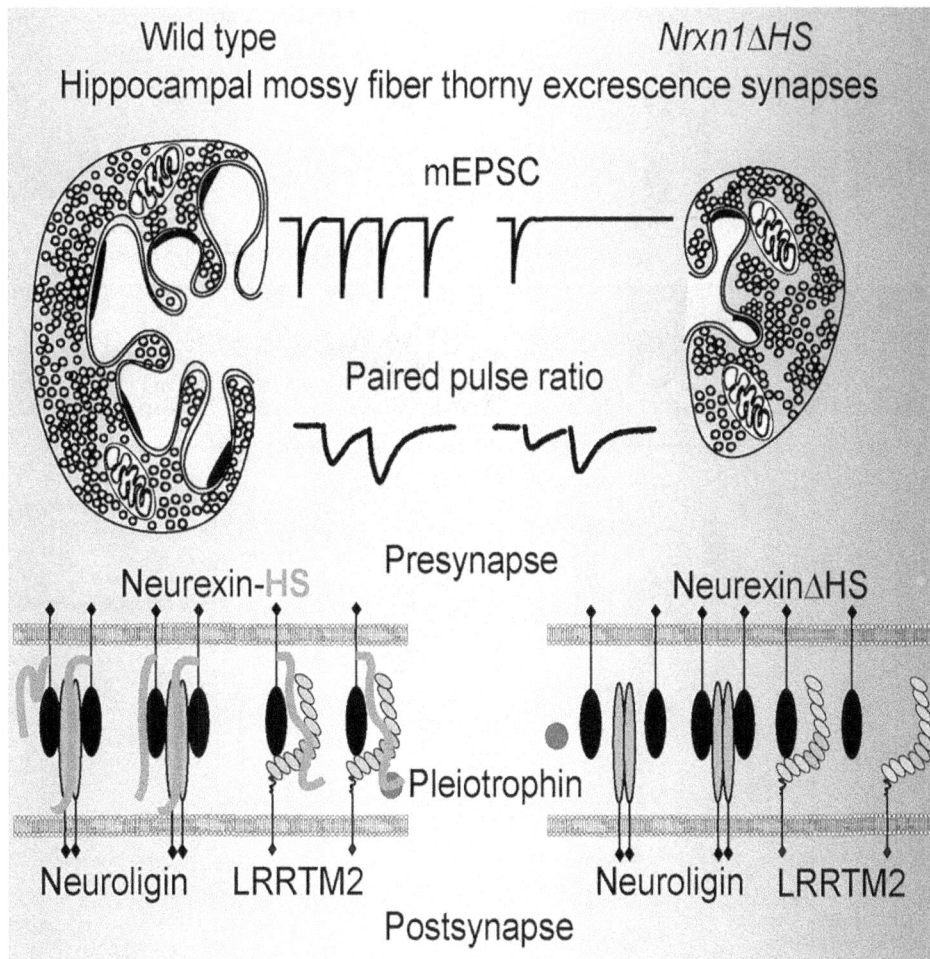

Figure 24.1 Graphical abstract from *Cell*

art representations of chemical syntheses and molecular structures were far simpler than graphical abstracts produced for contemporary digital infrastructures. New Internet-enabled affordances allowed editors to publish abstracts differently, and they began expecting scientists to produce visual summaries.

Although some journals allow authors to repurpose article figures as graphical abstracts, many explicitly prohibit this practice. And for good reasons. Figure 24.2 is one of nine figures supporting the argument that heparin sulfate participates in synaptic organization; however, these visualizations are not designed to facilitate browsing or circulation. Even experts would need time to unpack the figures and their captions (e.g. Figure 24.2's 500-word caption). The caption-less graphical abstract (Figure 24.1) visually synthesizes the entire argument. Such syntheses require scientists to think differently about the composition and rhetorical functions of visuals.

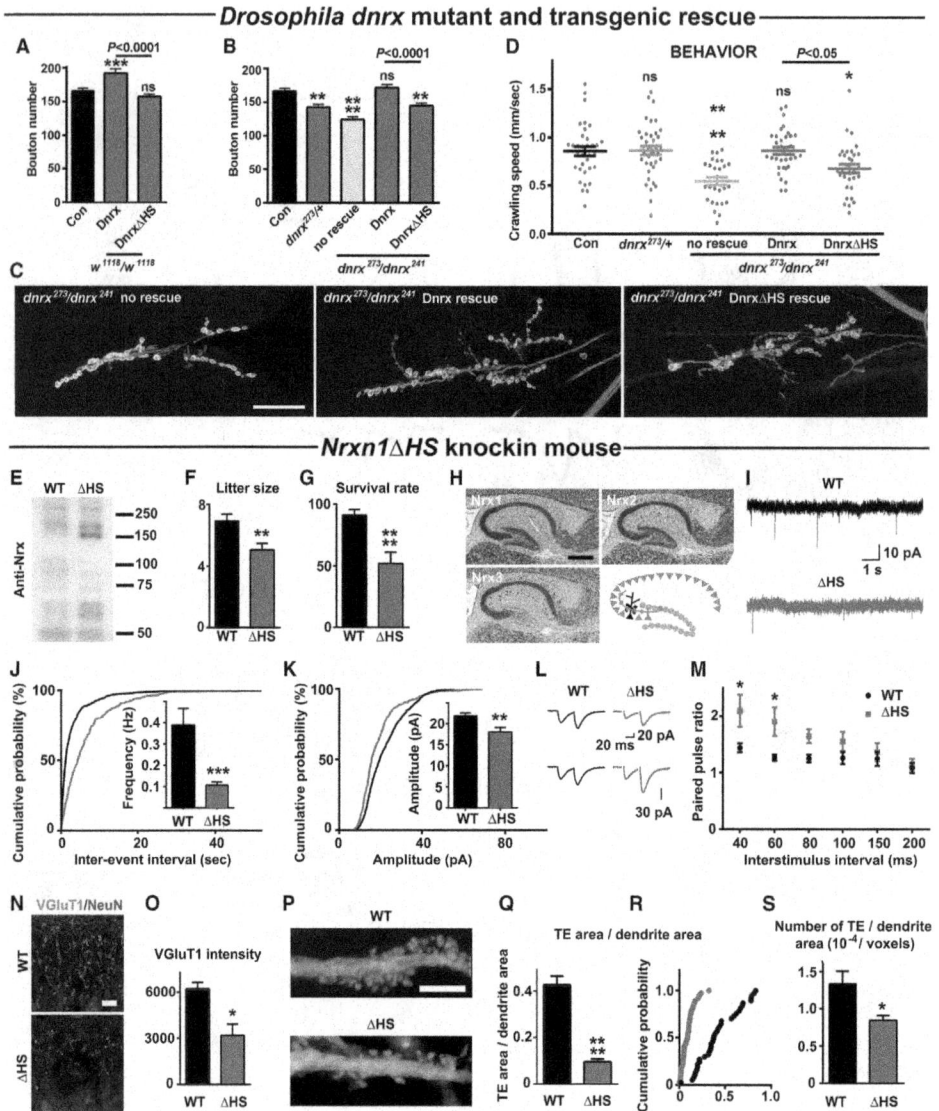

Figure 24.2 Figure 6 (of 9) included in Zhang et al. (2018)

Productive archives for unpacking rhetorical expectations of graphical abstracts include editorial advice published when graphical abstracts emerged as important new forms. For example, the American Chemical Society (2012) began providing "do this, not that" advice for creating what it calls "table of contents" graphics. The guidance emphasizes clear type, appropriate fonts, and providing informative details – but not too many details. Other guidance emphasizes a comprehensive understanding of visual arguments and digital tools for creating them. In 2010, *Cell* offered before-and-after examples to demonstrate effective design strategies;

for example, the visuals in Figure 24.3 summarize an argument about proton channel Hv1 and sperm-cell motility. Of the "after" version, the editors explain:

- The image's components have been reoriented to tell the story from left to right.
- Some arrows and text were removed for simplicity.
- The color palate was softened.
- The paper's take-away message and new findings ("Activation of Hv1") were set as the focal point of the abstract. (p. 3)

Though accurate, this description does not fully capture the visual literacy skills demonstrated in the revision, which include understanding vectors, small multiples, colors as cohesive ties, and strategies for making an image less "abstract" in the aesthetic sense. The changes also reflect relatively sophisticated digital composing skills, and for *Cell's* editors, scientists needed to have those skills or hire someone who does. *Cell's* early graphical-abstract documentation included links to an abstract design service, and submission guidelines for some journals still link to such services.

Submission guidelines also articulate expectations of viewership and use, which have changed over time and vary between journals. When introducing a graphical-abstract policy for a print-only publication in 1997, *Phytochemistry* imagined graphical abstracts would be used by "the average reader of *Phytochemistry*." In 2010, *Cell's* editors noted their potential "to encourage browsing, promote *interdisciplinary* scholarship, and help readers quickly identify which papers are most relevant" (emphasis added). In 2012, the American Chemical Society expected graphical abstracts to "capture the reader's attention" and "give the reader a quick visual impression of the essence of the manuscript without providing specific results" (p. 1). For *The Journal of Clinical Investigation* (2018), a graphical abstract "visually represents the primary findings of an article" and "appears at the top of the online version of the article" (para. 1). For *Applied Surface Science* (2018), it "should summarize the contents of the article in a concise, pictorial form designed to capture the attention of a wide readership online" (p. 8). Considered

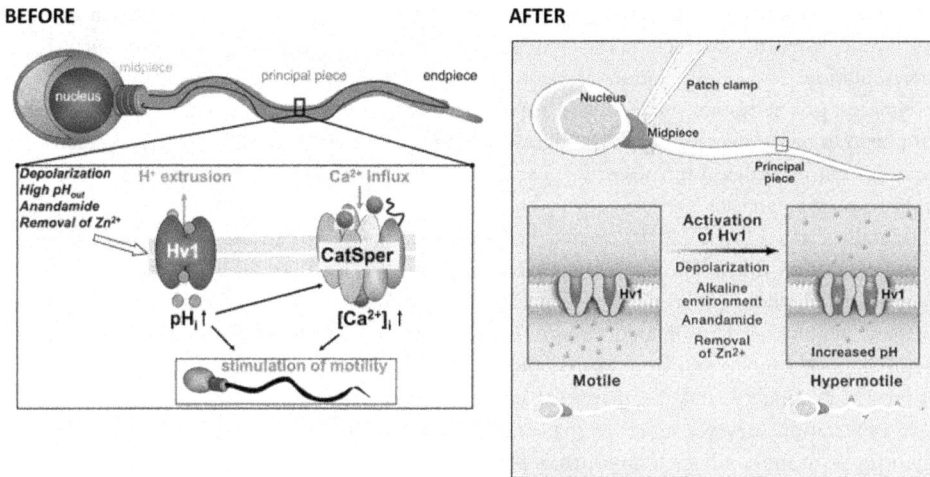

Figure 24.3 Before-and-after example demonstrating preferred design strategies for *Cell*

Source: "Cell Press Graphical Abstract Guidelines." © Elsevier, 2010. Reproduced with permission of Elsevier.

together, these editorial descriptions articulate several (possibly competing) functions: (1) to summarize – either to summarize the article's "essence" but not results or to summarize its "essence" while including key findings, and (2) to facilitate decisions about reading for browsers of journal issues or a wider online readership. Scholarship on graphical abstracts has examined both functions. Although graphical abstracts have not yet received extensive scholarly treatment, foundational work in writing studies, information science, and medical journals has provided insights into how graphical abstracts function as texts (artifact analysis) and how they circulate online (circulation analysis).

In the first study of graphical abstracts, Lane et al. (2015) argued that graphical abstracts operate simultaneously on four layers – visual composition, rhetorical purpose, modal relationship, and mediated context. Visual composition can be singular (one photograph, illustration, or diagram), stacked (multiple visual elements but no linking elements), relational (multiple visual elements with explicit links to signal relationships), or blended/hybrid (multiple visual elements combined; e.g. graphs with additional symbols or images incorporated to signal concepts). Rhetorical purpose can be topical-descriptive or argumentative – i.e. abstracts either introduce topics or present aspects of arguments. Modal relationships – i.e. how graphical abstracts (the visual mode) relate to texts (the verbal mode) – include redundant, complementary, supplementary, and stage-setting. Mediated contexts include dominant (appearing only with the title), parallel (appearing with other interpretive aids, such as textual abstracts), and subordinate (available but less salient than textual summaries).

Other early studies offered different typologies. For example, Yoon and Chung (2017) coded abstracts according to visualization type (table, chart, photo, combination), content (background, method, results, overview), and relation to the manuscript – new, duplicated (copied from the manuscript), modified (modification of a single figure), and integrated (multiple visualizations combined). Hullman and Bach (2018) identified four key categories of the "design space" of graphical abstracts (layout, depiction of time, text usage, and representational genre), and they plotted each on a continuum. For example, their layout continuum ranges from linear to single-image "free" arrangements – with zig-zag, forking, nesting, parallel, orthogonal, and centric layouts in between. Their itemized taxonomies led Hullman and Bach to identify three regularly occurring design patterns: process illustrations (illustrations with linear or forking layout, labels to name elements, and arrows to connect individual stages/elements), result representations (data visualizations presenting results often contextualized within a larger representation of a temporal process), and parallel layouts ("parallel pictures laid out horizontally" for purposes of comparison" [p. 14]). Significantly, these templates combine visual-element taxonomies with rhetorical purposes; e.g. parallel layouts "were frequently associated with survey and review type articles" (p. 14).

Hendges and Florek (2019) approached graphical abstracts through genre analysis and a user-expectation study. Their genre analysis coded abstracts for layout (Hullman and Bach's categories), originality (similar to Yoon and Chung's "relation to the manuscript"), and "nature of images," which they plot on a continuum from less specialized (illustration) to most specialized (symbolic notation). For all categories, they found "lack of uniformity in the genre in terms of layout, originality, and the nature of images within and across disciplines" (p. 77). Their user-expectation survey confirmed that expectations vary across disciplines and indicated that scientists want more advice than journals provide.

Arguably, findings from previous analyses can be synthesized into the questions listed in Box 24.1, though it includes other questions – about special *topoi*, figural patterns, and the influence of social and material constraints – derived from this chapter's case studies.

**Box 24.1   Productive questions for artifact and infrastructure
analysis of graphical abstracts**

- *Representation*: What does the graphical abstract represent?

    - *Rhetorical function*: Is it just announcing the topic or presenting an argument about that topic?
    - *Scope*: Does it provide an overview of the article, or does it focus on content from specific sections, such as methods or results?

- *Construction*: How is the graphical abstract constructed?

    - *Component sources*: Is the abstract novel, or is it related to the article's other visuals?
    - *Component types*:

        - What visual content elements are used (data visualizations, photographs, diagrams, etc.)?
        - What visual linking elements are used (space, arrows, enumeration, etc.)?
        - What textual elements are used (labels, legends, notes, etc.)?

    - *Arrangement*:

        - *General component arrangement*: Which visual patterns are used to compose the graphical abstract (e.g. Hullman and Bach's layout continuum)?
        - *Rhetorical arrangement*: Are specific form-meaning pairings used in summarizing the argument?
        - General design templates (Hullman and Bach)
        - Field-specific arrangements or special topoi
        - Figural patterns

- *Intertextual connections*: How does the graphical abstract relate to the abstracted text?

    - Relationship with other summary elements (Lane et al.)
    - Relationship with other visual elements (Yoon and Chung)

- *Social constraints*: How does the graphical abstract meet form and content expectations?

    - *Disciplinary conventions*: Does the abstract reflect discipline-specific design strategies?
    - *Venue expectations*: How are the expectations of the journal's style guide or submission instructions reflected by the abstract?

- *Material constraints*: How might publishing infrastructures influence the abstract's content, design, or rhetorical function?

Circulation studies of graphical abstracts consider how they move through and between discourse communities, are used by audiences, and influence reading behaviors. Most circulation studies have appeared in medical journals, where graphical abstracts are often called "visual abstracts" and follow an "infographic"/"slide" format (e.g. Figure 24.4). Some medical journals even provide branded PowerPoint templates to promote consistent designs (e.g. *Neurosurgery*, 2020).

An early circulation study by Ibrahim et al. (2017) tested whether including visual abstracts in tweets affects how people interact with vascular surgery articles. They found that tweeting

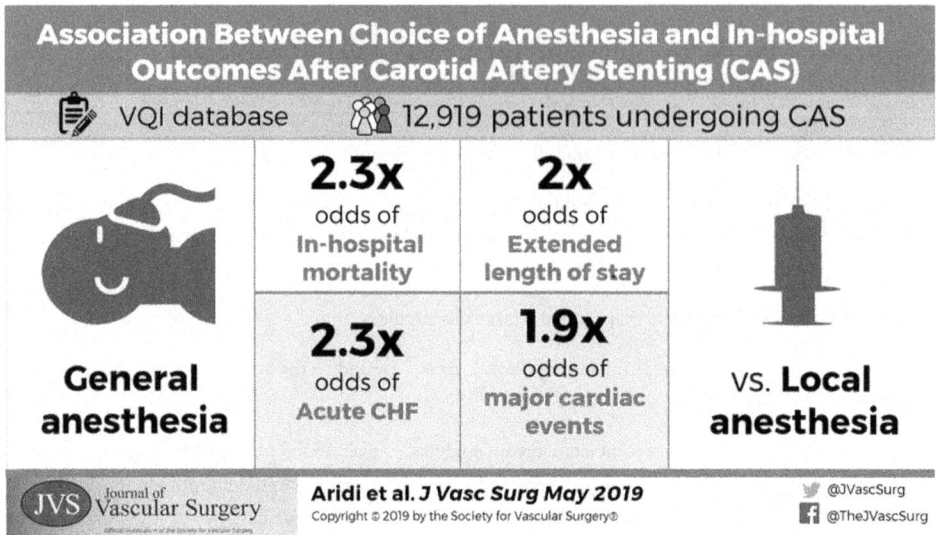

Figure 24.4   Visual abstract in an infographic format from *Journal of Vascular Surgery*

Source: Dakour-Aridi et al. (2019). Reproduced in accordance with the article's Creative Commons "Attribution 4.0 International (CC BY 4.0)" license.

graphical abstracts with titles resulted in an almost eightfold increase in impressions, more than an eightfold increase in retweets, and nearly a threefold increase in article visits when compared to tweeting an article title alone. Similar studies by Koo et al. (2019) and Lindquist and Ramirez-Zohfeld (2019) confirmed that tweets containing visual abstracts significantly outperformed text-only tweets with regard to key Twitter metrics.

Another set of studies compared graphical abstracts to other summary strategies. For example, Chapman et al. (2019) compared three strategies for disseminating surgical research to health care professionals and non–health care workers via Twitter: plain-English abstracts, visual abstracts, and standard tweets. They found that health care professionals accounted for 95% of engagements with surgical-research tweets and that visual abstracts "attracted significantly more engagements than plain English ones" (p. 1613). Bredbenner and Simon (2019) compared graphical abstracts to video abstracts, plain-language summaries, and textual abstracts, but they considered "engagement" beyond social media metrics. They found video abstracts and plain-language summaries produced higher levels of comprehension, feelings of understanding, and enjoyment. However, they also identified the importance of text in visual–verbal interactions, with participants often wanting "words or even a small description to go with the image" (p. 15). Bredbenner and Simon acknowledge that their use of graphical abstracts with little explanatory text might have influenced this finding.

When paired with early work on modal contexts (Lane et al., 2015), these circulation studies suggest another set of key questions; however, the questions in Box 24.2 also include questions emerging from the case studies to be described here. Specifically, many social media circulation studies have not gone beyond assessing tweets and retweets quantitatively; however, qualitative approaches to social media artifacts can also yield insights for authors and instructors of scientific communication.

Box 24.2   Productive questions about graphical-abstract circulation

- *Local circulation*: How does the journal's publishing infrastructure promote/inhibit circulation?

  - *Electronic infrastructure*: How many clicks does it take to see actionable graphical abstracts on the journal site?
  - *Print infrastructure*: How is the graphical abstract integrated into the article's "print" format (i.e. its static PDF)?

- *Social media circulation*: How has the abstract circulated through social media infrastructure?

  - *Agents*: Publishers, authors, readers, vendors, etc.
  - *Purposes*: Information sharing, self-promotion, journal promotion, application, humor, celebration, etc.

- *Reception and uptake*: How do users/readers react to graphical abstracts?

Although graphical abstracts have appeared on digital platforms for more than ten years, pedagogical concerns have not been much addressed. In 2015, Lane et al. described the lack of effective advice available for authors, noting that with "little guidance, and very little available discussion or critique, authors must invent their graphical abstracts largely in a vacuum" (p. 5). Since then, few new resources on making rhetorically effective abstracts have appeared. To date, only one textbook addresses them at length – Patience et al. (2015). However, its advice focuses on the technical features for creating images that display well online. Ibrahim's "A Primer on How to Create a Visual Abstract" (2016) offers sound design advice but only for the slide-style abstracts preferred by some medical journals. Guinda's (2017) classroom-based study of engineering students found that students' "'natural digital-native graphicacy' is conservative as to the medium, format and type of representation, but versatile regarding particular meanings, although not always unambiguous or register-appropriate" (p. 61). For Guinda, graphical abstracts are a promising teaching tool because they foreground "the ever-growing promotionalism that has colonised the dissemination of science and technology" while providing students with contexts for "learning technical or scientific disciplinary contents, acquiring a transductional repertoire (visual and/or multimodal), and reflecting on the possibilities granted by affordances, communicative situations, disciplines and cultures" (p. 86). Indeed, graphical abstracts can help students see nuances of multimodal argumentation within and across disciplines. Thus the remainder of this chapter demonstrates an analytical procedure that could help students and teachers of scientific communication approach graphical abstracts systematically and rhetorically.

Case studies: *Cell* and *Applied Surface Science*

The following case studies demonstrate methods of artifact analysis, infrastructure analysis, and circulation analysis. Study goals included assessing taxonomies from previous studies, comparing abstracts and infrastructures from different disciplines, and gaining a better understanding of the current state of graphical abstract use with regard both to design and to circulation. The following research questions emerged:

- What aspects of the models developed on the earliest examples of graphical abstracts remain relevant for current practices?
- How do delivery platforms affect the usability of graphical abstracts?
- Do specific disciplines have favored practices regarding graphical abstracts?
- Why do people tweet and retweet graphical abstracts?

With goals established, I sought as cases journals that reflected established graphical-abstracting practices, that could allow for meaningful cross-disciplinary comparison, and that had some form of social media presence. Thus, I chose to examine two journals from different disciplines with relatively long histories of circulating graphical abstracts online through portal pages and Twitter: *Cell* and *Applied Surface Science*.

Cell began publishing digital graphical abstracts in 2010. *Applied Surface Science* began publishing digital graphical abstracts with select articles in 2013; by 2015, most online versions of its articles included them. Though Elsevier publishes both journals, they differ in their expectations for graphical abstracts (Table 24.1). *Cell*'s expectations are more explicit and, as later analysis shows, seem to influence composing practices. *Applied Surface Science*'s less explicit guidelines allow for greater variability – especially in how graphical abstracts relate to other visuals.

The two journals also differ in their social media strategies. *Cell* is extremely active on Twitter. For example, within a week of releasing an issue on September 6, 2018, *Cell* tweeted abstracts from all but one article of the issue. In contrast, *Applied Surface Science* does not have its own Twitter feed, but tweets of its recent graphical abstracts appear in the Elsevier Physics feed.

For each journal, I examined abstracts from one issue from 2018 – a time when graphical abstracts were expected by both journals. Many studies of graphical abstracts have focused on early examples, and I wanted to check those earlier findings against later examples. In hindsight, I might have chosen an even later point, as social media activity for *Applied Surface Science* was minimal in 2018 but has since increased. Nonetheless, the 2018 time point provides a snapshot on graphical-abstract usage of two journals from different disciplines with established practices for publishing graphical abstracts. I only examined abstracts for research articles and not nonabstract images of commentary pieces. All 17 research articles in *Cell* from September 6, 2018, included graphical abstracts; 25 of 27 articles from the September 1 issue of *Applied Surface Science* included them. These cases provided (1) enough examples for meaningful qualitative analysis and (2) manageable datasets for comparing multiple analytical frames. For teaching contexts, instructors should select cases to support their goals; e.g., students might examine abstracts from multiple journals in the same field to identify discipline-specific conventions.

After establishing the scope of the cases, I populated an Airtable database with publication information, links to articles, and links to a full-sized version of each graphical abstract. (The scraping tool Webscrapper.io expedited this process.) I also arranged thumbnails of all abstracts on a PowerPoint slide for each issue, which allowed for comparisons within a single visual field. I then examined each graphical abstract, read each textual abstract, examined other visuals to check for repurposing, and gathered related tweets. I coded for the following features.

Artifact features

- The features of Lane et al.'s model: Visual Composition, Rhetorical Purpose, Modal Relationship, and Mediated Context

Table 24.1 Graphical abstract guidelines for *Applied Surface Science* and *Cell*

From *Applied Surface Science*, "Author Information Pack"

Graphical Abstract

A graphical abstract is mandatory for this journal. It should summarize the contents of the article in a concise, pictorial form designed to capture the attention of a wide readership online. Authors must provide images that clearly represent the work described in the article. Graphical abstracts should be submitted as a separate file in the online submission system. Image size: please provide an image with a minimum of 531×1328 pixels (h × w) or proportionally more. The image should be readable at a size of 5×13 cm using a regular screen resolution of 96 dpi. Preferred file types: TIFF, EPS, PDF or MS Office files. You can view Example Graphical Abstracts on our information site. Authors can make use of Elsevier's Illustration Services.

From "*Cell* Press Graphical Abstract Guidelines"

OVERVIEW

The graphical abstract is one single-panel image that is designed to give readers an immediate understanding of the take-home message of the paper. Its intent is to encourage browsing, promote interdisciplinary scholarship, and help readers quickly identify which papers are most relevant to their research interests.

TECHNICAL REQUIREMENTS

- Size: The submitted image should be 1200 pixels square at 300dpi.
- Font: Arial, 12–16points. *Smaller fonts will not be legible online*
- Preferred file types: TIFF, PDF, JPG
- Content: the abstract should consist of *one single panel*
- A note about color: Effective use of color can enhance the graphical abstract both aesthetically and by directing the reader's attention to focal points of interest. Authors are encouraged to select colors that are consistent with and complementary to the colors used on the Cell Press website. Heavily saturated, primary colors can be distracting.

UNIQUENESS AND CLARITY

The graphical abstract should:

- Have a clear start and end, "reading" from top-to-bottom or left-to-right
- Provide a visual indication of the biological context of the results depicted (subcellular location, tissue or cell type, species, etc.)
- Be distinct from any model figures or diagrams included in the paper itself
- Emphasize the new findings from the current paper without including excess details from previous literature
- Avoid the inclusion of features that are more speculative (unless the speculative nature can be made apparent visually)
- Not include data items of any type; all the content should be in a graphical form

KEEP IT SIMPLE

The graphical abstract should also:

- Use simple labels
- Use text sparingly
- Highlight one process or make one point clear
- Be free of distracting and cluttering elements

- Hullman and Bach's design-space features (Layout, Time, Text, and Representational Genre) and design templates (Process Diagram, Results Visualization, and Parallel Illustration)
- Yoon and Chung's visual and intervisual features: Visualization Type, Content, and Relation to the Manuscript
- Rhetorical figuration: Presence and type (*antithesis, antimetabole,* etc.)
- Any seemingly discipline-specific strategies

Infrastructure features

- Salience in electronic versions of articles
- Salience in "print" (PDF) versions
- Number of clicks to view a usable graphical abstract

Circulation features

- Tweet statistics: How many times was the article tweeted?
- Tweet sources: Who tweeted it (publisher, author, readers, vendors, etc.)?
- Tweet purpose: What does the tweet suggest about the poster's purpose?

Though productive, this time-consuming analytical process might be excessive for some instructors' or authors' goals. Nevertheless, the higher-level categories – artifact, infrastructure, and circulation analysis – should help instructors and authors identify productive frames for their purposes.

Findings about previous schema

Typologies from previous scholarship offered useful entry points for analyzing graphical abstracts; however, some categories seem more settled in 2018 than they were earlier. For example, only 1 of 43 abstracts expressed Lane et al.'s "topical-descriptive" rhetorical function. All others presented aspects of the argument and did not just introduce a topic. Similarly, only two of four options were consistently expressed for Yoon and Chung's "content" category: overview (10 of 42) or results (30 of 42). Likewise, the majority of abstracts only use text for labels (33 of 42). Textual annotations/commentaries appeared infrequently (2 of 42), which is not surprising given *Cell*'s directive for sparse text. However, even without that directive, abstracts for *Applied Surface Science* used text primarily for labels and occasionally for legends.

Working with earlier coding schemes also revealed insights about future uses of various typologies. For example, Hullman and Bach's layout typology offers useful vocabulary for design patterns; however, many abstracts apply multiple patterns; e.g. parallel forking processes. Thus their terms are descriptive but not necessarily categorical. Similarly, Yoon and Chung's "Content" category did not always code categorically; e.g. 2 of 42 abstracts included methods and results without providing a full overview.

Findings about disciplinary practices

Comparing Figure 24.1 (from *Cell*) and Figure 24.5 (from *Applied Surface Science*) illustrates differences in graphical abstracting practices between journals from different fields. (Note: Figure 24.5 is not from the sample but a later high-resolution example available under a Creative Commons license; nevertheless, it demonstrates the design practices typical of the sample.) First, they differ in their relation to other visuals. No aspect of this *Cell* abstract appears in the article. Similarly, 13 of 17 *Cell* abstracts in the sample are entirely "new" visuals, only three are "modified" versions of figures, and only one "integrates" visual components from several in-text visuals. In contrast, Figure 24.5 – an "integrated" graphical abstract – includes parts of three figures. This design practice is the most common in *Applied Surface Science* (13 of 25 abstracts). Three *Applied Surface Science* abstracts were "duplicates" of other figures, four were "modified" (e.g. labels removed but otherwise unchanged), and only five were novel. This

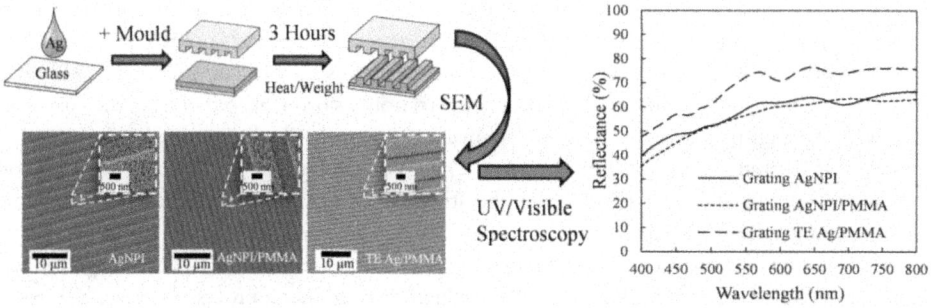

Figure 24.5   Graphical abstract from *Applied Surface Science*

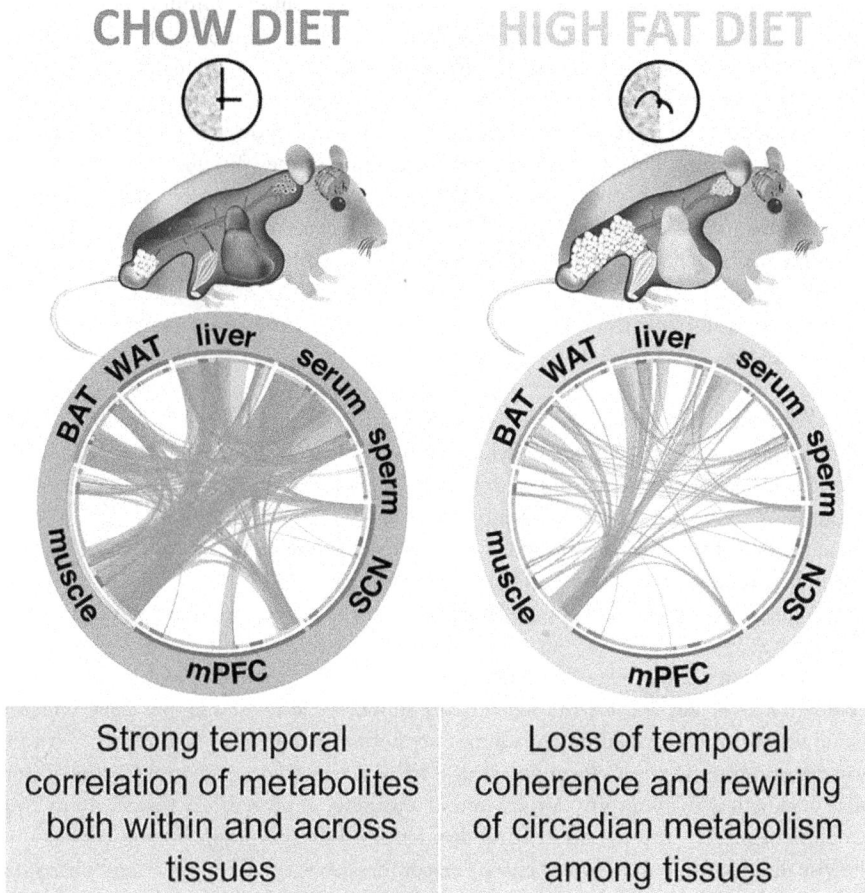

Figure 24.6   A graphical abstract that relies on visual and verbal antitheses to communicate its argument about circadian rhythms. The antitheses are even more striking in the original color image, which is available at https://doi.org/10.1016/j.cell.2018.08.042

difference might be attributed to the different guidelines of these journals. *Cell* discourages using figures appearing elsewhere; *Applied Surface Science* is silent on this point.

Second, these abstracts differ in the types of visual elements used to compose them. Although the *Cell* abstract includes some visualized data removed from its original context (e.g. the curves for mEPSC and paired-pulse ratio), its representational genre is primarily diagrammatic, like most *Cell* abstracts. Of 17 abstracts, all but 2 are diagrams. Only 6 of 25 abstracts from *Applied Surface Science* are purely diagrammatic; the others are combinations of data graphics, illustrations, micrographs and/or photographs. Four are micrographs elaborated by diagrams or micrographs, three depict structures at different scales (e.g. a micrograph shows a zoomed-in view of part of a diagram), and 12 are multivisual composites like Figure 24.5, which incorporates a process diagram, a series of micrographs (with inset micrographs of higher magnification), and a data visualization.

Another key difference is the relative frequency of design practices reliant on specific visual rhetorical figures. As Fahnestock (2002) demonstrated, rhetorical figures are far more than stylistic flourishes. They are form-meaning pairings that develop, epitomize, and reinforce lines of reasoning that scientists deploy verbally and visually. For example, throughout the history of science, scientists have deployed antithesis – a rhetorical figure composed of pairs of opposing terms (e.g. "To err is human; to forgive divine"). Textually, such strategies often appear in abstracts and titles – the compressed spaces where the rhetorical templates of figures can be especially impactful. Figural logics expressed visually can also be powerfully persuasive resources that circulate within and beyond scientific discourse communities (Buehl, 2016).

In *Cell*, authors regularly epitomize single-difference causal arguments with antitheses; e.g. Figure 24.1 presents differences between wild-type and modified genes through text, illustration, patterns of visualized data, and schematic diagrams. Similarly, Figure 24.6 deploys multiple oppositions to argue that "repertoires of tissue metabolism are linked and gated to specific temporal windows and how this highly specialized communication and coherence among tissue clocks is rewired by nutrient challenge" (Dyar et al., 2018, p. 1571):

- CHOW Diet (balanced energy diet) versus High Fat Diet (nutrient challenge)
- Functional clock (normal circadian rhythms) versus broken clock (disrupted circadian rhythms)
- Healthy mouse tissues versus unhealthy mouse tissues
- Numerous connections between tissues versus fewer connections
- Blue heading font and background shading versus yellow heading font and background shading (reproduced as darker and lighter gray in Figure 24.6)
- "Strong temporal correlation of metabolites" versus "Loss of temporal coherence"

Even if readers don't know what "temporal correlation of metabolites" involves, the pattern of binary differences is clear. The same figural logic supports the revised Figure 24.3, which uses visual and verbal antitheses to differentiate motility before and after the activation of Hv1. Such patterns are common in the *Cell* sample, with 8 of 17 abstracts using vertical parallel layouts to visualize antithetical relationships. Visual–verbal antitheses appeared less frequently in *Applied Surface Science* abstracts, and the antithetical distinctions were less emphatic and typically made in data visualizations (e.g. line or bar charts) or combinations of micrographs and charts. Only three abstracts present stark visual antitheses like those in *Cell*. Although some abstracts in both journals deploy other visual rhetorical figures as design strategies for demonstrating change or progression, none are used as frequently as antithesis, especially in *Cell*. However, it might be overreaching to claim these differences as discipline or journal specific. Additional research is needed on larger samples, with more journals, and across additional disciplines.

Findings about infrastructure

While browsing journal issues for this project, I often found that the resolution of graphical abstracts on online contents pages was too low to facilitate reading decisions. Approaching that frustration systematically through *Cell* and *Applied Surface Science* revealed that graphical abstracts have seemingly been more effective as spreadable social media than as browsing aids. However, recent updates to digital infrastructures might be changing this situation.

In 2018, many graphical abstracts presented on portal pages were not legible enough to influence reading behavior. If readers visited the website for *Applied Surface Science*, they would encounter a list of titles, and appearing beneath each title was a link marked "article preview" (step 1). Clicking that link (step 2) took these readers to the textual abstract, and they could then click another link to view a graphical abstract (step 3). But images were often of such low resolution (illegible text, blurry elements), they could not communicate an effective summary of the argument. To view a legible abstract, readers had to open the article (step 4), scroll past the highlights and textual abstract (step 5), and then click a link to download a high-resolution image (step 6). Although there were fewer steps for *Cell*, readers still needed to access the full article and scroll past textual summaries before being able to view an actionable graphical abstract. In short, too many steps were involved for these artifacts to facilitate browsing.

Updated infrastructures seem to be making better use of graphical abstracts. For example, by 2020, the default contents page for *Applied Surface Science* revealed graphical abstracts (thereby removing steps 1–3). However, some abstracts still fail to function as browsing aids because the resolution is too poor for legible text. For example, in the May 1, 2021, issue, labels on 4 of 21 abstracts on the contents page were illegible, though high-resolution versions could be accessed from the html article. *Cell* no longer includes any low-resolution graphical abstracts on contents pages; however, they are prominent both on articles' html versions (right below the highlights and summary) and PDF versions (on a separate cover page before any other summary element). *Applied Surface Science* does not include graphical abstracts in its PDFs. The larger point: even journals from the same publisher differ in how graphical abstracts are deployed within online infrastructures. Scientists and future scientists should study venue-specific constraints to create appropriate abstracts.

Findings about social media practices

Studies of graphical-abstract circulation have focused on the frequency of tweets and retweets of abstracts and other summarizing genres. Although interested in frequency, I also wanted to make additional qualitative assessments; therefore, I also coded tweets for source and purpose and noted whether tweets included graphical abstracts.

Tweet frequencies were uneven across the sample. By 2018, *Cell* actively distributed graphical abstracts through its Twitter feed, while *Applied Surface Science* tweeted less frequently through the Elsevier Physics feed. Unsurprisingly, *Cell* articles generated more tweets. Besides the original posts (17 tweets from *Cell*), the articles from September 6, 2018, were part of 239 other tweets. Articles from *Applied Surface Science* appeared in only two tweets – neither by the journal.

As Table 24.2 shows, the largest source of tweets were individual Twitter users seemingly without any connection to the article ($N = 163$), followed by aggregators, which include bots automatically retweeting content on specific themes ($N = 43$). Other sources include individuals with explicit links to the authors (colleagues offering congratulations), commercial interests (vendors, other journals), and other institutions (e.g. university and government labs).

For purpose, I coded any tweet without another distinct purpose as "information sharing" – by far the largest type (221 tweets). Most tweets consisted of only article titles and links without any additional commentary, though some included additional text: e.g. "An interesting study on [topic]" or "This is so cool!" Forty-nine of these tweets included graphical abstracts. Although any tweet linking to an article might be considered "information sharing," 37 tweets indicated additional purposes – article promotion, product promotion, celebration, collegial support, humor, argument, and memorializing deceased colleagues. (See Table 24.3.) Such tweets occurred less frequently than information sharing; nevertheless, they demonstrate that social media users circulate scientific content for various purposes. However, no clear pattern emerged of graphical abstracts being tied to specific purposes.

Finally, social media analysis demonstrates that graphical abstracts function as spreadable media. Although a majority (184 of 258) did not include graphical abstracts, nearly a third did (17 publisher tweets, plus 57 others). Further research on tweeting behavior could reveal whether including graphical abstracts in retweets is purposeful or coincidental.

Table 24.2 Sources of tweets

Source	Total tweets	Tweets containing graphical abstracts
Other individual	163	41
Aggregator	43	8
Publisher	17	17
Lab	15	2
Vendor	5	2
Colleague	4	1
Other journal	3	0
Author	2	2
Government lab	2	0
Library	2	0
Author's institution	1	0
Institutional repository	1	1
Total	258	74

Table 24.3 Purposes of tweets

Purpose	Total tweets	Tweets containing graphical abstracts
Information sharing (default)	221	49
Promotion (including 17 from *Cell*)	22	18
Argument	4	1
Celebration	3	2
Humor	3	2
Support	3	1
Memorial	1	0
Product promotion	1	1
Total	258	74

Conclusion

This chapter reviewed the history and rhetorical purposes of graphical abstracts and examined how researchers from different fields have approached them. Case studies of *Cell* and *Applied Surface Science* demonstrated a systematic way to approach graphical abstracts for research and teaching purposes. Given the sustained interest and increasing relevance of graphical abstracts, they should become a salient topic in scientific writing curricula.

References

American Chemical Society. (2012) 'Guidelines for table of contents/abstract graphics', http://pubsapp. acs.org/paragonplus/submission/toc_abstract_graphics_guidelines.pdf? (Accessed 1 May 2021)

Applied Surface Science. (2018) 'Author information pack', www.elsevier.com/journals/applied-surface-science/0169-4332?generatepdf=true (Accessed 1 May 2021)

Bredbenner, K., & Simon, S.M. (2019) 'Video abstracts and plain language summaries are more effective than graphical abstracts and published abstracts', *PloS one*, vol. 14, no. 11, p. e0224697, DOI: https://doi.org/10.1371/journal.pone.0224697

Buehl, J. (2016) *Assembling arguments: Multimodal rhetoric and scientific discourse*, Columbia, SC: University of South Carolina Press.

Cell Press. (2010) 'Cell Press graphical abstract guidelines', www.cell.com/pb/assets/raw/shared/figure-guidelines/GA_guide.pdf (Accessed 1 May 2021)

Chapman, S.J., Grossman, R.C., FitzPatrick, M.E.B., & Brady, R.R.W. (2019) 'Randomized controlled trial of plain English and visual abstracts for disseminating surgical research via social media', *British Journal of Surgery*, vol. 106, no. 12, pp. 1611–1616, DOI: 10.1002/bjs.11307

Dakour-Aridi, H. et al. (2019) 'Association between the choice of anesthesia and in-hospital outcomes after carotid artery stenting', *Journal of Vascular Surgery*, vol. 69, no. 5, pp. 1461–1470, DOI: 10.1016/j.jvs.2018.07.064

Dyar, K.A. et al. (2018) 'Atlas of circadian metabolism reveals system-wide coordination and communication between clocks', *Cell*, vol. 174, no. 6, pp. 1571–1585, DOI: 10.1016/j.cell.2018.08.042

Fahnestock, J. (2002) *Rhetorical figures in science*, Oxford: Oxford University Press.

Guinda, C.S. (2017) 'Semiotic shortcuts: The graphical abstract strategies of engineering students', *HERMES-Journal of Language and Communication in Business*, vol. 55, pp. 61–90, DOI: https://doi.org/10.7146/hjlcb.v0i55.24289

Hendges, G.R., & Florek, C.S. (2019) 'The graphical abstract as a new genre in the promotion of science', in M. Luzon & C. Perez-Llantada (eds.), *Science communication on the internet: Old genres meet new genres*, Amsterdam: John Benjamins Publishing Company, pp. 59–79.

Hullman, J., & Bach, B. (2018) 'Picturing science: Design patterns in graphical abstracts', *Diagrammatic Representation and Inference*, conference proceedings, International Conference on Theory and Application of Diagrams, Edinburgh, pp. 183–200, 18–22 June.

The Journal of Clinical Investigation. (2018) "Graphical abstracts", www.jci.org/kiosks/publish/graphical (Accessed 1 May 2021)

Ibrahim, A.M. (2016) 'A primer on how to create a visual abstract', *SurgeryRedesign.com*, www.SurgeryRedesign.com/resources (Accessed 1 May 2021)

Ibrahim, A.M., Lillemoe, K.D., Klingensmith, M.E., & Dimick, J.B. (2017) 'Visual abstracts to disseminate research on social media: A prospective, case-control crossover study' *Annals of Surgery*, vol. 266, no. 6, pp. e46-e48, DOI: 10.1097/SLA.0000000000002277

Koo, K., Aro, T., & Pierorazio, P.M. (2019) 'Impact of social media visual abstracts on research engagement and dissemination in urology', *The Journal of Urology*, vol. 202, pp. 875–877, DOI: 10.1097/JU.0000000000000391

Lane, S., Karatsolis, A., & Bui, L. (2015) 'Graphical abstracts: A taxonomy and critique of an emerging genre', *Proceedings of the 33rd Annual International Conference on the Design of Communication*, SIGDOC'15, July 16–17, Limerick, Ireland, pp. 1–9, DOI: https://doi.org/10.1145/2775441.2775465

Lindquist, L.A., & Ramirez-Zohfeld, V. (2019) 'Visual abstracts to disseminate geriatrics research through social media', *Journal of the American Geriatrics Society*, vol. 67, no. 6, pp. 1128–1131, DOI: 10.1111/jgs.15853

Marcus, E. (2010) '2010: A publishing odyssey', *Cell*, vol. 140, no. 1, p. 9, DOI: 10.1016/j.cell.2009.12.048

McGrath, F. et al. (2020) 'Structural, optical, and electrical properties of silver gratings prepared by nano-imprint lithography of nanoparticle ink', *Applied Surface Science*, vol. 537, p. 147892, DOI: https://doi.org/10.1016/j.apsusc.2020.147892

Neurosurgery. (2020) '*Neurosurgery* instructions for authors', https://academic.oup.com/neurosurgery/pages/Author_Guidelines

Patience, G.S., Boffito, D.C., & Patience, P. (2015) *Communicate science papers, presentations, and posters effectively*, London: Academic Press.

Yoon, J., & Chung, E. (2017) 'An investigation on graphical abstracts use in scholarly articles', *International Journal of Information Management*, vol. 37, no. 1, pp. 1371–1379, DOI: https://doi.org/10.1016/j.ijinfomgt.2016.09.005

Zhang, P., Lu, H., Peixoto, R.T., Pines, M.K., Ge, Y., Oku, S., Siddiqui, T.J., Xie, Y., Wu, W., Archer-Hartmann, S., Yoshida, K., Tanaka, K.F., Aricescu, A.R., Azadi, P., Gordon, M.D., Sabatini, B.L., Wong, R.O.L., & Craig, A.M. (2018) 'Heparan sulfate organizes neuronal synapses through neurexin partnerships', *Cell*, vol. 174, no. 6, pp. 1450–1464, DOI: 10.1016/j.cell.2018.07.002

25

DOING SCIENCE DIGITALLY

The role of multimodal interactive representations in scientific rhetoric

Lauren E. Cagle

Introduction

Nontextual forms of representation are critical elements of contemporary communication among scientists. As Gross et al. put it in their landmark study of the scientific article genre, "it is impossible to conceive of the argumentative practices in twentieth-century science without their visual representations (tables and figures)" (2002, p. 201). Visual representation's prominence is underscored by scientists' own descriptions of their reading practices, which often involve using figures as a primary tool for engaging with the scientific literature (DeVasto, Chapter 22 of this volume; Hutto, 2008, p. 112; Pain, 2017).

Thanks to the expanding affordances of digital scientific publishing, the last several decades have seen an increasing use of multimodal interactive representations (MIRs). I use the word "representation" rather than "visualization" to acknowledge the use of exclusively nonvisual modes to represent scientific knowledge or data. For example, the *Journal of Sound and Vibration* publishes sound clips in articles (2020); while a sound clip's user interface may be visual, the clip represents scientific data orally. Moreover, the use of multimodal affordances to build accessibility into communicative elements – such as alt text for images – means that even primarily visual representations may be accessed by users in other modes (e.g. aurally with a screenreader or haptically with Braille output).

Scientists themselves have noted an increasing use of MIRs and argued for their value (Vivanco et al., 2019; Perkel, 2018; Dang et al., 2018; Newe, 2016; Weissgerber et al., 2016; Barnes et al., 2013). *Nature* technology editor Jeffrey M. Perkel argues that static two-dimensional visualizations are limited in their ability to support robust uptake, critique, and reproduction of scientific knowledge, as they "are divorced from the underlying data, which prevents readers from exploring them in more detail by, for instance, zooming in on features of interest" (2018, p. 133). In other words, he writes, "static figures are just one perspective on the data" (2018, p. 134), whereas interactive visualizations allow readers to dig into data themselves.

Yet the move to MIRs is hardly a straightforward one, as evidenced by the many research articles, workshop proceedings, and other resources for producing, hosting, and circulating such representations (see McMahon, 2010; Barnes et al., 2013; Takeda et al., 2013; Newe, 2016). As a genre, MIRs are in a stage of active innovation (Miller, 2016), complicating efforts to capture in a single chapter the full breadth of MIRs' production and use, from manipulatable

 DOI: 10.4324/9781003043782-29

views of MATLAB figures ('Interactive journal figures', 2014) to three-dimensional molecule renderings (Dang et al., 2018) to complex network simulations (Wu et al., 2016). Moreover, the topic of MIRs invites attention not only to such visualizations as a way of *communicating* science *after* it is done but also as a way of *doing* science in the first place (Wickman, 2013; Fernández-Fontecha et al., 2018). Put simply, visualization can "facilitate insight and discovery" (Bayer, 1991, p. 224), for example, when advances in microscopy enable molecular biologists to "look deep into solid tissue and figur[e] out the dynamics of what's going on" (John White quoted in Eisenstein, 2006, n.p.). And, of course, the boundaries between these two uses – to *communicate* and to *do* science – are porous, in part because both call on scientists to *make arguments* about science, whether in the pages of a journal or in the more ephemeral contexts of research group conversations, conference Q&As, and other kairotic spaces (Price, 2011).

In the following sections, I first define multimodality and interactivity, then provide a recent history of MIRs' development and incorporation into scientific texts. I then analyze MIRs through two rhetorical frameworks: (1) the classical canons of Western rhetoric and (2) extant taxonomies of scientific visualization. In closing, I suggest future research directions for rhetorical studies of MIRs.

Defining multimodality and interactivity

Defining multimodality and interactivity will allow us to distinguish MIRs from the broader category of scientific visualizations, which are often already multimodal, due to the incorporation of both visual elements (e.g. Cartesian graphs, photographs, maps, and shape-based markup such as arrows) and typographic text (e.g. legends, captions, and axes labels). Moreover, the inclusion of a graphic, such as a figure, in an otherwise textual genre makes that genre itself already multimodal. In scientific communication, multimodal genres are the norm, not the exception (Lemke, 1998; Mohan et al., 2007; Gross, 2011). Lemke explains, "the 'concepts' of science . . . are semiotic *hybrids*, simultaneously and essentially verbal, mathematical, visual-graphical, and actional-operational. The actional, conversational, and written textual genres of science are historically and presently, fundamentally and irreducibly, *multimedia genres*" (1998, p. 87, emphasis in original). Given that scientific communication is already integrally multimodal, the following definitions provide us tools for cleaving out MIRs as distinct objects of analysis.

Defining multimodality

Multimodality is a well established concept in rhetoric, semiotics, and communication. Jewitt, Bezemer, and O'Halloran write, "If a 'means for making meaning' is a 'modality', or 'mode', as it is usually called, then we might say that the term 'multimodality' was used to highlight that people use multiple means of meaning making" (2016, p. 2), including nonlinguistic means. Kress writes that in fields such as sports, to which "the reach of *speech* or *writing* simply does not extend," "semiotic-conceptual work has to be and is done by means of other modes" (2010, p. 15). These other modes include "speech; still image; moving image; writing; gesture; music; 3D models; action; colour" (Kress, 2010, p. 28). Modes are not neatly distinguishable from one another: moving images consist of still images in quick succession; 3-D models may digitally render gesture; on-screen action may be in color and voiced-over.

Figure 25.1 illustrates the multimodality common to contemporary scientific visualizations. The figure includes two conventional graphic forms of representation: (1) connected hexagons and lines representing chemical structures and (2) a series of vertically oriented bars in a Cartesian

Chart 1. Halide Effect on the Reaction Outcome

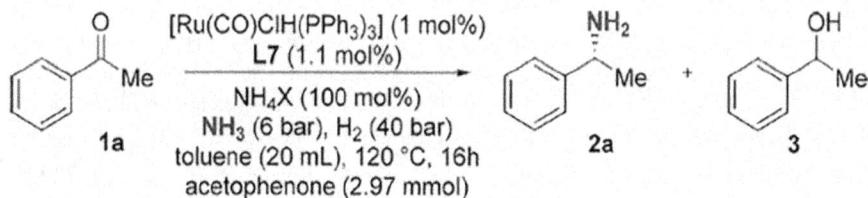

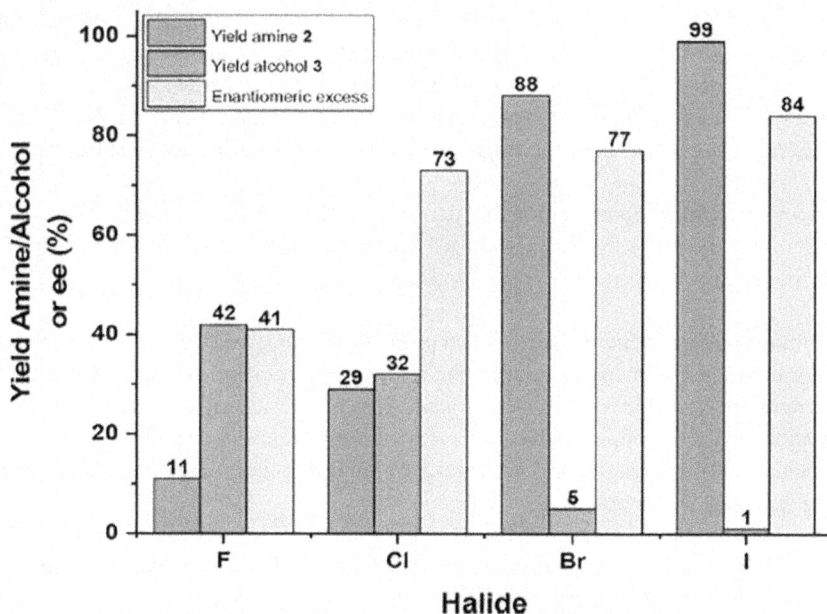

Figure 25.1 Multimodality common to contemporary scientific visualizations

Source: Copyright 2018 American Chemical Society. Reprinted with permission from Gallardo-Donaire, J. et al. (2018) 'Direct Asymmetric Ruthenium-Catalyzed Reductive Amination of Alkyl–Aryl Ketones with Ammonia and Hydrogen', *Journal of the American Chemical Society*, 140(1), pp. 355–361. doi: 10.1021/jacs.7b10496.

plane representing a dataset broken down by nominal class. This figure relies on multimodality to both meet genre conventions and convey information: skeletal formulae are connected by symbols (the arrow and the plus sign) and identified by textual abbreviations, and the bar graph's shapes are interpretable due to axis labels, a legend, and color contrasts.

Following from this example, my definition of multimodal in the context of scientific visualization and representation adapts Kress's argument:

> Multimodal scientific visualizations are intentionally designed data or concept representations in which semiotic-conceptual work is done by means of at least two modes, of which one is visual.

Notably, digitality is excluded from this definition, as multimodality does not require any of its modes to be digital (Cedillo and Elston, 2017, p. 7; Shipka, 2009, p. 347). For example, a printed scientific poster might be described orally by its author while being simultaneously interpreted into sign language. That said, a multimodal scientific *visualization* is a more specific thing than the event just described, as visualizations are bounded spatially rather than or in addition to temporally. So a multimodal scientific visualization isn't just any combination of one visual plus one other mode; it is specifically a combination of these modes intended to represent a dataset or concept within a bounded space (and perhaps time, as with video), whether that space is on a projector screen, a monitor, or a printed page.

Defining interactivity

In semiotic studies broadly, interactivity, or 'interaction,' is often used to describe the functioning of multimodal texts, in which modes 'interact' with one another to make meaning (see Kress and van Leeuwen, 1996; Fortune, 2005; Gross, 2009). In this chapter, I define *interactivity* somewhat differently, more in line with scholarship from the loosely connected fields of technical communication, information sciences, human–computer interaction, design, user experience, and experience architecture:

> Interactive scientific representations are data or concept representations which have an inherent and intentionally designed capacity to be changed in some way by a human user for a variety of purposes.

This definition treats interactivity as an inherent property of MIRs, meaning MIRs are interactive, whether or not users are interacting with them at any given moment. By locating the property of interactivity in the representation itself, rather than in a relational space between user and representation, I can distinguish MIRs' interactivity from the more general way in which *all* representations collaboratively produce meaning through audiences' interactive practices, such as reading, listening, and watching.

My definition is also indebted to Rawlins and Wilson's (2014) typology of "interactive data displays" (IDDs), which uses the balance of designer/user agency to theorize a continuum of interactivity with different levels ranging from complete designer agency (resulting in static data displays) to complete user agency (resulting in unconstrained user control of all data display elements). Industrial ecologists Vivanco et al. hit on a very similar interest, describing interactive visualization design as a process of "balancing user freedom with *the desire to communicate clear messages*" (2019, p. 525, emphasis added).

As with multimodality, references to "digitality" are absent from this definition. Since anything we interact with in order to access underlying processes and information is an interface, from books to fast food drive-through speakers, we can easily imagine a communicative role for nondigital scientific interfaces, such as the hands-on science museum demo and the wet lab. That said, MIRs are most typically IDDs embedded in digital scientific articles or stand-alone database interfaces.

Defining multimodal interactive representations

Combining my definitions of multimodal and interactive representations results in the following definition for MIRs:

> In scientific communication, multimodal interactive representations are intentionally designed data or concept representations which have an inherent and intentionally

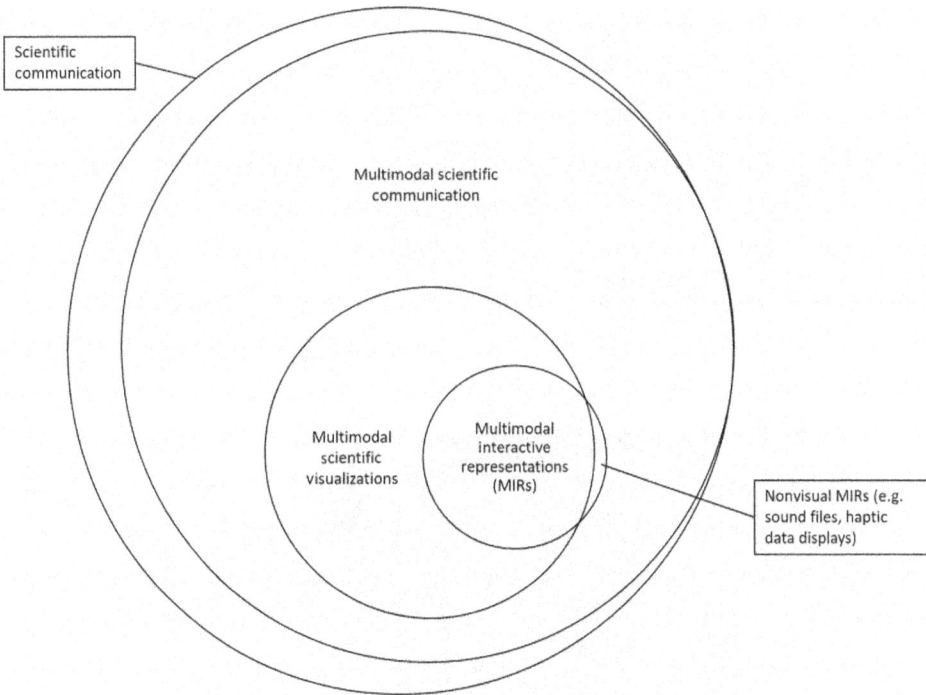

Figure 25.2 Conceptual model of MIR classification within the total set of scientific communication

designed capacity to be changed in some way by a human user for a variety of purposes and in which semiotic-conceptual work is done by means of at least two different modes.

As Figure 25.2 shows, the very presence of a multimodal visualization or MIR in a piece of scientific communication automatically makes that scientific communication itself multimodal, so very little of contemporary scientific communication falls outside the category of multimodality. Using the definition of MIRs, we can drill down within the broad category of scientific communication to focus on multimodal visualizations generally and MIRs specifically, including to see how the latter occasionally *doesn't* include visual modes.

To further illustrate my definition, I turn to two examples of MIRs: (1) a figure in a published peer-reviewed article and (2) a "data portal" enabling users to generate figures from a massive (and growing) dataset. The former, shown in Figure 25.3, is an example of what Rawlins and Wilson call a "Zoom and Pan" IDD.

While limited, the figure's interactivity is inherent to it, as it has built-in "zoom in," "zoom out," and "reset image size" hyperlinks, making it interactive independently of any user's ongoing interaction. Just to clearly explicate the difference I'm getting at, imagine that a reader opens this figure in a web browser, then uses the *browser's* zoom function to take a closer look. While the user would be *interacting* with the figure, the figure *itself* wouldn't be interactive independently of its context of use. So while its interactivity is minimal, the embedded zoom controls do make this figure an MIR.

A much less constrained MIR is the Allen Brain Atlas, a "data portal" to a gargantuan brain science dataset. Technically, the data portal comprises multiple 'atlases' and datasets, including a downloadable "Brain Explorer 2" software package which can be used to "view the human brain anatomy and gene expression data in 3-D" (Allen Institute for Brain Science, 2020) (see Figure 25.4).

Unlike Figure 25.3, which shows a relatively stable semiotic formation, Figure 25.4 shows us a portal into a wide variety of semiotic formations, whose precise combinations of modes, content, and design are constrained but not preordained. Thus data portals such as the Allen Brain Atlas differ from the MIR pictured in Figure 25.3 in three important ways: (1) they provide access to a much more varied dataset (which may or may not be technically "bigger,"

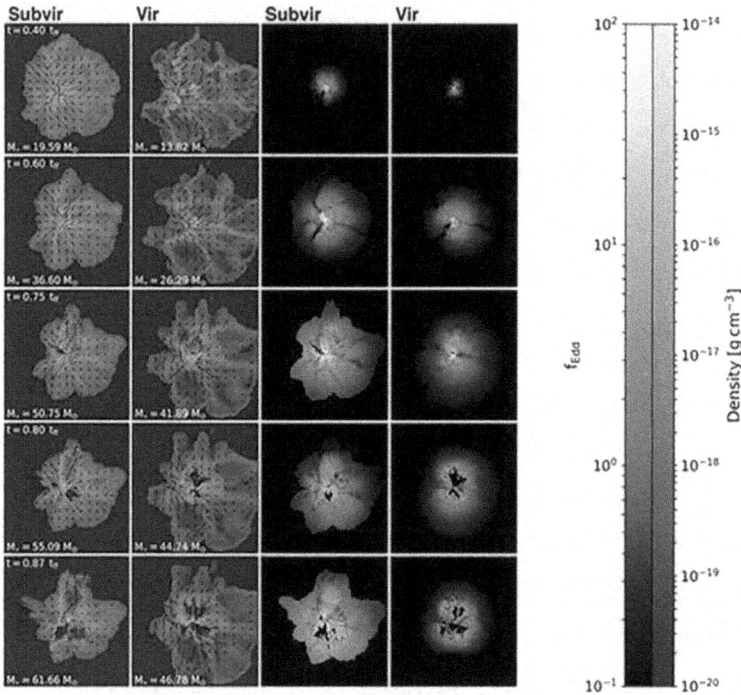

Figure 25.3   Density slices along the *y-z* plane of a collapsing, turbulent prestellar core into a massive stellar system for runs *SubVir* (far left column) and *Vir* (center left column) with velocity vectors overplotted. The velocity vectors are scaled as v in units of km s^{-1}, and we only over-plot vectors for densities $\geq 5 \times 10^{-20}$g cm^{-3}, The two right columns show the corresponding slices of the Eddington ratios ($f_{EM} = F_{WD}/F_{PM}$) for runs *SubVir* (center right column) and *Vir* (rightmost column). Each panel is (50 kau)^2, with the center of each panel corresponding to the location of the most massive star that has formed. The time of the simulation and mass of the most massive star are given in the upper left corner of the far left panels and the lower left corner of the panels in the two left columns, respectively

Source: Copyright AAS. Reprinted with permission from Rosen, A. L. et al. (2019) 'Massive-Star Formation via the Collapse of Subvirial and Virialized Turbulent Massive Cores', *The Astrophysical Journal*, vol. 887, no. 2, p. 108. doi:10.3847/1538-4357/ab54c6.

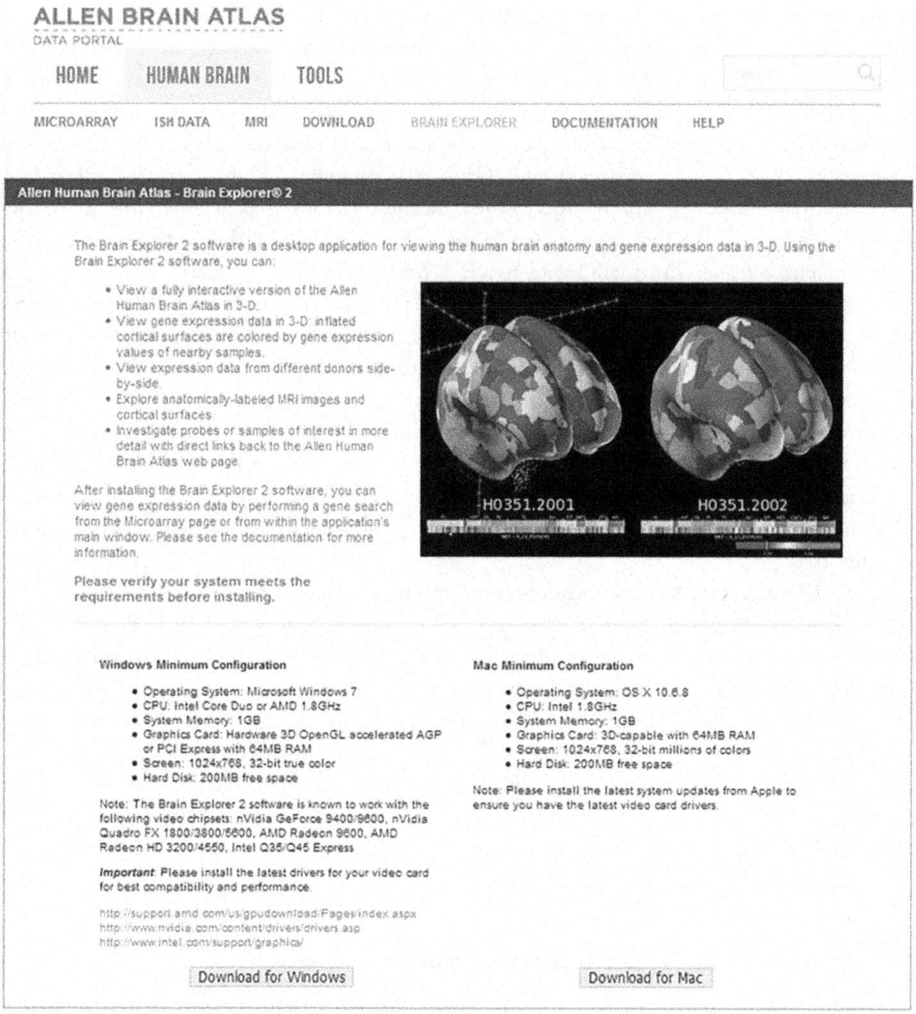

ALLEN BRAIN ATLAS
DATA PORTAL

HOME   HUMAN BRAIN   TOOLS

MICROARRAY   ISH DATA   MRI   DOWNLOAD   BRAIN EXPLORER   DOCUMENTATION   HELP

Allen Human Brain Atlas - Brain Explorer® 2

The Brain Explorer 2 software is a desktop application for viewing the human brain anatomy and gene expression data in 3-D. Using the Brain Explorer 2 software, you can:

- View a fully interactive version of the Allen Human Brain Atlas in 3-D.
- View gene expression data in 3-D: inflated cortical surfaces are colored by gene expression values of nearby samples.
- View expression data from different donors side-by-side.
- Explore anatomically-labeled MRI images and cortical surfaces.
- Investigate probes or samples of interest in more detail with direct links back to the Allen Human Brain Atlas web page.

After installing the Brain Explorer 2 software, you can view gene expression data by performing a gene search from the Microarray page or from within the application's main window. Please see the documentation for more information.

Please verify your system meets the requirements before installing.

H0351.2001   H0351.2002

Windows Minimum Configuration

- Operating System: Microsoft Windows 7
- CPU: Intel Core Duo or AMD 1.8GHz
- System Memory: 1GB
- Graphics Card: Hardware 3D OpenGL accelerated AGP or PCI Express with 64MB RAM
- Screen: 1024x768, 32-bit true color
- Hard Disk: 200MB free space

Note: The Brain Explorer 2 software is known to work with the following video chipsets: nVidia GeForce 9400/9600, nVidia Quadro FX 1800/3800/5600, AMD Radeon 9600, AMD Radeon HD 3200/4550, Intel Q35/Q45 Express

Important: Please install the latest drivers for your video card for best compatibility and performance.

http://support.amd.com/us/gpudownload/Pages/index.aspx
http://www.nvidia.com/content/drivers/drivers.asp
http://www.intel.com/support/graphics/

Mac Minimum Configuration

- Operating System: OS X 10.6.8
- CPU: Intel 1.8GHz
- System Memory: 1GB
- Graphics Card: 3D-capable with 64MB RAM
- Screen: 1024x768, 32-bit millions of colors
- Hard Disk: 200MB free space

Note: Please install the latest system updates from Apple to ensure you have the latest video card drivers.

Download for Windows   Download for Mac

Figure 25.4 Screenshot from the Allen Human Brain Atlas

Source: Copyright 2010 Allen Institute for Brain Science. Allen Human Brain Atlas. Available from human.brain-map.org. Screenshot printed with permission from the Allen Human Brain Atlas (https://human.brain-map.org/static/brainexplorer).

depending on the amount of data encoded in Figure 25.3); (2) they provide far more options for manipulating those data, including limiting which datasets and pieces of datasets are visualized at any given time; and (3) they are primarily intended for *doing* science rather than *communicating* it to other scientists. The key point, though, is that *both* the data portal and the zoom-able astronomical figure are MIRs; the category offers a broad umbrella.

A brief recent history of MIRs

While a history of MIRs could span centuries, I'll begin with predictions made about them in the 1990s and 2000s. (For a more extensive history of scientific data visualization, see DeVasto,

Chapter 22 of this volume.) Thanks to late twentieth-century increases in computing power and digital affordances, science communication researchers anticipated MIRs' increased ubiquity and utility well before the development of the web. Thirty years ago, Nancy L. Bayer predicted that "video and other media that can visually communicate visualizations will no doubt become more frequently used for publishing research results that were obtained with scientific visualization" (1991, p. 225). A decade later, Gross et al. opined that "visual images will undergo a . . . transformation: there will be links to the data and methods used to generate figures; there will be visual images that move and make sounds; and there will be three-dimensional images that the scientist-reader can manipulate to view from different perspectives" (2002, p. 233). Five years on, J. S. Mackenzie Owen argued that "the sciences can benefit from simulations, animations, stereo photographs, 3-D visualization, etc. that are made possible by the digital format" (2007, p. 44). Despite the anticipated benefits of MIRs, however, only 27% of pre-2002 online scientific publications contained any kind of multimedia (e.g. video, audio, software, and animations) (Mackenzie Owen, 2007, p. 144). This scarcity may have resulted from a scarcity of tools for creating MIRs; since 2002, however, dozens of tools for generating and circulating MIRs have been developed across scientific domains. Table 25.1 lists tools for creating MIRs and identifies the output types they produce, as well as the modalities they support for inputs and outputs.

Many MIR tools build on one another; some can be used in coordination (e.g. by composing static images in Biorender, then animating them in Adobe Illustrator), while others support formal integration (e.g. cellPACK, a specialized version of autoPACK, has sanctioned plugins for Blender and Maxon Cinema4D). In some cases, such integration is what allows tools to layer additional modalities onto outputs (e.g. Tilemap and ESRI integration enabling 3-D mapping). Many of these tools are built using the same languages (e.g. Python and R), making open source tools particularly appealing to scientists with specific use cases. Additionally, a number of guides have been published for teaching scientists how to design and create MIRs (e.g. Yau, 2011; Frankel and DePace, 2012; Ferster, 2013; Grant, 2018; Tominski and Schumann, 2020). Efforts to promote MIRs' creation and use seem to be going strong.

Rhetorical implications of MIRs

Pinning down the rhetorical implications of MIRs is the work of many scholars and extends far beyond this single chapter. Moreover, the kind of review or overview approach appropriate to a handbook chapter is complicated for MIRs by the fact that there isn't an established vocabulary for them. Yet there are certainly existing rhetorical frameworks we could apply to MIRs to begin shaping them into recognizable objects of rhetorical attention and to uncover areas and questions for further rhetorical investigation. To forward that work, I offer here two rhetorical frameworks for making sense of MIRs:

1   The five Western canons of rhetoric
2   Established taxonomies for scientific visualization, with a focus on taxonomies developed by Gross et al. (2002), Desnoyers (2011), and Rawlins and Wilson (2014)

I'll use two examples of MIRs to illustrate the applications of both these frameworks. The first example draws from Reaxys, a chemistry database that "consist[s] of deeply excerpted compounds and related factual properties, reaction and synthesis information as well as bibliographic data, navigated and displayed via an actionable interface" (Elsevier, 2020). The database can be queried either through a "quick search" or a "query builder." While both query

Table 25.1 Tools for creating multimodal and interactive representations

Tool	Output(s)	2D visuals	3D visuals	Audio	Haptic	Animated	Interactive	VR
Adobe Illustrator	General 2-D graphics	x				x		
Adobe Photoshop	General 2-D graphics	x				x		
Autodesk Maya	3-D graphics		x			x		
BioBlender	Protein visualizations			x		x		
Biorender	Life sciences figures	x						
Blender	3-D graphics			x		x		
Bokeh	Data visualizations	x					x	
CartoDB	Maps	x	x			x	x	
cellPACK project	Cell-scale structures		x					
Cytoscape	Protein interaction networks	x					x	
D3	Data visualizations	x				x	x	
EpiVis	Epidemiological time-series data	x					x	
Esri ArcGIS	Maps	x	x			x	x	x
geojson.io	Maps	x					x	
GIMP	General 2-D graphics	x				x		
GRASS GIS	Maps	x	x			x	x	
htmlwidgets	Data visualizations	x					x	
Inkscape	General 2-D graphics	x						
ipywidgets	User interfaces for data manipulation	x	x				x	
Jmol	Molecular structures	x	x			x	x	x
Keynote	General 2-D graphics	x				x		
Map Warper	Maps	x					x	
Mapbox Studio	Maps	x	x			x	x	x
matplotlib	Data visualizations	x	x			x	x	
Maxon Cinema 4D	3-D graphics		x			x		
Molecular Flipbook	Molecular models		x			x		
MS PowerPoint	General 2-D graphics	x				x		
Neatline	Maps and timelines	x				x	x	
Plotly's Dash for Python	Web apps for data visualization	x		x			x	
pygal	Vector graphics	x					x	
PyMOL	Molecular structures	x	x			x		
Python Molecular Viewer (PMV)	Molecular structures	x	x			x	x	
QGIS	Maps	x	x			x	x	
Shiny for R	Interactive web apps	x	x				x	
SimpleMap	Maps	x					x	
SimpleMappr	Maps	x						
Tableau	Data visualizations	x	x			x	x	
UCSF ChimeraX	Molecular structures	x	x			x	x	x
Visual Molecular Dynamics (VMD)	Biomolecular systems		x		x	x	x	x

functions include traditional text-input forms, these interfaces also allow users to literally *draw* their queries via user-friendly multimodal and interactive graphical user interfaces (GUIs) (see Figure 25.5).

Search results are presented from a variety of potential database object types, such as substances, reactions, and documents. The substances and reactions results include interactive visualizations of relevant structures. Figure 25.6 shows search results for "atenolol," including 121 substance and 159 reaction results.

Clicking the magnifying glass on the top result (Figure 25.6a) triggers a pop-up structure visualization (Figure 25.6b). Clicking "View in 3D" yields a manipulatable structure that can, among other things, be rotated along the x-, y-, and z-axes in three-dimensional space (Figure 25.6c). In other words, Reaxys queries return MIRs, which can be used to conduct chemical research, find and order necessary chemical substances, and prepare research results for publication.

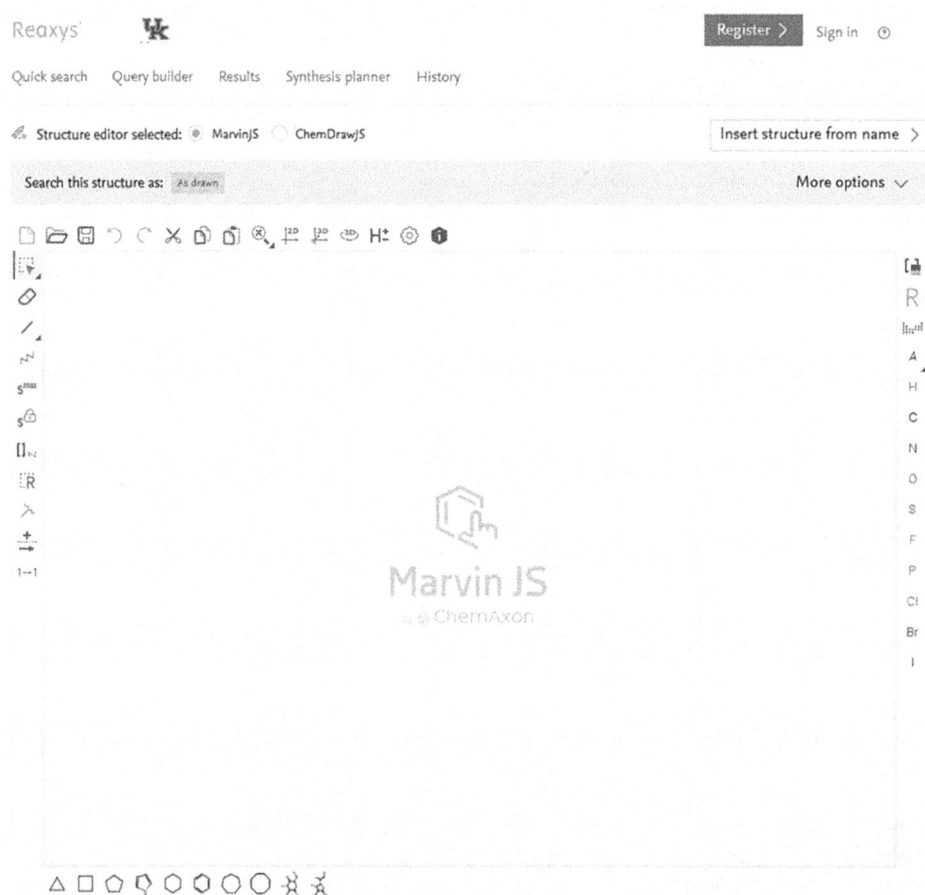

Figure 25.5 Screenshots from the Reaxys database

Source: Printed in accordance with the STM Permissions Guidelines (www.stm-assoc.org/intellectual-property/permissions/permissions-guidelines/). Elsevier, the publishing company which owns Reaxys, is an STM Guidelines signatory. Available from www.reaxys.com/.

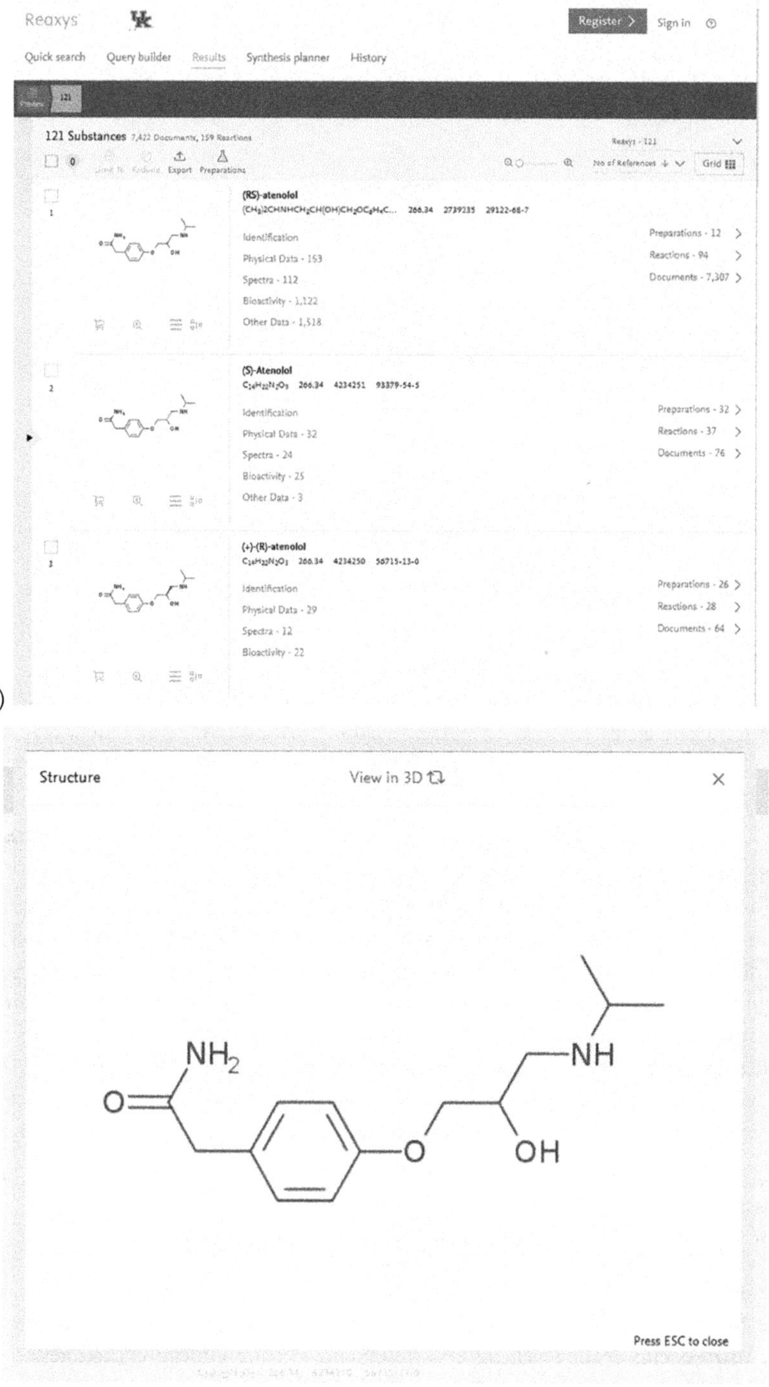

(a)

(b)

Figure 25.6   Screenshots from the Reaxys database

Source: Printed in accordance with the STM Permissions Guidelines (www.stm-assoc.org/intellectual-property/permissions/permissions-guidelines/). Elsevier, the publishing company which owns Reaxys, is an STM Guidelines signatory. Available from www.reaxys.com/.

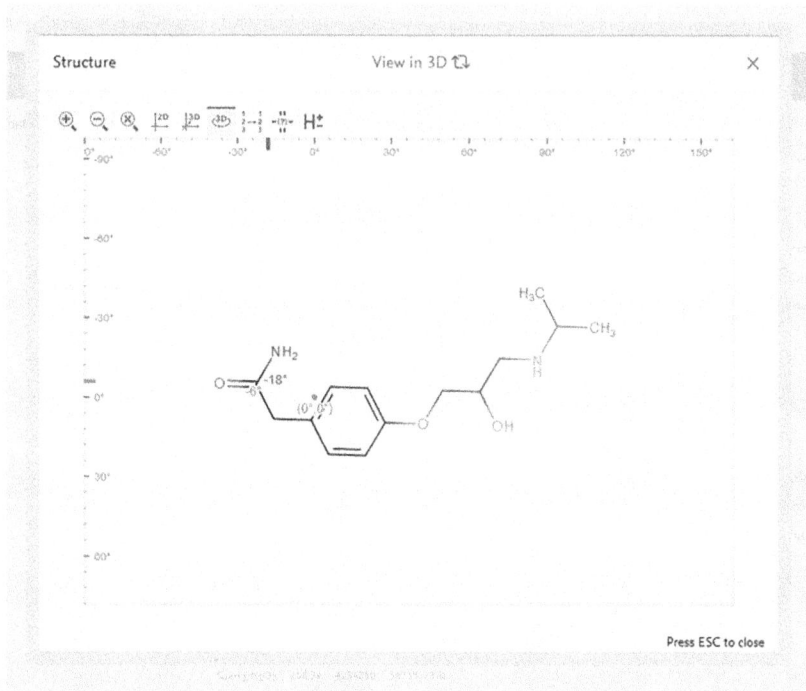

(c)

Figure 25.6   (Continued)

My second MIR example is a multilayered interactive online map produced by the Kentucky Geological Survey (KGS) (Figure 25.7).

This multimodal interactive GUI is likely more familiar to nonexperts than Reaxys's GUI, thanks to the ubiquity of online mapping services such as Google Maps. The KGS Interactive Map's affordances go well beyond such popular services, however. For example, the "Layers" menu shown in Figure 25.7 allows users to map specific datasets, narrowing their focus and manipulating the map in targeted ways. In short, it's possible for ten users to 'create' ten entirely different maps from this MIR, using a variety of both common (zoom, drag, pan, feature-of-interest pop-ups) and not so common (layer management) tools. This map is actually many maps, populated by many datasets. Multimodality and interactivity allow many maps to layer into one.

MIRs and the canons of rhetoric

The canons of classical Western rhetoric are useful analytical tools because they direct us to consider the lifecycle of a rhetorical object. Such a canon-based analysis might go as follows:

Invention occurs at the levels both of doing and of communicating the science. Being able to literally *see* data differently thanks to accessible MIR composition tools could very well lead to entirely new scientific discoveries. Reaxys's multimodal search options provide one rather mundane example of how this might occur; drawing chemical structures allows the researcher to input them representationally to find the same structure under different names, as when a structure is commercially licensed in some jurisdictions but not in others. Rhetoricians might

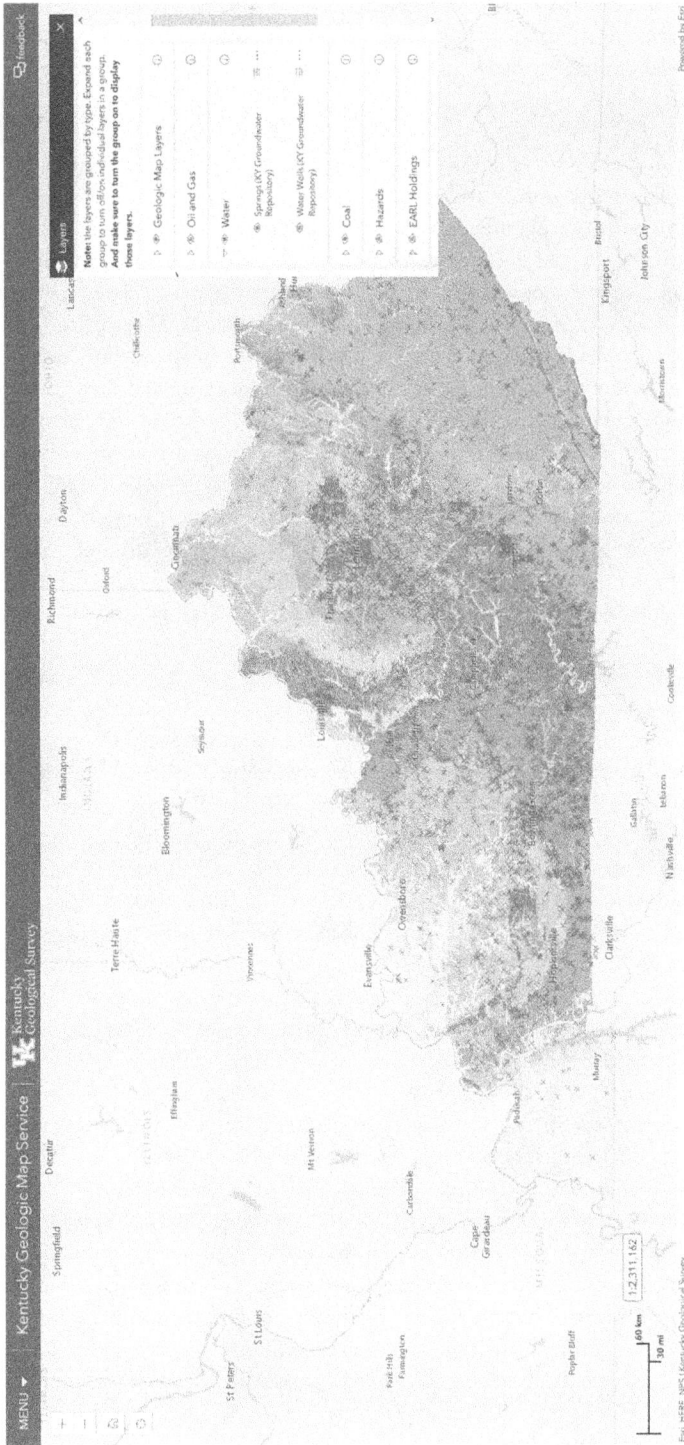

Figure 25.7 Screenshot of the Kentucky Geologic Map Service

Source: Printed with permission from the Kentucky Geological Survey, n.d. (www.uky.edu/KGS/). The Kentucky Geologic Map Service uses the Esri "World Topographic Map" as a base map. Available from https://kgs.uky.edu/kygeode/geomap/.

ask how MIRs contribute to scientists' ability to *do* science. (E.g. has Reaxys's MIR search function prompted experiments that otherwise wouldn't have happened?) We might also ask how MIRs affect which science scientists choose to do and communicate, as the ability to represent knowledge in new ways might prompt scientists to pursue new kinds of knowledge. (E.g. does layer-based mapping prompt scientists to pose research questions about data that can be easily mapped?)

As for **arrangement** and **style**, there are organizational and stylistic components both (1) to the internal design of an MIR and (2) to its external integration into genres of scientific communication. The canons of arrangement and style prompt us to consider not only the spatial arrangement and visual styling of 2-D representations but also, say, the temporal arrangement and oral styling of multimodal representations, not to mention the experience architecture and responsive element styling of multimodal interactive representations. Rhetoricians might ask how three-dimensional objects make meaning in two-dimensional space; how color, temporality, size, and other semiotic modes shape scientific communication; and how users' engagement with these modes and also with movement, buttons, links, and other interactive elements shapes their experiences. Delving a layer deeper, and connecting arrangement and style to invention, we might also ask how the tools for composing MIRs are designed. This question is a central question in technical communication, as technologies are never neutral but rather built, arranged, and styled to accommodate and promote particular users and outcomes; Reaxys seems to be built for chemists, while the user base for the KGS Interactive Map seems to be at least somewhat more diverse, given the more familiar interface and the presence of instructional text.

Of course, invention, arrangement, and style may well be constrained by publication outlets' specific technological capacities, which can be understood through the canons of **memory** and **delivery**. For both Reaxys and the KGS Interactive Map to work, there has to be sufficient memory to store the data that populate them. At a more restrained scale, such as the restricted interactivity of Figure 25.3's astrophysics visualization, MIRs still require their publication outlets to store the electronic encoding data making them up, to provide sufficient processing power to enable timely user interactions, and to present them, repeatedly, to multiple users, potentially all at once, via different screens.

Rhetoricians might then ask, for example, how the affordances of MIR memory and delivery might obscure or unveil scientific knowledge. The canon of memory, as argued by Johnson (2020), is now vested in machines as much as in human rhetors, if not more so. The scientific publication ecology has become a series of what Johnson terms "memory infrastructures," which "do not merely document pieces of a past. . . . [T]hey anchor, shape, and compose remembering and forgetting" (2020, p. 15). Which scholarly knowledge is anchored, shaped, and composed is driven by the memory's capacity to store and (re)present MIRs. Delivery is, of course, also driven by technological affordances; where and when MIRs appear is a function of where and when the technological capacity exists to make them appear.

And, as always, the canons are not extricable from one another. As delivery has shifted from handprinted letters to press-printed publications to pixel-based screen presentations, our expectations of scientific visualizations' style have also shifted. No longer does reproducing a data-rich map require laborious hand-copying. I can copy and paste an image from a digital map in less time than it took me to type this sentence. Thus do ever more complex and sophisticated MIRs become the standard for scientific publishing, and in a clear change of stylistic expectations driven by changes in delivery, maps *cannot* be hand-drawn if they are to be taken seriously as scientific representations.

Applying rhetorical taxonomies to scientific multimodal and interactive representations

Several established scientific visualization taxonomies might apply to MIRs. For example, Gross et al. (2002, pp. 64, 200) categorize visualizations based on type and function, whereas Johnson and Hertig, drawing from McGill, categorize based a combination of purpose and audience: (1) data analysis with researchers, (2) communication with peers, (3) education of students, and (4) outreach to the public (2014). Inspired by Linnean principles and organized according to types of information and sign content, Desnoyers's (2011) more complex taxonomy combines *type of data display* with *type of data* to establish distinct categories. Desnoyers, however, explicitly rules interactive and complex visuals out of bounds for his taxonomy, as "taxonomies can only consider singular cases" (2011, p. 124).

Nonetheless, MIR data display types are sometimes tied to specific types of data, such as protein–protein interactions, genome marks (see Marx, 2015), and "the branching structure of plant root systems" (Perkel, 2018, n.p.), so Desnoyers' taxonomy could be inspirationally useful at least. Of course, even if some data types, such as the chemical structures displayed in Reaxys or the 2-D representation of geographic sites on the KGS map, have well established forms of visual representation, many other data types can be displayed in a variety of ways. The proliferation of MIR types makes it a challenge to incorporate them into Desnoyers's taxonomy for precisely the reasons he describes, though; if multimodality or interactivity means that users are essentially engaging with or producing a series of representations rather than a single one, his taxonomy would have to extend to new categories capturing such stacking of representations. Essentially, Denoyers's taxonomy takes a static view of what counts as a representational scientific object in need of categorization, so it's an uneasy fit at best with MIRs.

To find a more multimodal and interaction-friendly taxonomy, I turn to Rawlins and Wilson (2014), whose taxonomy of "interactive data displays" (IDD) was previously discussed; their key classification features are levels of interactivity and type of agency. Classifying via interactivity and agency levels lets us make meaningful distinctions between the relatively simple interface controls of the Reaxys query results and the much more complex and option-rich interface of the KGS Interactive Map. Rawlins and Wilson's taxonomic schema holds perhaps the greatest promise for developing a taxonomy that can capture the full spectrum of MIRs as discussed in this chapter.

MIRs largely overflow the bounds of established taxonomies, raising the question of whether MIRs call for extensions or amendments to these existing taxonomies or whether they call on us to develop a whole new taxonomic schema accounting for their particularities. While developing and proofing a full MIR taxonomy is beyond the scope of this chapter, it would make a very useful project for a rhetorician interested in further developing scholarship in this area.

Future rhetorical research on MIRs

I close with some suggestions for future rhetorical research directions on MIRs. Some of the many issues I touch on barely or not at all in this chapter include:

- the role of scientific discipline in the development and adoption of MIRs;
- the publication issues raised by MIRs, such as who is responsible for integrating them into journals or other publication outlets, how they factor into peer review, and what long-term storage solutions exist for them;

- the need to ensure full accessibility of both the tools for composing MIRs and the MIRs produced through that composition process;
- the relationships among multimodality, interactivity, and accessibility as coproducing characteristics of scientific representations;
- the rhetorical figures and strategies enabled by MIRs, following Fahnestock's (2003) arguments about tables and the role of "verbal and visual parallelism" in scientific argument; and
- the functioning of MIRs as kinds of evidence, inspired by distinctions such as Perelman and Olbrechts-Tyteca's opposition of an example (meant to establish a rule) and an illustration (meant to "strengthen adherence to a known or accepted rule") (1971, p. 357).

In this chapter, I have established that multimodal and interactive representations (MIRs) are increasingly common in a variety of intrascientific communication contexts and have provided definitions to help begin identifying and classifying MIRs. Rhetoricians of science and science communication scholars will be well served by developing a basic understanding of this broad category of emerging genres and by contributing to our knowledge of it through further research.

References

Allen Institute for Brain Science. (2020) 'Allen brain map', https://portal.brain-map.org/

Barnes, D.G. et al. (2013) 'Embedding and Publishing Interactive, 3-Dimensional, Scientific Figures in Portable Document Format (PDF) Files', *PLOS ONE*, 8(9), p. e69446. doi: 10.1371/journal.pone.0069446.

Bayer, N.L. (1991) 'Scientific Visualization: New Opportunities for Technical Communicators', *Technical Communication*, 38(2), pp. 223–227.

Cedillo, C.V., & Elston, M.M. (2017) 'The Critique of Everyday Life; Or, Why We Decided to Start a New Multimodal Rhetorics Journal', *Journal of Multimodal Rhetorics*, 1(1), p. 8.

Dang, J. et al. (2018) 'Smart access to 3D structures', *Nature Reviews Chemistry*, 2(7), pp. 95–96. doi: 10.1038/s41570-018-0021-y.

Desnoyers, L. (2011) 'Toward a taxonomy of visuals in science communication', *Technical Communication*, 58(2), pp. 119–134.

Eisenstein, M. (2006) 'Something to see', *Nature*, 443(7114), pp. 1017–1019. doi: 10.1038/4431017a.

Elsevier. (2020) 'Reaxys', https://www-reaxys-com.ezproxy.uky.edu/#/about-content (Accessed 22 October 2020).

Fahnestock, J. (2003) 'Verbal and Visual Parallelism', *Written Communication*, 20(2), pp. 123–152. doi: 10.1177/0741088303020002001.

Fernández-Fontecha, A. et al. (2018) 'A multimodal approach to visual thinking: the scientific sketchnote', *Visual Communication*, 18(1), pp. 5–29. doi: 10.1177/1470357218759808.

Ferster, B. (2013) *Interactive visualization: Insight through inquiry*, Cambridge, MA: MIT Press.

Fortune, R. (2005) '"You're not in Kansas anymore": Interactions among semiotic modes in multimodal texts', *Computers and Composition*, 22(1), pp. 49–54. doi: 10.1016/j.compcom.2004.12.012.

Frankel, F., & DePace, A.H. (2012) *Visual strategies: A practical guide to graphics for scientists & engineers*, New Haven: Yale University Press.

Grant, R. (2018) *Data visualization: Charts, maps, and interactive graphics*, Boca Raton: Chapman and Hall, CRC Press, 1st edition.

Gross, A.G. (2009) 'Toward a Theory of Verbal – Visual Interaction: The Example of Lavoisier', *Rhetoric Society Quarterly*, 39(2), pp. 147–169. doi: 10.1080/02773940902766755.

Gross, A.G. (2011) 'A Model for the Division of Semiotic Labor in Scientific Argument: The Interaction of Words and Images', *Science in Context*, 24(4), pp. 517–544. doi: 10.1017/S0269889711000214.

Gross, A.G., Harmon, J.E., & Reidy, M.S. (2002) *Communicating science: The scientific article from the 17th century to the present*. Cary: Oxford University Press, Inc.

Hutto, D. (2008) 'Graphics and Ethos in Biomedical Journals', *Journal of Technical Writing & Communication*, 38(2), pp. 111–131. doi: 10.2190/TW.38.2.b.

'Interactive journal figures' (2014) *Materials Today*, 15 January. Available at: www.materialstoday.com/computation-theory/news/interactive-journal-figures/ (Accessed: 9 April 2020).

Jewitt, C., Bezemer, J., & O'Halloran, K. (2016) *Introducing multimodality*, London: Routledge.

Johnson, G. T., & Hertig, S. (2014) 'A guide to the visual analysis and communication of biomolecular structural data', *Nature Reviews Molecular Cell Biology*, 15(10), pp. 690–698. doi: 10.1038/nrm3874.

Johnson, N.R. (2020) *Architects of memory: Information and rhetoric in a networked archival age*, Tuscaloosa: The University of Alabama Press (Rhetoric, culture, and social critique).

Journal of Sound and Vibration (2020) 'Author Information Pack'. Elsevier. Available at: www.elsevier.com/wps/find/journaldescription.cws_home/622899?generatepdf=true (Accessed: 2 October 2020).

Kentucky Geological Survey. (n.d.) 'KGS geologic map information service', *WWW Document*, https://kgs.uky.edu/kygeode/geomap/ (Accessed 29 February 21)

Kress, G.R. (2010) *Multimodality: A social semiotic approach to contemporary communication*, London and New York: Routledge.

Kress, G.R., Leeuwen, T. V., & Leeuwen, D. (1996) *Reading images: The grammar of visual design*, London: Psychology Press.

Lemke, J. (1998) 'Multiplying meaning: Visual and verbal semiotics in scientific text', in J. Martin & R. Veel (eds.), *Reading science: Critical and functional perspectives on scientific discourse*, London: Routledge.

Mackenzie Owen, J.S. (2007) *The scientific article in the age of digitization*, Dordrecht: Springer, Information Science and Knowledge Management, vol. 11.

Marx, V. (2015) 'Visualizing epigenomic data', *Nature Methods*, 12(6), pp. 499–502. doi: 10.1038/nmeth.3409.

McMahon, B. (2010) 'Interactive publications and the record of science★', *Information Services & Use*, 30(1–2), pp. 1–16. doi: 10.3233/ISU-2010-0607.

Miller, C.R. (2016) 'Genre Innovation: Evolution, Emergence, or Something Else?', *The Journal of Media Innovations*, 3(2), pp. 4–19. doi: 10.5617/jmi.v3i2.2432.

Mohan, B. et al. (2007) 'Multimodal scientific representations across languages and cultures', in T.D. Royce & W.L. Bowcher (eds.), *New directions in the analysis of multimodal discourse*, New York: Psychology Press, pp. 275–298.

Newe, A. (2016) 'Enriching scientific publications with interactive 3D PDF: an integrated toolbox for creating ready-to-publish figures', *PeerJ Computer Science*, 2, p. e64. doi: 10.7717/peerj-cs.64.

Pain, E. (2017) 'How to (seriously) read a scientific paper', *Science*, www.sciencemag.org/careers/2016/03/how-seriously-read-scientific-paper

Perelman, C., & Olbrechts-Tyteca, L. (1971) *The new rhetoric: A treatise on argumentation*, translated by J. Wilkinson & P. Weaver, Notre Dame, IN: University of Notre Dame Press.

Perkel, J.M. (2018) 'Data visualization tools drive interactivity and reproducibility in online publishing', *Nature*, vol. 554, no. 7690, pp. 133–134, doi: 10.1038/d41586-018-01322-9.

Price, M. (2011) *Mad at school: Rhetorics of mental disability and academic life*, Ann Arbor: University of Michigan Press (Corporealities).

Rawlins, J.D., & Wilson, G.D. (2014) 'Agency and Interactive Data Displays: Internet Graphics as Co-Created Rhetorical Spaces', *Technical Communication Quarterly*, vol. 23, no. 4, pp. 303–322, doi: 10.1080/10572252.2014.942468.

Shipka, J. (2009) 'Negotiating Rhetorical, Material, Methodological, and Technological Difference: Evaluating Multimodal Designs', *College Composition and Communication*, 61(1), pp. 343–366.

Takeda, K. et al. (2013) 'Enhancing research publications using Rich Interactive Narratives', *Philosophical Transactions of the Royal Society A: Mathematical, Physical and Engineering Sciences*, vol. 371, p. 20120090, doi: 10.1098/rsta.2012.0090.

Tominski, C., & Schumann, H. (2020) *Interactive visual data analysis*, Boca Raton: CRC Press, 1st edition.

Vivanco, D.F. et al. (2019) 'Interactive Visualization and Industrial Ecology: Applications, Challenges, and Opportunities', *Journal of Industrial Ecology*, vol. 23, no. 3, pp. 520–531, doi: 10.1111/jiec.12779.

Weissgerber, T.L. et al. (2016) 'From Static to Interactive: Transforming Data Visualization to Improve Transparency', *PLOS Biology*, vol. 14, no. 6, p. e1002484, doi: 10.1371/journal.pbio.1002484.

Wickman, C. (2013) 'Observing Inscriptions at Work: Visualization and Text Production in Experimental Physics Research', *Technical Communication Quarterly*, vol. 22, no. 2, pp. 150–171, doi: 10.1080/10572252.2013.755911.

Wu, H. et al. (2016) 'MUFINS: multi-formalism interaction network simulator', *NPJ Systems Biology and Applications*, vol. 2, no. 1, pp. 1–10, doi: 10.1038/npjsba.2016.32.

Yau, N. (2011) *Visualize this: The flowing data guide to design, visualization, and statistics*, Indianapolis, IN: Wiley, 1st edition.

26

SCIENCE COMMUNICATION, VISUAL RHETORIC, AND *EBIRD*

The role of participatory science communication in fostering empathy for species

Amy D. Propen

Introduction

Public participation in scientific research projects, often referred to as "citizen science," constitutes an important component of science communication practice; citizen science allows scientists to accumulate "vast amounts of data" and build "a constituency for conservation . . . that is often required to bring about action in the political arena" (Bonter and Hochachka, 2009). Here, it is worth noting that this chapter understands *science communication* and *scientific communication* as related but different; that is, science communication and scientific communication "may share some common generalities ('science'), but the audience, purpose, and style are quite different for each." As Cristina Hanganu-Bresch and Kelleen Flaherty help describe, science communication is geared toward making scientific discoveries "intelligible to a lay (or general) audience," while scientific communication is more geared toward an audience of "specialists" who don't typically require the same level of explanations of basic concepts or background information (2020, p. 3). This chapter focuses primarily on the work of citizen science projects for advancing science communication in the public sphere. *eBird*, for instance, managed by the Cornell Lab of Ornithology, is "among the world's largest biodiversity-related science projects, with more than 100 million bird sightings contributed each year" ("About *eBird*"). Those who participate in *eBird* are typically birders who communicate scientific information by uploading their photos and other media to the *eBird* website, thereby adding to "a global database of bird populations" (National Audubon Society, 2019). Participatory projects like *eBird* have clear implications for citizen science and thus science communication, as birders can play a role in shaping knowledge about bird species around the globe, especially through the use of visual information.

In this chapter, I first provide some background about *eBird* and describe how participants can upload their data and accompanying photos to the *eBird* website or online app created by the Cornell Ornithology Lab. Please note that *eBird* allows users to upload multimedia data including photos, videos, and sound clips; however, for the purposes of this chapter, I focus on the ways that

DOI: 10.4324/9781003043782-30

visual information and photographs specifically constitute an important component of science communication. Next, I consider how *eBird* helps illustrate the connections between citizen science and science communication. Subsequently, I explore how *eBird*, specifically its photographic content, not only constitutes an important mode of science communication but, in doing so, also helps foster empathy for vulnerable species (Gruen, 2015) and helps demonstrate the geographical concept of "caring at a distance" (Silk, 1998). Finally, in demonstrating how *eBird* constitutes a productive mode of science communication, this chapter also considers the broader implications of scientific communication and visual rhetoric for contributing to species and habitat conservation at a time when such efforts are more necessary than ever before.

eBird: the communicative power of observation for knowledge-making

eBird is a large-scale, data-driven science communication project founded on the idea that "every birdwatcher has unique knowledge and experience" (eBird, 2020a). The project is managed by the Cornell Lab of Ornithology and includes collaborations with "hundreds of partner organizations, thousands of regional experts, and hundreds of thousands of users" (eBird, 2020a). *eBird* is essentially an open-access database with a user-friendly web interface that allows birdwatchers to submit information about "when, where, and how they went birding, and then fill out a checklist of all the birds seen and heard during the outing" (eBird, 2020a). The ability to share these data allows anyone from around the globe to locate birding hot spots any time of year (eBird, 2017). The accompanying *eBird* mobile app likewise allows users to summarize and submit their observations while on a hike or in the field. Birders can manage their personalized, location-based lists, upload photos and audio recordings, view others' photos and recordings, and even see real-time maps of where certain species have recently been spotted.

It is important to note that the goal of this chapter is not to embark on a usability study of the *eBird* site, per se; rather, I make note of *eBird*'s relative ease of use in order to help account for the connections between citizen science, science communication, and the necessary communicative structures that help empower participants and public audiences who might be well versed in ornithological knowledge but not always embracing of new technologies. That is, birdwatching and tech savviness are not necessarily synonymous with one another; thus *eBird*'s success over time and across audiences is arguably a testament to a web interface that allows even the most tech-weary participants to easily create an account and submit their data. Moreover, *eBird*'s success is also indicative of a community's desire to participate in citizen science, share information about science and conservation, and learn from one another – such a desire for knowledge-building has implications across audiences and for various types of research endeavors. To this end, by explicitly positioning *eBird* as "rewarding for birders, scientists, conservationists, and ultimately, the birds themselves," the Cornell Lab of Ornithology highlights some of its perceived, primary audiences for whom *eBird* holds significance, including bird species themselves ('Introduction to *eBird*'). The idea that science communication projects can span audiences and even species, can be rewarding for those involved, and can have implications beyond scientific observation or practice also speaks to some of the connections between and definitions of citizen science and science communication.

eBird as citizen science

eBird helps illustrate some of the important ways that citizen science integrates or interacts with professional science communication. From a practical standpoint, *citizen science* references "the emerging practice of using digital technologies to crowdsource information about natural phenomena" (Wynn, 2017, p. 2). This practice of using web-based technologies to crowdsource information often involves volunteers, or "citizen scientists," who are integral to the process of gathering field data, although such participants do not typically analyze the data themselves (Wynn, 2017, p. 2). Put more concisely, citizen science is defined "as public participation in scientific research projects, usually by volunteers who collaborate with scientists and researchers to increase scientific knowledge. The volunteers who participate in citizen science have varying levels of expertise" (Van Vliet and Moore, 2016, p. 13). As rhetoric of science scholar James Wynn describes, such descriptions of citizen science "situate the activity in a century-and-a-half-long tradition of citizens volunteering for governments and scientific institutions to gather information about natural phenomena like weather and astronomical events" (2017, p. 2). However, the main difference between citizen science and "volunteer projects from previous periods" is that "citizen science is unique because it capitalizes on the existence of widespread computing resources and the Internet that connects them" (Wynn, 2017, p. 3). *eBird* constitutes a project of citizen science through the ways that it leverages an open access database, web resources, and the Internet to allow citizen participants to upload their birding data and observations, which may then be used by other citizen participants and scientists alike. The work of citizen participants to crowdsource scientific information using digital technologies constitutes one of the main connections between citizen science and science communication.

eBird as science communication

The definition of *science communication* draws upon several related terms and ideas, which Burns et al. (2003) argue are nuanced and not necessarily synonymous with one another. Contributing to this complexity is the fact that science communication projects often involve a range of stakeholders, such as scientists, mediators, decision makers, various public audiences, the scientific community, and science practitioners. Again, a main point of connection between citizen science and science communication involves the ways that public audiences, especially "participants," become "directly or indirectly involved in science communication" through their involvement in projects like *eBird*, for instance (Burns et al., 2003, p. 184). As mentioned earlier, *eBird* understands knowledge-building about bird species as rewarding for multiple types of audiences. Burns et al. describe similar ideas in their general definition of science communication: "Science communication aims to enhance public scientific awareness, understanding, literacy, and culture by building . . . responses in its participants" that are related to "awareness, enjoyment, interest, opinion-forming, and understanding of science" (2003, p. 198). As they further describe, "Science communication also provides skills, media, activities, and dialogue to enable the general public, mediators, and science practitioners to interact with each other more effectively" (Burns et al., 2003, p. 199). In the case of *eBird*, it is arguably the use of multimodal communication, especially visual information, that helps to provide much of the "dialogue" that enables citizen participants to interact effectively.

Science communication projects often forward arguments and build knowledge through the use of visual and other multimodal information. In other words, projects like *eBird* can communicate *rhetorically* through the use of information such as sound clips, maps, photographs, and other illustrative materials. Because it is beyond the scope of this chapter to address all of *eBird*'s multimodal affordances, I focus specifically on its use of photos to help build knowledge within this global birding community.

The field of *rhetoric* and its related, predominant methodical approach of *rhetorical criticism* is typically concerned with the study of texts and discourses to achieve "a greater understanding of human action" (Segal, 2005, p. 2). Subsequently, rhetorical criticism may then involve critiques of textual or visual artifacts, such as photographs or maps, for instance, or even physical spaces like museum exhibits, and how they shape our understanding of events or impact discourse. Based on these discursive processes, we may thus "choose to communicate particular ways based on what we have discovered" (Foss, 2009, p. 3). Communication scholar Cara Finnegan notes that *visual rhetoric* in particular is concerned with visual texts, their reliance on cultural contexts, and how they shape knowledge-making (2004). Finnegan also points out that visual rhetoric "should recognize the influence of visual artifacts and practices, but also place them in the contexts of their circulation in a discursive field conceived neither as exclusively textual nor exclusively visual (Finnegan, 2004, p. 245). *eBird* indeed makes use of visual artifacts such as photographs, but, consistent with Finnegan's point, such photos are typically contextualized with additional data and surrounding text.

Science communication endeavors and thus citizen science projects such as *eBird* often rely on the interplay of visual, textual, and other multimodal information, and visual communication is an integral but hardly the sole component of participatory projects like *eBird*, which rely on digital interfaces to communicate with public audiences. *Digital rhetoric*, then, which accounts for knowledge-building through the use of multimodal communication, often focuses on "the application of rhetorical theory . . . to digital texts and performances"; it is also concerned with the "affordances and constraints" of artifacts of new media as well as their "potential for building social communities" (Eyman, 2012). As an artifact of digital rhetoric, *eBird* constitutes a technology that reflects public interest in building knowledge about species. As communication scholar Charles Bazerman has described, "technology, as a human-made object, has always been part of human needs, desires, values, and evaluation, articulated in language and at the very heart of rhetoric" (1998, p. 383). In the case of *eBird*, the use of visual information reflects current cultural discourses of what counts as valuable and legitimate communication about species in science and conservation projects.

"Every bird counts": science communication as aligned with conservation practice

In addition to its significance for science communication and the global birding community, *eBird* is a valuable tool for conservation science. Prior to the development of *eBird*, there was no way to fully understand the "population-level movements of many species" (eBird, 2017). *eBird* data combine sightings with "remote-sensed habitat data from NASA," which thus allows for the generation of "species distribution models that provide an unparalleled view of where and when birds are in the landscape," thus informing "better conservation strategies" and, as *eBird* puts it, "opening the door to the future of bird conservation" (eBird, 2017).

eBird's interest in bird conservation and their informal motto that "every bird counts" not only speaks to their interest in data-driven observations and sightings but also implicitly aligns with the main principles underpinning compassionate conservation (eBird, 2017). *Compassionate conservation* is a recent movement in conservation practice and policy that is informed by the idea that *every* individual animal body matters when it comes to conservation efforts. Broadly speaking, a compassionate conservation approach would advocate for viewing species as individuals, and it would thus take individual species into consideration within conservation policy (see Bekoff, 2010, 2014). Compassionate conservation is, at its core, grounded in an ethic of "do no harm." Part of the goal of compassionate conservation is to prompt and foster a critical awareness of how we understand our relationships with nonhuman animals and to foster an ethic of peaceful coexistence in which decision making about conservation practice and policy is grounded in a consideration of the most compassionate choice for all beings. By understanding the movements of individual species as contributing to "information that allows all of us to understand the movements and needs of birds at global scales," the Cornell Lab of Ornithology implicitly leverages *eBird* in ways that are aligned with both science communication and compassionate conservation practice ("Introduction to *eBird*"). Moreover, a compassionate conservation ethic is not incompatible with ideas about what philosopher Lori Gruen (2015) refers to as "entangled empathy."

eBird as fostering empathy

This chapter argues that science communication projects like *eBird* are capable of participating in conservation efforts and that knowledge-making related to science and conservation practice has the potential for engendering empathy for species. In distinguishing between sympathy and empathy, Gruen (2015) writes that "sympathy involves maintaining one's own attitudes and adding them to a concern for another. Sympathy for another is felt from the outside, the third person perspective. . . . Empathy, however, recognizes connection with and understanding of the circumstances of the other" (pp. 44–45). *eBird*, for instance, can be said to help foster empathy for bird species not only through the interplay of text and image that contributes to knowledge-making about species but also through the "awareness, enjoyment, interest, opinion-forming, and understanding of science" that Burns et al. suggest are often components and outcomes of science communication (2003, p. 198).

Gruen likewise acknowledges that attempts at empathy are sometimes incomplete and "in need of revisions. However, the goal is to try to take in as much about another's situation and perspective as possible" (2015, p. 45). *eBird*, as a project of conservation-related science communication, provides a forum and a context for the knowledge-building that is necessary for empathy to take shape. Similar to compassionate conservation, entangled empathy focuses on "another's experiential wellbeing," and so this kind of empathy tends to lead to action based on an assessment of what seems to be the best or most compassionate choice in helping to pursue the well-being of another (Gruen, 2015, p. 51).

eBird as a mode of "caring at a distance"

In the case of *eBird*, however, participants learn about bird species through a web-based interface and through the use of digital information. Pursuing the well-being of bird species may then take the form of conservation-based research strategies that happen as a result of community-based knowledge-building and participants' direct involvement in science communication projects. One birdwatcher may have initially viewed a bird in person and then photographed that bird and submitted the data; subsequently, another participant on the other side of the world may then learn about that bird through the shared knowledge and experience of the other birdwatcher. All the while, the Cornell Lab of Ornithology gathers and archives these data to inform their own efforts at protecting bird species; such communication, outreach, and research may happen from a relative distance, but the immediate impacts for bird species, related to habitat preservation and knowledge of migration patterns, for instance, are no less important.

To participate in this more remote and mediated knowledge-building about species is aligned with geographer John Silk's concept of *caring at a distance* (1998). Silk is interested in how "mass media articulate with other kinds of interaction and action to produce new forms of care and caring 'at a distance'" (1998, p. 166). The concept of "caring at a distance" extends the notion of "care and caring" from the more traditional face-to-face encounter to more distanced interaction. Silk argues that "mass media and electronic networks play a significant part in extending the range of care and caring beyond the traditional context of shared spatio-temporal locale and our 'nearest and dearest' to embrace 'distant others'" (1998, p. 179). Caring at a distance for bird species who might not reside in a birdwatcher's immediate geographic region but are still of interest to that birder represents a way of connecting with distant species who are likely affected by practices of human development and who may benefit from conservation research associated with the efforts of the Cornell Ornithology Lab. In this way, caring at a distance involves a mode of seeing that is once removed from immediate observation but is still associated with the remote learning, curiosity, and community knowledge-building that can take shape through the practices of participatory science communication.

As we consider the ways that *eBird* constitutes an important mode of science communication, in particular by fostering empathy for bird species and by enabling a kind of caring at a distance, it is helpful to take a look at a specific example in which visual information works in conjunction with surrounding text and context to help shape knowledge about a particular species of shorebird called the American Avocet.

Science communication and visual rhetoric in context: the American Avocet

Every bird species documented on *eBird* has its own web page that users can find when searching for that species. This main page includes a large photo of the bird, under which appears a brief, textual description of the species. Under the textual description, users will find "statistics" for the bird, including the number of photos and audio clips uploaded by users, and a scrollable range map. As users scroll farther down this main page, they will find more media from which to choose, in the form of "top photos," "top audio," and "top video," with the opportunity to click on "view all" for any of these categories. In short, every bird species documented on *eBird* is associated with a wealth of searchable material.

In order for *eBird* users to upload a photo or related birding data, they must first have an active *eBird* account, and they may then add information about the location of their specific sighting. A record of the sighting is then created, and users may essentially build on that record to upload additional media, such as photos or sound clips, for instance.

When searching for the American Avocet on *eBird*, for example, users will come across the main page for this bird and learn that this species is a "[d]istinctive large shorebird with a long, thin upturned bill and lean neck." Additionally, users will learn that this shorebird has "[b]old black-and-white wings" that are visible throughout the year; during the summer months, adults have a "buffy-orange wash" on their head. These birds frequently visit wetlands where, in order to find food, they swing their heads "back-and-forth in shallow water to catch small invertebrates" (eBird, 2020b), see Figure 26.1.

The photo in the figure is one of many representative images of the American Avocet available for users to view on *eBird*. Some Avocet photos depict the shorebird wading through shallow water to feed; other photos depict the bird standing still or swimming. It is important to note that while the photo in the figure appears on its own in *this* chapter, visitors to the eBird site will find that the photo is also contextualized with the name(s) of the photographer(s) or *eBird* user(s) who uploaded the image, in this case Hal and Kirsten Snyder; the possible date on which the photo was taken; the location where the photo was taken, in this case, "Civic Center Plaza, San Francisco, California, United States"; and any "behaviors" associated with the bird in the photo – in this case, "flying" (Snyder, moonbeampublishing.com; Macaulay Library, "American Avocet *Recurvirostra americana*"). It is possible for *eBird* users to upload other types of information as well, including information about the bird's age or sex, if known, or other notes. The result of this vast collection of birding data is a richly informative and aesthetically robust database that represents not only a wealth of knowledge about bird species but also the ongoing knowledge-building associated with participatory science communication. Moreover, the combination of statistical data with visual information also contributes to a holistic picture of each bird species.

While aggregate data about the American Avocet inform our knowledge about this species, the photo of the bird in Figure 26.1 arguably complements existing data and contributes to our curiosity about and intrinsic appreciation for this creature. In this photo, the Avocet is pictured flying near San Francisco's Civic Center Plaza. The bird's gaze meets the camera but does not quite seem to look directly at the camera; rather, we are presumably more aware of this bird than it is of us. Its long legs are sloped ever so slightly downward – indicative, perhaps, of either a recent takeoff or an intended landing. The photo depicts a full portrait of the bird's body, at a full profile, thus providing a clear view of the sheer, undeniable beauty of this creature in flight. The high resolution of the image itself and its production value, also allows viewers to get a glimpse of the texture of the Avocet's clean and interlocking feature structure – ideal for stable flight as well as buoyancy in the water. The angle of the sunlight creates a subtle silhouette along the top of the bird's body, in part because the bird has bold black-and-white wings and shoulders to begin with. Finally, the original, color version of this photo conveys the soft, blue-gray tones of the often hazy, overcast sky, typical of the Bay Area.

While photographs are considered to be a credible and arguably indispensable aspect of science communication, the work of photos to convey a specific and static representation of the world has not gone without critique. Some philosophers of science and scholars of feminist science studies have critiqued the work of photography for its tendency to present a mediated view of the world, one that reifies specific ways of knowing and seeing (see Haraway, 1991;

Figure 26.1   American Avocet (*Recurvirostra americana*)
Source: Hal and Kirsten Snyder, moonbeampublishing.com (used with permission).

Harding, 1996; Hayles, 1999). For instance, as Gregory Ulman describes, photography requires "a representing subjectivity – the photographer; an object of representation such as a landscape, plant, or animal; and a medium of representation such as a digital photograph" (2015, p. 27). The very practice of making a photograph then "separates its subject visually and temporally from its surroundings, translates it into a different physical form . . . and rejoins it with other photographic representations" (Ulman, 2015, p. 27). In other words, photography has the capacity to convey a specific version of the world that it purports to represent (Propen, 2018). In doing so, it may be viewed as defining or influencing our experience in certain ways, or "redefining to some degree viewers' relationships with and capacity to experience the object(s) depicted in the photograph by influencing how we expect a landscape or animal or wildflower to appear" (Ulman, 2015, p. 27).

In the case of the Avocet photo, the bird is indeed shown in its natural environment, and the placement of the photo on *eBird* then creates an additional context in which the photo itself becomes part of a database of representations of this bird – the Avocet is then grouped with its kin, but in a way that redefines our relationship with this bird by understanding the bird as contributing to a database of mediated, visual information and knowledge about this species. This is part of the knowledge-making work of science communication, and yet such work is paradoxically complex in its communicative power.

Such critiques of representationalism and the power of images to reify our relationships with bodies, objects, and environments are also described by Quinn R. Gorman (2009), who further acknowledges how photographs may represent the physical environment in ways that perpetuate realist and social constructionist perspectives. Gorman problematizes the communicative work of nature photography in describing those who advocate for "scientific or literary realism who believe that with the proper care there can be a more or less mimetic relationship between signs and 'natural' referents, [and] social constructionists who believe in the inevitably arbitrary, cultural, and constructed character of signs and sign systems, however carefully crafted or functional" (2009, p. 241). This perpetuates a challenge, he says, whereby we "allow the world to be captured by the discourse of realism, which asserts the competence of representational mimesis to reproduce the world in words, or we allow it to be captured by the discourse of textuality, which claims that the interests of a natural Other are inevitably and utterly invaded by our own cultural baggage" (Gorman, 2009, p. 241). Instead, Gorman acknowledges a more nuanced and less polarized view, whereby we might understand photography as providing "the ground for a representational ethics that resists the very *possibility* of a complete capture of the natural," even though it cannot provide a wholly "unobstructed view" (Gorman, 2009, p. 242, emphasis in original). Conceptualizing nature photography in this more nuanced way supports an understanding of visual rhetoric as participating in the practices of knowledge-building that are part and parcel of a science communication grounded more in care and curiosity, one that builds community through "awareness, enjoyment, interest, opinion-forming, and understanding of science" that underpin participatory projects of science and conservation. (Burns et al., 2003, p. 198).

Conclusion and implications

Gorman tells us that some photos "exert an attraction based upon their ability to move a viewer" (Gorman, 2009, p. 245). The pull exerted by such photos speaks to the visual, rhetorical power of images. The Avocet photo is arguably one such image. The photo clearly and simply conveys the beauty of this species and its ability, given optimal circumstances, to thrive in its natural habitat. Moreover, to understand the actual photo of this bird as contributing to a global knowledge base, perpetuated and sustained by birders who care about the conservation of species and habitat, speaks to the ways that science communication projects can foster empathy and caring at a distance. As such, the artifacts of science communication and citizen science, in this case photos, can help give substance to "the cultural meaning and categories" that allow *eBird* participants "to take substantive, positive steps toward change" (Gorman, 2009, p. 254).

In this way, visual culture and imagery can then play a role "in extending the scope of beneficence beyond our 'nearest and dearest' to embrace distant others" (Silk, 1998, p. 165). *eBird* provides a credible framework for citizen science, or participatory science communication, which taps into visual and digital rhetoric in order to reach diverse audiences, build knowledge about bird species, and subsequently argue for the necessity of species and habitat conservation. By providing birders and researchers with a forum in which they can share information and learn about the circumstances of vulnerable bird species around the globe, *eBird* may be understood as helping to foster empathy for bird species not only through the visual and digital rhetorics that contribute to knowledge-making about species but also by fostering the kinds of awareness, enjoyment, and interest that are necessary for sustained, participatory science communication. These projects of conservation-based science communication can ultimately help in pursuing the well-being of another – a necessary part of empathetic practice in an age of climate change. Finally, to understand *eBird* as an important project of participatory science communication helps demonstrate that how we represent nonhuman animals discursively and visually reflects our cultural understandings of the relationships between humans, nonhumans, and the environment; as such, how "we talk or write about animals, photograph animals, think about animals, imagine animals – represent animals – is in some very important way deeply connected to our cultural environment" (Rothfels, 2002, p. xi). In the case of *eBird*, such empathetic practice plays out through research about habitat conservation, such that migration routes, ecosystems, and ultimately the population of species may be maintained. While the combined efforts of *eBird* participants may happen from a distance, the implications of participatory science communication for bird species are no less important.

Acknowledgments

I would especially like to thank Hal and Kirsten Snyder of moonbeampublishing.com for permission to use their photo of the American Avocet in this chapter. Thank you also to the staff at the Cornell Lab of Ornithology for answering my many questions. Finally, I am very appreciative of the feedback and helpful suggestions provided by the editors of this volume.

References

Bazerman, C. (1998) 'The production of technology and the production of human meaning." *Journal of Business and Technical Communication*, vol. 12, no. 3, pp. 381–387, DOI:10.1177/1050651998012003006

Bekoff, M. (2010) *The animal manifesto: Six reasons for expanding our compassion footprint*, Novato: New World Library.

Bekoff, M. (2014) *Rewilding our hearts: Building pathways of compassion and coexistence*, Novato: New World Library.

Bonter, D.N., & Hochachka, W. (2009) 'A citizen science approach to ornithological research: Twenty years of watching backyard birds', *Proceedings of the Fourth International Partners in Flight Conference: Tundra to Tropics*, pp. 453–458.

Burns, T.W., D.J. O'Connor, and S.M. Stocklmayer. (2003) 'Science communication: A contemporary definition,' *Public Understanding of Science*, vol. 12, pp. 183–202, DOI:10.1177/09636625030122004

eBird. (2017) 'Introduction to eBird', www.youtube.com/watch?time_continue=164&v=-t-0xAjxakw&feature=emb_logo (Accessed 18 August 2020) eBird. (2020a) 'About eBird', https://ebird.org/about (Accessed 12 August 2020)

eBird. (2020b) 'American Avocet *Recurvirostra Americana*', https://ebird.org/species/ameavo (Accessed 21 August 2020)

Eyman, D. (2012) 'On digital rhetoric', www.digitalrhetoriccollaborative.org/2012/05/16/on-digital-rhetoric/ (Accessed 17 August 2020)

Finnegan, C. (2004) 'Review essay: Visual studies and visual rhetoric." *Quarterly Journal of Speech*, vol. 90, pp. 234–256, www.mendeley.com/research/review-essay-visual-studies-visual-rhetoric

Foss, S.K. (2009) *Rhetorical criticism: Exploration & practice*, Long Grove: Waveland Press.

Gorman, Q.R. (2009) 'Evading capture: The productive resistance of photography in environmental representation', in S.I. Dobrin & S. Morey (eds.), *Ecosee: Image, rhetoric, nature*, Albany: State University of New York Press, pp. 239–256.

Gruen, L. (2015) *Entangled empathy: An alternative ethic for our relationships with animals*, New York: Lantern Books.

Hanganu-Bresch, C., & Flaherty, K. (2020) *Effective scientific communication: The other half of science*, New York: Oxford University Press.

Haraway, D.J. (1991) *Simians, cyborgs, and women: The reinvention of nature*, New York: Routledge.

Harding, S. (1996) 'Science is "good to think with"', in A. Ross (ed.), *Science wars*, Durham: Duke University Press, pp. 16–28.

Hayles, K.N. (1999) *How we became posthuman: Virtual bodies in cybernetics, literature, and informatics*, Chicago: University of Chicago Press.

Macaulay Library, 'American Avocet *Recurvirostra Americana*', https://search.macaulaylibrary.org/catalog?req=true&mediaType=p&region=United%20States%20(US)&regionCode=US&userId=USER131436 1&user=Hal%20and%20Kirsten%20Snyder (Accessed 21 August 2020)

National Audubon Society. (2019) 'Great backyard bird count should be "finchy" and fun: Red-breasted Nuthatches and possibly Evening Grosbeaks will be highlights in North Carolina', *Audubon*, https://nc.audubon.org/press-release/great-backyard-bird-count-should-be-finchy-and-fun (Accessed 14 January 2020).

Propen, A. (2018) *Visualizing posthuman conservation in the age of the anthropocene*, Columbus: The Ohio State University Press.

Rothfels, N. (2002) 'Introduction', in N. Rothfels (ed.), *Representing animals*, Bloomington: Indiana University Press, pp. vii–xv.

Segl, J.Z. (2005) *Health and the rhetoric of medicine*, Carbondale: University of Southern Illinois Press.

Silk, J. (1998) 'Caring at a distance', *Ethics, Place and Environment*, vol. 1, no. 2, pp. 165–182, DOI: https://doi.org/10.1080/1366879X.1998.11644226

Snyder, H., & Snyder, K. (n.d.) 'American Avocet *Recurvirostra Americana*,' moonbeampublishing.com, photo used with permission.

Ulman, L.H. (2015) 'Beyond nature photography: The possibilities and responsibilities of seeing,' in S. Rust, S. Monani, & S. Cubitt (eds.), *Ecomedia: Key issues*, New York: Routledge, pp. 27–45.

Van Vliet, K., & C. Moore (2016) 'Citizen science initiatives: Engaging the public and demystifying science', *Journal of Microbiology and Biology Education*, pp. 13–16, DOI: 10.1128/jmbe.v17i1.1019

Wynn, J. (2017) *Citizen Science in the Digital Age: Rhetoric, Science, and Public Engagement*. Tuscaloosa, University of Alabama Press

PART 4

Scientific communication pedagogy

27

EMERGING PRACTICES IN SCIENCE COMMUNICATION PEDAGOGY

Kate Maddalena

Introduction

Curriculum in science writing and science communication is crucial in terms of the university's role in preparing students for work in the world as well as preparing students to be critically engaged citizens of the world. These pedagogies comprise both scientific communication, or communication within scientific communities of expertise and science communication, or communicating science with public audiences of varying levels of engagement and expertise. The speedup in the conversion from research into circulating information of various qualities and stages of completeness – often information that is not yet knowledge – has made the need for improved science literacy across highly diverse public audiences painfully clear (Liu, 2009; Sharon and Baram-Tsabari, 2020). A dearth of public science literacy is only one problem; public demand for practical applications of science, directly tied to questions about the warrant for public funding of science, is also increasing. Direct accountability of scientific research to public stakeholders is a good thing but only insofar as public science literacy is maintained and improved upon. A combination of decreasing public science literacy and increased public accountability of science could spell problems for the scientific endeavor (Alcoreza, 2021).

The pedagogy of science communication seeks farsighted solutions to these ongoing problems. It has long been the work of writers and communicators to make and remake meaning of knowledge produced by the sciences (Miller, 1979; Gross, 1994; Fahnestock, 1986). It is now the work of scientists and researchers themselves to become more versatile writers and communicators. Critical writing pedagogies belong in both worlds; the departmental divides between the skills of science and the skills of the humanities are logistically problematic at best and philosophically dangerous at worst (see Maddalena and Reilly, 2017; Goodwin 2019 for examples). First and foremost, it is important to *teach writing and communication as a basic requirement of any science degree*, and it is important to *teach science writing, rhetoric of science, history of science, and/or science and technology studies as a basic requirement of any humanities degree*.

Since the advent of the Internet, communities, contexts, modes, and genres have been destabilized, complicated, and reshuffled, in some cases more than once over the course of a decade. A college student entering a career in communication must be able to understand the shifting territory and leverage of their skills in it. Contemporary pedagogies across most disciplines have thus become what we like to call "student focused." In the spirit of

DOI: 10.4324/9781003043782-32

student-focused teaching, I frame this entry as three interrelated sections about three kinds of student and the way each relates to science itself as well as various audiences for science communication. These three categories of science communicator are neither discrete nor clear-cut. One main theme pervades all three, however: students of science communication need to work closely with scientists themselves (and, in some cases, *as* scientists or social scientists themselves) in productively *embedded* contexts.

The academic scientific communicator

Academic science writing as an advanced service course grew out of the rise of rhetoric and composition as a curricular phenomenon (Crowley, 1995). The study of writing and writing pedagogy was coming into its own as a field, and science writing was an obvious opportunity both as an object of study (for rhetoric of science) and as needful of a formalized pedagogy (Connors, 2004). Penrose and Katz's (1998, 3rd edition 2009) text *Writing in the Sciences* supports such advanced-level pedagogy. The course is typically outside the required curriculum, and students use an elective credit to get writing support and enrichment. The benefits of such courses are manifold; they bring writers from multiple disciplines together and, if taught well, allow for students to see the rhetorical underpinnings of science writing practices that transcend disciplines, as well as to critique some entrenched practices that may not serve any rhetorical purpose. The composition pedagogy model also teaches emerging scientists an orientation to the writing process: a set of habits and practices that they can continue to develop over their careers. Workshop settings and peer review sessions prepare them to be good collaborative coauthors and journal reviewers, as well. At the graduate level, research institutions have also developed a best practice of offering formal writing support in the form of composition workshops for "occluded genres" like theses and dissertations: highly technical genres with specific genre expectations that the writer is only expected to produce once in their career (Autry and Carter, 2015). Since Penrose and Katz, other texts continue to strive to show science writers how to write publishable prose. For freshman-level science writing (and beyond), Gerald Graff and Kathy Birkenstein's much loved *They Say/I Say* (2006) continues to show students the importance of framing academic prose (and, indeed, all nuanced arguments) as a conversation with peers. Joshua Schimel's (2012) *Writing Science* also offers an excellent, diagram-based framework. There are also excellent publicly available online resources for academic science writers, most notably Scitable, a writing resource published by the journal *Nature*. Scitable, because it is a resource produced for scientists by an eminent scientific journal, makes an excellent text for an upper-level science writing course. Finally, the best new resources for science writers recognize that science is not monolithic. As I will mention in the conclusion of this chapter, the future of technical science writing instruction is field-specific. The best work is work in context. Texts that speak to the specificity of writing tasks in scientific fields began emerging in earnest in the 2010s; two excellent examples are *Writing Papers in the Biological Sciences* (McMillan and McMillan, 2012) and *Writing for Today's Health Care Audiences* (Bonk, 2015).

But there are problems with teaching science writing for academic audiences in a class devoted to technical and specialized genres, even when adjusted to allow for field-specific contexts. Training academic science writers in academic genres tends to reify genre conventions and may unwittingly perpetuate dogmatic discursive gatekeeping that compromises the scientific process and/or excludes nonexperts from conversations of expertise. Some genre conventions have become so habitual and expected that they actively instantiate bad cultural habits in academic culture (Herndl, 1993). John Swales's (1990) oft cited genre study of academic prose is an excellent example of this phenomenon. Swales's study pinpointed, among other things, that a successful rhetorical "move" in academic prose is to "establish a niche" or "indicate a gap" in research that your research can then fill. The approach has become widely known as the C.reate a R.esearch S.pace, or CARS, model. The pervasive belief that research must break brand-new ground or fill a previously unoccupied gap is noticeably weakening what should be a solid base of replicated results, especially in the social sciences and medical fields (Loken and Gelman, 2017; Loken, 2019). Teaching Swales's "moves" uncritically may produce successful grant proposals, but it may also unwittingly undermine a field's more noble epistemological functions.

Teaching communicative habits as "the way it's done" in scientific fields can also exacerbate problems in outreach and public communication. Academic writing pedagogy, like academic publishing, has lagged quite a bit behind the explosion of new funding-seeking, knowledge-making, and peer-and-public-engaging genres that have proliferated online in the past decade (Mehlenbacher, 2019). A dependency on genre conventions also contributes to a tendency for scientists and researchers to be unable to "recognize and react to the rhetorical situation for [academic, technical] writing" (Lerner, 2007). The conventional modularity of research articles enables strategic reading, a "cherry-picking" approach to consumption that can privilege results over nuance and rhetorical context (Sollaci and Pereira, 2004). Treating writing in the sciences as specialized and therefore privileged maintains a highly problematic deficit model of public communication that rhetoricians of science have described for decades and that scientists themselves can't seem to shake (Miller, 2001; Sturgis and Allum, 2004). Other nuanced issues such as weight-of-evidence misrepresentation (Kortencamp and Basten, 2015) and manufactured controversy (Ceccarelli, 2013) are habits of scientific communities for rhetorical reasons similar to the Swales gap, and they tend to balloon when they emerge in public discourse. Jargon can also become a point of public controversy; "stem cell" (Leydesdorff and Hellsten, 2005), "climate change" (Nisbet, 2009), and "tipping point" (Russill, 2009) are all terms that originated in technical discourse and got repurposed in public conversation for differing purposes.

The key, then, is to teach even academic scientific communication as a critical and creative endeavor that responds to a unique context, not just as the task of fitting content to a set genre mold. Eric Loken (2019) takes the replication crisis as a creative opportunity for the field of psychology, "not really a scientific crisis, because the awareness is bringing improvements in research practice, new understandings about statistical inference and an appreciation that isolated findings must be interpreted as part of a larger pattern." The challenges of ossified genre conventions should be seen as a similar opportunity for teachers and students of science communication. Students should learn to re-see genre as rhetorical genre studies first defined it: as social action (Miller, 1984). To return to the example of the literature review and finding a niche, critical approaches to teaching the genre are available and compelling. Alfonso Montuori (2005) posits a critical and "creative inquiry" approach to the literature review that helps students use the genre to position their own identity in a community of scholarship. Colleen A. Reilly (Maddalena and Reilly, 2017) uses Montuori's approach in her innovative professional science writing program as a lynchpin project for a semester course. New approaches to the literature review as genre are emerging from the sciences themselves, as well. Christopher Luederitz et al. (2016) teach the literature review in their sustainability sciences classrooms as a way for students to take ownership in research in a transdisciplinary field. *Science Communication in Theory and Practice* (Stocklmayer et al.), an edited collection published in 2001, describes scientists and newly minted science communicators grappling with newly transdisciplinary issues. Along these same lines, Gross and Harmon's (2010) *The Craft of Scientific Communication* is a genre-based approach to successful publishing and publicly communicating, pitched to the academic communicator.

The new gold standard for teaching scientific communication is to embed communication experts in science courses. In this way, experts in writing and rhetoric can introduce the critical and creative composition processes in the field or in the lab, where they ostensibly should be happening anyway. Embedded approaches strive to bust the myth of doing writing separately from doing science – as some kind of extra task that occurs after the fact. Scholars and teachers of rhetoric, writing, and/or communication have found an embedded or teamed course model rewarding for both teacher and student (see Druschke and McGreavy, 2016; Cagle, 2017; Kraft et al., 2019). It should be noted that embedded approaches require considerable commitments of time, resources, and energy from the scholars and teachers involved as well as their institutions; most of the projects cited here were grant-funded projects. It should also be noted that the prospect of scholars of rhetoric joining a teaching and/or research team of scientists as embedded experts has raised questions of disciplinary territory and concerns about adequate critical distance, as well (see Ceccarelli, 2013; Druschke, 2014; Cagle, 2017). The embedded/teamed approach to science communication pedagogy may also best facilitate positive learning outcomes for the next two kinds of students treated here: public science communicators and practitioners of public science.

The science communicator/science journalist

Historically, models of public science communication have upheld the science/public divide, or the "deficit model," by seeing public communication as the purview of journalism and education. But while traditional journalism has struggled to maintain itself as a profession at the beginning of the twenty-first century (and liberal democracy has struggled to maintain itself as a result), journalism in science and technology, after a stormy rebirth, is in the midst of a "golden age" (Hayden and Hayden, 2018). The golden age of public science communication owes its transition partly to a vibrant and volatile community of bloggers who engage one another, researchers, and various publics on personal and professional blogs and social media networks (for a much more detailed genre analysis of science blogging, see Mehlenbacher, 2019). It is impossible to define the contemporary science communicator in simple terms, and it is perhaps disingenuous to categorize them separately from practitioners of public science as defined in the next section. But I mean to describe the professional science communicator who has supplanted the science journalist here. Basically, in the early 2000s, several science-minded writers made their names in a brand-new landscape, formed professional networks that gained the attention of the academic science community (Mehlenbacher, 2019), and leveraged book deals and permanent positions at high-profile publications. Science communication is now a booming field for professional writers employed by labs and by popular and academic publications, and it has enriched existing networks of professional communicators like public information officers, public relations professionals, and journalists. The common trait of these professionals is that their work is to render technical language accessible to nonexpert audiences, a skill that rhetorician Jeanne Fahnestock (1986) called "accommodation."

Though some science communicators are trained in journalism or PR, many are not. The American Academy of Arts and Sciences' Public Face of Science Initiative recently published a report detailing "priorities for the future" that includes an appendix on science communication skills. The appendix calls out the "disjointed and disconnected nature of practical advice surrounding science communication training efforts" and says that "research and evaluative literatures can expand to better support the utility and application of these communication skills" (AAAS, 2020, Appendix A). In other words, research is needed in pedagogies of the new science communication. Many successful courses in long-form writing about science for public audiences concentrate on workshopping narrative nonfiction and professional development that is student driven and tailored to the individual. Two invaluable resources for such workshops are the *Best American Science and Nature Writing* series (a series by Houghton Mifflin, edited by prominent scientists who are also communicators) and the *Best American Science Writing* series (published by Harper Collins). Also, most scientists think critically about science communication; the field has produced its own pedagogy, and a wide array of instructional texts are available online (see "Public Science" in this chapter for recommendations).

There is also a robust culture of professional development within the science communication, or "#SciComm," community. One best practice for writing and communication teachers in the field is to provide students with guided access to the community itself at the Public Library of Science (PLOS) Scicomm forum (scicomm.plos.org), through the American Association for the Advancement of Sciences's (AAAS) devoted science communication resources (aaas.org > programs > communicating science), at the Scicomm national annual conference (scicommcon. org) or in one of the regional chapters of the international organization; many such chapters are student led and based on college campuses. Finally, students who truly want to speak this community's language and begin to converse with its members in its most common genre must find the community on Twitter: the #scicomm hashtag is a place to start, but finding and following consistent, clear voices is an invaluable critical reading skill for the twenty-first century.

One of the clear voices in the SciComm professional community is Ed Yong, a British journalist who is now a writer for *The Atlantic* and the magazine's preeminent science perspective. Yong is a self-made journalist who achieved great success by developing a charismatic, nuanced, intellectual, and prolific voice. Yong (2010) has published a resource in the form of a list of science communicators and their own stories of how they got started in the field: "On the Origin of Science Writers." The list is an excellent professional development resource for students; it showcases the professional agility required to negotiate the always changing terrain of twenty-first-century journalism, but it also reassures readers of the never straight path to rewarding work in science communication. In Yong's case, he ended up with a relatively conventional job in science journalism at an independent publication. Such positions are rare. There is a much larger and still growing need for the kind of work described in the next section, as an embedded communicator in a public science project. A trend in continuing professional development for working journalists, even those at more traditional posts, is embedded fellowships at research institutions with working scientists (Valenti and Tavana, 2005).

Public science

Science communication has been taught and conceived of thus far as a version of "popular science," science accommodated for nonexpert publics and "popularized" by professionals who are experts in writing and communication (Bucchi and Trench, 2008). But scientists themselves see a growing need for science that already speaks to and enlists stakeholders: "[t]here is broad agreement [among highly published scientists] that the science community would benefit from additional science communication training and that deficit model thinking remains prevalent" (Besley and Tanner, 2011). The heading for this section breaks with the pattern of the headings that precede it for a reason; public science is an emerging collaborative practice rather than a singular authorial identity. Practitioners of public science may be scientists, writers, or both, but they are engaged in projects that specify public communication as a goal from the outset. Practically, this means that research grants are written and courses are taught with writing, rhetoric, and public communication/outreach as a line in the budget and a learning outcome on the syllabus. Communicators are embedded, contributing members of a public science project: a lab, a department, a collective, etc. Holly Menninger, director of Public Science at North Carolina State University in the United States, helped coconceive of models of public science in the Rob Dunn Lab, an early example of a lab supporting working public science projects. Reflecting on public science models, Menninger stresses that public science helps career researchers move past the deficit model and "think creatively – and more authentically – about how they engage the public" (Menninger, 2015).

Public science was first imagined by ecologists and environmentalists who knew that research in conservation had to integrate human communities and therefore public communication into its research designs (Cox, 2012; Robertson and Hull, 2001). Public science has since worked to break down the pervasive and persistent idea among scientists that skills in internal, expert-to-expert communication are superior to skills in communicating to nonexperts (Chilvers, 2013; Dijkstra and Gutteling, 2012; Gieryn, 1983; Mogendorff et al., 2016). Though not all public science is citizen science, public science pedagogies draw from and work alongside programs in citizen science to allow stakeholders to contribute to datasets, directly benefit from findings, and even work to set research agendas (see Cooper, 2016). Because public science is already working in interface with nonexpert local communities, it is accountable to them, giving the evolving pedagogy of public science a practical means by which to challenge the problematic elitist politics of most STEM fields as they stand. Such critique- and public-oriented agenda setting is often most successfully reimagined in undergraduate and early graduate classrooms: "[e]xperts-in-training who have not yet completely internalized the discursive patterns and professional identities that help reproduce the hegemony of technical-scientific expertise may be of crucial importance here" (Mogendorff et al., 2016, p. 47).

Many textbooks and dynamic, actively maintained teaching resources emphasize science communication with a public science orientation. *Ethics and Practice in Science Communication* (Priest et al., 2018), an academic collection of essays about specific communicative contexts and their ethical stakes, is an excellent text for graduate-level public science with an eye to affecting policy. Several more accessible, trade-type books address working scientists with practical advice in situated argumentation. *Escape from the Ivory Tower* by Nancy Baron (2010) and *Championing Science* by Roger Aines and Amy Aines (2019) are excellent examples of the "practical advice" genre. All of these texts address the newly dominant modalities of sound, video, and the visual in some way, but texts that address visual communication deserve mention here as well: *Introduction to Data Visualization & Storytelling* (Berengueres et al., 2019) pairs excellent guidance in designing visuals with examples and narrative frames from the emergent field of data journalism, and *Data Visualization: Principles and Practice* (Telea, 2014) is a more in-depth and technical handbook for digital designers. For dynamic, up-to-date examples and critiques of data interpretation, teachers can connect students to *FlowingData*, an early blog by Nathan Yau copyrighted and extended in 2007. Several broader online teaching resources are also excellent. *The Open Notebook*, a nonprofit founded in 2010 as a resource for science journalists, has evolved into a hub for science communicators negotiating the social ecology of research, journalism, and the Internet. Finally, in 2020, the MIT Knight School of Journalism published an open-access *Science Editing Handbook* that leads students of contemporary science communication through the multiple, interrelated, and overlapping contexts of institutional science and science journalism, with an emphasis on the emerging practice of public science.

Public science also offers the opportunity for students to actively study science communication with empirical methodologies. In 2009, Matthew Nisbet and Dietram Scheufele called for "a more scientific approach to science communication, i.e., one that is less exclusively driven by intuition, personal experience, or traditional ways of 'doing communication,' and more by an empirical understanding of how modern societies make sense of and participate in debates over science and emerging technologies." More recently, voices from the field of science and technology studies have made a similar call for embedded and cosupportive scholarship (Felt and Davies, 2020). Likewise, the aforementioned AAAS report on science communication devotes a special section to "The Crucial Role of the Social and Behavioral Sciences" in understanding the reception and impact of science communication (AAAS, 2020, p. 25). The public science model, which gives communicators a voice in research design and often includes social science methodologies alongside natural and physical science research approaches, is well suited to these calls.

Unresolved issues and future directions

At the time of this writing, we are living through a historical moment that requires deep and considered reflection about the ways of life that the generations before us took for granted. The need for trained science communicators and scientifically literate global citizens will surely only increase in the years to come. Faith Kearns's (2021) book, *Getting to the Heart of Science Communication: A Guide to Effective Engagement*, gives a comprehensive and prescient snapshot of the field as it stands; Kearns offers humane and justice-oriented approaches to some of its most wicked problems. Science writers and communicators may need to respond to contexts we can't yet imagine, but a few trends that suggest the best pedagogical practices of the future seem clear: (1) the need for widespread, embedded, "post-critical" work, (2) the need for more effective ways for science communication to articulate a politics in order to escape being "captured" and politicized, (3) a concerted effort to address science communication in global and indigenous contexts rather than only Western and postindustrial contexts, and (4) the development of a detailed and agile pedagogy of writing and communication around health, public health, and medicine.

Scholars who study the rhetoric of science must teach scientific and science communication, and they must find ways to do so in more direct contact with science itself and with the diverse publics who are science's stakeholders. At the same time, the field of rhetoric must maintain its critical lens (see Depew, 2014; Goodwin, 2014; Goodwin et al., 2014; Vernon, 2014). The best way to do this is to include communication and writing in science curricula by hiring science communication experts into science departments and/or working with colleagues in communication and rhetoric to design truly interdisciplinary degree programs that require upper-level courses in writing and communication. The bellwethers for this practice have been the environmental sciences (see Druschke, 2014; Druschke and McGreavy, 2016).

Directly following this point, science and technical communication pedagogy must refuse to avoid communicative work's inherently political facets and commit to making its politics more transparent. Such critical transparency must start in the classroom. A transparent politics of science communication helps the public stage swing back toward nuance. One aspect of our classrooms' politics must be a commitment to self-critique and active change in terms of identity, representation, and access. Culturally entrenched institutions are awakening to their own culpability in systematic oppression of underrepresented groups. Academic discourse, perhaps especially discourse in the hard sciences, uses supremacist attitudes toward language and economic and geographic exclusion to reproduce a largely racist, sexist, gender-ignorant, ableist, and privileged culture (Canfield et al., 2020; Rao et al., 2014). Embedded pedagogies of public science, because they ask social science questions about stakeholders and set public agendas for research at the outset, have the potential to intervene and affect much needed change in academic culture at the ground level.

Demographic diversity and access across social difference are related to geopolitically determined access to science communication pedagogy as well. Future teachers of science communication need to find ways to stress access and collaboration that is global as well as indigenous awareness that is local. Globally minded science communication will address issues of collaboration across borders and cultures in spite of geopolitical limitations and divides between postindustrial and "developing" countries (du Plessis, 2008). Indigenous-oriented research will expand a definition of knowledge to include first peoples' "perspectives, ideas, values, and opinions into a values-policy nexus" and give us "insight and foresight to manage natural resources sustainably"(Cooper, 2016, pp. 173–174).

Finally, teachers of science communication should begin to specialize in and teach courses on the communication of health and medicine in embedded contexts, taking environmental science programs as a model, especially at the undergraduate level. Public health is already public science, in that the social scientific aspects of public communication are treated as important questions for research. But the SARS-CoV2 pandemic (and several exacerbating public health crises) in the United States and much of the developed world by extension is evidence of a vicious cycle of public disenfranchisement and medical illiteracy (see Alcoreza, 2021). Graduate work in public health, especially in Integrative Graduate Education and Research Traineeship (IGERT) programs, can be bellwethers for embedded/coproductive pedagogies of writing in health and medicine at the undergraduate level.

References

Aines, R.D., & Aines, A.L. (2019) *Championing science: Communicating your ideas to decision makers*, Berkeley: University of California Press, 1st edition.

Alcoreza, O.B. (2021) 'Science Literacy in the Age of (Dis)Information: A Public Health Concern', *Academic Medicine*, 96(2) p. 12, doi: 10.1097/ACM.0000000000003848

American Academy of Arts and Sciences. (2020) *The public face of science in America: Priorities for the future*, New York: American Academy of Arts and Sciences.

Autry, M.K., & Carter, M. (2015) 'Unblocking occluded genres in graduate writing: Thesis and dissertation support services at North Carolina State university', in *Composition forum*, New York: Association of Teachers of Advanced Composition, vol. 31.

Baron, N. (2010) *Escape from the ivory tower: A guide to making your science matter*, Washington, DC: Island Press.

Berengueres, J., Fenwick, A., & Sandell, M. (2019) *Introduction to data visualization & storytelling: A guide for the data scientist*, New York: Independently Published.

Besley, J.C., & Tanner, A.H. (2011) 'What science communication scholars think about training scientists to communicate', *Science Communication*, 33(2), 239–263, DOI: https://doi.org/10.1177/10755 47010386972

Bonk, R.J. (2015) *Writing for today's healthcare audiences*, Toronto: Broadview Press.

Bucchi, M., & Trench, B. (eds.). (2008) *Handbook of public communication of science and technology*, London: Routledge.

Cagle, L.E. (2017) 'Becoming" forces of change": Making a case for engaged rhetoric of science, technology, engineering, and medicine'. *Poroi*, 12(2), 3, DOI:10.13008/2151–2957.1260

Canfield, K.N., Menezes, S., Matsuda, S.B., Moore, A., Mosley Austin, A.N., Dewsbury, B.M., Smith, H.M. et al. (2020) 'Science communication demands a critical approach that centers inclusion, equity, and intersectionality', *Frontiers in Communication*, 5, 2, DOI: https://doi.org/10.3389/fcomm.2020.00002

Ceccarelli, L. (2013) 'To whom do we speak? The audiences for scholarship on the rhetoric of science and technology', *Poroi*, 9(1), 7, DOI: 10.13008/2151–2957.1151

Chilvers, Jason. (2013) 'Reflexive engagement? Actors, learning, and reflexivity in public dialogue on science and technology', *Science Communication*, *35*(3), 283–310, DOI: https://doi.org/10.1177/1075 547012454598

Connors, R.J. (2004) 'The rise of technical writing Instruction in America', in J. Johnson-Eilola & S.A. Selber (eds.), *Central works in technical communication*, Oxford: Oxford University Press, illustrated edition.

Cooper, C. (2016) *Citizen science: How ordinary people are changing the face of discovery*, New York: Abrams.

Cox, R. (2012) *Environmental communication and the public sphere*, London: Sage Publications.

Crowley, Sharon. (1995) 'Composition's ethic of service, the universal requirement, and the discourse of student need', *Journal of Advanced Composition*, *15*(2), 227–239.

Depew, D. (2014) 'Introduction to Volume 10.1', *Poroi*, *10*(1), Article 1, 5, DOI: https://doi.org/10.13008/2151-2957.1186

Dijkstra, A.M., & Gutteling, J.M. (2012) 'Communicative aspects of the public-science relationship explored: Results of focus group discussions about biotechnology and genomics', *Science Communication*, *34*(3), 63–391, DOI: https://doi.org/10.1177/1075547011417894

Druschke, C.G. (2014) 'With whom do we speak? Building transdiscipinary collaborations in rhetoric of science', *Poroi*, 10(1), 10, DOI: https://doi.org/10.13008/2151-2957.1175

Druschke, C.G., & McGreavy, B. (2016) 'Why rhetoric matters for ecology', *Frontiers in Ecology and the Environment*, 14(1), 46–52, https://doi.org/10.1002/16-0113.1

du Plessis, H. (2008) 'Public communication of science and technology in developing countries', *Handbook of Public Communication of Science and Technology*, p. 213.

Fahnestock, J. (1986) 'Accommodating science: The rhetorical life of scientific facts', *Written Communication*, vol. 3, no. 3, pp. 275–296, doi:10.1177/0741088386003003001

Felt, U., & Davies, S.R. (eds.) (2020) *Exploring science communication: A science and technology studies approach*, London: Sage Publications Limited.

Gieryn, T.F. (1983) 'Boundary-work and the demarcation of science from non-science: Strains and interests in professional ideologies of scientists', *American Sociological Review*, *48*(6), 781–795, DOI: https://doi.org/10.2307/2095325

Goodwin, J. (2014) 'Introduction: Collaborations between scientists and rhetoricians of science/technology/medicine', *Poroi*, 10(1), Article 5, 5, DOI: https://doi.org/10.13008/2151-2957.1176

Goodwin, J. (2019) 'Sophistical refutations in the climate change debates', *Journal of Argumentation in Context*, 8(1), 40–64, DOI: https://doi.org/10.1075/jaic.18008.goo

Goodwin, J., Dahlstrom, M.F., Kemis, M., Wolf, C., & Hutchison, C. (2014) 'Rhetorical resources for teaching responsible communication of science', *Poroi*, 10(1), Article 7, 5, DOI: https://doi.org/10.13008/2151-2957.1179

Graff, G., & Birkenstein, C. (2006) *They say, I say*, New York: WW Norton.

Gross, A.G. (1994) 'The roles of rhetoric in the public understanding of science', *Public Understanding of Science*, 3(1), 3–23, DOI: https://doi.org/10.1088/0963-6625/3/1/001

Gross, A.G., & Harmon, J.E. (2010) *The craft of scientific communication*, Chicago: University of Chicago Press.

Hayden, T., & Check Hayden, E. (2018) 'Science journalism's unlikely golden age', *Frontiers in communication*, 2, 24, DOI: https://doi.org/10.3389/fcomm.2017.00024

Herndl, C.G. (1993) 'Teaching discourse and reproducing culture: A critique of research and pedagogy in professional and non-academic writing', *College Composition and Communication*, 44(3), 349–363, DOI:10.2307/358988

Kearns, F. (2021) *Getting to the heart of science communication*, Washington, DC: Island Press.

Kortencamp, K.V., & Basten, B. (2015) 'Environmental science in the media: Effects of opposing viewpoints on risk and uncertainty perceptions', *Science Communication*, *37*(3), 287–313, DOI: https://doi.org/10.1177/1075547015574016

Kraft, J., Walck-Shannon, E., Reilly, C., & Stapleton, A.E. (2019) 'Integrating case studies into graduate teaching assistant training to improve instruction in biology laboratory courses', bioRxiv, 594853.

KSJ Science Editing Handbook. (2020) 'KSJ handbook', https://ksjhandbook.org/

Lerner, N. (2007) 'Laboratory lessons for writing and science', *Written communication*, 24(3), 191–222, DOI: https://doi.org/10.1177/0741088307302765

Leydesdorff, L., & Hellsten, I. (2005) 'Metaphors and diaphors in science communication: Mapping the case of stem cell research', *Science communication*, 27(1), 64–99, DOI: https://doi.org/10.1177/1075547005278346

Liu, X. (2009) 'Beyond Science Literacy: Science and the Public', *International Journal of Environmental and Science Education*, 4(3), 301–311.

Loken, E. (2019) 'The replication crisis is good for science', *The Conversation*, https://theconversation.com/the-replication-crisis-is-good-for-science-103736

Loken, E., & Gelman, A. (2017) 'Measurement error and the replication crisis', *Science*, 355(6325), 584–585, DOI: 10.1126/science.aal3618

Luederitz, C., Meyer, M., Abson, D.J., Gralla, F., Lang, D.J., Rau, A.L., & von Wehrden, H. (2016) 'Systematic student-driven literature reviews in sustainability science – an effective way to merge research and teaching', *Journal of Cleaner Production*, 119, 229–235, DOI: 10.1016/j.jclepro.2016.02.005

Maddalena, K., & Reilly, C.A. (2017) 'Dissolving the divide between expert and public', in H. Yu & K.M. Northcut (eds.), *Scientific communication: Practices, theories, and pedagogies*, New York: Taylor and Francis.

McMillan, V.E., & McMillan, V. (2012) *Writing papers in the biological sciences*. London: Palgrave Macmillan.

Mehlenbacher, A. (2019) *Science communication online: Engaging experts and publics on the internet*, Columbus, OH: The Ohio State University Press.

Menninger, H. (2015) 'Authentic public science', *NC State University, College of Science News.*

Miller, C.R. (1979) 'A humanistic rationale for technical writing', *College English*, 40(6), 610–617, DOI: https://doi.org/10.2307/375964

Miller, C.R. (1984) 'Genre as social action', *Quarterly Journal of Speech*, 7, 151–167, DOI: https://doi.org/10.1080/00335638409383686

Miller, S. (2001) 'Public understanding of science at the crossroads', *Public understanding of science*, 10(1), 115–120, DOI: https://doi.org/10.3109/a036859

Mogendorff, Karen, te Molder, Hedwig, van Woerkum, Cees, & Gremmen, Bart. (2016) 'Turning experts into self-reflexive speakers: The problematization of technical-scientific expertise relative to alternative forms of expertise', *Science Communication*, 38(1), 26–50, DOI: https://doi.org/10.1177/1075547015615113

Montuori, A. (2005) 'Literature review as creative inquiry: Reframing scholarship as a creative process', *Journal of Transformative Education*, 3(4), 374–393, DOI: https://doi.org/10.1177/1541344605279381

Nisbet, M.C. (2009) 'Communicating climate change: Why frames matter for public engagement', *Environment: Science and Policy for Sustainable Development*, 51(2), 12–23, DOI: https://doi.org/10.3200/ENVT.51.2.12-23

Nisbet, M.C., & Scheufele, D.A. (2009) 'What's next for science communication? Promising directions and lingering distractions', *American journal of botany*, 96(10), 1767–1778, DOI: 10.3732/ajb.0900041

The Open Notebook. (2010) 'The open notebook', www.theopennotebook.com/ (Accessed 31 January 2021)

Penrose, A.M., & Katz, S.B. (1998) *Writing in the sciences*, New York: St. Martin's Press.

Priest, S., Goodwin, J., & Dahlstrom, M.F. (eds.) (2018) *Ethics and practice in science communication*, Chicago: University of Chicago Press.

Rao, K., Ok, M.W., Bryant, B.R. (2014) 'A review of research on universal design educational models', *Remedial and Special Education*, 35(3), 153–166, DOI: https://doi.org/10.1177/0741932513518980

Robertson, D.P., & Hull, R.B. (2001) 'Beyond biology: Toward a more public ecology for conservation', *Conservation Biology*, 15(4), 970–979. www.jstor.org/stable/3061316

Russill, C., & Nyssa, Z. (2009) 'The tipping point trend in climate change communication', *Global environmental change*, 19(3), 336–344, doi:10.1016/j.gloenvcha.2009.04.001

Schimel, J. (2012) *Writing science: How to write papers that get cited and proposals that get funded*, New York: Oxford University Press.

Sharon, A.J., & Baram-Tsabari, A. (2020) 'Can science literacy help individuals identify misinformation in everyday life?' *Science Education*, 104(5), 873–894. https://doi.org/10.1002/sce.21581

Sollaci, L.B., & Pereira, M.G. (2004) 'The introduction, methods, results, and discussion (IMRAD) structure: A fifty-year survey', *Journal of the Medical Library Association*, 92(3), 364–367.

Stocklmayer, S., Gore, R., & Bryant, C. (eds.) (2001) *Science communication in theory and practice*, Springer, vol. 14, 1st edition, www.springer.com/gp/book/9781402001307

Sturgis, P., & Allum, N. (2004) 'Science in society: Re-evaluating the deficit model of public attitudes', *Public understanding of science*, 13(1), 55–74, DOI: https://doi.org/10.1177/0963662504042690

Swales, J. (1990) *Genre analysis: English in academic and research settings*, Cambridge: Cambridge University Press.

Telea, A.C. (2014) *Data visualization: Principles and practice*, London: A K Peters, CRC Press, 2nd edition.

Valenti, J.M., & Tavana, G. (2005) 'Report: continuing science education for environmental journalists and science writers: in situ with the experts', *Science Communication*, 27(2), 300–310, DOI: https://doi.org/10.1177/1075547005282474

Vernon, J.L. (2014) 'Leveraging rhetoric for improved communication of science: A scientist's perspective', *Poroi*, *10*(1), Article 11, 6, DOI: https://doi.org/10.13008/2151-2957.1181

Yau, N. (2007) 'Flowing data', https://flowingdata.com/

Yong, E. (2010) 'On the origin of science writers', *Discover Magazine*.

28

PEDAGOGIES FOR SUPPORTING GLOBAL SCIENTISTS' RESEARCH WRITING

James N. Corcoran and Karen Englander

Plurilingual EAL scientists' strengths, needs, and challenges

Scientists who use English as an additional language (EAL) are often viewed as inherently flawed academic communicators, ostensibly limited by their lack of advanced English language writing proficiency (Curry and Lillis, 2019; Hultgren, 2019). In this chapter, we challenge such a *deficit perspective* by presenting a pedagogical approach for supporting EAL scientists' research writing in targeted ways that recognizes both their strengths as plurilingual communicators as well as the distinct challenges they face in a complex, metric-heavy science world where English is privileged. Readers will note throughout this chapter that we have mindfully chosen to use the term *plurilingual EAL* rather than *ESL*, as the term *plurilingual* signals the ability of a person to use multiple languages for various purposes in a variety of contexts (Cenoz and Gorter, 2020; García and Otheguy, 2020; Lau and Van Viegen, 2020; Marshall and Moore, 2018). This is certainly the case for the scientists referred to in this chapter, who often use multiple languages for scholarly publication purposes.

For several decades now, scientific research has been designated as a commodity that is counted and ranked in a knowledge economy (Altbach, 2013; OECD, 1996). Scientific products (i.e. research articles) are valued commodities that are counted and ranked for global visibility. Across disciplines, value is awarded to research (and researchers) published in selected journals, i.e. those included in indexes such as the Web of Science (formerly Thomson Reuters, now curated by Clarivate Analytics) or Scopus. Though robust publication occurs in other languages (Curry and Lillis, 2017; Gotti, Chapter 13 in this volume; Hamel, 2013; Salö, 2017), more than 90% of the journals included in these indexes are published in English (Liu, 2017). In the knowledge economy, publication in indexed journals is recognized at three scales: the individual researcher, their institution of higher education, and the country where they work (Englander and Uzuner-Smith, 2013). Research production is tallied by funding bodies, resulting in incentives at the national and institutional levels for plurilingual EAL scientists to put their efforts either primarily or exclusively toward achieving these publication outcomes (Lei and Jiang, 2019; Luo and Hyland, 2019). Nowhere is the dominance of English more apparent than in the natural sciences (Montgomery, 2013; Ware and Mabe, 2015), although the social sciences

DOI: 10.4324/9781003043782-33

and humanities are by no means exempt (Flowerdew and Li, 2009; Gea-Valor et al., 2014). And nobody is more acutely impacted by the dominance of English than plurilingual EAL scientists (Flowerdew, 2019; McKinley and Rose, 2018; Politzer-Ahles et al., 2016). Thus conversations about international scientific communication should be carried out with a recognition of the global, neoliberal, asymmetrical market of knowledge production in which scientists work (Demeter, 2019; Lillis and Curry, 2010; Nygaard and Bellanova, 2017). In this chapter, we consider challenges that are most acute or amplified for plurilingual EAL scientists, be they in natural or social science disciplines, while reflecting upon pedagogical interventions that might effectively and equitably support them.

Over the past decade, a body of research emanating from the fields of applied linguistics, education, and writing studies has outlined many of the challenges plurilingual scholars across disciplines face when writing up their work for publication in indexed journals. Some of this work has focused on plurilingual EAL scientists in center nations such as Canada, New Zealand, or the United States, where English is the dominant local language (Fazel, 2019; Huang, 2010; Simpson et al., 2016). However, the majority of the world's scientists live in the global periphery (Wallerstein, 1991) or semiperiphery (Bennett, 2014; Sheridan, Chapter 8 in this volume), where English is not the local language. Research from this emerging field, *English for research publication purposes (ERPP)*, has charted the experiences and challenges of plurilingual EAL scientists across non-Anglophone geographical regions, including Europe (e.g. Arnbjörnsdóttir and Ingvarsdóttir, 2017), Asia (e.g. Mu, 2020), Africa (e.g. Kwanya, 2020), the Middle East (e.g. Gholami and Zeinolabedini, 2015), and Latin America (e.g. Monteiro and Hirano, 2020). For more global perspectives, see also edited volumes on ERPP by Cargill and Burgess (2017), Corcoran et al. (2019), Curry and Lillis (2017) and Habibie and Hyland (2019).

This expanding body of work in the burgeoning field of ERPP has contributed to heightened awareness of the issues faced by plurilingual EAL scientists. It clarifies many of the acute challenges faced by plurilingual EAL scientists across disciplines as they attempt to meet increasing expectations for publishing their work in indexed international (i.e. English-medium) scholarly journals. Salient challenges include (but are not limited to) identifying appropriate journals; positioning their research as valuable to an international readership; attending to journal-specific author guidelines; lexicogrammatical accuracy; intercultural rhetorical differences; section-specific discursive expectations (e.g. linking introduction/discussion sections); attending to reviewer and editor feedback; forming and maintaining networks; obtaining sufficient research equipment and resources, among other issues. Many of these challenges are shared among all scientists, regardless of their first language, discipline, or geographical location (Hultgren, 2019). However, recent research has shown that plurilingual EAL authors operate in a world of science that is unequal, where Anglophone, center-based scientists have an inherent competitive advantage (Curry and Lillis, 2019; Hanauer and Englander, 2013). Thus advanced scholarly writing for plurilingual EALs, particularly those located in the global peripheries, is more difficult and requires distinctive support and interventions (Cargill, 2019). This work has also highlighted the need for dispensing with monolingual, deficit understandings or conceptions of these scholars as deficient or flawed users of English (Cook, 2001; Flowerdew and Ho Wang, 2016; Lin, 2016). In fact, plurilingual EALs are better understood as complex, multi-/plurisompetent users of languages, with scholarly voices and identities linked to their multiple linguistic resources (Cenoz et al., 2017; Coste et al., 2009). We draw on these plurilingual conceptions and orientations, advocating for international publishing that has the potential to welcome diverse scientific voices, ways of knowing, and research agendas (Gentil, 2018; Kubota, 2019; Sousa Santos and Menezes, 2020; Sugiharto, 2020).

Over the past decade, we, the authors, have added to the growing body of ERPP research, providing several key findings stemming from our investigations into Latin American scientists' experiences with scholarly writing for publication:

1 there is an additional, quantifiable burden experienced by scientists when writing research in an additional language (Hanauer and Englander 2011; Hanauer et al., 2019);
2 there are specific reasons why scientists choose to put their efforts toward publishing in English versus Spanish (Englander et al., forthcoming);
3 there are specific, salient challenges scientists face when attempting to publish their work in English (Corcoran, 2015); and
4 there are detrimental individual and collective beliefs about English and science (e.g. good science is in English; good English = good science) that serve to maintain unequal relations of power in scientific knowledge production (Corcoran, 2019; Corcoran and Englander, 2016).

Moreover, our findings illustrate that Latin American scientists, like their peers around the world, write research in English in order to achieve professional advancement, win institutional/federal monies, and connect with an international audience. Interestingly, many of these scientists *also* publish their work in Spanish/Portuguese-language journals, often in order to connect with a domestic audience more impacted by their research (Corcoran, 2019). These findings raise a particular paradox: scientists are often incentivized to publish their work in international, English-language journals for prestige and financial reward rather than in national-language journals where their work may contribute to solving problems of regional and national concern. Finally, though quantitative data are elusive on this front, qualitative data suggest that pedagogical interventions can improve EAL scientists' confidence, genre awareness, and overall ability to navigate the research article submission and review process (Corcoran, 2017; Englander and Corcoran, 2019).

Based on these findings and our experiences with the trials and tribulations of plurilingual scientists in Mexico, Ecuador, and Brazil, we propose a targeted, culturally responsive, and flexible pedagogical approach for supporting plurilingual EAL scientists' research writing. In so doing, we look to challenge the status quo in scientific knowledge production – while also supporting the broader advancement of science. Our approach is mindfully oriented to readers of this chapter who might adapt it for their particular disciplinary audience and global context, wherever that might be.

English for research publication purposes (ERPP): critical versus pragmatic pedagogies

Scientists, concerned with supporting research writing, have endeavored to help novice scholars with books that describe intuitive writing practices. Well known examples include *Scientists Must Write* by Robert Barrass (2002), *Eloquent Science* by David Schultz (2009), and the continually updated *How to Write and Publish a Scientific Paper*, first published by Robert Day (1979) and more recently coauthored with Barbara Gastel (2016). These books, while certainly useful to EAL scientists, often do not take advantage of the research writing knowledge and expertise emanating from the field of applied linguistics. Over the past decade, applied linguists with expertise in diverse aspects of the discipline have published books designed to support plurilingual scientists' English-language research writing. Seminal works in this area include those informed by discourse and genre analysis (Swales and Feak, 2012; Matarese, 2013); corpus studies (Cargill and O'Connor, 2013; Glasman-Deal, 2009); social writing practices (Curry and Lillis, 2013; Paltridge and Starfield, 2016); and critical applied language studies (Englander and Corcoran, 2019). This applied work has contributed immensely to the burgeoning field of English for research publication purposes, inspiring pedagogical innovation and research agendas. As reported in the recent state-of-the-art review by Li and Flowerdew (2020), instructors in locations around the world draw on this work in creating programs for EAL scientists and scholars.

Straightforward, pragmatic books on how to write science, while meeting an immediate need, typically ignore or underplay issues of power, access, and context that often pervade the experiences of plurilingual EAL scientists. They invariably advise scientists to simply identify what an English-language paper looks like and then replicate it – an *identify and replicate* approach. Courses aimed at supporting international scientists often take on an approach aligned with some of the how-to books just listed. The courses typically cover two main topics, in descending importance:

1   the structure and organization of a typical scientific paper as Introduction, Methods, Results, and Discussion (IMRD/IMRAD) and the rhetorical moves (Swales, 1990) made in each section, including sentence-level lexical and grammatical accuracy;
2   the publishing process from journal selection and manuscript submission to publication.

Such pragmatic pedagogies are often quite helpful in improving scientists' awareness of disciplinary writing conventions as well as the journal submission and review process, while potentially improving plurilingual EAL scientists' English language proficiency. There is great benefit in achieving these learning outcomes. However, pragmatic pedagogies may also overlook issues of language, power, and identity that are central to the experiences of plurilingual EAL scientists attempting to achieve publication in English language journals. By ignoring these issues, such approaches potentially reinforce or exacerbate the unequal relations in the production and dissemination of science.

Based on our direct participation in several English for research publication purposes (ERPP) programs in Latin America, we draw on extant theory and empirical findings to present here an effective and equitable approach to supporting plurilingual scientists: *Critical Plurilingual Pedagogies* (Englander and Corcoran, 2019). Several themes central to our critical, plurilingual approach, as outlined next, are often overlooked in pragmatic ERPP approaches. Rather than *identify and replicate*, we recommend an *identify and situate* approach, where scientists identify the characteristics of English language scientific research articles and then situate those characteristics based on their own historical and social (including linguistic) knowledge and experience. This situated approach not only validates plurilingual scientists' knowledge and experience but also promotes awareness of a broader range of choices and greater sense of agency in the production and dissemination of their work. An identify and situate pedagogy comprises two main tenets: *Critical Genre and Language Awareness* and *Critical Publishing Practices and Processes*. In the following subsections, we present examples of how an identify and situate approach can be implemented, with suggested readings that may be useful for those looking to adapt our approach for their classroom.

Identify and situate: critical genre and language awareness

Common pragmatic approaches to teaching effective writing for publication emphasize identification of genre-specific norms, structures, and rhetorical patterns of disciplinary writing (e.g. general-to-specific presentation of ideas; typical citation patterns). Aspiring authors are encouraged to examine published papers in their disciplines and desired journals in order to develop an understanding of how papers are organized into sections, which information should be included where, and what common phrases (e.g. "as shown in figure," "in agreement with") are used to accomplish particular rhetorical moves (Swales and Feak, 2012). This enhanced genre awareness will undoubtedly serve scientists well. Further, pragmatic approaches often focus on improving English language proficiency, with a focus on lexicogrammatical accuracy. Increased lexicogrammatical accuracy is most certainly a desirable learning outcome. However, hyperfocus on replicating dominant Anglophone norms can serve to cast plurilingual EAL scientists as deficient or inadequate rather than as language-resource rich. Our approach challenges an uncritical acceptance of normative genre and language conventions in a number of ways:

- Examining changes in the format of science papers over time (from Newton through Darwin to post–World War II), noting that IMRD/IMRAD only became entrenched as a "rigidly conventionalized" structure structural model of research articles in the 1970s (Atkinson, 1996, p. 347) – the mid-twentieth century – and highlighting new, evolving genres. By understanding that the research article is an ever evolving genre in response to reader and publisher needs and communicative technologies, we provide more agency to the scientific author (see Bondi, Chapter 14 in this volume). Further, see Hedges and Florek (2019), Buehl (Chapter 24 in this volume), and Cagle (Chapter 25 in this volume) for discussions of the emerging digital scientific genres.
- Discussing and reflecting upon problematic statements such as this one from a former editor of *Science*:

  > If you see people making multiple mistakes in spelling, syntax and semantics . . . you have to wonder whether, when they did their science, they weren't making similar errors of inattention.
  >
  > *(Bloom, cited in Gibbs, 1995, p. 97)*

  This comment reveals a monolingually oriented response to plurilingual EAL manuscripts, where inaccurate writing is conflated with bad science. By considering the validity of such a position, scientists can examine their own ideas about writing in an additional language.
- Identifying the role and positioning of the scientist as an active agent in the work while examining cultural and disciplinary variation. How much *I* and *we* are used in particular sections of a science paper varies with discipline and language, revealing different traditions of establishing the researcher's voice, which can empower plurilingual writers. For example, see Lorés-Sanz (2011) for an examination of how an author's voice is constructed in research papers.
- Emphasizing the value of intelligibility over linguistic accuracy. Preoccupation with correct prepositions or the distinction between "which" and "that" can require expensive language corrector services and delay manuscript submission; such detail rarely causes a manuscript to be rejected, as long as the content is intelligible. For example, see Rozycki and Johnson (2013), where they present nonstandard Englishes in award-winning engineering research articles.
- Analyzing and promoting diversity of expression. Many researchers have identified distinct rhetorical similarities and differences between research papers in English and other languages (e.g. Spanish, Russian, etc.). EAL scientists can perform their own first-language and discipline-specific comparisons in order to determine suitability of language use. Such

writing decisions support scientists maintaining their plurilingual voice. For example, see Pérez-Llantada (2012) or Bennett (2014), who report on hybrid discourses, i.e. published papers that display characteristics of the writer's first language in English-language papers.

Identify and situate: critical publishing practices and processes

Pragmatic approaches to supporting plurilingual EAL scholars often provide information on the manuscript submission process (e.g. managing submission portals, journal instructions to authors), the review and revision process, and the text-based challenges of scientific writing for publication in a top-down, prescriptive manner. Enhanced understanding of these often occluded submission and review processes are extremely valuable to scientists, particularly novice ones. However, pragmatic approaches often overlook the social practices of research writing, including topics such as the unequal allocation of power in science production, actual and perceived notions of "bias" in scientific adjudication, and/or the possibility of publishing in multiple languages and for distinct audiences. Bringing a critical lens to these topics creates the space for scientists to more fully examine their experiences with publishing, while developing sustainable social practices of research writing that will serve them well throughout their academic trajectories. We suggest implementing an identify and situate approach that acknowledges scientists' multifaceted, multilingual social and political contexts in several ways:

- Examining and reflecting on the impact of publishing choices by discussing what is gained and what is lost by publishing scientific work in English rather than in other languages. Validating the research work done in different regions and published in other languages can empower plurilingual EAL scientists. For example, see "The Hidden Bias of Science's Universal Language" (Huttner-Koros, 2015) in the mainstream magazine *The Atlantic*.
- Exploring the different content and audiences who could be reached by a piece of research if it is published in a national/regional language versus in English. For example, see Curry and Lillis (2013, pp. 15–21) for how to critically evaluate the impact of publishing choices. For a more provocative position, consider Philip Altbach's (2013) argument that plurilingual scholars have a responsibility to disseminate their research in local languages in order to demonstrate their commitment to issues of local and/or national importance.
- Reflecting on the agency and relations of power between authors and language *correctors* (e.g. editors and translators), journal reviewers and editors. Discussing scientists' perceptions of fairness in editorial review and effective means of maintaining the integrity of their work can empower plurilingual EAL scientists. For example, see McKinley and Rose (2018) for a discussion of language instructions in author guidelines, and Saposnik et al. (2014) for quantitative analysis of peripheral versus center country manuscript acceptance.
- Examining the systemic challenges faced by plurilingual scientists who live outside the English-dominant countries. Scientists from a variety of disciplines have published their firsthand experiences, which can be referenced for discussion. For example, see Clavero (2011), Fregonese (2017), or Umakantha (1997).
- Exploring strategies for building and maintaining networks of scientists and others (e.g. language editors) who can support a broad range of meaningful research and publishing activities. Such networks can be indispensable for collaborations, introductions to disciplinary influencers, access to funding and publishing opportunities, and sustainable publishing outcomes. For example, see Curry and Lillis's suggestion of "critical" strategies for plurilingual scientists (2013).

As we hope is evident, our critical plurilingual approach to pedagogy for English for research publication purposes is not a blueprint. Rather, it is an approach to scientific research writing that values the linguistic resources of plurilingual scientists, examines the relations of power inherent in science publishing, and acknowledges the identity and situated contexts involved in performing and communicating scientific research. We emphasize that all pedagogical support for scientists must be determined locally, "taking into account the historical, socio-economic, political, and linguistic concerns, constraints and pressures of those contexts" (Englander and Corcoran 2019, p. 223). In other words, one pedagogical size should not fit all.

Charting a path toward a more diverse, equitable scientific landscape

In this chapter, we have outlined some state-of-the-art research in a new subfield called English for research publication purposes (ERPP). This field includes emerging descriptions, evaluations, and propositions of pedagogical approaches to supporting the research writing of plurilingual EAL scientists from across disciplines. One such approach is our research- and theory-driven *Critical Plurilingual Pedagogies*. This approach is meant to be adapted and adopted to respond to local scientists' advanced disciplinary literacy needs and desires. As mentioned previously, we are adamant that the responsibility of creating a more equitable landscape of scientific knowledge production rests not only with those producing science but also with those adjudicating and supporting such production. Thus, we leave you, the reader, with a set of three principles that may guide your future practices:

1 *Think globally, act locally*: Support for plurilingual EAL scientists should be guided by local needs. Supports might include a range of faculty- or institution-based initiatives, including scholarly writing (for publication) courses, writing groups, vetted lists of trustworthy editors/translators, etc. We encourage the local development of expertise for conducting pedagogical interventions rather than hiring experts from abroad.

2 *Encourage diversity of expression*: Those responsible for pedagogical support should validate plurilingual EAL scientists' use of their multiple linguistic resources. Emphasis on diverse ways of knowing, doing, and communicating should occur at the textual and pedagogical levels (e.g. delivery of course content in a language other than English even when the desired scientific product will be in English). At the textual level, such an approach should emphasize the value of communicative intelligibility over prescriptive accuracy in production and adjudication of scientific texts. Emphasizing the potential affordances of plurilingual knowledge production (e.g. where multiple languages are used to "do" science) can lead to a richer scientific landscape where epistemological and ontological diversity advances science in ways a homogeneous Anglophone world of science cannot.

3 *Empower EAL scientists*: Institutions, faculties, and departments should recognize and validate scientists' desires to publish in multiple languages (i.e. in regional and national peer-reviewed journals) in order to convey research findings to appropriate stakeholders. ERPP courses should provide space for scrutiny of institutional and national evaluation policies that affect scientists' lives. Research institutions and course designers should provide opportunities and strategies for scientists to develop networks of international and domestic colleagues who can facilitate scientific production/engagement.

In sum, we argue that there is an ethical responsibility to reduce the inequitable burden borne by plurilingual EAL scientists in a world of knowledge production dominated by English. This can be accomplished in three important ways: (1) allocating additional institutional time for scientists to produce/revise their research works; (2) providing accessible institutional research writing resources; and (3) reflecting upon our language adjudication/support practices. Fundamentally, we believe that enacting these changes to our research support practices can not only lead to greater equity in knowledge production but also the advancement of science.

References

Altbach, P.G. (2013) 'Advancing the national and global knowledge economy: The role of research universities in developing countries', *Studies in Higher Education*, vol. 38, no. 3, pp. 316–330, https://doi.org/10.1080/03075079.2013.773222

Arnbjörnsdóttir, B., & Ingvardsdóttir, H. (2017) 'Issues of identity and voice: Writing English for research purposes in the semiperiphery', in M.J. Curry & T. Lillis (eds.), *Global academic publishing: Policies, perspectives and pedagogies*, Clevedon: Multilingual Matters, pp. 73–87.

Atkinson, D 1996, 'The Philosophical Transactions of the Royal Society of London, 1675–1975: A sociohistorical discourse analysis', *Language in Society*, vol. 25, no. 3, pp. 333–371, www.jstor.org/stable/4168717

Barrass, R. (2002) *Scientists must write*, New York: Routledge.

Bennett, K. (2014) 'Introduction: The political and economic infrastructure of academic practice: The "semiperiphery" as a category for social and linguistic analysis', in K. Bennett (ed.), *The semiperiphery of academic writing*, London: Palgrave Macmillan, pp. 1–9.

Cargill, M. (2019) 'The value of writing for publication workshops', in P. Habibie & K. Hyland (eds.), *Novice writers and scholarly publication: Authors, mentors, gatekeepers*, London: Palgrave Macmillan, pp. 195–214.

Cargill, M., & Burgess, S. (eds.) (2017) *Publishing research in English as an additional language: Practices, pathways and potentials*, Adelaide: University of Adelaide Press.

Cargill, M., & O'Connor, P. (2013) *Writing scientific research articles*, West Sussex: Wiley-Blackwell, 2nd edition.

Cenoz, J & Gorter, D 2020, 'Pedagogical translanguaging: An introduction', *System*, vol. 92, pp. 1–7, DOI: https://doi.org/10.1016/j.system.2020.102269

Cenoz, J., Gorter, D., & May, S. (eds.) (2017) *Language awareness and multilingualism*, Cham: Springer.

Clavero, M 2011, 'Unfortunately, linguistic injustice matters', *Trends in Ecology and Evolution*, vol. 26, no. 4, p. 156, DOI:https://doi.org/10.1016/j.tree.2011.01.011

Cook, V 2001, 'Using the first language in the classroom', *Canadian Modern Language Review*, vol. 57, no. 3, pp. 402–423, DOI: https://doi.org/10.3138/cmlr.57.3.402

Corcoran, J.N. (2015) 'English as the international language of science: A case study of Mexican scientists' academic writing for publication', PhD thesis, University of Toronto, Toronto.

Corcoran, J.N. (2017) 'The potential and limitations of an English for research publication purposes course for Mexican scholars', in M.J. Curry & T. Lillis (eds.), *Global academic publishing: Policies, practices, and pedagogies*, Clevedon: Multilingual Matters, pp. 242–255.

Corcoran, J.N. (2019) 'Addressing the "bias gap": A research-driven argument for critical support of plurilingual scientists' research writing', *Written Communication*, vol. 36, no. 4, pp. 538–577, https://doi.org/10.1177/0741088319861648

Corcoran, JN & Englander, K 2016, 'A proposal for critical-pragmatic pedagogical approaches to English for research publication purposes', *Publications*, vol. 4, no. 1, pp. 1–10, DOI:10.3390/publications4010006

Corcoran, J.N., Englander, K., & Muresan, L. (eds.) (2019) *Pedagogies and policies on publishing research in English: Local initiatives supporting international scholars*, New York: Routledge.

Coste, D., Moore, D., & Zarate, G. (2009) *Plurilingual and pluricultural competence*, Strasbourg: Council of Europe.

Curry, M.J. & Lillis, T. (2013) *A scholar's guide to getting published in English: Critical choices and practical strategies*, Clevedon: Multilingual Matters.

Curry, MJ & Lillis, T 2019, 'Unpacking the lore on multilingual scholars' publishing in English: A discussion paper' *Publications*, vol. 7, no. 27, pp. 1–14, DOI:10.3390/publications7020027

Curry, M.J., & Lillis, T. (2017) 'Problematizing English as the privileged language of global academic publishing', in M.J. Curry & T. Lillis (eds.), *Global academic publishing: Policies, practices, and pedagogies*, Clevedon: Multilingual Matters, pp. 1–22.

Day, R.A. (1979) *How to write and publish a scientific paper*, Santa Barbara: Greenwood Press.

Demeter, M 2019, 'The world-systemic dynamics of knowledge production: The distribution of transnational academic capital in the Social Sciences', *Journal of World-Systems Research*, vol. 25, no. 1, pp. 111–144, DOI: https://doi.org/10.5195/jwsr.2019.887

Englander, K., & Corcoran, J.N. (2019) *English for research publication purposes: Critical plurilingual pedagogies*, New York: Routledge.

Englander, K., Carrasco, A., Kent, R., & Corcoran, J.N. (forthcoming) 'Modelar, atender o resistir a las políticas académicas de publicación: Respuestas disciplinarias en instituciones mexicanas de educación superior', in R.E. Hamel (ed.), *Políticas del lenguaje en América Latina*, Berlin: De Gruyter.

Englander, K & Uzuner-Smith, S 2013, 'The role of policy in constructing the peripheral scientist in the era of globalization', *Language Policy*, vol. 12, no. 3, pp. 231–250, DOI:10.1007/s10993-012-9268-1

Fazel, I. (2019) 'Writing for publication as a native speaker: The experiences of two anglophone novice scholars', in P. Habibie & K. Hyland (eds.), *Novice writers and scholarly publication: Authors, mentors, gatekeepers*, London: Palgrave Macmillan, pp. 79–98.

Flowerdew, J 2019, 'The linguistic disadvantage of scholars who write in English as an additional language: Myth or reality?', *Language Teaching*, vol. 52, no. 2, pp. 249–260, doi:10.1017/S0261444819000041

Flowerdew, J & Ho Wang, S 2016, 'Author's editor revisions to manuscripts published in international journals', *Journal of Second Language Writing*, vol. 32, pp. 39–52, DOI: https://doi.org/10.1016/j.jslw.2016.03.004

Flowerdew, J & Li, Y 2009, 'English or Chinese? The trade-off between local and international publication among Chinese academics in the humanities and social sciences', *Journal of Second Language Writing*, vol. 18, no. 1, pp. 1–16, DOI: https://doi.org/10.1016/j.jslw.2008.09.005

Fregonese, S 2017, 'English: Lingua Franca or Disenfranchising?', *Fennia-International Journal of Geography*, vol. 195, no. 2, pp. 194–196, DOI: https://doi.org/10.11143/fennia.67662

García, O & Otheguy, R 2020, 'Plurilingualism and translanguaging: commonalities and divergences', *International Journal of Bilingual Education and Bilingualism*, vol. 23, no. 1, pp. 17–35, DOI: https://doi.org/10.1080/13670050.2019.1598932

Gastel, B., & Day, R.A. (2016) *How to write and publish a scientific paper*, Cambridge: Cambridge University Press, 8th edition.

Gea-Valor, ML, Rey-Rocha, J & Moreno, AI 2014, 'Publishing research in the international context: An analysis of Spanish scholars' academic writing needs in the social sciences', *English for Specific Purposes*, vol. 36, pp. 47–59, DOI:10.1016/j.esp.2014.05.001

Gentil, G 2018 'Modern Languages, Bilingual Education, and Translation Studies: The Next Frontiers in WAC/WID Research and Instruction?', *Across the Disciplines*, vol. 15, no. 3, pp. 114–129, DOI: https://doi.org/10.37514/ATD-J.2018.15.3.16

Gholami, J & Zeinolabedini, M 2015, 'A diagnostic analysis of erroneous language in Iranian medical specialists' research papers', *The Journal of Tehran University Heart Center*, vol. 10, no. 1, pp. 58–67.

Gibbs, WW 1995, 'Lost science in the third world', *Scientific American*, vol. 273, no. 2, pp. 92–99, DOI:10.1038/scientificamerican0895-92

Glasman-Deal, H. (2009) *Scientific research writing for non-native speakers of English*, Oxford: World Scientific Publishing.

Habibie, P., & Hyland, K. (eds.) (2019) *Novice writers and scholarly publication: Authors, mentors, gatekeepers*, London: Palgrave Macmillan.

Hamel, R.E. (2013) 'Language policy and ideology in Latin America', in R. Bailey, R. Cameron & C. Lucas (eds.), *The Oxford handbook of sociolinguistics*, Oxford: Oxford University Press, pp. 609–628.

Hanauer, DI & Englander, K 2011, 'Quantifying the burden of writing research articles in a second language: Data from Mexican scientists', *Written Communication*, vol. 28, no. 4, pp. 403–416, DOI: https://doi.org/10.1177/0741088311420056

Hanauer, D.I., & Englander, K. (2013) *Scientific writing in a second language*, Anderson: Parlor Press.

Hanauer, D.I., Sheridan, C., & Englander, K. (2019) 'Linguistic injustice in the writing of research articles in English as a second language: Data from Taiwanese and Mexican researchers', *Written Communication*, vol. 36, no. 1, pp. 136–154, https://doi.org/10.1177/0741088318804821

Hedges, G. & Florek, C.S. (2019) 'The graphical abstract as a new genre in the promotion of science', in M.J. Luzón & C. Pérez-Llantada (eds.), *Science communication on the internet: Old genres meet new genres*, Amsterdam: John Benjamins, pp. 59–80.

Huang, JC 2010, 'Publishing and learning writing for publication in English: Perspectives of NNES PhD students in science', *Journal of English for Academic Purposes*, vol. 9, no. 1, pp. 33–44, DOI: doi:10.1016/j.jeap.2009.10.001

Hultgren, AK 2019, 'English as the language of academic publication: On equity, disadvantage, and "non-nativeness" as a red herring', *Publications*, vol. 7, no. 2, pp. 1–13, https://doi.org/10.3390/publications7020031

Huttner-Koros, A. (2015) 'The hidden bias of science's universal language', *The Atlantic*, 21 August, www.theatlantic.com/science/archive/2015/08/english-universal-language-science-research/400919/

Kubota, R 2019, 'Confronting epistemological racism, decolonizing scholarly knowledge: Race and gender in applied linguistics', *Applied Linguistics*, vol. 41, no. 5, pp. 712–732, doi:10.1093/applin/amz033

Kwanya, T 2020, 'Publishing and perishing? Publishing patterns of information science academics in Kenya', *Information Development*, vol. 36, no. 1, pp. 5–15, DOI: https://doi.org/10.1177/0266666918804586

Lau, S.M.C. & Van Viegen, S. (2020) 'Plurilingual pedagogies: An introduction', in S.M.C. Lau & S. Van Viegen (eds.), *Plurilingual pedagogies: Critical and creative endeavors for equitable language (in) education*, Cham: Springer, pp. 3–22.

Lei, J. & Jiang, T. (2019) 'Chinese university faculty's motivation and language choice for scholarly publishing', *Ibérica*, vol. 38, pp. 51–74.

Li, Y., & Flowerdew, J. (2020) 'Teaching English for research publication purposes: A review of language teachers' pedagogical initiatives', *English for Specific Purposes*, vol. 59, pp. 41–59, https://doi.org/10.1016/j.esp.2020.03.002

Lillis, T. & Curry, M.J. (2010) *Academic writing in a global context: The politics and practices of publishing in English*, London: Routledge.

Lin, A.M. (2016) *Language across the curriculum & CLIL in English as an additional language (EAL) contexts: Theory and practice*, New York: Springer.

Liu, W 2017, 'The changing role of non-English papers in scholarly communication: Evidence from Web of Science's three journal citation indexes', *Learned Publishing*, vol. 30, no. 2, pp. 115–123, DOI: https://doi.org/10.1002/leap.1089

Lorés-Sanz, R 2011, 'The construction of the author's voice in academic writing: The interplay of cultural and disciplinary factors', *Text & Talk*, vol. 31, no. 2, pp. 173–193, DOI: https://doi.org/10.1515/text.2011.008

Luo, N & Hyland, K 2019, '"I won't publish in Chinese now": Publishing, translation and the non-English speaking academic', *Journal of English for Academic Purposes*, vol. 39, pp. 37–47, DOI: https://doi.org/10.1016/j.jeap.2019.03.003

Marshall, S & Moore, D 2018, 'Plurilingualism amid the panoply of lingualisms: addressing critiques and misconceptions in education', *International Journal of Multilingualism*, vol. 15, no. 1, pp. 19–34, DOI: https://doi.org/10.1080/14790718.2016.1253699

Matarese, V. (ed.) (2013) *Supporting research writing*, Oxford: Chadros Publishing.

McKinley, J., & Rose, H. (2018) 'Conceptualizations of language errors, standards, norms and nativeness in English for research publication purposes: An analysis of journal submission guidelines', *Journal of Second Language Writing*, vol. 42, pp. 1–11, doi:10.1016/j.jslw.2018.07.003

Monteiro, K & Hirano, E 2020, 'A periphery inside a semi-periphery: The uneven participation of Brazilian scholars in the international community', *English for Specific Purposes*, vol. 58, pp. 15–29, DOI: https://doi.org/10.1016/j.esp.2019.11.001

Montgomery, S.L. (2013) *Does science need a global language? English and the future of research*, Chicago: University of Chicago Press.

Mu, C. (2020) *Understanding Chinese multilingual scholars' experiences of writing and publishing in English: A social-cognitive perspective*, Stuttgart: Springer.

Nygaard, L., & Bellanova, R. (2017) 'Lost in quantification: Scholars and the politics of bibliometrics', in M.J. Curry & T. Lillis (eds.), *Global academic publishing: Policies, perspectives and pedagogies*, Clevedon: Multilingual Matters, pp. 23–36.

OECD. (1996) *The knowledge-based economy*, Paris: OECD.

Paltridge, B., & Starfield, S. (2016) *Getting published in academic journals: Navigating the publication process*, University of Michigan Press, Ann Arbor.

Pérez-Llantada, C. (2012) *Scientific discourse and the rhetoric of globalization: The impact of culture and language*, London: Continuum.

Politzer-Ahles, S, Holliday, JJ, Girolamo, T, Spychalska, M & Harper Berkson, K 2016, 'Is linguistic injustice a myth? A response to Hyland (2016)', *Journal of Second Language Writing*, vol. 34, pp. 3–8, DOI: 10.1016/j.jslw.2016.09.003

Rozycki, W & Johnson, NH 2013, 'Non-canonical grammar in best paper award winners in engineering', *English for Specific Purposes*, vol. 32, no. 3, pp. 157–169, DOI: https://doi.org/10.1016/j.esp.2013.04.002

Salö, L. (2017) *The sociolinguistics of academic publishing: Language and the practices of homo academicus*. New York: Palgrave Macmillan.

Saposnik, G, Ovbiagele, B, Raptis, S, Fisher, M & Johnston, SC 2014, 'The effect of English proficiency and research funding on acceptance of submitted articles to *Stroke* journal', *Stroke*, vol. 45, pp. 1862–1868, DOI: https://doi.org/10.1161/STROKEAHA.114.005413

Schultz, D.M. (2009) *Eloquent science*, Boston: American Meteorological Society.

Simpson, S., Caplan, N., Cox, M., & Phillips, T. (2016) *Supporting graduate writers: Research, program, and curriculum design*, Ann Arbor: University of Michigan Press.

Sousa Santos, B. & Menezes, M.P. (eds.) (2020) *Knowledges born in the struggle: Constructing the epistemologies of the Global South*, London: Routledge.

Sugiharto, S. (2020) 'Enacting the locus of enunciation as a resistant tactic to confront epistemological racism and decolonize scholarly knowledge', *Applied Linguistics*, April 19, pp. 1–7, https://doi.org/10.1093/applin/amaa023

Swales, J.M. (1990) *Genre analysis: English in academic and research settings*, Cambridge: University of Cambridge Press.

Swales, J.M. & Feak, C.B. (2012) *Academic writing for graduate students*, Ann Arbor: University of Michigan Press, 3rd edition.

Umakantha, N. (1997) 'Beyond the language barrier', *Nature*, 27 Feb, vol. 385, no. 764, https://doi.org/10.1038/385764c0

Wallerstein, I. (1991) *Geopolitics and geoculture*, Cambridge: Cambridge University Press.

Ware, M., & Mabe, M. (2015) *The STM report: An overview of scientific and scholarly journal publishing*, The Hague: International Association of Scientific, Technical and Medical Publishers.

29

THRESHOLD CONCEPTS IN SCIENTIFIC WRITING LITERACY

What citizens and scientists need to know about scientific writing

Gwendolynne Reid

Introduction

Through public engagement efforts such as citizen science and open access initiatives, laypeople are more likely to encounter original scientific reports, scientific databases, and scientists communicating with one another online, alongside traditional popularizations of science. Brian Trench (2008, p. 185) calls this the "inside-out" nature of contemporary scientific communication, observing how new media have rendered "the back-stage preparation" of science "visible to the prospective spectators of the front-stage performance" (p. 187), a state of affairs that means science's "uncertainties and contests can no longer be hidden from public view" (p. 195). Referencing the often conflicting sources encountered online, Trench notes that it "takes an above-average internet literacy to distinguish these different types of information and informant from each other" (p. 193). More specifically, we can safely say that it takes an above-average scientific literacy to make sense of public scientific knowledge-making in such cases.

The premise of this chapter is that explicit knowledge about scientific writing contributes to a holistic view of scientific literacy that includes procedural knowledge about how science works and that this is increasingly important in contemporary media environments. This builds on the work of others who have argued for the importance of textual literacy in scientific literacy (Hand et al., 2003; Norris and Phillips, 2003; Yore et al., 2003; Zerbe, 2007; Gigante, 2014). With the backstage of science increasingly accessible to a range of audiences, scientists need to consider how their work will be read and used by science communicators and members of the public. In addition, scientists are also often educators, whether in classrooms or through science outreach efforts, and as such need to consider how best to support scientific literacy among the nonspecialists they engage with. Beginning with the rationale for explicitly including textual literacy in scientific literacy, the chapter then explains threshold concepts as distinct from core concepts and describes four such threshold concepts that can function as portals toward deeper understanding of scientific writing: (1) scientific writing is central to scientific inquiry; (2) scientific writing is rhetorical; (3) scientific genres serve distinct purposes in scientific genre ecosystems; and (4) scientific writing and language are contested and dynamic. The chapter

 DOI: 10.4324/9781003043782-34

concludes by discussing the implications of this expanded understanding for scientific literacy and science communication, including common concerns around social and rhetorical views of science.

Putting literacy back into scientific literacy

Scientific literacy, a phrase coined in 1958, has been a concept since at least the turn of the twentieth century (Feinstein, 2010). Definitions of scientific literacy vary widely enough to earn criticism about its usefulness (DeBoer, 2000): one study identified 74 distinct definitions (Norris et al., 2014). At its core, however, scientific literacy refers to what the average person needs to know about science to function in society for both personal and social purposes (Lederman et al., 2013). While common measures of scientific literacy have historically focused on propositional knowledge – lists of common facts, such as "the center of the Earth is very hot" (National Science Board, 2020, p. 24) – modern understandings are holistic, including procedural knowledge of how science works and of its social nature. The annual *Indicators* report by the National Science Foundation (NSF), for example, now includes four process questions to measure understandings of scientific inquiry (National Science Board, 2020). This procedural knowledge has also expanded from the simplified "scientific method" to a model of scientific inquiry that includes "the iterative and social nature of scientific work" and emphasizes "argumentation and model building in addition to the formulation and testing of hypotheses" – scientific literacy is not simply content knowledge but includes interrelated dimensions such as "understanding of scientific practices," "epistemic knowledge," "identifying and judging scientific expertise," and even "dispositions and habits of mind" (National Academies, 2016, pp. 29, 32).

While not typically included in definitions of scientific literacy, a small body of research has made the case that textual literacy plays a role in scientific literacy (Hand et al., 2003; Norris and Phillips, 2003; Yore et al., 2003). Among the dimensions they include, the National Academies of Sciences, Engineering, and Medicine (NASEM) list "foundational literacies" as part of scientific literacy but define these in a limited way as "textual literacy, visual literacy, and understanding of graphs and charts" that "should be distinguished from 'disciplinary literacy,' which is the knowledge and skill required to understand the specific forms of specialized texts commonly used in a discipline" (p. 32). Norris and Phillips (2003, p. 225) review a number of arguments that similarly recognize the role of textual literacy in scientific literacy but restrict this to being able to understand popularizations of science, such as newspaper articles. In contrast, Norris and Phillips themselves emphasize the constitutive role of textuality in science and the importance of "fundamental literacy" to scientific literacy: "Reading and writing are inextricably linked to the very nature and fabric of science . . . [T]ake them away and there goes science" (p. 226). Acknowledging that some science can be learned without reading, they yet argue that "a failure to learn how to read scientific text points to a failure to understand science" (p. 237). Similarly, an interdisciplinary group called the Island Group, of which Norris and Phillips were members, has argued that, in achieving the goal of "Science Literacy for All," educators and researchers need to take seriously the "language practices of the research laboratories and classrooms" and that "these uses of language must be seen as a legitimate part of scientific literacy" (Hand et al., 2003, p. 614).

Along with science educators, writing experts have explored how their work could contribute to scientific literacy, for example by collaborating with science educators on writing pedagogy through writing programs. Perrault (2012), for example, describes a general education course on popular writing about science and technology that asks students to examine how popularizations represent theories of the public-science relationship and requires students to learn about scientific writing. Gigante (2014) describes a course for future scientists that seeks to make explicit otherwise tacit knowledge about "how science works," especially scientific communication, so that students might later contribute to communicative efforts in public engagement with science. Both Perrault and Gigante build on Zerbe's (2007) work arguing for scientific literacy as an important part of university first-year writing. While often framed as disciplinary literacy potentially irrelevant to composition, scientific texts, Zerbe argues, are a dominant cultural discourse and part of the reading and writing students should critically engage with in first-year writing to develop as "literate, informed students and citizens" (p. 5). Drawing a connection between scientific literacy and literacy studies, Zerbe describes three models evident in the research on literacy writ large – autonomous, critical, and ideological – and argues that scientific literacy instruction should be informed by the ideological model. These models move from the most basic, context-free understanding of literacy to the most contextual and multidimensional: an autonomous model focuses on basic literacy skills like decoding texts that assumes meaning is fixed and that skills and meanings will transfer easily across contexts; a critical model emphasizes the contextual, interpretive dimension of texts and literacy, encouraging students to read beyond denotative meaning to also examine intention, assumptions, etc.; the ideological model includes the first two but further emphasizes the culturally negotiated nature of literacies as social practices embedded in other social systems that always include struggles over meaning, identity, and power. In this view, literacy practices are never neutral and always involve struggles over which literacies and meanings will dominate. It is worth noting that while work from science education and literacy-related fields complement each other in useful ways, the taxonomy Zerbe outlines doesn't always map neatly onto how terms like "critical" are used across these conversations. Calls for "critical scientific literacy" sometimes overlap with what Zerbe calls an ideological approach, which includes attention to science's dominant role in society and the role of conflict and ideology within science as a cultural institution.

Priest's (2013, p. 143) case for critical scientific literacy, for example, is premised on the idea that "nonscientist citizens in our science- and technology-oriented society have a right and a responsibility to engage in discussions on what to make of science and how best to govern it." Toward that goal, Priest argues that citizens need some understanding of the sociology and philosophy of science in order to make sense of competing scientific truth claims and emerging science (p. 144). Crucially, she notes that social knowledge is also part of how scientists evaluate scientific claims outside their own fields and participate in broader conversations about science. Whether scientist or layperson, we all use "heuristic cues" like author credentials, journal reputation, and peer review status to evaluate scientific claims outside our expertise (p. 139). As issues like climate change and the COVID-19 pandemic illustrate, a balance must be struck between trust and healthy skepticism for science to realize its full epistemic and societal potential. In this chapter, I argue that familiarity with scientific writing, including disciplinary literacy, is, like the examples Priest lists, a form of social knowledge that can help citizens make better sense of scientific claims as well as engage critically with the role of science in society. This knowledge can serve as additional "heuristic cues" but also provides concrete insight into scientific knowing. While always important, the exponential growth of scientific publication and the disintermediation of science in new media (Trench, 2008) alongside current science-relevant crises add a special urgency to considering the role of writing in scientific literacy.

Threshold concepts in scientific writing literacy

Rather than list features and facts about scientific writing, the discussion that follows describes four threshold concepts (TCs) that can function as portals toward deeper understanding of scientific writing and engagement with research on it. This approach is premised on the idea that threshold concepts can be used pedagogically to achieve a more robust scientific literacy. First, a word on the distinction between threshold and core concepts: while studying economics education, education researchers Meyer and Land (2006) found that certain concepts were key to further learning in the discipline, forming doorways to further learning. To learn them, students entered liminal states or "thresholds" that could feel unsafe and "troublesome" (cited in Cousin, 2006). Once through this threshold, students' knowledge was transformed, and they progressed in ways that would be impossible without this new understanding. To identify threshold concepts, Meyer and Land offer five characteristics: threshold concepts are transformative, irreversible, integrative, bounded, and troublesome. By this, Meyer and Land mean that threshold concepts transform both the learner and their knowledge; that this transformation is difficult to unlearn; that the concepts make visible otherwise hidden connections; that the concepts are discrete and definable; and that the concepts are often counterintuitive and uncomfortable. Among other benefits, identifying threshold concepts in a field can help educators focus on key curricular "jewels" that will foster deep, lasting, transformative learning (Cousin, 2006).

Since Meyer and Land defined them, threshold concepts have been used in a number of fields, such as computer science (Eckerdal et al., 2006) and science education (McKinnon and Vos, 2015). For writing researchers, the most influential example is Adler-Kassner and Wardle's (2015) edited collection *Naming What We Know*, recently updated with the collection *(Re) Considering What We Know* (2020). Their 2015 collection identifies five threshold concepts in writing studies. The threshold concepts that follow complement these five, representing threshold concepts from fields connected with both writing studies and science: rhetoric of science, professional writing, writing in the disciplines, and science communication. Each concept represents a shared understanding about scientific writing that is critical to further learning in these areas and that is important for nonspecialist understandings of science. For those versed in one or more of these fields, these concepts may be common sense, a reaction common for specialists as their worldview has already been transformed and is difficult to unlearn. For others, these concepts may be "troublesome," a quality Perkins defines as "counter-intuitive, alien (emanating from another culture or discourse), or seemingly incoherent" (cited Cousin, 2006, p. 4). For scientists-as-communicators, these concepts offer useful meta-knowledge about scientific communication as well as thresholds they might guide students and nonspecialist audiences toward in expanding their scientific literacy. Neither exhaustive nor sequential, this list is a starting point.

Scientific writing is central to scientific inquiry

Scientific writing literacy is predicated on the understanding that scientific writing is central to scientific inquiry. Written texts are central to the entire process of doing science and are both epistemic and constitutive. In their influential ethnography, Latour and Woolgar (1979, p. 52) observed that the laboratory resembled "a system of literary inscription." Similarly, Wickman's (2010) laboratory ethnography noted the centrality of the notebook, mediating the entire process of inquiry from contingent laboratory-based interpretations to those grounded in the discipline. Based on his analysis of scientific texts, Jay Lemke (1998, p. 87) has argued that scientific concepts are "semiotic hybrids," inseparably and simultaneously "verbal, mathematical, visual-graphical, and actional-operational." While orality and even gesture play important roles in scientific meaning-making, many of the modes required for these hybrids cannot exist outside of textuality, a point made by Norris and Phillips (2003, p. 231). "Scientific theory," they write, "cannot exist outside of text altogether" as such attempts "quickly [run] into insurmountable shortcomings of expressive power, memory, and attention" and the need for textual elements such as graphs, diagrams, and equations. While we can represent and learn scientific ideas orally, this knowledge is "parasitic" to that built through text as it "would not have existed, been preserved, and inherited" without text (p. 231).

For scientists, TC1 – *scientific writing is central to scientific inquiry* – may be both intuitive and troublesome as one study found that many scientists recognize writing as an aid to the clarity of their scientific ideas but not as integral to knowledge construction (Yore et al., 2004, p. 359). For those who study scientific writing, however, writing is seen as inseparable from scientific thought and social interaction and is generally framed through a constructivist lens that stresses science as a social endeavor. Understanding writing in this way can help scientists embrace their writing practice as the activity of doing science rather than as an onerous adjunct to it. For public audiences, this understanding can moderate naïve understandings of scientific knowledge as existing independently of the human enterprise of science and as a wholly certain and objective one. Public audiences who do not understand the central role scientific writing plays in science are less likely to be attuned to the interpretive and interactive processes at work in the texts they encounter, whether that writing is the formal "frontstage" or less finalized "backstage" of science. Understanding writing as integral to scientific inquiry helps nonspecialists understand that when they read a scientific report, they are reading science-in-process – scientists actually "doing" science by interacting with one another – and pay closer attention to the way scientists write about the phenomenon under study and what this indicates about the stage of the inquiry and interpretation of evidence. Lederman et al. (2013, p. 140) argue that the scientifically literate should understand the tentative nature of scientific knowledge, as well as the distinctive functions of observations, inferences, and theories. Understanding scientific writing as constitutive provides a concrete mechanism for noticing and understanding these distinctions.

Scientific writing is rhetorical

Another threshold concept (TC2) – *scientific writing is rhetorical* – may be more troublesome due to popular understandings of rhetoric as persuasion that may deceive. When those who study communication invoke rhetoric, however, they are referring to the discipline focused on the study and art of using symbols – often language – to think and act with others. Drawing on Aristotle, Maher (2020, pp. 104–105) posits that "rhetoric itself is persistently troublesome because it deals in probability" – the art of rhetoric developed to negotiate issues without certain answers – "and therefore not only offers no assurance of truth but challenges the very nature of truth." Science has developed as a way to produce true knowledge about the world and so rhetoric may seem incompatible with that enterprise. A social understanding of science, however, underscores the role of argument in producing knowledge that is validated by the scientific community and the fact that much of this knowledge is developed through inference and therefore probable. Lederman et al.'s (2013) argument that scientific literacy includes understanding that "scientific knowledge is tentative (subject to change)" is another way to say that it is probable, based on the best evidence available at a given moment. Analyses of scientific communication have demonstrated the deeply rhetorical nature of science (Gross et al., 2002). In his work on disciplinary writing, linguist Hyland (2000, p. 11) defined disciplines, both scientific and humanistic, as "the contexts in which disagreement can be deliberated." A rhetorical approach to scientific writing provides a lens for understanding how that deliberation takes place. Studies in the rhetoric of science offer another clue to why this threshold concept is elusive for many: scientific writing persuades through style and presentation in addition to arguments and evidence, but this style often appears as a nonstyle and therefore unrhetorical (though writing outside of style is not actually possible). In a study of biology proposal writing, for example, Myers (1985, p. 220) finds that these must "persuade without seeming to persuade." Yet while scientific style appears to be a nonstyle, analyses demonstrate that it employs recognizable stylistic patterns one would not necessarily expect, such as rhetorical figures like antithesis, and that these influence scientific thought (Fahnestock, 2002).

Though initially troubling, understanding scientific writing as rhetorical provides the necessary basis for examining the norms scientific communities have developed about what is scientifically persuasive, how to deliberate disagreements, and how to use language and other symbolic means to think and act together. While the concept (TC1) that *scientific writing is central to scientific inquiry* draws attention to the important work that writing performs in scientific knowledge production, it does not convey its persuasive dimension or the importance of other modes, such as orality (TC2). Acknowledging the persuasive dimension of science makes visible the rhetorical tradition human beings have developed to produce knowledge about the world that is more certain than hunches or superstition. By acknowledging this, we can see and study the communicative practices scientists have developed to support a specific type of reasoning about a world that is not self-evident or fully accessible to the senses. For scientists-as-communicators, this understanding affords greater sensitivity to how they participate in this rhetorical tradition. For members of the public, this understanding can provide greater sophistication about how the scientific community produces, shares, and negotiates claims and evidence about our world. As Zerbe (2007, p. 98) points out, this can also lead to ideological literacy about science that facilitates informed deliberation about the role of science in our larger culture.

Scientific genres serve distinct purposes in scientific genre ecosystems

A third threshold concept (TC3) – *scientific genres serve distinct purposes in scientific genre ecosystems* – relates to the notion that scientific rhetorical traditions have developed over time. Over centuries, scientists have repeatedly confronted similar communicative problems and developed conventionalized ways to address them (e.g. experimental reports, case studies, meta-analyses). In his work on the experimental article, Bazerman (1988, p. 62) defines genre as "a socially recognized, repeated strategy for achieving similar goals in situations socially perceived as being similar." As scientists over history have employed similar strategies for achieving similar goals, these strategies have become regularized, developing regular patterns. Genres have a dialectical relationship with their communities, both shaped by the community's activities, values, and structures but also shaping and constituting the community – not simply "responding to the emerging regularities of the rhetorical universe" but also "helping indeed to create that rhetorical universe" (p. 48). A corollary of this understanding of genre is that scientific communication consists of many genres, each fulfilling a different role for the scientific community. Researchers have described the relationships among genres using various terms (e.g. system, network). I use the term "ecosystem" here to foreground the notion that "genres act together to play particular roles in the communication of a community, much as particular species fill particular niches in a biological ecosystem" (Casper, 2016, p. 77). In addition, the metaphor of the ecosystem conveys that varied actors interact with one another – genres interact with other genres but also with media, tools, and social systems. It is important not to overly systematize how we understand genres: individual genre performances vary, with individuals innovating regularly. This is one way genres evolve, and so we can only say that genres like the experimental report are "stabilized-for-now" (Schryer, 1993, p. 204).

For scientists-as-communicators, this threshold concept may trouble the notion that a uniform, unchanging "good scientific writing" exists and add increased sensitivity to the distinct purposes of scientific genres as well as to how research genres may interact with public-facing genres, like science news articles and science videos, or with changes in media, technologies, and processes. For nonspecialists, awareness that scientists use discrete genres for distinct purposes and at distinct stages of inquiry enhances understanding of the scientific processes that produce, refine, vet, and validate scientific claims, potentially fostering greater sensitivity to the degree of certainty in scientific claims they encounter.

Scientific writing and language are contested and dynamic

While scientists use language rhetorically to produce knowledge (TC2), they also debate the conventions of that language and their genres (TC4). Genres include conventionalized relationships between patterns of reasoning, symbolic systems, tools, media but also between *people* – scientists, editors, research participants, policymakers, laypeople. These conventions are often the product of struggle and are inherently ideological. Histories of the experimental article, for example, have underscored the role of journal editors in its evolution (Bazerman, 1988). A study of medical journals from 1935 to 1985 demonstrates how these shifted from topical structures to the IMRAD structure (Introduction, Methods, Results, Discussion), a shift editors encouraged to aid peer reviewers and readers with a predictable, modular format (Sollaci and Pereira, 2004). Another struggle: scientists have long debated how to represent human beings in their genres. An analysis of medical publications on the unethical Tuskegee syphilis study illustrates the dangers of dehumanizing terms such as "host" or "syphilitic," which may have helped maintain the medical community's complacency for so long (Solomon, 1985). New relationships with the public, such as in citizen science, have similarly raised questions about how to acknowledge these relationships (Hunter and Hsu, 2015). Negotiations related to changing media are also common (Pérez-Llantada and Luzón, 2019). Passive voice and first person are familiar sites of debate: over time, passive voice has gained ground as scientific writing has emphasized "the objects and processes of the natural world, the methods and materials of the laboratory" and as "abstract nouns have increasingly occupied the subject position" (Gross et al., 2002, p. 166). Some defend first person, however, as a more economical and accurate representation of scientific inquiry. And we cannot ignore struggles over the primacy of English in scientific publication and whether this represents the development of a "lingua franca" or "linguistic hegemony and cultural imperialism" (Hyland, 2009, p. 180).

For scientists-as-communicators and laypeople, this threshold concept opens up a view of science as itself a literary achievement with a fascinating and worthwhile history. More importantly, this concept affords a view of scientific writing both as it is and *as it could be*, encouraging ethical and innovative thinking about communicative choices. Scientific genres like the experimental report may have a 350-year history and strong forces maintaining its stability, but choices exist even within these structures and authorial innovation and resistance have occurred all along the way. Realizing this can help communicators evaluate which conventions are worth resisting for ethical or communicative reasons. Combined with a rhetorical, relational understanding of scientific communication, this concept encourages attention to how communicative choices may empower or disempower others. For laypeople, this threshold concept draws attention to how scientists have honed their rhetorical tradition as part of honing their inquiry and how relationships between science and the world are not inevitable but rather a product of ongoing negotiation.

Implications for scientific literacy and science communication

If scientific writing literacy plays a role in scientific literacy and these threshold concepts are worth emphasizing, what implications does this have for scientific literacy and science communication practice? First of all, one strategy that some science communicators already employ would be to explicitly talk about writing when communicating about science. For example, when communicating about a study, not simply sharing findings, but also sharing the paper's genre and what this means. Along with this, communicators might note the stage of inquiry this represents and what types of responses might be hoped for or expected (e.g. replication, a randomized controlled trial) or even the types of responses and debate that occurred earlier, either during the research and publication process or perhaps in the prior literature. Rather than eliminating hedges like "may" or "suggests" and replacing them with more certain terms like "proved," communicators might maintain and explain them, not as signs of weak evidence but rather as examples of precision.

In the "inside-out" world of contemporary science communication, however, we cannot expect (or desire) that laypeople are only accessing popularized science through professional intermediaries. Recognizing that access goes beyond the technical ability to locate a file, many open access publications are experimenting with connected genres, such as the lay summary, to render scientific writing more accessible. These, however, still mediate the science for readers, who may wish to try reading it themselves. Open access publications might also add reading guides for nonspecialists, such as on common genres, stages of publication, certainty indicators, or common features of important genres. With scientific publication continuing to evolve, these can be useful for those unfamiliar with more recent practices or genres. It is also worth noting that new media have facilitated public engagement with science much earlier than publication, such as citizen science projects. These interactions provide opportunities for explicit talk about writing, such as showing early writing from a lab notebook or discussing how new literature or peer feedback is refining the inquiry and paper. These reinforce how the scientific rhetorical tradition is designed to strengthen science.

While open access to science has been generally embraced as a common good, emphasizing the social nature of science through its writing presents a number of concerns. Some worry that laying open the messy backstage of scientific inquiry and argument might lead to the erosion of expert authority or to a cynical relativism. Bad faith actors can exploit this to sow uncertainty for ulterior motives, as in the case of climate change. Latour (2004, p. 226), for example, quoted a 2003 Republican strategist arguing the need, on the issue of global warming, "to continue to make the *lack of scientific certainty* a primary issue." Reasserting authority and keeping the backstage of science invisible, however, is not necessarily the answer to these challenges. In her analysis of how scientific values and argumentative norms like fairness, open-mindedness, and freedom of inquiry can be exploited to "manufacture" scientific controversy for public audiences, Ceccarelli (2011, p. 213) warns against de-emphasizing scientific debate (or social construction) as scientists end up "unwittingly confirming the very charge leveled against them: that they are a closed-minded orthodoxy conspiring to silence the opposition." Instead, Ceccarelli recommends science communicators "acknowledge that debate is important to science" but that "debate on this particular subject has already taken place in scientific forums and been decided against dissenters . . . who have not yet offered a persuasive case" (p. 213). These issues have reached a crisis point during the post-truth moment. Some media scholars have even asked whether lessons on critical thinking and media literacy may actually have backfired (boyd, 2018). The emerging answer to this conundrum, for science and other institutions like journalism, however, is not to retreat to comfortable authority but rather to support complex, multipronged approaches to epistemological education (Barzilai and Chinn, 2020).

Ultimately, calls for a holistic scientific literacy such as this one are calls for a both/and approach: members of the public need to know both when to trust experts and scientific truth claims and when to be critical. Pseudoscience, bad faith actors, and harmful uses of science exist and "even genuine experts can be wrong, self-interested, or both" (Priest, 2013, p. 140). Citizens have a right to science – both inquiry and knowledge – and to deliberative decision making about its role in society. Additional social knowledge about how science works, such as knowledge about its writing, can refine the heuristic cues citizens use to assess claims about science. More importantly, it can promote deeper epistemological awareness. This may sometimes lead to accepting scientists' claims and proposals and at other times rejecting them and proposing alternative applications of or relationships with science. The response, however, will be better positioned as a dialogic partnership with a chance of bettering science and society.

References

Adler-Kassner, L., & Wardle, E. (eds.) (2015) *Naming what we know: Threshold concepts of writing studies*, Logan: Utah State University Press.

Adler-Kassner, L., & Wardle, E. (eds.) (2020) *(Re)considering what we know: Learning thresholds in writing, composition, rhetoric, and literacy*, Logan: Utah State University Press.

Barzilai, S., & Chinn, C.A. (2020) 'A review of educational responses to the "post-truth" condition: Four lenses on "post-truth" problems', *Educational Psychologist*, vol. 55, no. 3, pp. 107–119, DOI: https://doi.org/10.1080/00461520.2020.1786388

Bazerman, C. (1988) *Shaping written knowledge: The genre and activity of the experimental article in science*, Madison: University of Wisconsin Press.

boyd, d. (2018) 'You think you want media literacy . . . do you?, Data & society: Points', https://points.datasociety.net/you-think-you-want-media-literacy-do-you-7cad6af18ec2

Casper, C. (2016) 'The online research article and the ecological basis of new digital genres', in A.G. Gross & J. Buehl (eds.), *Science and the internet: Communicating knowledge in a digital age*, London: Routledge, pp. 77–98.

Ceccarelli, L. (2011) 'Manufactured scientific controversy: Science, rhetoric, and public debate', *Rhetoric & Public Affairs*, vol. 14, no. 2, pp. 195–228, DOI: 10.1353/rap.2010.0222

Cousin, G. (2006) 'An introduction to threshold concepts', *Planet*, vol. 17, no. 1, pp. 4–5, DOI: https://doi.org/10.11120/plan.2006.00170004

DeBoer, G.E. (2000) 'Scientific literacy: Another look at its historical and contemporary meanings and its relationship to science education reform', *Journal of Research in Science Teaching*, vol. 37, no. 6, pp. 582–601, https://doi.org/10.1002/1098-2736(200008)37:6 < 582::AID-TEA5 > 3.0.CO;2-L

Eckerdal, A. *et al.* (2006) 'Putting threshold concepts into context in computer science education', *ACM SIGCSE Bulletin*, vol. 38, no. 3, pp. 103–107, DOI: https://doi.org/10.1145/1140123.1140154

Fahnestock, J. (2002) *Rhetorical figures in science*, Oxford: Oxford University Press.

Feinstein, N. (2010) 'Salvaging science literacy', *Science education*, vol. 95, no. 1, pp. 168–185, DOI: https://doi.org/10.1002/sce.20414

Gigante, M.E. (2014) 'Critical science literacy for science majors: Introducing future scientists to the communicative arts', *Bulletin of Science, Technology & Society*, vol. 34 no. 3–4, pp. 77–86, DOI: https://doi.org/10.1177/0270467614556090

Gross, A.G., Harmon, J.E., & Reidy, M.S. (2002) *Communicating science: The scientific article from the 17th century to the present*, Oxford: Oxford University Press.

Hand, B.M., Alvermann, D.E., Gee, J., Guzzetti, B.J., Norris, S.P., Phillips, L.M., Prain, V., & Yore, L.D. (2003) 'Message from the "Island Group": What is literacy in science literacy?', *Journal of Research in Science Teaching*, vol. 40, no. 7, pp. 607–615, DOI:10.1002/tea.10101

Hunter, J., & Hsu, C.-H. (2015) 'Formal acknowledgement of citizen scientists' contributions via dynamic data citations', in R.B. Allen, J. Hunter, & M.L. Zeng (eds.), *Digital libraries: Providing quality information. 17th international conference on Asia-Pacific digital libraries*, Seoul: Springer, pp. 64–75.

Hyland, K. (2000) *Disciplinary discourses: Social interactions in academic writing*, London: Pearson, Longman.

Hyland, K. (2009) *Academic discourse: English in a global context*, New York: Continuum.

Latour, B. (2004) 'Why has critique run out of steam? From matters of fact to matters of concern', *Critical inquiry*, vol. 30, no. 2, pp. 225–248, DOI: https://doi.org/10.1086/421123

Latour, B., & Woolgar, S. (1979) *Laboratory life: The construction of scientific facts*, Beverly Hills: Sage.

Lederman, N.G., Lederman, J.S., & Antink, A. (2013) 'Nature of science and scientific inquiry as contexts for the learning of science and achievement of scientific literacy', *International Journal of Education in Mathematics Science and Technology*, vol. 1, no. 3, pp. 138–147.

Lemke, J.L. (1998) 'Multiplying meaning: Visual and verbal semiotics in scientific text', in J.R. Martin & R. Veel (eds.), *Reading science: Critical and functional perspectives on discourses of science*, London: Routledge, pp. 87–113.

Maher, J.H. (2020) 'Rhetoric as persistently "troublesome knowledge": Implications for disciplinarity', in L. Adler-Kassner & E. Wardle (eds.), *(Re)considering what we know: Learning thresholds in writing, composition, rhetoric, and literacy*, Logan: Utah State University Press, pp. 94–112.

McKinnon, M., & Vos, J. (2015) 'Engagement as a threshold concept for science education and science communication', *International Journal of Science Education, Part B*, vol. 5, no., 4, pp. 297–318, DOI: https://doi.org/10.1080/21548455.2014.986770

Meyer, J.H.F., & Land, R. (eds.) (2006) *Overcoming barriers to student understanding : threshold concepts and troublesome knowledge*, London: Routledge.

Myers, G. (1985) 'The social construction of two biologists' proposals', *Written Communication*, vol. 2, no. 3, pp. 219–245, DOI: https://doi.org/10.1177/0741088385002003001

National Academies of Sciences, Engineering, and Medicine. (2016) *Science literacy: Concepts, contexts, and consequences*, Washington, DC: National Academies Press.

National Science Board. (2020) *Science and technology: Public attitudes, knowledge, and interest*, NSB-2020-7, National Science Foundation, https://ncses.nsf.gov/pubs/nsb20207/introduction

Norris, S.P., & Phillips, L.M. (2003) 'How literacy in its fundamental sense is central to scientific literacy', *Science education*, vol. 87, no. 2, pp. 224–240, DOI: https://doi.org/10.1002/sce.10066

Norris, S.P., Phillips, L.M., & Burns, D.P. (2014) 'Conceptions of scientific literacy: Identifying and evaluating their programmatic elements', in *International handbook of research in history, philosophy and science teaching*, New York: Springer, pp. 1317–1344.

Pérez-Llantada, C., & Luzón, M.J. (eds.) (2019) *Science communication online: Connecting traditional and new genres*, Amsterdam: John Benjamins.

Perrault, S. (2012) 'Course Description: UWP 011: Popular Science & Technology Writing', *Composition Studies*, vol. 40, no. 2, pp. 112–133, www.jstor.org/stable/compstud.40.2.0112

Priest, S. (2013) 'Critical science literacy: What citizens and journalists need to know to make sense of science', *Bulletin of Science, Technology & Society*, vol. 33, no. 5–6, pp. 138–145, DOI: https://doi.org/10.1177/0270467614529707

Schryer, C.F. (1993) 'Records as genre', *Written communication*, vol. 10, no. 2, pp. 200–234, DOI: https://doi.org/10.1177/0741088393010002003

Sollaci, L.B., & Pereira, M.G. (2004) 'The introduction, methods, results, and discussion (IMRAD) structure: A fifty-year survey', *Journal of the Medical Library Association*, vol. 92, no. 3, pp. 364–371.

Solomon, M. (1985) 'The rhetoric of dehumanization: An analysis of medical reports of the Tuskegee syphilis project', *Western journal of speech communication*, vol. 49, no. 4, pp. 233–247, DOI: https://doi.org/10.1080/10570318509374200

Trench, B. (2008) 'Internet: Turning science communication inside-out?', in M. Bucchi & B. Trench (eds.), *Handbook of public communication of science and technology*, London: Routledge, pp. 185–198.

Wickman, C. (2010) 'Writing material in chemical physics research: The laboratory notebook as locus of technical and textual integration', *Written Communication*, vol. 27, no. 3, pp. 259–292, DOI: https://doi.org/10.1177/0741088310371777

Yore, L.D., Bisanz, G.L., & Hand, B.M. (2003) 'Examining the literacy component of science literacy: 25 years of language arts and science research', *International journal of science education*, vol. 25, no. 6, pp. 689–725, DOI: https://doi.org/10.1080/09500690305018

Yore, L.D., Hand, B.M., & Florence, M.K. (2004) 'Scientists' views of science, models of writing, and science writing practices', *Journal of research in science teaching*, vol. 41, no. 4, pp. 338–369, DOI: https://doi.org/10.1002/tea.20008

Zerbe, M.J. (2007) *Composition and the rhetoric of science: Engaging the dominant discourse*, Carbondale: SIU Press.

30

LANGUAGE AS A REALIZATION OF SCIENTIFIC REASONING IN SCIENTIFIC TEXTS AND ITS IMPORTANCE FOR PROMOTING SECONDARY SCHOOL STUDENTS' DISCIPLINARY LITERACY

Moriah Ariely and Anat Yarden

Introduction

Communicating in a written or spoken form is a fundamental practice of science and has become a major educational goal in science education (National Research Council [NRC], 2012). Goldman and Bisanz (2002) suggested that communication of scientific information in our society has three main roles: communicating among scientists, popularizing information generated by the scientific community, and providing scientific education. Communication of scientific information is achieved mainly through scientific texts that can be classified into two main groups: academic texts and popular texts. Academic texts include research articles (primary scientific literature, PSL) and academic textbooks, while popular texts include, among others, media news articles.

Writing of all science texts, whether academic or popular, is a social act that is constrained by genre norms, social values, and implicit beliefs about the audience, all of which shape the content of the text (Bazerman, 1983). Scientists write differently in professional journals than they do in popular science magazines, since language use is always acquired and licensed by specific social and historically shaped practices, representing the values and interests of distinctive groups of people (Gee, 1999). Since different text genres serve different goals, they serve the needs of different discourse communities. Research articles and popular articles, for example, are produced in different social contexts and have different communicative functions (Goldman and Bisanz, 2002). Research articles are commonly used by the scientific community, while popular articles are commonly used by the general public. Accordingly, each discourse community (scientists or the general public) shares a common set of norms for interaction and a language that marks the community (Swales, 2001).

DOI: 10.4324/9781003043782-35

In this chapter, we focus on the language of scientific research articles and its functionality for producing and organizing scientific knowledge. Next we discuss a pedagogical approach for learning about the communicative functions of scientific research articles and for promoting students' disciplinary literacy.

The scientific discourse – a functional view

Disciplines differ in how they generate, communicate, evaluate, and renovate knowledge, and these differences are realized in the way that content experts use language in their social-cognitive practices (Fang and Schleppegrell, 2008). Accordingly, discipline-specific texts are organized linguistically to accomplish particular communicative purposes (Schleppegrell, 2002, 2004). It is widely recognized that different disciplines have their own jargon. However, it is less recognized that grammar is also different across disciplines and even across different text-genres in the same discipline (Fang, 2012b).

Systemic functional linguistics (SFL) is an approach for reasoning grammatically about language. Language is viewed as a system of options, in which the choice of lexicogrammatical items reflects different social functions (Halliday and Matthiessen, 2014). Analysis of texts using the SFL approach has shown that different text genres, even in the same discipline, include different linguistic features (Fang and Schleppegrell, 2010; Fang, 2012b). Thus the structure and grammar of research articles provide the semiotic means to build arguments throughout the article, reflecting the ways in which scientists do, explain, theorize, organize, and challenge science (Fang, 2012b; Lemke, 1990; Martin, 1993; Norris and Phillips, 2008; Suppe, 1998; van Lacum et al., 2012).

The research article (PSL) is a means of communication among scientists. These articles are written by scientists for other scientists in a specific field of research. PSL articles function to persuade readers of their knowledge claims (Yarden et al., 2015; Yore et al., 2004). Accordingly, the discourse of scientific papers is an argumentative discourse aimed at persuading the scientific community to accept new knowledge and arguments presented in them. By accepting these arguments, they become part of the "scientific knowledge" or "facts" upon which there is a consensus within the relevant discipline (Latour, 1987; Livnat, 2010). Since the scientific article is a written product, authors' persuasive efforts are necessarily linguistic and textual in nature (Livnat, 2010).

From a textual point of view, PSL articles are structured and written in a way that serves the logic of the reasoning contained in the text, which enables the authors to argue their claims in the article as accepted in the discipline (Yarden, 2009; Yarden et al., 2015). An examination of more than 1,000 data-based research articles revealed that articles from different disciplines have a common organizational structure and a variety of speech acts through which authors create an argumentative structure (Suppe, 1998).

PSL articles are structured in a canonical manner (Yarden, 2009); namely, they follow the Introduction-Methods-Results-Discussion (IMRD) structure. Each section of the PSL article has a different rhetorical role (Swales, 2001). The sections of the article serve the purpose of being accepted by the scientific community: the introduction situates the research within already accepted previous work and shows how it is continuous with it; the methods are shown to match the requirements of quantitative science; the results try to convince the readers of the validity of the research; and finally, the discussion tries to convince the readers that the research has an explicatory power (Parkinson, 2001).

From a linguistic point of view, the scientific discourse is often referred to as a "register," a constellation of lexical and grammatical features that characterize particular uses of language

(Schleppegrell, 2002). Scientific texts can be recognized by the combined effect of several clusters and features and, more importantly, by how those clusters and features are related throughout the text (Fang, 2005; Snow, 2010). As a register, the scientific discourse contains a unique lexicon, semantics, and structure, which enables the scientists to conduct specialized kinds of cognitive and semiotic work (Fang, 2005; Martin, 1993). The scientific discourse contains a high density of information-bearing words and grammatical processes, such as extended noun phrases, embedded clauses, and nominalizations, that compress complex ideas into a few words (Osborne, 2014). In addition, the appearance of objectivity is carefully constructed by the use of grammatical conventions in the discourse of science such as hedging and passivation. These features are functional in building the argument throughout the text.

Let us demonstrate the functionality of these grammatical features in scientific texts, with the following two examples taken from a PSL article (Davoodi-Semiromi et al., 2010):

1   In this study, we conjugated CTB with two major malarial vaccine antigens, AMA1 and MSP1, in tobacco and lettuce chloroplasts and **female BALB/c mice were immunized orally with tobacco transplastomic leaves or subcutaneously with purified or enriched vaccine antigens**. Different groups of immunized mice and control were challenged with cholera toxin and sera were evaluated against the malarial parasite. <u>Oral immunization</u> provided both mucosal and systemic immunity while subcutaneously immunized mice developed systemic immunity.

(p. 225)

2   Our data suggest that only antigen-specific CTB-IgM antibody level significantly changed after CT challenge but not any other antibodies tested.

(p. 229)

First, we refer to nominalization, a grammatical process in which an element having the nature of an action or a process is given a nominal form. Through nominalizations, the scientific language theorizes concrete life experiences into abstract entities, which can then be further examined and critiqued (Fang, 2005). In the first example, "oral immunization" (underlined) is the nominalized form of the action of immunizing the mice orally (marked in bold) that is presented earlier in the paragraph. As shown in the example, nominals are functional since they can synthesize or abstract previously presented information into entities and are therefore effective in developing arguments in the text (Fang, 2005). Nominalization can also serve to avoid the fact that some agent is involved, thus creating objectivity.

Second, technicality in scientific texts is the use of terms or expressions with a specialized field-specific meaning (Wignell et al., 1993). Generally, technical terms (nouns or adjectives) are names of objects or phenomena or verbs describing a unique activity in the discipline ("AMA1," "MSP1," "chloroplast," "cholera toxin," "CTB-IgM antibodies," etc.). Technical terms can also derive from nominalizations as this is a most powerful resource for creating taxonomies (categories and subcategories) (Wignell et al., 1993). In disciplinary texts, technical terms derive their meaning from being organized into taxonomies (i.e. technical taxonomies), with many layers of organization built into them (Halliday, 1993). In the first example, the term "tobacco transplastomic leaves" has three layers of organization (Figure 30.1).

Another example is the term "antigen-specific CTB-IgM antibody" from the second example, which has four layers of organization.

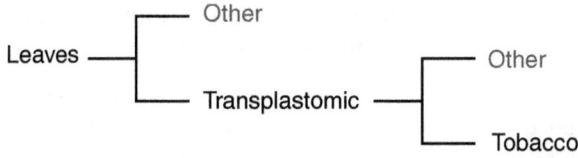

Figure 30.1 Example of technical taxonomies in the technical term "tobacco transplastomic leaves"

Third, authors of research articles establish their objectivity through the use of technical vocabulary, declarative sentences, and a passive voice (Fang, 2005). These grammatical resources enable the realization of an assertive author who is the knowledgeable expert who provides accurate and objective information (Fang, 2005; Schleppegrell, 2002). Using the passive voice, the actor (the participant performing the action) can be omitted, allowing the action (and not the actor) to be at the focus of attention (Fang, 2005). For example, in the following sentence from the first example, the action (the immunizing of mice) is at the focus of attention.

"... female BALB/c mice were immunized orally with tobacco transplastomic leaves or subcutaneously with purified or enriched vaccine antigens." [The passive verb is underlined.]

Another significant communicative resource that plays a critical role in PSL articles is hedging. Hedging is an expression of uncertainty concerning the factuality of statements and an indication of the author's deference to the readers (Yarden, 2015). In science, propositions are presented with caution and precision. Thus hedges play a critical role in gaining ratification for claims from a powerful peer group by allowing the author to present statements with appropriate accuracy, caution, and humility. Therefore, it is helpful in negotiating the perspective from which conclusions can be accepted (Hyland 1996, 1998). In the second example, in the previous page, there are two types of hedging: (1) "Our data," which functions to limit the generalizability of the claim by acknowledging personal responsibility, and (2) the verb "suggest," which limits the writer's commitment to the statement made in the text.

Finally, in scientific writing, more content words (verbs, adjectives, and some adverbs) are packed into the clause (lexical density; Halliday, 1993). The high density of information is partly achieved by the use of longer and more complex noun phrases and partly by nominalizations and technical terms, as previously explained. Complex noun phrases are useful in constructing scientific definitions, as they allow information to be integrated and concentrated and enable the author to provide concise but also complete and accurate descriptions of technical concepts (Fang, 2012a). An example occurs in the following sentence from the first example: "different groups of immunized mice and control were challenged with cholera toxin" (the noun phrase is underlined).

The use of scientific language as described in this section is mostly apparent in the texts that scientists produce to communicate their findings to other members of their community – the PSL articles. Therefore, scientific texts are not merely tools for the storage and transmission of scientific knowledge; they are "constitutive parts of science," and they are essential vehicles for the expression of scientific thoughts (Norris and Phillips, 2003). Nevertheless, these features make the reading of PSL articles a challenging task. Therefore, in science education, PSL articles require suitable adaptation in order to be employed by novice readers.

Using scientific text in secondary school to promote disciplinary literacy

Disciplinary literacy is concerned with the strategies of reading and writing in a specific discipline. It is grounded in the belief that reading and writing are integral to disciplinary practices and that disciplines differ not only in content but also in the ways this content is produced, communicated, and critiqued (Shanahan and Shanahan, 2008, 2020). Disciplinary literacy emphasizes the way of thinking in a discipline, such as what is important to pay attention to or what level of confidence the field has in the knowledge it produces (Shanahan and Shanahan, 2020). Therefore, promoting scientific disciplinary literacy is important for *all students* in order for them to take their place in society as literate consumers of scientific knowledge.

Language as central to disciplinary literacy

Students engage in more advanced literacy tasks as they move from primary school to middle school and high school. As the tasks evolve, their language is structured in a more condensed and technical way. Thus secondary school students are expected to read language that presents formal and technical knowledge which is distant from their everyday lives (Schleppegrell, 2004, 2006). According to the recent U.S. framework for K–12 science education, by twelfth grade students should develop the ability to extract the meaning of scientific text, and, in high school, this practice should be further developed by providing students with more complex texts (National Research Council [NRC], 2012).

Previous studies have shown that specialized academic language in general and scientific language in particular pose great challenges for students' comprehension of scientific texts (e.g. Fang and Schleppegrell, 2008). Written science adopts a wide range of linguistic features that are seldom used in everyday informal spoken interaction. The transition from reading instruction that focuses on developing reading skills in elementary grades to reading that supports knowledge gain in academic subjects is often difficult. One reason for this difficulty is that students are usually not familiar with expository, specialized, and technical language that characterizes academic writing (Fang, 2006). Students who are not familiar with the specialized meaning-making and grammatical resources of scientific writing are likely to experience significant difficulties when reading and writing science texts (Fang, 2005). Therefore, for students to become scientifically literate, they must learn to cope with the specialized language of science (Fang, 2005), and use language in new ways that are specific to the discipline (Schleppegrell, 2004).

Adapted Primary Literature (APL)

What is APL?

Adapted primary literature (Yarden et al., 2001) refers to an educational genre specifically designed to enable the use of PSL articles in high school. The adaptation process of APL articles includes several defined steps (Yarden et al., 2015); the canonical structure (IMRD) and the writing style of the article are maintained, while matching its content and the complexity with students' prior knowledge and assumed cognitive abilities (Yarden et al., 2001). Since APL articles represent the structure, linguistic norms, epistemic standards, and content found in PSL articles (Ariely et al., 2019; Yarden et al., 2015), they can be used to promote important aspects of scientific literacy that are harder to achieve using other text genres such as textbooks or popular articles.

Two decades of research on the use of APL articles in science education have shown that these articles can help students improve their understanding of inquiry, active learning, and integration of knowledge (Falk et al., 2008) and enable students to pose questions that reveal a higher level of thinking and uniqueness (Brill and Yarden, 2003). Students' understanding of the nature of scientific inquiry and their ability to criticize scientific research improved after reading APL articles, compared to students who read popular articles (Baram-Tsabari and Yarden, 2005). APL articles were also found to be useful in promoting students' understanding of scientific and mathematical reasoning and argumentation (Norris et al., 2009) and for supporting the discourse of science and the promotion of disciplinary literacy (Koomen et al., 2016).

APL as an apprenticeship genre

Theories of situated cognition view learning as enculturation, an act of taking on the behaviors and worldview of a culture or knowledge domain that may be achieved through engaging in the authentic activities of the culture (Brown et al., 1989). In other words, learning means participating in the activities and practices of the community (Lave and Wenger, 1991). The socialization into the community of practice occurs through apprenticeship. A newcomer learns its ways of knowing by participating in the ways of doing that define a community (Brown et al., 1989; Lave and Wenger, 1991). According to Lave and Wenger (1991), discourse has an important role in learning. Namely, participating in a community of practice means learning to talk the way full participants of the community talk. Thus an "apprenticeship genre" is a genre that can encourage socialization into disciplinary communities (Carter et al., 2007).

It was previously claimed that an APL article is somewhat closer to a PSL article than to a popular article, since it represents the structure, claims, content, and uncertainty that are found in PSL articles (Yarden et al., 2015). Indeed, a linguistic analysis of an APL article compared to a PSL and a popular article revealed that the adaptation of the APL article lowers the lexical complexity, while at the same time retaining the main linguistic features of a research article, and the authenticity of the scientific writing (Ariely et al., 2019). Namely, analysis of the APL's grammar compared to a PSL article, and a popular article on the same topic revealed that key grammatical features such as nominalizations, passivation, and technicality are mostly retained in the APL (but not in the popular article). At the same time, the APL language is less complex and less dense compared to the PSL, suggesting that the APL article is easier to read. Thus the APL article is linguistically more closely related to the PSL article than to the popular article in all the previously mentioned grammatical features (Ariely et al., 2019). A summary and comparison of the main attributes characterizing the PSL, APL, and popular text genres in eight different dimensions are presented in Table 30.1.

As presented in Table 30.1, the authors and the target audience of APL articles are different from those of PSL articles. However, APL and PSL articles are similar in the main text type, the organizational structure, the content and the presentation of science. Concerning the main purpose of each genre and the functionality of the language in each text genre, the PSL and APL articles share the same function of construing specific realms of scientific knowledge and beliefs, but the adaptations in the APL article's language make the text less complex and thus probably more readable for high school students. This is not the case for popular articles which are different from both PSL and APL articles in all parameters, including the functionality of language (Table 30.1).

Table 30.1 Various attributes characterizing three text genres: PSL, APL, and popular articles

	PSL	*APL*	*Popular*
Main purpose★	Having claims accepted by the scientific community (Hyland, 1998; Myers, 1989)	Enable the use of scientific research articles in schools, as a model of scientific reasoning and communication (Yarden et al., 2001)	Communicating scientific findings to nonscientists (Norris and Phillips, 1994; Parkinson and Adendorff, 2004)
Authors	Scientists (Myers, 1989; Yore et al., 2004)	Science educators and scientists (Norris et al., 2009; Yarden et al., 2001)	Science journalists (Nwogu, 1991)
Target audience	Scientists (Myers, 1989; Yore et al., 2004)	Students (Yarden et al., 2001)	General public (Nwogu, 1991)
Main text type	Argumentative (Hyland, 1998; Jiménéz–Aleixandre and Federico-Agraso, 2009; Suppe, 1998)	Argumentative (Norris et al., 2009)	Varying (expository, narrative, argumentative) (Jiménéz-Aleixandre and Federico-Agraso, 2009; Penney et al., 2003)
Organizational structure	Canonical (Suppe, 1998; Swales, 2001)	Canonical (Baram-Tsabari and Yarden, 2005; Yarden et al., 2001)	Noncanonical (Nwogu, 1991)
Content	Evidence to support conclusions (Suppe, 1998)	Evidence to support conclusions (Falk and Yarden, 2009); Yarden et al., 2001)	Facts with minimum evidence (Jiménéz-Aleixandre and Federico-Agraso, 2009)
Presentation of science	Uncertain (Suppe, 1998)	Uncertain (Falk and Yarden, 2009; Yarden et al., 2001)	Various degrees of certainty (Penney et al., 2003)
Functionality of language★	Construing specific realms of scientific knowledge and beliefs (Fang, 2005; Martin, 1993)	Construing specific realms of scientific knowledge and beliefs in a form that is more readable and interpretable to students (Ariely et al., 2019)	Reporting about new scientific findings in a form that is interpretable by nonscientists (Norris and Phillips, 1994; Parkinson and Adendorff, 2004)

Source: Following Yarden et al., 2015.

★ Attributes that did not appear in Yarden et al., 2015.

Given these features and affordances of using APL in science education, we suggest that APL be used as a model of scientific communication and for promoting disciplinary literacy among school students. Since linguistic features can reflect beliefs about knowledge, the APL article can serve as an apprenticeship genre for learning the unique features of scientific language, reasoning, and communication.

Implications

Science is a socially situated practice in which scientists' values for what counts as good questions, appropriate methods, and good answers are constructed and negotiated within particular scientific disciplines and communities (Sandoval and Morrison, 2003). It is through participation in discipline-specific practices (such as reading, writing, talking, etc.) that disciplinary knowledge is used, shared, critiqued, refined, and expanded. Therefore, disciplinary enculturation requires the learning of language patterns that construct the knowledge, values, and worldview of the discipline (Hynd-Shanahan, 2013).

To develop students' disciplinary literacy, teachers should teach their students how to read specialized disciplinary texts by highlighting the structural and linguistic differences between scientific text genres and by discussing with their students how these differences reflect the context and communicative purposes of the genre. However, even today the vast majority of texts that students read in the science classroom are texts obtained from textbooks, popular research articles from the media, or review articles from the Internet (Sung et al., 2015). These texts often do not reflect the core attributes of authentic scientific reasoning and are antithetical to the epistemology of authentic science (Chinn and Malhotra, 2002). "Messy" classroom environments in which students engage with complex evidence, such as multistep, complex experimental designs, are more authentic in reflecting the kinds of reasoning valued for scientific literacy (Duncan et al., 2018). This calls for a careful consideration of texts and materials used for science education, which should include texts that more authentically reflect scientific reasoning and communication. In addition, a more disciplinary approach for reading instruction is needed, one that shifts the focus from solely the texts' content to the language of the text and its functionality.

Functional language analysis has shown that disciplinary texts are constructed in patterns of language and offers teachers a set of practical tools for engaging students in systematically analyzing the language patterns and discussing the meanings of these patterns in disciplinary texts (Fang, 2012b). This can be done by providing students with activities such as comparing paragraphs taken from an APL article and a popular article; identifying differences in the language of different scientific text genres; analyzing the role grammatical features such as nominalizations, extended noun phrases, passivation, and technicality in each text; and subsequently discussing the differences and the functionality of language in each genre (Ariely and Yarden, 2018). Recognizing the role of language in construing knowledge and value can enable teachers to help students recognize the specialized patterns of language in the texts they read (Fang and Schleppegrell, 2010). In this way, teachers can advance their students' awareness of the language of the discipline and facilitate their enculturation into the scientific discourse community. We also suggest that teachers themselves adapt research articles while being aware of the form of language they use when adapting the article. According to Brown et al. (1989) engaging in an authentic activity is the only way learners can gain access to the standpoint that enables practitioners to act meaningfully and purposefully. Science teachers are, in most cases, not scientists. Therefore, they have little or no experience in writing scientific research articles. However, since the APL and PSL articles were found to have several shared features, writing an APL article can be considered a legitimate peripheral participation (Lave and Wenger, 1991), enabling teachers to engage in the kinds of writing that full participants (namely scientists) do.

Conclusions

A science literacy linguist asserts that "from the perspective of functional linguistics, learning the specialized language of science is synonymous with learning science" (Fang, 2005, p. 337).

Having language awareness, namely, being aware of the ways meanings are created in the text has an important consciousness-raising potential for readers (Hyland, 2007). Teachers who operate on the notion that there is nothing particularly distinctive about the genres in which science is communicated may fail to mentor their students in the necessary literacy practices which would help them read in science (Osborne, 2014). Therefore, understanding "the structure of the genres and the grammar of technicality" (Martin, 1993, p. 202) are important for both students and teachers. It will allow them to recognize the nature and function of genres specific to the discipline and to use the author's intent as a frame for a critical response (Osborne, 2014).

To conclude, engaging with APL as an apprenticeship genre can advance teachers' and students' awareness of the scientific argumentative discourse and facilitate their enculturation into the scientific discourse community.

References

Ariely, M., Livnat, Z., & Yarden, A. (2019) 'Analyzing the language of an adapted primary literature article: Towards a disciplinary approach of science teaching using texts', *Science & Education*, vol. 28, no. 1–2, pp. 63–85, doi: org/10.1007/s11191-019-00033-5

Ariely, M., & Yarden, A. (2018) 'Using authentic texts to promote disciplinary literacy in biology', in K. Kampourakis & M.J. Reiss (eds.), *Teaching biology in schools*, New York and Abingdon: Routledge, pp. 204–215.

Baram-Tsabari, A., & Yarden, A. (2005) 'Text genre as a factor in the formation of scientific literacy', *Journal of Research in Science Teaching*, vol. 42, no. 4, pp. 403–428, doi: org/10.1002/tea.20063

Bazerman, C. (1983) 'Scientific writing as a social act: A review of the literature of the sociology of science', in P.V. Anderson, J.R. Brockman, & C. Miller (eds.), *New essays in technical and scientific communication: Research, theory, practice*, Farmingdale, NY: Baywood Publishing Company, Inc, pp. 156–184.

Brill, G., & Yarden, A. (2003) 'Learning biology through research papers: A stimulus for question-asking by high-school students', *Cell Biology Education*, vol. 2, no. 4, pp. 266–274, doi: org/10.1187/cbe.02-12-0062

Brown, J.S., Collins, A., & Duguid, P. (1989) 'Situated cognition and the culture of learning', *Educational Researcher*, vol. 18, no. 1, pp. 32–42.

Carter, M., Ferzli, M., & Wiebe, E.N. (2007) 'Writing to learn by learning to write in the disciplines', *Journal of Business and Technical Communication*, vol. 21, no. 3, pp. 278–302, doi: org/10.1177/1050651907300466

Chinn, C.A., & Malhotra, B.A. (2002) 'Epistemologically authentic inquiry in schools: A theoretical framework for evaluating inquiry tasks', *Science Education*, vol. 86, no. 2, pp. 175–218, doi: org/10.1002/sce.10001

Davoodi-Semiromi, A., Schreiber, M., Nalapalli, S., Verma, D., Singh, N.D., Banks, R.K., Chakrabarti, D., & Daniell, H. (2010) 'Chloroplast-derived vaccine antigens confer dual immunity against cholera and malaria by oral or injectable delivery', *Plant Biotechnology Journal*, vol. 8, no. 2, pp. 223–242, doi: org/10.1111/j.1467-7652.2009.00479.x

Duncan, R.G., Chinn, C.A., & Barzilai, S. (2018) 'Grasp of evidence: Problematizing and expanding the next generation science standards' conceptualization of evidence', *Journal of Research in Science Teaching*, vol. 55, no. 7, pp. 907–937, doi: org/10.1002/tea.21468

Falk, H., Brill, G., & Yarden, A. (2008) 'Teaching a biotechnology curriculum based on adapted primary literature', *International Journal of Science Education*, vol. 30, no. 14, pp. 1841–1866, doi: org/10.1080/09500690701579553

Falk, H., & Yarden, A. (2009) '"Here the scientists explain what i said." Coordination practices elicited during the enactment of the results and discussion sections of adapted primary literature', *Research in Science Education*, vol. 39, no. 3, pp. 349–383, doi: 10.1007/s11165-008-9114-9

Fang, Z. (2005) 'Scientific literacy: A systemic functional linguistics perspective', *Science Education*, vol. 89, no. 2, pp. 335–347, doi: org/10.1002/sce.20050

Fang, Z. (2006) 'The language demands of science reading in middle school', *International Journal of Science Education*, vol. 28, no. 5, pp. 491–520, doi: org/10.1080/09500690500339092

Fang, Z. (2012a) 'The challenges of reading disciplinary texts', in T.L. Jetton & C. Shanahan (eds.), *Adolescent literacy in the academic disciplines: General principles and practical strategies*, New York: Guilford Press, pp. 34–68.

Fang, Z. (2012b) 'Language correlates of disciplinary literacy', *Topics in Language Disorders*, vol. 32, no. 1, pp. 19–34, doi: 10.1097/TLD.0b013e31824501de

Fang, Z., & Schleppegrell, M.J. (2008) 'Language and reading in secondary content areas', in *Reading in secondary content areas: A language-based pedagogy*, Ann Arbor: University of Michigan Press, pp. 1–17.

Fang, Z., & Schleppegrell, M.J. (2010) 'Disciplinary literacies across content areas: Supporting secondary reading through functional language analysis', *Journal of Adolescent & Adult Literacy*, vol. 53, no. 7, pp. 587–597, doi: org/10.1598/JAAL.53.7.6

Gee, J.P. (1999) 'Discourses and social languages', in *An introduction to discourse analysis: Theory and method*, London and New York: Routledge, vol. 8, issue 3, pp. 11–39.

Goldman, S.R., & Bisanz, G.L. (2002) 'Toward a functional analysis of scientific genres: Implications for understanding and learning processes', in J. Otero, J.A. Le'on & A.C. Graesser (eds.), *The psychology of science text comprehension*, Mahwah, NJ: Erlbaum, pp. 19–50.

Halliday, M.A.K. (1993) 'Some grammatical problems in scientific English', in M.A.K. Halliday & J.R. Martin (eds.), *Writing science: Literacy and discursive power*, Pittsburgh: University of Pittsburgh Press, pp. 69–85.

Halliday, M.A.K., & Matthiessen, C. (2014) *Halliday's introduction to functional grammar*, London and New York: Routledge, 4th edition.

Hyland, K. (1996) 'Writing without conviction? Hedging in science research articles', *Applied Linguistics*, vol. 17, no. 4, pp. 433–454, doi: org/10.1093/applin/17.4.433

Hyland, K. (1998) *Hedging in scientific research articles*, Amsterdam and Philadelphia: John Benjamins Publishing Company, doi: org/10.1075/pbns.54

Hyland, K. (2007) 'Genre pedagogy: Language, literacy and L2 writing instruction', *Journal of Second Language Writing*, vol. 16, no. 3, pp. 148–164, doi: org/10.1016/j.jslw.2007.07.005

Hynd-Shanahan, C. (2013) 'What does it take? The challenge of disciplinary literacy', *Journal of Adolescent & Adult Literacy*, vol. 57, no. 2, pp. 93–98, doi: org/10.1002/JAAL.226

Jiménez-Aleixandre, M.P., & Federico-Agraso, M. (2009) 'Justification and persuasion about cloning: Arguments in Hwang's paper and journalistic reported versions', *Research in Science Education*, vol. 39, no. 3, pp. 331–347, doi: 10.1007/s11165-008-9113-x

Koomen, M.H., Weaver, S., Blair, R.B., & Oberhauser, K.S. (2016) 'Disciplinary literacy in the science classroom: Using adaptive primary literature', *Journal of Research in Science Teaching*, vol. 53, no. 6, pp. 847–894, doi: org/10.1002/tea.21317

Latour, B. (1987) *Science in action: How to follow scientists and engineers through society*, Cambridge, MA: Harvard University Press.

Lave, J., & Wenger, E. (1991) *Situated learning: Legitimate peripheral participation*, Cambridge: Cambridge University Press.

Lemke, J.L. (1990) *Talking science: Language, learning, and values*, Norwood, NJ: Albex Publishing.

Livnat, Z. (2010) 'Impersonality and grammatical metaphors in scientific discourse: The rhetorical perspective', *Lidil. Revue de Linguistique et de Didactique Des Langues*, vol. 41, pp. 103–119, doi: org/10.4000/lidil.3015

Martin, J.R. (1993) 'Literacy in science: Learning to handle texts as technology', in M.A.K. Halliday & J.R. Martin (eds.), *Writing science: Literacy and discursive power*, Pittsburgh: University of Pittsburgh Press, pp. 166–202.

Myers, G. (1989) 'The pragmatics of politeness in scientific articles', *Applied Linguistics*, vol. 10, no. 1, pp. 1–35, doi: org/10.1093/applin/10.1.1

National Research Council (NRC) (2012) *A framework for K-12 science education: Practices, crosscutting concepts, and core ideas*, Washington, DC: The National Academies Press.

Norris, S.P., & Phillips, L.M. (1994) 'Interpreting pragmatic meaning when reading popular reports of science', *Journal of Research in Science Teaching*, vol. 31, no. 9, pp. 947–967, doi: org/10.1002/tea.3660310909

Norris, S.P., Macnab, J.S., Wonham, M., & de Vries, G. (2009) 'West Nile virus: Using adapted primary literature in mathematical biology to teach scientific and mathematical reasoning in high school', *Research in Science Education*, vol. 39, no. 3, pp. 321–329, doi: 10.1007/s11165-008-9112-y

Norris, S.P., & Phillips, L.M. (2003) 'How literacy in its fundamental sense is central to scientific literacy', *Science Education*, vol. 87, no. 2, pp. 224–240, doi: org/10.1002/sce.10066

Norris, S.P., & Phillips, L.M. (2008) 'Reading as inquiry', in R.A. Duschl & R.E. Grandy (eds.), *Teaching scientific inquiry: Recommendations for research and implementation*, Rotterdam, NL: Sense Publishers, pp. 233–262, doi: org/10.1163/9789460911453_018

Nwogu, K.N. (1991) 'Structure of science popularizations: A genre-analysis approach to the schema of popularized medical texts', *English for Specific Purposes*, vol. 10, no. 2, pp. 111–123, doi: org/10.1016/0889-4906(91)90004-G

Osborne, J. (2014) 'Teaching scientific practices: Meeting the challenge of change', *Journal of Science Teacher Education*, vol. 25, no. 2, pp. 177–196, doi: org/10.1007/s10972-014-9384-1

Parkinson, J. (2001) 'Popular and academic genres of science: A comparison, with suggestions for pedagogical applications', Unpublished PhD Thesis, University of Natal, Durban.

Parkinson, J., & Adendorff, R. (2004) 'The use of popular science articles in teaching scientific literacy', *English for Specific Purposes*, vol. 23, no. 4, pp. 379–396, doi: org/10.1016/j.esp.2003.11.005

Penney, K., Norris, S.P., Phillips, L.M., & Clark, G. (2003) 'The anatomy of junior high school science textbooks: An analysis of textual characteristics and a comparison to media reports of science', *Canadian Journal of Science, Mathematics and Technology Education*, vol. 3, no. 4, pp. 415–436, doi: org/10.1080/14926150309556580

Sandoval, W.A., & Morrison, K. (2003) 'High school students' ideas about theories and theory change after a biological inquiry unit', *Journal of Research in Science Teaching*, vol. 40, no. 4, pp. 369–392, doi: org/10.1002/tea.10081

Schleppegrell, M.J. (2002) 'Linguistic features of the language of schooling', *Linguistics and Education*, vol. 12, no. 4, pp. 431–459.

Schleppegrell, M.J. (2004) *The language of schooling: A functional linguistics perspective*, Mahwah, NJ and London: Lawrence Erlbaum Associates.

Schleppegrell, M.J. (2006) 'The challenges of academic language in school subjects', in I. Lindberg & K. Sandwall (eds.), *Spraket och kunskapen: att lra pa sitt andrasprak i skola och högskola*, Göteborg: Göteborg universitet institutet för svenska som andraspråk, pp. 47–69.

Shanahan, C., & Shanahan, T. (2020) 'Disciplinary literacy', in J. Patterson (ed.), *The SATsuit and classroom practice: English language arts/literacy*, New York: College Board, pp. 91–125.

Shanahan, T., & Shanahan, C. (2008) 'Teaching disciplinary literacy to adolescents: Rethinking content-area literacy', *Harvard Educational Review*, vol. 78, no. 1, pp. 40–60, doi: org/10.17763/haer.78.1.v62444321p602101

Snow, C.E. (2010) 'Academic language and the challenge of reading for learning about science', *Science*, vol. 328, no. 5977, pp. 450–452, doi: 10.1126/science.1182597

Sung, Y., Wu, M., Chen, C., & Chang, K. (2015) 'Examining the online reading behavior and performance of fifth-graders: Evidence from eye-movement data', *Frontiers in Psychology*, vol. 6, pp. 1–15, May, doi: org/10.3389/fpsyg.2015.00665

Suppe, F. (1998) 'The structure of a scientific paper', *Philosophy of Science*, vol. 65, no. 3, pp. 381–405, doi: org/10.1086/392651

Swales, J. (2001) *Genre analysis: English in academic and research settings*, Cambridge, New York and Melbourne: Cambridge University Press, 1990 1st ed.

van Lacum, E.B., Ossevoort, M., Buikema, H., & Goedhart, M. (2012) 'First experiences with reading primary literature by undergraduate life science students', *International Journal of Science Education*, vol. 34, no. 12, pp. 1795–1821, doi: org/10.1080/09500693.2011.582654

Wignell, P., Martin, J.R., & Eggins, S. (1993) 'The discourse of geography: Ordering and explaining the experiential world', in M.A.K. Halliday & J.R. Martin (eds.), *Writing science: Literacy and discursive power*, London: The Falmer Press, pp. 136–165.

Yarden, A. (2009) 'Reading scientific texts: Adapting primary literature for promoting scientific literacy', *Research in Science Education*, vol. 39, no. 3, pp. 307–311, doi: org/10.1007/s11165-009-9124-2

Yarden, A., Brill, G., & Falk, H. (2001) 'Primary literature as a basis for a high-school biology curriculum', *Journal of Biological Education*, vol. 35, no. 4, pp. 190–195, doi:org/10.1080/00219266.2001.9655776

Yarden, A., Norris, S.P., & Phillips, L.M. (2015) *Adapted primary literature,* Dordrecht: Springer, doi: org/10.1007/978-94-017-9759-7

Yore, L.D., Hand, B.M., & Florence, M.K. (2004) 'Scientists' views of science, models of writing, and science writing practices', *Journal of Research in Science Teaching*, vol. 41, no. 4, pp. 338–369, doi: org/10.1002/tea.20008

31

PROFESSIONAL DEVELOPMENT IN SCIENCE COMMUNICATION FOR PRACTICING SCIENTISTS

The role of science communication training programs in shaping participating scientists' skills

Yael Barel-Ben David and Ayelet Baram-Tsabari

Introduction

Scientists' ability to communicate their research and its importance to diverse publics, stakeholders, and peers outside their discipline is crucial. However, most scientists are not trained in science communication to do so. To address this need, science communication training programs have proliferated within academic and professional settings during the last 20 years (Mulder et al., 2008).

Several approaches govern the practice of science communication: the public understanding of science (PUS) approach and dialogical approach. PUS claims that the public's diminished support for science is due to a lack of knowledge about science (Sturgis and Allum, 2004). It focuses mainly on educational activities by scientists such as public lectures and "open science days" to inform the public. PUS-inspired science communication training programs mostly involve training in how to deliver lectures and to convey the scientific message in a coherent and accessible way.

By contrast, dialogical approaches (Sturgis and Allum 2004; Trench, 2008; Brossard and Lewenstein, 2010; Eagleman, 2013; Gouyon, 2016) such as public engagement with science (PES) emphasize feedback, contextual knowledge, consultation, and two-way conversations between scientists and society (Trench, 2008; Eagleman, 2013; Lewenstein, 2015; Gouyon, 2016). This change in focus does not mean abandoning previous skills such as standing in front of a camera or handling difficult questions. These are complemented by the use of storytelling tools and encouraging a two-way dialogue with the audience in addition to empathy and trust building strategies (Bray et al., 2012; Bucchi and Trench 2014; Aurbach et al., 2019).

There are many valid ways to assess the effectiveness of science communication training programs (Baram-Tsabari and Lewenstein, 2016; Barel-Ben David and Baram Tsabari, 2020;

DOI: 10.4324/9781003043782-36

Nisbet and Scheufele, 2009). We present a pre- and postassessment of scientists' output in three science communication training programs to shape writing skills and discuss the learning processes the scientists underwent.

The diversity of science communication training programs

When addressing the public, scientists need to adapt their message to fit an unfamiliar medium and genre by following new rules, skills, and practices (National Academies of Sciences Engineering and Medicine, 2017). The QUEST[1] project estimated there are roughly 117 science communication courses in 19 European countries (Villa, 2019) that vary in terms of length, skills, goals, framework, and design. Some address oral communication while others focus on writing for the media. Training programs can be oriented toward different audiences such as scientists, reporters, students, or practicing science communicators. This makes it possible to cater to different needs but at the same time complicates the systematic evaluation of their effectiveness (Baram-Tsabari and Lewenstein, 2017b; Bray et al., 2012; Brownell et al., 2013b; Fischhoff and Scheufele, 2013; Miller and Fahy, 2009; Mulder et al., 2008; Sevian and Gonsalves, 2008). Scientists acknowledge the importance of communicating their scientific work to the public and have positive attitudes toward these programs (Besley, 2015; Besley et al., 2015; Besley and Tanner, 2011; Besley and Nisbet, 2013; McCann et al., 2015).

Ample research has been conducted on science communication skill development during science communication training programs (e.g. Baram-Tsabari and Lewenstein 2017a, 2017b; Besley et al., 2016; Dudo and Besley, 2016), the importance of scientists having these skills (Leeming, 2017; Bankston and McDowell, 2018; Vollbrecht et al., 2019) and which skills promote scientists' public engagement (Mercer-Mapstone and Kuchel, 2015, 2017). These studies suggest scientists should communicate their science in an accessible manner, for example by reducing the use of jargon and promoting trust and empathy. Therefore, training should provide key skills such as promoting trust and dialogue, using clear and vivid language, tailoring the message to the audience, etc. (Aurbach et al., 2019; Baram-Tsabari and Lewenstein, 2017a, 2012; Brownell et al., 2013a; Mercer-Mapstone and Kuchel 2015; Metcalfe and Gascoigne 2009; Montgomery, 2017; National Academies of Sciences Engineering and Medicine, 2017; Rakedzon and Baram-Tsabari, 2016, 2017; Sevian and Gonsalves, 2008). Recent articles have emphasized the role of core science communication noncontext-specific skills, such as knowing your audience, the use of adequate language, narrative structure, metaphors and analogies, personal connection, and nonverbal communication (Aurbach et al., 2019; Davies et al., 2019; Sanders, 2019; Stevens et al., 2019; Altman et al., 2020; Joubert, 2020).

Methodology

Training programs

We examined (1) semester-long academic courses, (2) one- to two-day workshops in Hebrew led by the first author, and (3) one- to two-day workshops in English led by the Alan Alda Center (AAC), as described in Table 31.1.

All the training programs were held between 2017 and 2019.

Academic courses: Two optional semester-long graded academic credit courses open to both graduate and undergraduate students from all fields of study in a STEM university in Israel were examined: an introduction to science communication course and an advanced practicum course. The introductory course consisted of an overview of science communication theory

Table 31.1 Science communication training programs involved in the study

Type of training program	Academic courses (Hebrew)	Workshops in Israel (Hebrew)	Alan Alda Center workshops in Israel and in the United States (English)
Number of training programs	2	7	9
Number of participants	5–21	14–28	24–32
Length	13 sessions of 3 academic hours each	One or two days (6–12h)	One or two days (8–16h)
Participants	Graduate and undergraduate students	Researchers, scientists, and students in all research fields	Researchers, scientists, and students in STEM and health fields in academy, industry, and government branches
Venue	University	University and industrial sector	University, industrial sector, and government

and emphasized practical experience in written genres such as blogs, press releases, interviews, and oral presentations such as three-minute science talks (modeled after the FameLab[2] competition), televised interviews, and short science topic clips (Baram-Tsabari and Lewenstein, 2017a). The advanced-level practicum course was aimed at communicating science through an apprenticeship in five leading science communication venues in Israel. Participants practiced writing about research and scientific topics. Professional staff and the course instructors guided the students on producing scientific content for the public.

Workshops: The workshops dealt with science communication skills such as reducing jargon, awareness of the 'curse of knowledge', connection to people's daily lives and the use of narrative, humor, and analogies when communicating science. They emphasized message distilling, originating from journalistic practice, and connecting empathically by using improvizational theater methods to communicate science in an accessible way to different audiences (Alda, 2010; Bernstein, 2014). Whereas the AAC workshop defines itself as an immersive practice-oriented workshop, the workshops in Israel started with an introduction to science communication and presented research-based information relevant to the workshop and participants. The workshops in Israel used a team-taught, informal, in-person pedagogy to provide the participants with skills, strategies, and confidence in communicating science to the general public. The sample consisted of seven workshops hosted by the researcher in Israel (in university and industry settings) and nine additional workshops hosted by the AAC staff as described in Table 31.1.

We focused on assessment of science communication training programs through their written skills since writing categories (see Aurbach et al., 2019) are interchangeable with speech (e.g. style, grammar, voice and tense, compelling storytelling elements, awareness of jargon, etc.).

Key differences between training programs

Length of exposure: The workshop participants had short, highly intense sessions in a short time span, while course students were exposed to the information and practice over time and could take longer to absorb and contemplate both practice and theory (Table 31.1).

Orientation: The workshops were practical and heavily based on active participation of the scientists in communicating their science. The courses were academically driven and partially theoretical. However, students were given exercises on the science communication skills discussed during classes and were graded on them.

Demographics and science communication experience: Of the participants, about a quarter were early career scientists studying toward a master's degree (24.6%) or had a postdoctoral position (24%). Participants came from research fields spanning medicine and public health, urban planning, and the social sciences, but the majority were from the natural and life sciences (63%). Most participants had some science communication experience prior to the training program (see Table 31.2).

Research tools and data collection

The dataset was composed of questionnaires analyzed using a mixed methods approach.

Questionnaires: Online questionnaires were based on previously validated tools (Peters, 2012). Participants were asked to anonymously answer the online questionnaire before and after their training. There were 36 questions (open-ended, multiple choice, and Likert scale) on the pre-training questionnaire and 41 on the post-training questionnaire. The questionnaires focused on participants' reasons for taking part in science communication training programs and activities, as well as their attitudes regarding such activities.

Skills were assessed based on an open question asking participants to describe their research to a lay audience: "Using 150–200 words please describe your research to a nonspecialized audience (people without science background)" (see examples in Table 31.4). In total, 169 questionnaires were collected. The final database consisted of 118 valid pre-training

Table 31.2 Self-reported experience with science communication activities in Israeli versus U.S.- based scientists

Location	Israel		United States	
Sample	*Science communication training participants (n = 68) (2017–2019)*	*Scientists survey respondents (n = 606) (2015)**	*Science communication training participants (n = 107) (2017–2019)*	*Scientists' Pew report (2015)***
No experience	35.3%	42%	31.8%	No regular interactions 11%
Some experience (1–5 times)	45.6%	51.2%	36.5%	Occasional interactions 48%
Considerable experience (more than 5 times)	19.1%	6.8%	31.8%	Frequent interactions 41%

* Data based on the Reiss et al. (2016), survey of active Israeli scientists.

** Data based on PEW report (2015), a census survey of land grant North American–based scientists. Israeli scientists were asked to estimate the number of science communication activities they participated in during the last three years. In the census study of U.S.–based land grant scientists, participants were asked to estimate their participation over the last year.

Table 31.3 Coding scheme for assessing participants' skills

Cluster	Category (inter-rater reliability achieved for manually evaluated categories)	Measurement	Example
Clarity	Jargon use	De-Jargonizer	"By accelerating/shaking electrons one can make radiation, e.g. radio waves, X-rays, etc. . . . My research looks at new ways to accelerate electrons and analyzes any new properties of the radiation that may come. One such example is to look at the recoil on the electron when the radiation is emitted" [pre-training, academy]. (Common: 75%, 39; mid-frequency: 23%, 12; rare: 2%, 1; Score: 87)
	Readability	Flesch Reading Ease Flesch-Kincaid Grade Level	Using the previous example: Flesch Reading Ease – 42.7; Flesch-Kincaid Grade Level – 11.4
	Type of explanation (0.88)	Absent	"I work on enhancing the performance of modern computing systems" [pre-training, academy].
		Definition	"photons, the fundamental particles that light is made of" [pre-training, academy]
		Elucidating explanation	"Many of the applications we use in our daily life use machine learning algorithms and especially Artificial Neural Networks to perform tasks such as matching on-line commercials to users, tagging people on Facebook and much more" [pre-training, academy].
		Quasi-scientific	"Enzymes are Nature's workhorses" [post-training, academy].
		Transformative explanation	
Content	Science content (0.7)	Number of units – any stand-alone fact stated about scientific research or its results	"Most imaging technologies (MRI, CT, X-rays) provide information about anatomy and alterations in anatomy caused by disease. However, some diseases affect the function of an organ or structure but do not cause changes in anatomy. My research uses real-time imaging techniques to assess abnormal function. More specifically, I use video fluoroscopy (real time X-ray) to evaluate swallowing function in patients with difficulty swallowing" [pre-training, academy]. Coded as four science content units.

(Continued)

Table 31.3 (Continued)

Cluster	Category (inter-rater reliability achieved for manually evaluated categories)	Measurement	Example
	Nature of science (0.84)	Number of units – any indication of research dynamics, affordances, limitations, and human aspects counts as one unit	"It is important to understand the natural evolution of the planet to predict its evolution in the future and not just prepare for changes that are under way, but also make the optimal use of the planet's resources with the least amount of harming it and us" [post-training, academy]. Contains one NoS unit.
	Connection to everyday life (0.82)	Present/absent	Present: "when we look around us we see things that are very heterogeneous, very different, trees, and dogs, and cats, and every human is different, so it's hard to describe all of these things at once, using some kind of simple theory" [pre-training, academy].
	Impact or implications for the audience or society (0.79)	Present/absent	Present: "Our findings will be used by scientists and policymakers to regulate chemical waste and pollution for the health of the environment" [post-training, academy].
	Personal connection (1)	Absent	"Geometric group theorists study the connection between abstract algebraic objects, groups, and geometry" [pre-training, academy].
		Declarative	"My research focuses on developing such methods" [pre-training, academy].
		Essential	"the research that I am most passionate about is using the patient's own stem cells to regenerate the retina" [pre-training, academy].
Knowledge organiza-tion	Framing* (0.79)	Social progress	"My research explores how human activities influence biodiversity, but also how people can benefit from biodiversity conservation" [pre-training, academy].
		Economic development/ competitiveness	"The market of IoT is exponentially increasing. It has been predicted that by 2020 more than 20 billion devices will be connected to the internet" [pre-training, industry].
		Morality and ethics	"and the practice removes massive amounts of biodiversity from the sea, contributing to damaged marine ecosystems and a financial loss for local and indigenous small-scale fishers who depend on these ecosystems for the livelihoods" [pre-training, government].

Cluster	Category (inter-rater reliability achieved for manually evaluated categories)	Measurement	Example
		Science and technology uncertainty	"But the challenge is that unlike playing chess, the behavior of physical system is usually far more complicated, nonlinear, uncertain, and more importantly, we often do not have adequate data, in terms of both quantity and quality, for learning" [post-training, academy].
		Public accountability	"With this information in hand, people can better protect their natural and water resources for future generations" [pre-training, government].
	Scaffolding explanation – a way of building understanding incrementally, helping the audience to follow the logical development of more complicated ideas (0.83)	Present/absent	"The microbiome is a term used to describe the trillions of bacteria that live in the human gut. The bacteria in our gut are a big part of our health. These bacteria are making nutrients, using nutrients, making toxins, and processing toxins, and because there are so many of them, which bacteria are there, and what they are doing can influence how healthy or sick we are. In fact, they may even dictate how long we live. When researchers give old fish the gut bacteria of young fish, the old fish live a lot longer" [pre-training, academy].
Style	The use of metaphors and analogies (1)	Present/absent	Present: "Imagine you have many heavy packages that you must transfer from a nearby building to the building you are in. This will be a complicated task and at the end you will be very tired. We have a similar problem in the computer architecture world" [post-training, academy].
	Narrative (0.8)	Present/absent	Present: "The current issue is that people used to experience nature much more in the past than now, they used to climb trees, go fishing . . . when they were children, while now children are spending most of their time indoors, in front of screens, instead of going out. If people go out more often and interact more with nature, they are more likely to feel part of the natural world, and in turn, to care about it, and ultimately, to protect it" [post-training, government].

(Continued)

Table 31.3 (Continued)

Cluster	Category (inter-rater reliability achieved for manually evaluated categories)	Measurement	Example
	Humor (1)	Present/absent	Present: "If only that worker could spend this time working instead! And of course he would be less tired too. A possible solution to this problem is to let the worker work from home. No more time and energy spent on getting to work!" [pre-training, academy].

Source: Based on a framework suggested by Baram-Tsabari and Lewenstein, 2012.

★ Several categories suggested by Baram-Tsabari and Lewenstein (2012) had no representation in our sample; hence they were excluded from the table. These include 'Framing: Pandora's box/Frankenstein's monster/runaway science, Middle way/alternative path, Conflict and strategy, and Science versus nonscience.

questionnaire-generated texts, and 39 post-training texts, of which only 27 could be paired with a corresponding pre-training text.

Data analysis

COMPUTERIZED ASSESSMENT TOOLS

One of the open-ended questions asked the participants to describe their research to nonscientists. The English texts produced ($n = 97$) were assessed for accessibility to readers in terms of vocabulary, length, and sentence complexity via the following tools:

1   *The De-Jargonizer*: An automated jargon identification program to improve and adapt vocabulary use to different audiences (Rakedzon et al., 2017). The program categorizes the words into three levels: high frequency/common words, midfrequency/normal words, and jargon, rare and technical words. The higher the score, the clearer and more jargon-free it is. An accessible text should contain no more than 2% of rare and unfamiliar words (Rakedzon et al., 2017).

2   *Flesch Reading Ease*: Assesses readability. The analysis is performed by the Microsoft Word readability option tool and can be used on any Word document to assess the readability, based on word and sentence length. The score ranges from 0 to 100. A desirable score is in the range of 60 to 70 and reflects an eighth-grade-level text (Flesch, 1948).

3   *Flesch-Kincaid Grade Level*: An analysis performed by the Word readability option tool that can be used on any Word document to assess the appropriateness of the grade level of a text, based on students' reading level. The score represents the grade level required to understand the text. A good Flesch-Kincaid Grade Level score would be 7–8 for a text to be very readable for most people (Flesch, 1948).

To determine whether there were significant differences for each category between the pre- and post-training texts, an independent samples *t*-test was conducted.

Each text was further coded according to 11 categories adapted from Baram-Tsabari and Lewenstein (2012) (Table 31.3).

Coding reliability

An intercoder reliability process was conducted with three peer coders in three consecutive rounds. In each round of coding, 10% of the database was randomly selected, using the Excel randomizer; on the third round, 20% were coded.

Coders did not know which training program the text was taken from or whether it was pre- or post-training. The selected items were then coded separately by the coder and compared to the researcher's coding. After each round, a deliberation phase refined the categories and codebook (Table 31.3).

Findings

Scientists' popular descriptions of their research before and after the training: The open-ended question in the questionnaires generated a database of 157 texts describing participants' research to lay audiences. Each text was assessed for readability using the computerized tools and manually to assess training impact on written skills. All the pre-training texts were aggregated to compare the means with all the post-training texts, as shown in Table 31.4.

Clarity: Overall, there was a significant improvement in the post-training text in all computerized measurements. The texts contained less jargon and were more accessible to readers than the pre-training text, showing improvement in clarity-related skills (e.g. reducing jargon, avoiding long and passive sentences, etc.). A significant improvement in all the readability measurements was found between the post- and pre-training texts (De-Jargonizer score: $p < 0.01$; Flesch Reading Ease score: $p < 0.01$; Flesch-Kincaid Grade Level: $p < 0.01$); on average, fewer words were used in the post-training texts.

Several other trends emerged between the pre- and post-texts although these did not reach significance:

Content: A decrease in the number of science content units from the pre- to the post-texts, from an average of 4.1 scientific content units to 3.1 Most of the post-training texts presented the relevance of the research to people's lives.

Knowledge organization: Most pre-training texts chose to frame science and technology in terms of uncertainty (38%), whereas the post-training texts tended to be framed to a greater extent in terms of social progress (51.3%). Another frame that was used was economics (7.6% pre; 10.3% post).

Style: There were almost no differences in any of the style categories, although the post-training texts were marginally lower in all categories than the pre-training texts (Table 31.4).

Paired pre- and post-texts

In the database, 27 pre- and post-training texts could be paired. Readability improved on all three factors, with a decrease in the number of words. Participants gave fewer explanations in their post- than in their pre-training texts but 'Connection to everyday life' and 'Impact or implications for audience or society' almost doubled in the post-training texts.

Table 31.4 Characteristics of scientists' popular texts written prior ($n = 118$) and after ($n = 39$) science communication training

Cluster	Category	Measurement	Text		Statistic
			Pre *(n = 118)*	Post *(n = 39)*	
Clarity	Jargon use	De-Jargonizer score	92.1	94.6	$p < 0.01$
	Readability	Flesch Reading Ease score	33.2	47.4	$p < 0.01$
		Flesch-Kincaid Grade Level	14.1	11.7	$p < 0.01$
		Number of words	167.1	132.6	$p < 0.05$
	Type of explanation	Absent	34.7%	46.2%	n.s.
		Definition	31.4%	20.5%	
		Elucidating explanation	18.6%	20.5%	
		Quasi-scientific	50.9%	70.7%	
		Transformative explanation	90.3%	50.1%	
Content	Science content	Average no. of units ± SD	40.1 ± 2.8	30.1 ± 3.0	$p = 0.056$
	Nature of science	Average no. of units ± SD	10.0 ± 1.1	00.9 ± 1.0	n.s.
	Connection to everyday life	% Present	30.5%	38.5%	n.s.
	Impact or implications for the audience or society	% Present	48.3%	59.0%	n.s.
	Personal connection	Absent	29.7%	23.1%	n.s.
		Declarative	52.5%	61.5%	
		Essential	17.8%	15.4%	
Knowledge organization	Scaffolding explanation	% Present	72%	66.7%	n.s.
	Framing	Absent	4.2%	2.6%	n.s.
		Social progress	41.5%	51.3%	
		Economic development/ competitiveness	70.6%	10.3%	
		Morality and ethics	40.2%	50.1%	
		Science and technology uncertainty	38.1%	30.8%	
		Public accountability	40.2%	00.0%	
Style	The use of metaphors and analogy	% Present	23.7%	23.1%	n.s.
	Narrative	% Present	44.9%	43.6%	n.s.
	Humor	% Present	14.4%	12.8%	n.s.

Discussion

Science communication training programs aim to support experts in communicating their science and expertise to nonexperts in "simple and compelling messages that are delivered clearly and with empathy" (Joubert, 2020, p. 3) to achieve a variety of goals. The ability to communicate in an understandable, accessible, and clear manner among experts and between experts and nonexperts has become a necessity for practicing academicians. Unfortunately, most academic curricula do not provide scientists with such skills. To fill this need, science communication training programs have emerged. But are these types of professional development effective in supporting scientists in acquiring the relevant skills?

We assessed the extent to which science communication training programs impacted participants' skills. Our findings reveal the skills manifested in participants' writing before and after training programs (Baram-Tsabari and Lewenstein, 2012) and paint a complex picture.

Some skills appear easier to learn than others. Writing skills categorized under 'Clarity', such as reducing the use of jargon and avoiding passive sentences to increase text readability, showed improvement from pre-training to post-training, indicating an improvement in participants' skills. However, more complex skills, such as providing vivid explanations and knowing your audience to be able to connect science to their everyday lives, did not show a difference. For example, we expected to see more use of humor and metaphors in the post-training texts since these are main concepts in the training programs. This might indicate that these concepts were not transferred to participants' writing or that they demand more practice and attention. This could be attributed to the subjective nature of the categories, even though good or high reliability was achieved on all coding categories. Since there is no 'gold standard' for science communication training, the skills aimed at supporting scientists in their science communication endeavors include setting communication goals, knowing your audience, and adjusting your message accordingly to nonverbal communication, and using plain language, compelling storytelling, and appropriate grammatical choices, to name a few (Aurbach et al., 2019; Sanders, 2019). The manual coding categories represented higher levels of skills than the computerized categories which were mainly technical and easier to apply (e.g. reduce the use of jargon, avoid passive sentences, shorten sentence length, etc.). They demanded more challenging skills (e.g. vivid explanations, knowing your audience to be able to connect your science to their everyday lives, building a story of the research using scaffolding and narrative, etc.) that might demand more capacity building to achieve. Thus developing science communication skills takes time and should be practiced over time. As suggested by Rodgers et al. (2018), learning effective science communication skills and putting them to use is a process.

Moreover, a genre-shift needs to be taken into consideration. Many science communication training programs emphasize verbal communication over written skills. Therefore, in these programs, the lion's share of the training is dedicated to practicing verbal communication, attention to tone, body language, etc. Although many of the main concepts, skills, and tools regarding experts' content are considered interchangeable between writing and speaking, clearly there are still differences. While writing can be spread over a period of time, allowing rewriting and feedback, speaking is an immediate action, and once you utter a word, it cannot be unsaid. Another aspect is the body language and intonation accompanying verbal communication that can support connecting to the audience for example. Here, we concentrated on the written skills the participants acquired during their participation in science communication training programs because writing is the most accessible form of communication scientists use when communicating with the public. Today, anyone can post a blog, write a Facebook post or a tweet, or sometimes even publish a news item on leading news sites. Second, focusing on writing allowed us to collect more data and use neutral computerized measures to evaluate the text levels. Nevertheless, a deeper examination of the interchangeability of these skills is needed.

Overall, whereas there is an impact of short-term, one-time interventions as Rakedzon and Baram-Tsabari (2017) found, such impact may be limited to clarity-related skills.

Limitations

In addition to the genre shift and the subjectivity of assessment categories, computerized tools have limitations for assessing readability such as the inability of the De-Jargonizer to distinguish between the meaning of words used in professional terminology and everyday life (e.g. 'values' or 'positive trend'; Somerville and Hassol, 2011). Thus some words that are highly professional in meaning in a specific context are regarded as common words and are not flagged. In general, the three computerized tools, although often used to determine levels of text readability, only address semantics, without referring to content or context. Taking these limitations into consideration, a triangulation approach was taken analyzing each text using the three computerized tools and a manual assessment. By doing so, we increased the external validity of our analysis.

Generalizability

Any training programs that can support experts' endeavors in sharpening their science communication skills are valuable. From the professional development point of view, we see that these training sessions have a quantifiable positive effect on skills. Hence continuing education credits should be offered to scientists to acquire tactics and methods to become more accessible to the public. Training program developers should choose realistic learning goals and adapt their program accordingly. Science communication trainers should not expect their participants to use humor and analogies woven into a storytelling arc by the end of a one-time intervention. Therefore, providing short interventions over time, allowing participants to absorb and execute their new skills and testing them in real-life situations might have a stronger effect on skills.

Notes

1 QUEST is a European Union's Horizon 2020 research and innovation funded program. Its focus is on assessing the effectiveness of and developing tools for communication and dialogue between science and the public.
2 FameLab is an international competition held at the Rotterdam Science Festival. It challenges participating scientists to present their science in three minutes addressing a heterogeneous crowd. Participants are judged on their clarity, coherence, and charisma.

References

Alda, A. (2010) 'In your own voice', in D. Kennedy & G. Geneva Overholser (eds.), *Science and the media*, Cambridge, MA: American Academy of Arts and Sciences, pp. 10–12.

Altman, K., Yelton, B., Hart, Z., Carson, M., Schandera, L., Kelsey, R.H., Porter, D.E., & Friedman, D.B. (2020) '"You gotta choose your words carefully": Findings from interviews with environmental health scientists about their research translation perceptions and training needs', *Journal of Health Communication*, vol. 5, no. 25, pp. 454–462, DOI: 10.1080/10810730.2020.1785060

Aurbach, E.L., Prater, K.E., Cloyd, E.T., & Lindenfeld, L. (2019) *Foundational skills for science communication: A preliminary framework*, White paper, University of Michigan, hdl.handle.net/2027.42/150489

Bankston, A., & McDowell, G.S. (2018) 'Changing the culture of science communication training for junior scientists', *Journal of Microbiology & Biology Education*, vol. 1, no. 19, DOI: 10.1128/jmbe.v19i1.1413

Baram-Tsabari, A., & Lewenstein, B.V. (2012) 'An instrument for assessing scientists' written skills in public communication of science', *Science Communication*, vol. 35, no. 1, pp. 56–85, DOI: https://doi.org/10.1177/1075547012440634

Baram-Tsabari, A., & Lewenstein, B.V. (2016) 'Assessment', in M.C.A. Van der Sanden & M.J. de Vries (eds.), *Science and technology education and communication: Seeking synergy*, Rotterdam, The Netherlands: Sense Publishers, pp. 161–184.

Baram-Tsabari, A., & Lewenstein, B.V. (2017a) 'Science communication training: What are we trying to teach?', *International Journal of Science Education, Part B*, vol. 3, no. 7, pp. 285–300, DOI: https://doi.org/10.1080/21548455.2017.1303756

Baram-Tsabari, A., & Lewenstein, B.V. (2017b) 'Preparing scientists to be science communicators', in *Preparing informal science educators*, Cham: Springer, pp. 437–471, DOI:10.1007/978-3-319-50398-1_22

Barel-Ben David, Y., & Baram Tsabari, A. (2020) 'Evaluating science communication training: Going beyond self-reports', in T. Newman (ed.), *Handbook of science communication training*, New York: Routledge, pp. 122–139.

Bernstein, R. (2014) 'Communication: Spontaneous scientists', *Nature*, vol. 505, no. 7481, pp. 121–123, DOI: 10.1038/nj7481–121a

Besley, J.C. (2015) 'What do scientists think about the public and does it matter to their online engagement?', *Science and Public Policy*, vol. 42, no. 2, pp. 201–214.

Besley, J.C., Dudo, A.D., & Storksdieck, M. (2015) 'Scientists' views about communication training', *Journal of Research in Science Teaching*, vol. 52, no. 2, pp. 199–220, DOI: https://doi.org/10.1002/tea.21186

Besley, J.C., Dudo, A.D., Yuan, S., & Abi Ghannam, N. (2016) 'Qualitative interviews with science communication trainers about communication objectives and goals', *Science Communication*, vol. 38, no. 3, pp. 356–38, DOI: https://doi.org/10.1177/1075547016645640

Besley, J.C., & Nisbet, M. (2013) 'How scientists view the public, the media and the political process', *Public Understanding of Science*, vol. 22, no. 6, pp. 644–659, DOI: https://doi.org/10.1177/0963662511418743

Besley, J.C., & Tanner, A.H. (2011) 'What science communication scholars think about training scientists to communicate', *Science Communication*, vol. 33, no. 2, pp. 239–263, DOI: 10.1177/1075547010386972

Bray, B., France, B., & Gilbert, J.K. (2012) 'Identifying the essential elements of effective science communication: What do the experts say?' *International Journal of Science Education, Part B: Communication and Public Engagement*, vol. 2, no. 1, pp. 23–41, doi: https://doi.org/10.1080/21548455.2011.611627

Brossard, D., & Lewenstein, B.V. (2010) 'A critical appraisal of models of public understanding of science', in L. Kahlor & P. Stout (eds.), *Communicating science: New agendas in communication*, New York: Routledge, pp. 11–39.

Brownell, S.E., Price, J.V., & Steinman, L. (2013a) 'Science communication to the general public: Why we need to teach undergraduate and graduate students this skill as part of their formal scientific training', *Journal of Undergraduate Neuroscience Education*, vol. 12, no. 1, pp. E6–E10.

Brownell, S.E., Price, J.V., & Steinman, L. (2013b) 'A writing-intensive course improves biology undergraduates' perception and confidence of their abilities to read scientific literature and communicate science', *Advances in Physiology Education*, vol. 37, no. 1, pp. 70–79, doi: 10.1152/advan.00138.2012

Bucchi, M., & Trench, B. (2014) *Routledge handbook of public communication of science and technology*, New York: Routledge, 2nd edition.

Davies, S.R., Halpern, M., Horst, M., Kirby, D.A., & Lewenstein, B. (2019) 'Science stories as culture: Experience, identity, narrative and emotion in public communication of science', *Journal of Science Communication*, vol. 18, no. 5, p. A01, https://doi.org/10.22323/2.18050201

Dudo, A., & Besley, J.J.C. (2016) 'Scientists' prioritization of communication objectives for public engagement', *PLoS One*, vol. 11, no. 2, p. e0148867, DOI: https://doi.org/10.1371/journal.pone.0148867

Eagleman, D.M. (2013) 'Why public dissemination of science matters: A manifesto', *The Journal of Neuroscience*, vol. 33, no. 30, pp. 12147–12149, DOI: 10.1523/JNEUROSCI.2556–13.2013

Fischhoff, B., & Scheufele, D.A. (2013) 'The sciences of science communication', *Proceedings of the National Academy of Sciences*, vol. 110, supplement 3, pp. 14031–14032, DOI: https://doi.org/10.1073/pnas.1312080110

Flesch, R. (1948) 'A new readability yardstick', *Journal of Applied Psychology*, vol. 32, no. 3, pp. 221–233, DOI: 10.1037/h0057532

Gouyon, J.-B. (2016)'1985, Scientists can't do science alone, they need publics', *Public Understanding of Science*, vol. 25, no. 6, pp. 754–757, DOI: 10.1177/0963662516650361

Joubert, M. (2020) 'From top scientist to science media star during COVID-19 – South Africa's Salim Abdool Karim', *South African Journal of Science*, vol. 116, no. 7–8, pp. 1–4, http://dx.doi.org/10.17159/sajs.2020/8450

Leeming, J. (2017) 'The next generation of science outreach', *NatureJobs Blog*, 14 April, blogs.nature.com/naturejobs/2017/04/14/the-next-generation-of-science-outreach

Lewenstein, B.V. (2015) 'Identifying what matters: Science education, science communication, and democracy', *Journal of Research in Science Teaching*, vol. 52, no. 2, pp. 253–262, https://doi.org/10.1002/tea.21201

McCann, B.M., Cramer, C.B., & Taylor, L.G. (2015) 'Assessing the impact of education and outreach activities on research scientists', *Journal of Higher Education Outreach and Engagement*, vol. 19, no. 1, pp. 65–78.

Mercer-Mapstone, L., & Kuchel, L. (2015) 'Teaching scientists to communicate: Evidence-based assessment for undergraduate science education', *International Journal of Science Education*, vol. 37, no. 10, pp. 1613–1638, https://doi.org/10.1080/09500693.2015.1045959

Mercer-Mapstone, L., & Kuchel, L. (2017) 'Core skills for effective science communication: a teaching resource for undergraduate science education core skills for effective science communication: A teaching resource for undergraduate science education', *International Journal of Science Education, Part B*, vol. 7, no. 2, pp. 181–201, doi:10.1080/21548455.2015.1113573

Metcalfe, J.E., & Gascoigne, T. (2009) 'Teaching scientists to interact with the media', *Issues*, vol. 87, pp. 41–44.

Miller, S., & Fahy, D. (2009) 'Can science communication workshops train scientists for reflexive public engagement? The ESConet experience', *Science Communication*, vol. 31, no. 1, pp. 116–126, doi:10.1177/1075547009339048

Montgomery, S.L. (2017) *The Chicago guide to communicating science*, Chicago, IL: University of Chicago Press, 2nd edition.

Mulder, H.A.J., Longnecker, N., & Davis, L.S. (2008) 'The state of science communication programs at universities around the world', *Science Communication*, vol. 30, no. 2, pp. 277–287, doi:10.1177/1075547008324878

National Academies of Sciences Engineering and Medicine. (2017) *Communicating science effectively: A research agenda*, Washington, DC: The National Academies Press.

Nisbet, M.C., & Scheufele, D.A. (2009) 'What's next for science communication? Promising directions and lingering distractions', *American Journal of Botany*, vol. 96, no. 10, pp. 1767–1778, doi:10.3732/ajb.0900041

Peters, H.P. (2012) 'Gap between science and media revisited: Scientists as public communicators', *Proceedings of the National Academy of Sciences*, vol. 110, supplement 3, pp. 14102–14109, doi:10.1073/pnas.1212745110

Pew Research Center. (2015) 'Public and scientists' views on science and society', www.pewinternet.org/2015/01/29/public-and-scientists-views-on-science-and-society

Rakedzon, T., & Baram-Tsabari, A. (2016) 'Assessing and improving L2 graduate students' popular science and academic writing in an academic writing course', *Educational Psychology*, vol. 37, no. 1, pp. 48–66, https://doi.org/10.1080/01443410.2016.1192108

Rakedzon, T., & Baram-Tsabari, A. (2017) 'To make a long story short: A rubric for assessing graduate students' academic and popular science writing skills', *Assessing Writing*, vol. 32, pp 28–42, https://doi.org/10.1016/j.asw.2016.12.004

Rakedzon, T., Segev, E., Chapnik, N., Yosef, R., & Baram-Tsabari, A. (2017) 'Automatic jargon identifier for scientists engaging with the public and science communication educators', *PLoS One*, vol. 12, no. 8, p. e0181742, https://doi.org/10.1371/journal.pone.0181742

Reiss, N., Baram-Tsabari, A., & Peters, H.P. (2016) 'Scientists' attitudes in consolidated and fragile science communication cultures', presented at the Public Communication of Science and Technology (PCST) Network conference, Istanbul, Turkey, 17–22 April.

Rodgers, S., Wang, Z., Maras, M.A., Burgoyne, S., Balakrishnan, B., Stemmle, J., & Schultz, J.C. (2018) 'Decoding science: Development and evaluation of a science communication training program using a triangulated framework', *Science Communication*, vol. 40, no. 1, pp. 3–32, https://doi.org/10.1177/1075547017747285

Sanders, S. (2019) 'Selling without selling out: How to communicate your science', *Science/AAAS Custom Publishing Office*, www.sciencemag.org/custom-publishing/webinars/selling-without-selling-out-how-communicate-your-science

Sevian, H., & Gonsalves, L. (2008) 'Analysing how scientists explain their research: A rubric for measuring the effectiveness of scientific explanations', *International Journal of Science Education*, vol. 30, no. 11, pp. 1441–1467, https://doi.org/10.1080/09500690802267579

Somerville, R.C.J., & Hassol, S.J. (2011) 'Communicating the science of climate change', *Physics Today*, vol. 64, no. 10, pp. 48–53, https://doi.org/10.1063/PT.3.1296

Stevens, S., Mills, R., & Kuchel, L. (2019) 'Teaching communication in general science degrees: Highly valued but missing the mark', *Assessment & Evaluation in Higher Education*, vol. 44, no. 8, pp.1163–1176, https://doi.org/10.1080/02602938.2019.1578861

Sturgis, P., & Allum, N. (2004) 'Science in society: Re-evaluating the deficit model of public attitudes', *Public Understanding of Science*, vol. 13, no. 1, pp. 55–74, doi:10.1177/0963662504042690

Trench, B. (2008) 'Towards an analytical framework of science communication models', in D. Cheng, M. Claessens, T. Gascoigne, J. Metcalfe, B. Schiele, & S. Shi (eds.), *Communicating science in social contexts: New models, new practices*, New York: Springer, pp. 119–135.

Villa, R. (2019) 'Where are science communication courses in Europe?' *QUEST*, questproject.eu/where-are-science-communication-courses-in-europe/

Vollbrecht, P.J., Frenette, R.S., & Gall, A.J. (2019) 'An effective model for engaging faculty and undergraduate students in neuroscience outreach with middle schoolers', *Journal of Undergraduate Neuroscience Education*, vol. 17, no. 2, p. A130.

32

GRAND CHALLENGES

A case study in the complications of and best practices for writing across the curriculum in scientific and technical communication classrooms

Jenna Morton-Aiken

Introduction

In *The Forgotten Tribe: Scientists as Writers*, Lisa Emerson reports that scientists struggle with their identities as writers. She writes, "Many of the scientists in this study struggled to articulate their understanding of writing, because writing – though utterly central to their careers – was something they didn't think about" (2016, p. 179). As many faculty know anecdotally, and as Emerson describes in a separate New Zealand study, STEM students often come to class "convinced that they can't write and they won't have to and shouldn't have to write" (2019, p. 169). Based on interviews with faculty and students, Emerson argues that STEM students absorb this attitude from earlier educational training that reinforces a false bifurcation between writing and science. Students in her study recall that the only writing done in science class was lab reports that Emerson writes "positioned writing as knowledge reporting and completely failed to model the complex, creative process of writing to make meaning" (2019, p. 176).

This chapter unpacks the challenges and opportunities I encountered while developing a Writing Across the Curriculum (WAC) program at Massachusetts Maritime Academy (MMA), a special-mission university with a focus on technical skills and hands-on learning experiences in STEM fields. With a learn-do-learn philosophy that underpins teaching within and beyond the traditional classroom, MMA students graduate well prepared to meet the technical and hard skill demands of their chosen profession. Formal feedback from employers and informal conversations with faculty, however, consistently reported that students needed more support in developing their writing and communication skills earlier and more often during their time at MMA. First, I will describe the PLIA framework (Populations, Landscapes, Infrastructure, and Assessment) that I developed in response to my work with the general education writing rubric at the University of Rhode Island (Foley-Schramm et al., 2018) and SciWrite@URI (Druschke et al., 2018; Morton-Aiken and Reynolds, 2018[1]). I will then situate MMA's context within the PLIA framework and demonstrate how those facets informed my WAC design choices.

DOI: 10.4324/9781003043782-37

WAC programs, sometimes more specifically designated as Writing in the Disciplines (WID), aim to bridge the gap between content and communication. David Russell explains WAC "connects us to one another in powerful ways" (2006, p. 14) because WAC programs are about building relationships, sharing ideas, and engaging with the *how* of making meaning in the world around us. It's not "a quick fix to the perceived 'problem of student writing'" (Association for Writing Across the Curriculum, n.d.) but instead can be a moment to help "faculty members make connections, with students and with each other" (2006, Maimon quoted in McLeod and Soven, 2006, p. 14). Moreover, WAC is also a chance to support professional development for practicing scientists because, as Emerson's senior scientist in genetics shares, "I tell my students that you may think you're a scientist – you're not – you're a writer who writes about science" (2016, p. 179) (for more, see Barel-Ben David and Baram-Tsabari, Chapter 31 in this volume).

In scientific communication, WAC work often means unpacking the ability to communicate science as the doing of science itself. Though WAC technically stands for Writing Across the Curriculum, perhaps a more meaningful approach is something more akin to writing across the campus, toward raising the profile, the value, and the practice of writing in all aspects of the institution. As the Statement of WAC Principles and Practices states:

> WAC refers to the notion that writing should be an integral part of the learning process throughout a student's education, not merely in required writing courses but across the entire curriculum. Further, it is based on the premise that writing is highly situated and tied to a field's discourse and ways of knowing, and therefore writing in the disciplines (WID) is most effectively guided by those with expertise in that discipline. WAC also recognizes that students come to the classroom with a wide range of literacy, linguistic, technological, and educational experiences, but that all students can learn to become more proficient writers.
>
> *(International Network of WAC Programs, 2014)*

What follows here is my attempt to do just that by using the PLIA framework to understand the context in which I am functioning so that I can support a culture of writing even in a space where it might not be considered a natural fit. Ultimately, the framework questions are intended to provoke planning conversations and considerations of a writing program, and to serve as a practical guide to accommodate local contexts. After all, while program designers all likely hope to implement best practices of writing programs, many will also "recognize that 'best' will mean productive and useful but also flexible and permeable – we need to offer ideas, not to standardize or stabilize approaches to teaching scientific communication" (Davis and Frost, 2017, p. 239). The four PLIA factors and related questions are not meant to be exhaustive but rather to serve as points of consideration that can inform the specific application of writing program practice and research at individual institutions.

Though challenging, writing programs and initiatives can also provide an exciting moment to support students in their growth as researchers, professionals, experts, thinkers, and do-ers. They can be conversation openers with faculty about their own writing practices, classroom pedagogy, and development as experts within their fields. They can also be a powerful platform to engage administrators about meaningful assessment practices, faculty support, program design, and student success. To do this, however, writing programs require careful consideration to match best practices from the field with pragmatic applications in a local context. My hope is that sharing both the PLIA framework and MMA's approaches will help illuminate a potential path forward at any institution.

The challenges of scientific writing

The work shared here builds on an ever increasing number of valuable resources on WAC/WID,[2] writing-intensive courses, writing to learn, and other resources developed to support writing at institutions of higher learning, particularly in scientific communication. Added to this are the interdisciplinary and institutional challenges of launching, sustaining, and assessing writing programs in general and scientific writing programs specifically.[3]

Even agreeing on terms can be difficult when STEM, science communication, scientific communication, and technical communication are often used interchangeably and inconsistently. Han Yu and Kathryn Northcut's introduction to their *Scientific Communication: Landscape, Theories, and Pedagogies* argues for the parsing of scientific/science communication as a distinct specialized and valuable branch of technical communication, though in their discussion of science communication versus scientific communication, they also note that "[u]ltimately, what is important is not semantic rigidity but a consensus in the way we understand science as social discourse: created and used by people in dynamic and unstable contexts, for varying and ever-changing reasons" (2017, p. 12). Most significantly, they close their introduction by "riffing on Neil deGrasse Tyson's infamous words: 'Science communication is important whether we want it to be or not'" (2017, p. 13).

Michael Carter addresses the challenges of writing in the disciplines, using meta-genres and meta-disciplines to unpack writing as a process to help "faculty understand their disciplines as ways of knowing, not just domains of declarative knowledge, and thus to see more readily how writing is related to knowing" (2007, p. 388). In other words, framing the ability to communicate how science is developed, reported, and understood as equally critical to the specialist knowledge itself: "the lab experience is a way of doing that is directed toward a way of knowing. It is primarily in writing the lab report, however, that doing becomes knowing" (Carter, 2007, p. 388). The ways of writing and knowing in science can be difficult to bridge even through WAC work because, as Joan Mullin explains, WAC people often come from English or rhetoric and composition backgrounds, and "cross-disciplinary programs may become codified through the disciplinary lens of one person and the field or group to which he or she belongs" (2008, p. 496). She stresses the importance of listening for the WAC "facilitators [to maintain] a certain disciplinary neutrality, a meta-awareness of their own frames" (Mullin, 2008, p. 496).

WAC can then become a moment to examine the communication habits and lenses of all disciplines. Neal Lerner, for example, addresses the disconnect of lab reports from the circulation of knowledge in scientific discourse communities by tracing the history of teaching college lab reports. He notes that the modern lab report "captures none of the dynamism of the experimental process and largely ignores the complexity of the rhetorical choices open to scientific writers" (2007, p. 215).[4] In fact, focusing on the *how* of communication in individual fields, disciplines, and programs can productively highlight the distinctions between habit, expertise, and genre in order to communicate more effectively with one another in all disciplines and writing situations.[5] Susanne Hall explains the need for this distinctive expertise even at the writing center level, arguing in Chapter 33, "Disciplinary Expertise," in this collection that STEM writers bring important and unique value to the writing center as discipline-specific tutors.

In my experience, the absence of writing in many courses is more oversight or avoidance than intentional omission. Many instructors learned how to communicate in their disciplines or professions by absorbing, internalizing, and applying writing conventions without necessarily understanding that they were learning how to write in that particular discourse community. These faculty are often subject experts and/or experienced professionals who have rarely had the benefit of studying pedagogy, let alone in the teaching of writing. Some also carry their own scars from English teachers past and are hesitant to talk about writing at all. Those scars matter because how teachers themselves feel about writing has a direct bearing on their ability to feel empowered when having conversations about writing, particularly when that involves teaching the writing itself (Troia et al., 2012; Bayat, 2014; Cremin and Oliver, 2016). The PLIA framework that follows illuminates such challenges and opportunities with faculty and institutions to help writing program facilitators navigate the way forward.

PLIA framework: four framing questions for program design

The PLIA framework, a series of questions grounded in four quadrants, is designed to prompt program developers who probably know about writing to consider the context of those who don't know about writing before proceeding. Though not exclusive to writing programs, these facets and questions intentionally touch on the fact that writing programs occupy a unique space within institutions because they can reach into every discipline and into every classroom. That doesn't, however, mean that they should. As Martha Townsend notes, "[I]f either faculty or administration is unwilling or disinterested, the WAC program will likely fail" (2008, p. 51). From my own experience, I recommend starting with folks already interested and, when possible, also offer free food.

Whether building new programs or evaluating existing ones, these four areas offer helpful starting points for where to start: Populations, Landscapes, Infrastructure, and Assessment. Though the questions are laid out in a linear format, they should be considered networked and nonhierarchical. Questions about participants that start in Populations, for example, might naturally lead to considerations of other issues such as exigency (in Landscape) and compensation (in Infrastructure), so I encourage readers to work through them in whatever order best suits individual circumstances.

Populations

- Who is your ideal participant? Why? Where do they "live" within your institution? How many participants do you need to gain traction across campus?
- Who is your realistic participant? How are they different from the ideal? Where do realistic participants "live" within your institution? Do you have existing relationships with the faculty/administrators of those institutional "homes"?
- What similar programs/initiatives exist on campus? How did they perform? What were the key elements of their success/challenges?
- How many participants do you want/need? How will you recruit? Will you need to be selective about accepting participants? How will you decide?
- Who are your faculty/administrative allies within individual institutional units and across the institution?
- Who are your faculty/administrative obstacles within individual institutional units and across the institution?

Landscape

- Why is this program a good idea? Why is now a good time to do it?
- What aspects of institutional culture can you use to your advantage? What aspects pose challenges to both launch and long-term viability?

- What will you be able to leverage from this program for your personal/professional development or progression?
- How does this program relate to the university/institutional unit/department's mission?
- How will you articulate the value of this program to natural supporters? How will you justify use of resources to (inevitable) challengers?
- Why will participants join the program? Why will participants stay/complete the program? Why will stakeholders encourage them to participate? How will stakeholders encourage and support participant success? How might stakeholders discourage or impede participation and/or completion?
- How long is the course of the program? What are the primary outcomes? What are the primary activities? What are the tangible/intangible participant takeaways?
- Who will carry out the primary labor? What compensation will they receive? Where is their institutional home? How will their institutional home support and enable their participation?

Infrastructure

- Where do you imagine the program's institutional home? Why is it the best choice? How realistic is your expectation of support?
- What funding and/or other support might this home provide? Where do you anticipate primary funding will come from? Is funding sustainable? Will different expenses need different sources of funding (faculty salary versus equipment versus student stipends, etc.)?
- Who will track budgets/expenses? Who will perform administrative tasks? Who needs compensations and how much?
- What elements make up the core of the program? Do these already exist, and/or can they be modified?
- What space will you need? What equipment/software? Other overheads? Do you anticipate resistance from other groups on campus regarding the use of communal resources?

Assessment

- What is your research question? What kinds of information help you answer that research question? What kinds of assessment tools help you gather data?
- Who will be in charge of assessment? Do you have an institutional assessment specialist available? How is your relationship with that person/office? Who has expertise in writing-specific assessment practices? Are you familiar with IRB (Internal Review Board) protocols? Are you certified in IRB research practices?
- What is the role of assessment on campus or in this program's academic home unit?
- How will you encourage participant engagement? How will you track participants across metrics? How will you ensure privacy and data security?
- How much time do you have to develop assessment tools before collection? What is the data collection timeline? How will you test/validate tools before data collection? When will you code/analyze the data?
- Who will administer the survey tools? Who will code the data? Who will do the statistical/computational analysis? Who will pay for assessment-specific labor?
- How will the results of this assessment be reported and potentially used for faculty, student, program, and/or institutional decision making?

These questions are intended to allow program designers to map a variety of possible pathways to success. Next, I use these categories to situate my WAC approaches within the larger MMA context.

Institutional profile: Massachusetts Maritime Academy

Established in 1891, MMA is the second oldest state maritime academy in the United States. Degree offerings include seven bachelor of science degrees, including maritime transportation, marine engineering, and international maritime business. While this chapter focuses on under-graduate education, MMA also offers three master of science degrees in related fields. Faculty in the three service departments (humanities, social science, and science and mathematics) support general education and courses required in the major like technical writing and business communication.

MMA population

MMA is full of bright, driven students interested in securing high-paying jobs upon graduation. They often come having had poor experience with writing in the past and/or feeling like writing serves no real purpose in their lives or professions. Some are happy to tell me this directly. Most are traditional college students; nearly all are unmarried and childless but have financial or other responsibilities at home. Most students are required to live on campus and participate in the Regiment of Cadets, a civilian but fully uniformed student body committed to the institutional values of discipline, knowledge, and leadership. Life in the Regiment includes military-like uniforms, chains of command, assigned watch shifts on the training ship *Kennedy*, and daily outside morning formation. Women and students of color are still highly underrepresented, and approximately 10% of all students go on to become commissioned military officers.

Faculty in the majors tend to be practitioners with impressive, specialized expertise and hands-on professional experience while faculty in the service departments are usually more traditionally trained academics. When asked in my own classes, students tend to report that they do little writing outside of humanities and social sciences, though many complaints about workload toward the end of the semester indicate otherwise. Like faculty at countless institu-tions, MMA instructors focus on content and are reluctant to integrate writing pedagogy that they often perceive coming at the cost of critical specialized material.

MMA landscape

Historically, MMA's focus has been on hard skills and experiential learning with little room in the curriculum for electives or writing in the disciplines. More recently, however, personnel changes at the administrative and faculty levels, as well as employer feedback, have made way for increasing cultural space on campus for conversations about writing. Despite this, when I was asked to launch the program, there was no established WAC culture, no English majors, no relationship with the academic resource center, and no budget.

MMA infrastructure

The WAC program lives in the Humanities Department, a service department with limited resources and until recent years, limited institutional influence. We now have two more tenure-track writing specialists, though all of us are currently pretenure. Students are required to take four humanities courses, but as mentioned, other graduation requirements leave little room or desire for more writing-specific courses. Faculty professional development of any kind is delivered through the Office for Institutional Effectiveness (OIE), another unit with limited resources and wide responsibilities. Students can find additional writing help at the Writing Resource Center, a unit overseen by the Assistant Dean of Academic Resources.

MMA assessment

The OIE works with faculty and administration to bring learning outcomes and teaching practices into alignment and to communicate data about these and other metrics to a wide variety of stakeholders. A hardworking office with limited resources, this would mean minimal support for writing program assessment development or analysis through absolutely no fault of the OIE.

As the then lone faculty member specializing in rhetoric and composition with reluctant faculty and a student body relatively hostile to writing, I would need to make do with expertise, sheer determination, and an intentional splash of cheerful naïveté to get WAC off the ground.

Making WAC work at Massachusetts Maritime Academy

MMA is uniquely shaped by the Regiment and by its roots as a maritime academy, influences that can be found everywhere on campus. I wanted to be clear that WAC would work *with* faculty and students, meeting them and respecting their interests and expertise to build a culture of writing together rather than dictating one at them. I leaned heavily into my answers for the Populations quadrant of the PLIA framework, allowing them to inform most of my plans when adapting best practices and approaches to MMA specifically. Acutely aware of the low priority of writing among most folks on campus, I decided to launch a three-prong strategy, targeting students, faculty, and institutional partners in separate but related low-stakes activities.

Working with students

MMA cadets are hands-on students who excel when they can break it, fix it, make it better, so I design writing classrooms centering around concrete tasks and writing knowledge and practices framed as professionalization. Accustomed to working collaboratively in the Regiment, group projects can become a source of energy and visible success when appropriately scaffolded. I make a point to remind them regularly class is a space to try it, fail, and then figure out how to do better next time. The experimental approach is something they have rarely encountered in previous writing environments, and in reflective writings to me, they consistently report how much they appreciate the emphasis and investment in their writing process.

I lean into the practical nature of the academy by giving students relevant assignments. In technical writing, they find and annotate job ads, develop resumes, and write letters of application. I start with these genres alongside best writing practices in order to establish credibility and practicality for the course, and the timing often results in co-ops or full-time employment secured with these materials by the end of the semester. We move to group work for assignments like memos and lab reports and then revise both individual and group work for the end of the semester.

I designed our support course, Applied Writing, with another focus. Students are required to take this class after failing the Writing Proficiency Exam (described next) and are often angry and likely embarrassed to have failed the exam. These students usually come with poor experiences with writing, and many see the course as punishment. Again, I leverage their passion for their professions, asking them to read, annotate, analyze, and synthesize articles in their own majors. They appreciate the freedom to pursue their goals even as we have ongoing conversations about writing practices like drafting, freewrites, peer review, and global revision and seem able to find agency in a writing space where they previously felt little to none.

All my classes include a final portfolio with revised drafts of the assignments produced over the entire term. They write short- and long-term reflections on the material and their growth as writers, elevating their metacognition to larger gains that last beyond the semester. Now aware of the rhetorical choices they make when they move from text and e-mail to memo and lab report, they feel prepared to meet the communication demands of their academic and professional lives.

Working with faculty

Faculty engagement at any institution can be challenging because even those interested in writing pedagogy are reluctant to add to already full schedules. My genuine recommendation to counter that reluctance is to befriend individuals in different departments, often through attending cross-campus activities and, again, to offer free food at WAC events. While these two elements won't convert skeptics, it often gives exhausted enthusiasts the extra momentum needed to attend formal events.

Starting without connections or a budget at MMA, I needed to leverage what resources I did have. I asked colleagues in my own department about folks in other departments who were already interested in writing and then sent individual e-mails inviting them to join me at a specific table during an administration-funded welcome back lunch. The idea was simply to meet one another and ask about their needs. Most were happy to attend and talk about writing. Since many faculty already eat on the Mess Deck (dining hall), I used that initial meeting to schedule biweekly conversations. Every other Friday, I came to these sessions ready to talk about specific topics but deliberately let them lead conversations, offering feedback and advice rather than mandates. In addition, I encouraged faculty to drop by my office or e-mail about writing so that talking about writing became a typical activity with a colleague rather than a high stakes encounter with an expert.

I also made a point to attend workshops and faculty lunchtime teaching conversations held by the OIE. In addition to strengthening my own pedagogy, I built relationships, shared writing pedagogies, and answered faculty questions and concerns in another low-key environment. More soft opening than splashy rollout, my strategy to build a community who supported writing slowly and casually was intentional. While MMA's culture that prioritized particular student credentials and experiences might not have much formal room for writing requirements, casual lunches and copier room conversations about writing didn't threaten the system in place. Instead, it allowed instructors to integrate what they were interested in where they could and leave the rest.

My approaches seemed to work: instructors began to experiment, mostly successfully, with writing pedagogy like reflective writing, peer review, and situated writing assignments. I received a grant from the OIE to run an asynchronous rubric-writing workshop over a summer with four of WAC participants, paying them and myself a small summer stipend. All reported increased confidence in rubric design, and most also reported beneficial changes to assignment design because of the program. The asynchronous and colleague-driven nature of the workshop had the extra benefit of modeling peer review and experimentation with the Eli Review platform in a guided and low-stakes environment.

Working with the institution

Though cultural shifts at the student and faculty level are critical, I believe institutional records and structures are needed to formalize and support momentum in writing programs. My past and present department chairs supported my plans, and I was fortunate my department also stood behind my push for the two new hires to be in rhetoric and composition. This support, along with the two new hires, meant we could build an official writing committee to expand the capacity of the Writing Program Administrator (also me) and begin to plan for a larger presence in the curriculum. We've brought our writing program outcomes into alignment with the Council of Writing Program Administrators recommended outcomes (2014) and continued to implement a departmental shift to genre-based writing, an approach that feels particularly well suited for our STEM community. We also continue to revise the protocols for the Writing Proficiency Exam, an on-demand two-hour writing task students must pass before graduating. While the nature of the writing exam – with its lack of drafting, peer review, or professionally situated writing situation – does not necessarily align with best practices like portfolio keeping, it still functions within the specific context of MMA, particularly by providing a credential on their records codifying the importance of writing in a quantitative-heavy environment.

Conclusion

I believe that WAC must be more about a cultivated shift in culture than adding more activities or requirements to succeed in the short or long term. These programs are about conversation and collaboration, not mandates or requirements and, above all, that means understanding the nature of the context in which the WAC program operates. I share the PLIA framework here because asking these questions helped me understand the complexity of the MMA environment and design accordingly.

What I have not addressed enough here, however, is how individual WAC directors can pivot toward specific circumstances. In closing, I share the additional questions and motivations that helped me build my WAC framework even after determining MMA's PLIA context:

- *Who are you as a practitioner?* I combine three approaches when developing strategies to work with both students and faculty. First, I give students and faculty agency in our interactions, positioning them as participants who are experts in other areas and are also capable of developing the skills and knowledge that they need to feel confident and competent as writers and teachers of writing. Second, I incorporate SciWrite@URI's three tenets of habitual writing, frequent peer review, and multiple genres for multiple audiences (Druschke et al., 2018) wherever possible. Third, I focus on writing assignments with realistic and specifically articulated rhetorical situations (with audience, purpose, and context) in genres relevant to student goals and learning outcomes.
- *What do you/they want to read/write?* Building on Art Young's "Assign only what you want to read" (2006, p. 32), I would advocate that faculty find a way to make assignments that students actually want to write. While students might not come to the classroom excited to write lab reports or instructions sheets, giving them tools, choices, and creativity in real-life applications provides them with the motivation to dig in and try it anyway. My students are in my class because they want to do something, and once they understand that learning

how to write will help them do *their* thing, they invest more time and energy than mere requirements for graduation could have inspired.

- *What campus changes can you leverage?* Critical personnel changes at MMA resulted in making room to expand the role and value of writing across campus, especially with new hires and promotions that put academics and hard skills on a more equal footing. These changes also created space to build better relationships with the Academic Resource Center on writing tutors and to partner with the OIE on faculty development workshops.
- *What do you bring as* you? Though I lacked personal networks and institutional memory, I was a fresh face who did not carry the baggage of battles from long ago. I could leverage my then recent graduate work, experience on general education writing rubric development, contributions to SciWrite@URI, and general enthusiasm into "nice to meet you" conversations that soon turned to "how can I help you with writing in your class" conversations. Far from a conscript myself, I genuinely believe that I could support students directly and indirectly through work in my classroom and writing pedagogy support in the classrooms of my colleagues.
- *How can you showcase the work?* As the assessment report writer for the department, I was able to showcase this work to the administration through the OIE's departmental assessment reports. Early feedback indicates that other departments are inspired by the CWPA-driven outcomes and are interested in implementation for their own program outcomes. I continue to work with interdisciplinary faculty on the institutional assessment committee, raising the profile of the assessment work within the department, as well as pushing for inclusion of writing assessment metrics in institutional surveys and reports. I've also presented on the work at conferences and have published in forums such as this collection.
- *How do you demonstrate respect for other ways of doing things?* This is an important one. For MMA students, this means telling them out loud, multiple times, that I don't expect them to be English majors, as well as encouraging them to work through how this immediate application will transfer beyond our current classroom. For faculty, it means respecting their expertise and finding ways to support assignment design and assessment and specifically to unpack professional expectations and genre conventions for writers still learning the ways of making meaning in this discipline.

Most of all, WAC at MMA meant finding small, productive, and sustainable ways to start talking about what it means to write in the academy, in the disciplines, in the profession, and in the world. Some faculty have integrated peer review while others have started integrating work journals instead of quizzes; those who have attempted such pedagogies report increases in student engagement, reading, and subject retention. They feel more confident teaching writing in their disciplines and are receiving more polished and enjoyable material from students. Even designing rubrics that address the specific expectations of a genre have helped both faculty and students understand what is being taught, learned, and modeled in written communication.

Situated in local context articulated through the PLIA framework, WAC at MMA seeks to find common ground among rhetoric and composition's best practices and MMA's learn-do-learn model. Though our program must still grapple with large issues like funding, oversight, labor, and tenure precarity, recognizing the points of congruence within the existing populations, landscapes, infrastructure, and assessment can still go a long way toward establishing the foundation of building a community of writers across the curriculum.

Notes

1 An early draft of this four-faceted framework was introduced at the 2018 International Writing Across the Curriculum conference with Dr. Nedra Reynolds and is shared here with permission of Dr. Reynolds.
2 For more information on WAC, see Colorado State University's WAC Clearinghouse and the Association for Writing Across the Curriculum websites
3 While this chapter focuses on undergrads, graduate students also face significant challenges of learning to write as scientists (Kuehne, 2014; Kuehne and Olden, 2015).
4 For more on case reporting genres, see Wickham's "Medical Case Report" and Mathison's "Scientific Letters."
5 For more on science communication pedagogy, see Maddalena's Chapter 27, "Emerging Practice," Mando's Chapter 21, "Scientist Citizen," and Reid's Chapter 29, "What Citizens and Scientists," in this volume.

References

Association for Writing Across the Curriculum (n.d.) 'Mission, association for writing across the curriculum', www.wacassociation.org/mission/ (Accessed 8 October 2020).

Bayat, N. (2014) 'The Effect of the Process Writing Approach on Writing Success and Anxiety', *Educational Sciences: Theory & Practice*, doi: 10.12738/estp.2014.3.1720.

Carter, M. (2007) 'Ways of Knowing, Doing, and Writing in the Disciplines', *College Composition and Communication*, 58(3), pp. 385–418.

Council of Writing Program Administrators. (2014) 'WPA outcomes statement for first-year composition (3.0), WPA outcomes statement for first-year composition (3.0)', Approved 17 July 2014, http://wpacouncil.org/positions/outcomes.html (Accessed 12 August 2016).

Cremin, T., & Oliver, L. (2016) 'Teachers as writers: a systematic review', *Research Papers in Education*, pp. 1–27, doi: 10.1080/02671522.2016.1187664.

Davis, C., & Frost, E.A. (2017) 'A rhetorical approach to scientific communication pedagogy in face-to-face and digital contexts', in H. Yu & K. Northcut (eds.), *Scientific communication: Practices, theories, and pedagogies*, New York: Routledge, pp. 239–257.

Druschke, C.G., Reynolds, N., Morton-Aiken, J., Lofgren, I.E., Karraker, N.E., & McWilliams, S.R. (2018) 'Better Science Through Rhetoric: A New Model and Pilot Program for Training Graduate Student Science Writers', *Technical Communication Quarterly*, vol. 27, no. 2, pp. 175–190, doi:10.1080/10572252.2018.1425735

Emerson, L. (2016) *The forgotten tribe: Scientists as writers*, University Press of Colorado Fort Collins, https://wac.colostate.edu/books/emerson/tribe.pdf

Emerson, L. (2019) '"I'm not a writer": Shaping the literacy-related attitudes and beliefs of students and teachers in STEM disciplines', in *Theorizing the future of science education research*, New York: Springer, pp. 169–187.

Foley-Schramm, A. Fullerton, B., James, E.M., & Morton-Aiken, J. (2018) 'Preparing Graduate Students for the Field: A Graduate Student Praxis Heuristic for WPA Professionalization and Institutional Politics', *WPA: Writing Program Administration*, 41(2).

International Network of WAC Programs. (2014) *Statement of WAC principles and practices – The WAC clearinghouse*, WAC Clearinghouse, https://wac.colostate.edu/principles/ (Accessed 8 October 2020).

Kuehne, L.M. (2014) 'Practical Science Communication Strategies for Graduate Students', *Conservation Biology*, 28(5), pp. 1225–1235, DOI: 10.1111/cobi.12305

Kuehne, L.M., & Olden, J.D. (2015) 'Opinion: Lay summaries needed to enhance science communication', *Proceedings of the National Academy of Sciences*, 112(12), pp. 3585–3586, doi: 10.1073/pnas.1500882112

Lerner, N. (2007) 'Laboratory Lessons for Writing and Science', *Written Communication*, 24(3), pp. 191–222, doi: 10.1177/0741088307302765.

McLeod, S.H., & Soven, M.I. (2006) *Composing a community: A history of writing across the curriculum*, Lauer Series in Rhetoric and Composition, West Lafayette: Parlor Press.

Morton-Aiken, J., & Reynolds, N. (2018) 'Working through interdisciplinary writing support program infrastructures', presentation at the International Writing Across the Curriculum Conference, Auburn, AL.

Mullin, J.A. (2008) 'Interdisciplinary Work as Professional Development: Changing the Culture of Teaching', *Pedagogy: Critical Approaches to Teaching Literature, Language, Composition, and Culture*, 8(3), pp. 495–508, doi: 10.1215/15314200-2008-008.

Russell, D.R. (2006) 'Introduction: WAC's beginnings: Developing a community of change agents', in S.H. McLeod & M.I. Soven (eds.), *Composing a community: A history of writing across the curriculum*, Lauer Series in Rhetoric and Composition, West Lafayette: Parlor Press, pp. 3–15.

Townsend, M. (2008) 'WAC program vulnerability and what to do about it: An update and brief bibliographic essay', *The WAC Journal*, 19, pp. 45–61.

Troia, G.A., Shankland, R.K., & Wolbers, K.A. (2012) 'Motivation Research in Writing: Theoretical and Empirical Considerations', *Reading & Writing Quarterly*, 28(1), pp. 5–28, doi: 10.1080/10573569.2012.632729.

Young, A. (2006) *Teaching writing across the curriculum*, Pearson Prentice Hall, https://wac.colostate.edu/books/landmarks/young-teaching

Yu, H., & Northcut, K.M. (2017) *Scientific communication: Practices, theories, and pedagogies*, London: Routledge.

33

DISCIPLINARY EXPERTISE, WRITING CENTERS, AND STEM WRITERS

Susanne Hall

Writers in the fields of STEM (science, technology, engineering, and mathematics)[1] can and should make use of the writing centers on their campuses. What resources they will find there will vary. In some writing centers, STEM writers seeking support with composing a technical research proposal will find themselves discussing the document with an undergraduate English major who has little prior experience with the content, genre, or purpose of such a document. In other centers, the writer might work with a professional writing specialist who holds a PhD in a STEM field and has written similar documents. This chapter seeks to explain the uses and limits of college writing centers for STEM writers, aiming to inform instructors and mentors, as well as writers, about the support that writing centers may offer. STEM instructors and students need to understand what kinds of disciplinary expertise their local writing center staff members have in order to make strategic use of the center's resources, and writing center directors and tutors should likewise be cognizant of the nature and limits of that expertise in order to work effectively with STEM writers.

The chapter begins with context on the nature of writing centers, which typically face the challenge of supporting many types of writers and writing. It proceeds from there into a presentation of theoretical and empirical research on the role of disciplinary expertise in the writing center tutorial. After establishing that this research suggests that tutors' disciplinary expertise can indeed be valuable in tutorial sessions, I then provide a specific vocabulary for more precisely discussing disciplinary expertise relevant to supporting STEM writers. Previous notions of disciplinary expertise in writing centers have been vague, and I suggest that a more precisely defined set of terms will help aid in important conversations about how to support STEM writers. I propose that disciplinary expertise in the following areas is relevant to supporting STEM writers: subject matter, literacy and numeracy, genre, rhetoric, writing process, and discourse communities.

The ideas presented here are based upon analysis of writing studies research, as well as my own experience directing the writing center at the California Institute of Technology (Caltech), which serves undergraduates, graduate students, postdoctoral researchers, and other advanced researchers in STEM fields.

A brief introduction to writing centers

In order for STEM instructors and writers to understand the uses and limits of the writing centers on their campuses, some understanding of the nature and variety of writing centers is

DOI: 10.4324/9781003043782-38

useful. Writing centers, generally speaking, seek to provide feedback, workshops, and other kinds of support for academic writers across many courses and contexts. While some centers may by design serve only a small subset of campus members, many centers strive to broadly support undergraduates, graduate students, postdoctoral researchers, faculty, and other members of the campus community, groups who are at different points in their development as writers. Furthermore, while some writing centers may focus on supporting writers only in certain curricula or disciplines, many seek to support writers across a wide variety of disciplines, who are writing for varied purposes and within different contexts.

Therefore, writing centers face the challenge of supporting diverse writers with many different needs and goals. In her article on the Graduate Writing Center at UCLA, writing center researcher Sarah Summers (2016) argues that "writing consultants [who are used to supporting undergraduates] often need new strategies to support graduate writers" (p. 121) and notes that many centers do not provide training in those strategies. Such training supplements and shapes the knowledge that writing center staff members bring into their work as tutors. One area in which training for writing center staff is often focused is disciplinary differences in writing. Historically, many writing centers have had a "generalist" staff model in which any member of the staff was trained to respond to any writer who might seek the center's support. In more recent years, more centers have added "specialists" to their staff who have specific knowledge or training to respond to particular types of writing.

I believe that disciplinary expertise helps tutors support STEM writers, not only because I have seen it happen on a daily basis in Caltech's writing center but also because the current research supports that idea. I will share that research here, starting with a key theoretical idea in the next section and followed, in the subsequent section, with empirical research.

Reading and responding to academic writing calls for disciplinary expertise

To explain why disciplinary expertise is valuable to tutoring STEM writers, I want to first share an idea from writing in the disciplines of expert David Russell (1995): the analogy of academic writing as "ball handling." Using activity theory, Russell presents an analogy "between games that require a particular kind of tool – a ball – and activity systems (disciplines, professions, businesses, etc.) that require a particular kind of tool – the marks that we call writing" (p. 57). As Russell explains, being skilled in one ball-based game, like soccer, has little inherent relationship to skill in another, like ping pong. Russell argues that there is "no autonomous, generalizable skill called ball using or ball handling that can be learned and applied to all ball games" (p. 57) because the context of the ball's use is so deeply important. Analogously, according to Russell, there is no general skill of writing that can be taught and directly transferred between disciplinary or professional activity systems.[2]

The idea of the generalist writing tutor can be understood via Russell's metaphor as a "ball-sports coach" who offers advice to people engaged in activities as different as baseball, billiards, and bocce ball. For novice college writers, such instruction can be useful, especially for students in some introductory writing courses that do not ask students to write in real-world, disciplinary writing genres. And even at more advanced levels, dialogue with an outsider to one's field can be enlightening. However, dialogue with a nonexpert is not necessarily more useful or a suitable replacement for dialogue with a disciplinary expert. Just as young soccer players benefit from having a coach who is a soccer expert as their skills develop, STEM writers who visit the writing center will benefit from being able to choose to work with tutors who have personal experiences reading, writing, and conducting research in their fields of study.

Empirical evidence for the value of disciplinary expertise in writing tutoring

Russell's work provides a theoretical justification for cultivating disciplinary expertise in writing center staff. What empirical evidence exists to support the value of disciplinary expertise in writing center tutors? I have found three empirical studies that compare writing center sessions led by nondisciplinary-expert tutors to those led by tutors with disciplinary expertise. These studies all strongly confirm that a tutor's disciplinary expertise (variously defined, about which more later) promotes more successful tutorial sessions. The studies focused on disciplinary expertise in literature (Kiedaisch and Dinitz, 1993), engineering (Mackiewicz, 2004), and history and political science (Dinitz and Harrington, 2014).[3] Tutorial success was judged by researchers as well as, in Kiedaisch and Dinitz (1993) and Dinitz and Harrington (2014), by expert faculty in the fields who later reviewed the sessions. The studies all have limits, with small sample sizes being a shared one. A summary of their key findings follows.

Focus of session

- Experts focus more on higher-order issues (the quality and clarity of ideas being expressed) and less on local issues (sentence-level clarity and correctness) than nonexperts (Kiedaisch and Dinitz, 1993; Dinitz and Harrington, 2014).
- Experts asked more questions about higher-order issues, pushing students to further think about their work (Kiedaisch and Dinitz, 1993; Dinitz and Harrington, 2014).

Nature of session

- Experts pushed back on a student's ideas, while nonexperts tended to accept them (Dinitz and Harrington, 2014).
- Experts had recursive sessions that looped back to previous topics when relevant, while nonexperts moved through papers in a linear manner (Dinitz and Harrington, 2014).
- Experts drew more general lessons for the session that the writer could transfer beyond the session into further learning (Dinitz and Harrington, 2014).
- An expert treated engineering writing as a real-world document rather than as a class assignment and focused more on purpose and audience (Mackiewicz, 2004).

Quality of guidance

- Nonexperts lacked genre knowledge and gave incorrect advice regarding the purpose of academic writing, wrongly encouraging writers working on informative writing to write in a persuasive mode (Mackiewicz, 2004).
- Nonexperts gave incorrect advice about local issues that were subject to disciplinary rules they did not know about (e.g. capitalization of key terms, informal versus formal tone) (Mackiewicz, 2004).
- Nonexperts failed to make the limits of their knowledge clear to writers, giving incorrect advice confidently, whereas experts were capable of marking areas of higher or lower confidence accurately for writers (Mackiewicz, 2004).

While there is a need for more empirical research on this topic, the extant studies strongly point toward the value of disciplinary expertise for a variety of reasons, including more in-depth discussions of writing, more dynamic and impactful discussions, and greater accuracy of advice.

Types of disciplinary expertise in STEM that are relevant to tutoring STEM writers

If we take seriously that disciplinary expertise among writing tutors can help them respond to writers' work, then we need to think more about what kinds of disciplinary expertise matter as well as why they matter. Existing studies of disciplinary expertise in tutors have different and relatively vague ways of defining that expertise, including by a tutor's major (Kiedaisch and Dinitz, 1993), by professional experience in a related field (Mackiewicz, 2004), and by having completed coursework in a discipline (Dinitz and Harrington, 2014). A more detailed exploration of types of disciplinary expertise that are relevant to tutoring STEM writers is useful for three reasons. First, it could aid future empirical research on this topic, which should address this issue in detail. Next, it can help shape tutor hiring and training practices, suggesting areas that training might address and acknowledging areas that only strategic hiring can address. Finally, it could help inform local conversations between writers, instructors/mentors, and writing center staff about what a particular writing center can and cannot offer STEM writers. Logistically, writing centers cannot offer every writer a tutor with the exact kinds of expertise that would most benefit them, and thus the best tutorial interactions will be those in which the nature and limits of a tutor's expertise are explicitly clear to both writer and tutor.

I have built my elaboration of STEM tutor expertise on writing-in-the-disciplines expert Anne Beaufort's (2007) model of the knowledge domains that expert writers draw on. Beaufort notes four such knowledge domains: subject matter knowledge, genre knowledge, rhetorical knowledge, and writing process knowledge. I will add one additional category that Beaufort does not consider: disciplinary literacies and numeracy. All five of these knowledge domains are understood to exist within a larger, sixth context of discourse community knowledge.

Subject matter knowledge

STEM subject matter knowledge has two important types: understanding of key concepts and experience with research methods.

Understanding of key concepts includes knowing overarching theories and specific ideas in STEM fields. For example, readers in astrophysics will understand the theory of special relativity as well as what a Lorentz transformation is. In much STEM writing, theories are tacit and assumed to already be understood by one's reader. Advice that suggests writers should explicitly present that knowledge to their readers would be misguided in many contexts.

Just as importantly, subject matter knowledge includes experience with research methods. This may be less obvious to tutors with humanistic training, where research methods often most essentially include locating the right texts and interpreting them in a particular way. For scientists and engineers, prior knowledge of the field's key research tools and techniques are particularly important to reading experimental work.

Tutors who lack knowledge of theories, key concepts, and methods of a field in which writers are working may have trouble reading and making sense of their writing. They will need to ask writers to define terms and explain assumed knowledge about theories and methods. Some amount of that explanation may be productive for a writer, but unless the writer is writing to an audience of nonexperts, a complete lack of subject matter knowledge may be a hindrance to effective STEM writing tutoring. Ideally, writing tutors working with STEM writers will have a grasp of major subject area theories and approaches even when they lack knowledge of specific concepts or methods a writer addresses.

Literacy and numeracy

Literacy is most simply the ability to read and write in a language such that one can understand and be understood. The related term "numeracy" (or "mathematical literacy") is the ability to understand numerical information and to use numbers to communicate and solve problems. Several unique literacies are foundational to reading and writing in STEM.

Scientific Englishes[4] *literacy* is a knowledge of the grammar, syntax, and styles of the Englishes typically used in STEM writing. The Englishes that appear in published STEM writing diverge from those of humanistic academic writing in noticeable ways. A reader comparing published general guides to academic writing with guides to STEM writing will note different advice about matters of grammar and syntax (e.g. advice about the use of passive voice). More complexly, citation styles and the role of intertextuality in STEM differs from the humanities. Even between and within STEM fields, varied scientific Englishes exist.

Quantitative literacy is crucial to a great deal of STEM writing, since much STEM work is built on a chassis of math. Most basically, the ability to understand common mathematical symbols and functions is a prerequisite to reading much STEM writing. Mathematical formulas are core meaning-making elements of many STEM texts.

Computational literacy is important, as the need to share and explain computer code and algorithms increasingly appears across STEM fields. This literacy entails an ability to read computer code and algorithms, identifying the meaningful structures that produce results. A reader for whom code and algorithms are illegible will struggle to read many documents in STEM.

Visual literacy in this context entails a familiarity with the ways STEM writers typically share data and other information. For example, an inability to understand how to read common tables, charts, and diagrams, such as the structural formula of a chemical compound or a circuit diagram, to name just two examples, will cause trouble reading and responding to writing in chemistry or electrical engineering, respectively. Visual communication is crucial in STEM.

Tutors who lack scientific Englishes, quantitative, computational, and visual literacies will have difficulty reading the most important parts of many STEM texts, much as an English speaker without training or experience reading Italian or Arabic would have trouble reading and responding to an essay that includes key passages in Italian or Arabic. Ideally, writing tutors working with STEM writers will possess a basic level of these literacies.

Genre knowledge

Each discipline tends to communicate in a number of genres that have developed across time to respond to communication needs that repeat themselves (Devitt, 1993). For example, STEM researchers repeatedly encountered the need to share the results of experiments with readers, and, over time, the genre of the Introduction-Methods-Results-and-Discussion (IMRAD) research article was developed as a relatively efficient way to do so.[5] A number of genres that are common in STEM fields are largely unused or used very differently in humanities fields. Examples include the structured abstract, IMRAD paper, lab report, grant proposal, poster presentation, progress report, and review paper.

Understanding a genre involves more than understanding its form. As genre theory expert Amy Devitt (1993) explains, genre is the result of social activity – of a repeated situation happening in the world that called the genre into existence and that leads to its continued existence and evolution. Posters exist because of the way some academic conferences work as well as the kinds of research participants at those conferences do. If conferences change, the poster may change or disappear. Ideally, writing tutors working with STEM writers will understand both the form and social context of key STEM genres, so that they can provide writers with feedback about how to successfully use a genre in a given context.

Rhetorical knowledge

Rhetorical knowledge is an understanding of how effective arguments are made in a particular context. In many genres of STEM writing, arguments are implied rather than explicit, especially when viewed from the point of view of a reader with training in the humanities. This rhetorical approach fits with a traditional ideology of empirical scientific research in which a researcher is discovering rather than creating knowledge. The typical IMRAD article proceeds by noting an interesting question, explaining a way the question was investigated, sharing collected data, providing an interpretation of that data, and noting limits of the project and areas for future work. Each one of these rhetorical moves involves the implication of argument but rarely the explicit statement of argument. Other common STEM genres, like the poster and conference presentation, likewise follow this rhetoric of implied argumentation.

Other STEM genres do foreground argument. Some STEM review articles argue for the best path for research in an area to take, while others are neutral. Letters written to journals to point out mistakes and oversights in published work are pointedly argument driven. Grant proposals typically include explicit arguments for the impact of research.

Beyond the consideration of whether arguments are presented explicitly or implicitly, the nature of reasoning and evidence is primarily quantitative in STEM writing. Grawe and Rutz (2009) (an economist who researches higher education and a writing studies researcher, respectively) define quantitative reasoning as "the habit of mind to consider the power and limitations of quantitative evidence in the evaluation and construction of arguments in personal, professional, and public life" (p. 3). Tutors who lack this habit of mind and the skills to execute such reasoning will have trouble identifying and understanding most arguments in STEM fields.

Tutors trained in the humanities may lack quantitative reasoning skills, as well as knowledge of how to present arguments to STEM readers in a convincing manner. Ideally, writing tutors working with STEM writers will be familiar with the variety of rhetorical approaches used in STEM fields.

Writing process knowledge

The process of producing writing is also deeply influenced by discipline. One unique aspect of the writing process in STEM is that making sense of data is often the key interpretative work of empirical STEM writing, and composing figures that show patterns in data is often the starting place for writing a research-based STEM article or report. Thus much STEM writing starts with an interpretative and communicative task that doesn't focus on alphabetic composition. Another key element of the process for some STEM writers is that writing is often done collaboratively. It thus involves ongoing dialogues and a complex division of labor that differs greatly from that of individually authored works.

413

The individual mathematician proving a theorem and the hundreds of authors contributing to an empirical research paper out of a research collaborative like the European Organization for Nuclear Research (CERN) have very different writing processes, as do their students. Ideally, writing tutors working with STEM writers will understand how diverse STEM writing processes are and how they may differ from humanistic writing processes. Developing skills for working with collaborative authorship situations, which are often challenging, will also be helpful.[6]

Discourse community knowledge

All these elements of disciplinary expertise – subject matter knowledge, literacy and numeracy, genre knowledge, rhetorical knowledge, and writing process knowledge – are constituted by the social discourse communities who read and compose STEM writing. As a writing professor, Charles Bazerman (1988) was charged to prepare college students to write across disciplines, and he quickly discovered that he could not understand the written texts of other disciplines without also understanding "the social and intellectual activity which the text was a part of" (p. 4). This realization started a line of inquiry that led to his landmark book, *Shaping Written Knowledge*, about the experimental STEM journal article. This world of social and intellectual activity described by Bazerman was further defined by linguist John M. Swales (1990), who called each of these worlds a discourse community. Swales explains that a discourse community may be geographically dispersed and socially diverse, but it nonetheless has "common goals, participatory mechanisms, information exchange, community specific genres, a highly specialized terminology and high level of expertise" (p. 29). Academic disciplines are one kind of discourse community – they share goals, hold conferences and publish journals, write in specific genres, have their own lexicon, and contain both expert and novice members. Membership in or familiarity with a discourse community helps writers understand not just how to address an audience but also *why* they are addressing that group in a particular way.

Disciplinary subject matter, literacies, genres, rhetoric, and writing processes all exist within the social context of the discourse community. Writing tutors who wish to support advanced STEM writers will benefit not only from having a knowledge of these discrete elements but also from having had experiences with the discourse community. Experiences like working in labs or other research settings (or even touring them), attending conferences, reading journals, and talking with experts in the field both formally and informally help inform a tutor's knowledge of how communication works within a discourse community.

Limitations of this vocabulary

I have based my articulation of the types of disciplinary expertise that are relevant to supporting STEM writers on existing research as well as hands-on experience. However, there is a need for much more research in this area. I hope that future empirical research on the topic of STEM expertise in writing centers will draw on and investigate these concepts. Such research may show some items in this list matter more than others, or it may reveal additional types of expertise I have overlooked here. My hope is that having this more comprehensive way of delineating and discussing different types of relevant STEM disciplinary expertise helps us move toward more robust support for STEM writers.

Making use of a writing center for STEM writing

I have suggested how meaningful disciplinary expertise can be for a writing center's ability to support STEM writers. However, even well resourced writing centers (and many are, unfortunately, resource challenged) will be unable to provide every writer with writing tutors who have all the kinds of disciplinary knowledge as just outlined. As I shared previously, many writing centers support writers in a wide variety of disciplines, and disciplinary expertise is not the only consideration they have in making decisions about hiring and training tutors. Furthermore, issues of both mission and funding will dictate the levels of experience with academic writing that staff have – centers may employ undergraduate tutors, graduate tutors, faculty tutors, other professional tutors, or some combination of these. The good news is that those who teach STEM writing can do a great deal to utilize and support the writing centers on their campuses, and the benefits will be theirs to reap. I recommend the following actions for STEM instructors and research mentors who wish to get the most out of their campus writing centers:

- *Introduce yourself to the writing center director:* Let the director know you would like to learn about the center's resources. During an introductory meeting, you can become familiar with the center's history and mission. You can also become acquainted with the current resources the center offers STEM students and faculty. Develop an understanding of the kinds of expertise staff do and do not have, using this chapter as a guide toward questions you might ask.
- *Once a relationship is established, inquire about a center's ability and willingness to develop course or lab-based partnerships or other customized resources:* Many writing centers have both informal and formal ways of offering targeted support for courses or labs with interested instructors or principal investigators. Such programs of support allow for collaborations that ensure your students make the most of the writing center's available resources.
- *Offer to support your writing center by participating in staff training:* Your willingness to share your own disciplinary expertise can be valuable professional development opportunity for writing center staff.
- *Use your power on campus to support your center's continued growth:* While advocating for appropriate funding on relevant committees is often especially needed, this is not the only way to help. Connecting a center's director and staff to colleagues in your department can help to build the relationships that will allow a center to recruit tutors from your field and better serve writers in your field.
- *Encourage your students to make use of the writing center and help set appropriate expectations for what kinds of help they can find there.*

For STEM writers seeking to get the best support from their campus writing centers, I suggest the following actions:

- *Begin a meeting with a writing center tutor by getting to know each other:* First, introduce yourself and your intellectual interests and goals. Follow sharing about yourself by asking your tutor about her or his level of your familiarity with your field and the genre of writing you are doing. For example, you might say, "Have you taken any courses in [my field]?" or "Have you ever written or read a review article?" From these opening questions, you can probe other, relevant kinds of disciplinary knowledge and experience I have elucidated in this article. The answers to these questions will help you understand the nature and limits of your writing center reader's expertise, which can then usefully shape how you focus your time in the remainder of the session.

- *Be clear about your goals*: In the vast majority of writing centers, tutors are interested in supporting a writer's goals for their work rather than in imposing their own goals onto you. If you arrive at a meeting with some clear goals you can share, you can also talk with the tutors about how confident they feel about their ability to support your goals.
- *See a response from a writing center tutor as what it is – a response from a single reader*: Well trained writing center tutors will never seek to present themselves as all-knowing authorities on your writing. Instead, they will see themselves as skilled, supportive readers who want to do all they can to help you pursue your goals. You should see them in the same way. Their feedback provides a useful set of data points about how you might revise, but it should be weighed against feedback from other readers as well as your own assessment of the work. Act on the feedback that is useful to you.

Writing centers thrive on collaborative learning, and by getting to know and supporting your local center, you will increase its ability to support STEM writers. This collaboration will help us continue moving toward a world in which all writing centers at institutions that educate STEM writers can fully support those writers.

Notes

1 This chapter will broadly address the ways writing centers support STEM writers, but it is worth acknowledging what a diverse group "STEM writers" is and the diverse writing practices they undertake. Where possible, I will delineate between varied groups yoked together with this term, but I lack the space to fully explore the different needs that those in fields as diverse as computer science, biology, and mathematics (to name just three examples) have as writers.
2 The main purpose of Russell's paper is to challenge models of the first-year writing class which assume such generalizable and easily transferrable writing skills do exist. By extension, his theory challenges the approach of writing centers that were built to support those writing curricula and which devalued disciplinary expertise as a result.
3 Two other empirical studies cover similar topics. Wong (1988) compares tutorial sessions led by disciplinary experts who either possess or lack specific subject area knowledge relevant to the paper. Smith (2003) examines the role of technical expertise in how writing instructors (not tutors) evaluate student writing in engineering.
4 Linguists and other writing experts have questioned whether the "standard English" that is typically assumed to be the dialect for academic writing is an actual coherent linguistic dialect or an ideological tool (Greenfield, 2011). For that reason, I use the term "Englishes" here to highlight the lack of one, definitive dialect, even as I do wish to suggest that there are differences between patterns of Englishes used by humanistic and scientific writers.
5 See Harmon and Gross's (2007) edited collection of science writing from the seventeenth to the twentieth centuries for an introduction to how genres and the rhetoric of science communication have changed since the scientific revolution. This history may be useful in training tutors to understand STEM genres.
6 See Coffey et al., 2017 for more on supporting team writing in the writing center.

References

Bazerman, C. (1988) *Shaping written knowledge: The genre and activity of the experimental article in science*, Madison, WI: University of Wisconsin Press, https://wac.colostate.edu/books/landmarks/bazerman-shaping/

Beaufort, A. (2007) *College writing and beyond: A new framework for university writing instruction*, Logan, UT: Utah State University Press.

Coffey, K.M., Gelms, B., Johnson, C.C., & McKee, H.A. (2017) 'Consulting with collaborative writing teams', *The Writing Center Journal*, vol. 36, no. 1, pp. 147–182. Available at: www.jstor.org/stable/44252641

Devitt, A.J. (1993) 'Generalizing about genre: New conceptions of an old concept', *College Composition and Communication*, vol. 44, no. 4, pp. 573–586, doi: 10.2307/358391

Dinitz, S., & Harrington, S. (2014) 'The role of disciplinary expertise in shaping writing tutorials', *Writing Center Journal*, vol. 33, no. 2, pp. 73–98.

Grawe, N., & Rutz, C. (2009) 'Integration with writing programs: A strategy for quantitative reasoning program development', *Numeracy*, vol. 2, no. 2, doi: http://dx.doi.org/10.5038/1936-4660.2.2.2

Greenfield, N. (2011) 'The "standard English" fairy tale', in *Writing centers and the new racism*, Logan, UT: Utah State University Press, pp. 33–60.

Harmon, J.E., & Gross, A.G. (2007) *The scientific literature: A guided tour*, Chicago: University of Chicago Press.

Kiedaisch, J., & Dinitz, S. (1993) ' "Look back and say 'so what'": The limitations of the generalist tutor', *The Writing Center Journal*, vol. 14, no. 1, pp. 63–74.

Mackiewicz, J. (2004) 'The effects of tutor expertise in engineering writing: A linguistic analysis of writing tutors' comments', *IEEE Transactions on Professional Communication. IEEE Transactions on Professional Communication*, vol. 47, no. 4, pp. 316–328, doi: 10.1109/TPC.2004.840485

Russell, D. (1995) 'Activity theory and its implications for writing instruction', in J. Petraglia (ed.), *Reconceiving writing, rethinking writing instruction*, Mahwah, NJ: Lawrence Erlbaum Associates, Inc., pp. 51–77.

Smith, S. (2003) 'The role of technical expertise in engineering and writing teachers' evaluations of students' writing', *Written Communication*, vol. 20, no. 1, pp. 37–80, doi: 10.1177/0741088303253570

Summers, S. (2016) 'Building expertise: The toolkit in UCLA's graduate writing center', *The Writing Center Journal*, vol. 35, no. 2, pp. 117–145.

Swales, J. (1990) 'The concept of the discourse community', in *Genre analysis: English in academic and research settings*, Boston: Cambridge University Press, pp. 21–32.

Wong, I.B. (1988) 'Teacher-student talk in technical writing conferences', *Written Communication*, vol. 5, no. 4, pp. 444–460.

INDEX

For Product Safety Concerns and Information please contact our EU
representative GPSR@taylorandfrancis.com
Taylor & Francis Verlag GmbH, Kaufingerstraße 24, 80331 München, Germany